PLANT-POLLINATOR INTERACTIONS

Plant–Pollinator Interactions

From Specialization to Generalization

Edited by Nickolas M. Waser and Jeff Ollerton

The University of Chicago Press Chicago and London

The University of Chicago Press, Chicago 60637
The University of Chicago Press, Ltd., London

Printed in the United States of America

15 14 13 12 11 10 09 08 07   2 3 4 5

ISBN: 0-226-87400-1 (paper)

Library of Congress Cataloging-in-Publication Data

Plant-pollinator interactions : from specialization to generalization / edited by Nickolas M. Waser and Jeff Ollerton.
p. cm.
Includes bibliographical references and index.
ISBN 0-226-87400-1 (pbk. : alk. paper)
1. Pollination by insects. 2. Pollination by animals. I. Waser, Nickolas Merritt, 1948– II. Ollerton, Jeff.
QK926.P52 2006
571.8'642—dc22
2005011907

⊗ The paper used in this publication meets the minimum requirements of the American National Standard for Information Sciences—Permanence of Paper for Printed Library Materials, ANSI Z39.48-1992.

To the memory of our predecessors who set the stage; to our contemporaries who opened new horizons; and to our successors who will resolve the resulting puzzles.

Contents

Preface

The genesis of this book goes back well over a decade to our independent interests in the meaning of floral form, the importance of different floral visitors for pollination, the origins and evolutionary diversification of plant-pollinator interactions, and related topics. In the early 1990s, our paths crossed when we and three other colleagues decided to write a paper linking these topics with patterns of specialization and generalization in pollination systems. After a gestation of three years we submitted this paper to the "Concepts" section of the journal *Ecology*, where it appeared in 1996.

Our 1996 paper urged pollination workers to take a second look at a common assumption that specialized plant-pollinator interactions were the norm and were the usual outcome of evolutionary change within angiosperm lineages. Ours was not a lone voice; others around the globe—for example, John Thompson and Pete Feinsinger in the United States and Carlos Herrera and Javier Herrera in Spain—had raised similar themes. Indeed, our target audience was not only professional pollination biologists but also ecologists and evolutionary biologists from other disciplines, whose views of plant-pollinator interactions may be colored by textbook examples of close coevolution of plants and pollinators, such as *Ficus*-fig wasp and *Yucca*-yucca moth relationships. Professional opinion of that paper seemed to range from "a stimulating synthesis" to "everyone knew this already" to "heresy!" suggesting that we had achieved our aim of getting people to talk and reconsider.

It appeared to the two of us that the late 1990s witnessed a renewed vigor in efforts to understand specialization and generalization in pollination interactions at levels from a single species to entire local communities to comparison across geographic regions. We began to see papers in journals and hear talks at meetings on topics that ranged from cognitive ecology of specialist versus generalist pollinators, to pollination interactions as networks, to interpretations of flowers as compromises for use of different pollinators, to nonadaptive diversification of floral traits, and on and on. A search of the ISI Web of Knowledge database for papers with "pollinat*" in the title suggests growth within the discipline (e.g., 901 papers in 2004 vs. 566 in 1991), some of which seems to involve new conceptual issues such as specialization and generalization, since searching for "pollinat*" and "generaliz*" together yields 16 papers from 1981 to 1996, compared to 74 from 1996 through July 2005.

By the millennium, we were convinced that enough new thinking and results were available that it would be fruitful to showcase them together in some form. We began to solicit colleagues to join in two endeavors. The first was a symposium titled "Specialization and Generalization in Plant–Pollinator Interactions," which we organized together with Regino Zamora from the Universidad de Granada. In proposing our symposium to the Ecological Society of America (ESA), we argued that an understanding of patterns of specialization and generalization in plants and pollinators is fundamental to the evolutionary ecology of this interaction, that the interaction is ubiquitous and important in virtually all terrestrial ecosystems, and that exploring such issues in pollination would be of interest to those working on other ecological interactions. The ESA agreed and sponsored our symposium in August 2002. Our second simultaneous endeavor was this book. The symposium format limited the number of voices that could be heard and the detail of each message. The remedy was to invite a larger group of workers to present, in writing, a greater variety and depth of perspectives on specialization and generalization in pollination systems.

This larger set of voices is, indeed, diverse. The present discussion of specialization and generalization is not a polarized debate with two sides (as the unfortunate dichotomy of language might suggest) but rather a vigorous exchange with diverse starting points and styles. By choosing those to invite as authors, we intended to represent this diversity and to reach beyond obvious, established workers to include those from many career stages and geographic regions. As a result, this book is a collaboration of graduate students, postdoctoral fellows, and untenured and tenured academics of 12 different nationalities. Of course, we still could not include all voices, and not all those invited could contribute; therefore, some themes are not as well represented as we might wish. On the whole, however, we believe that our attempt at inclusivity and diversity has resulted in a volume that captures the breadth and vigor of the discussion.

Not all of the authors of these chapters agree with one another's views (indeed, not all coauthors do!), but the disagreement is constructive rather than sterile, as readers will quickly see. Indeed, differences of opinion extend to the usage of the terms *specialization* and *generalization* themselves, and we have deliberately avoided defining these terms in the book. Our reasoning is that any proper attempt to discuss definitions and interpretations would constitute a chapter in itself and that such a chapter may imply that its authors have a corner on the "proper" definitions or wish to impose their view on other authors or readers of the book, neither of which is true. Better to let individual authors in this book define the terms, explicitly or implicitly, as they see fit!

A few words about the cover of the book are in order. Pollination interactions are characterized by their diversity, but to catch the eye a book needs a simple cover. Thus we must be content with two images to represent the great richness of plant–pollinator interactions and the complexity of studying them. The top image shows a species of barberry (*Berberis*) with a rather unspecialized floral morphology, being pollinated by a mason bee (*Osmia ribifloris*). This solitary bee is an "eclectic oligolege": it collects pollen from only a few plant species that are not closely related to one another. The bottom image depicts the montane larkspur *Delphinium barbeyi,* with a much more specialized floral morphology, being visited by a male rufous hummingbird (*Selasphorus rufus*), a generalist pollinator that does not match the flower's pollination syndrome.

Finally, we include some words on the layout of the book and how to use it. In our planning, we envisioned several themes that illustrate exciting recent advances, and we divided the book into sections representing those themes. Because what authors wrote often crosses our somewhat arbitrary conceptual boundaries, as one hopes and expects in a vigorous field, we warn the reader in advance that our placement of chapters into sections is imperfect. At the beginning of each section, except for the first (which comprises a single introductory chapter on the longer history of the ideas) and last (which comprises a capstone chapter extending some of the themes of the book from pollination to other mutualisms), we and others have provided brief introductions. As editors, we strove to induce authors to produce final chapters of similar length and level, the level being one that would stimulate those established in pollination biology, those entering the field, and those in allied fields, without wandering into excessive technical detail. The reader must judge how well we succeeded!

In closing, we extend our warmest thanks to a number of people. First and foremost, we thank our colleagues who wrote chapters for the book; their effort and commitment come through clearly in the quality of what you are about to read. Other valued colleagues generously provided external peer reviews of chapter manuscripts: Ruben Alarcón, Anita Diaz, Yoko Dupont, Andreas Erhardt, Dave Goulson, Dennis Hansen, Jason Hoeksema, J. Nathaniel Holland, Rebecca Irwin, Frances Knapczyk, Claire Kremen, Andrew Lack, Jane Memmott, Christine Müller, Paul Neal, Mary Price, Alastair Robertson, Ann Smithson, Karen Strickler, Ørjan Totland, and Neal Williams. We are likewise indebted to the many authors of the book who provided internal peer review of one another's chapters. Constructive criticism on the book's draft manuscript came from Christine Müller in Zürich, Beverly Rathcke in Ann Arbor, and one anonymous referee—we thank you! Jenny Townend's help in preparing the index was timely and invaluable. Our editor par excellence at the University of Chicago Press, Christie Henry, provided sage advice and guidance along with unfailing (but not uncritical!) support, and Jennifer Howard and Christine Schwab guided us expertly through final production of the book.

Last but not least, we thank Mary Price and Jenny Townend for adding those essential ingredients, encouragement and love, throughout the gestation of the book.

PART I

Introduction and History

CHAPTER ONE

Specialization and Generalization in Plant–Pollinator Interactions: A Historical Perspective

Nickolas M. Waser

und bey den Pflanzen, bey denen ihre allzu nahe Nachbarschaft, der Wind und Insekten zu einer widernatürlichen Vermischung täglich Gelegenheit geben.
(and with the plants, whose all too close proximity gives daily opportunity to wind and insects for an unnatural mixing.)
—Kölreuter (1761)

[Ein System] müßte auch die Beziehung zu den Bestäubern berücksichtigen, wenn es den Kern der Blütenbiologie überhaupt treffen soll.
([A system] also ought to consider the relationship to the pollinators, if it is at all to get at the essence of floral biology.)
—Vogel (1954)

The angiosperms, or flowering plants, comprise about one-sixth of all described species, and the insects almost two-thirds (Wilson 1992). These speciose groups thereby dominate the flora and fauna of Earth's land surface, and interactions between them are dominant elements of terrestrial ecosystems. The most obvious of these interactions is that between flowering plants and insect herbivores, an elaborate evolutionary arms race which fosters remarkable adaptations in both adversaries (Labandeira et al. 1994; Schultz 2002). Almost as obvious is the interaction between flowering plants and the insects that visit and pollinate their flowers.

Not all insects are pollinators, nor pollinators insects. Most pollinators are drawn from the insect orders Hymenoptera, Diptera, Lepidoptera, and Coleoptera, and these animals are joined by some species from other insect orders and from the vertebrates, in particular some birds and bats (Proctor et al. 1996). The majority of angiosperms (by some estimates the great majority; Nabhan and Buchmann 1997; Renner 1998) rely in whole or part on such animals for pollination, rather than on abiotic agents such as wind or water. Hence, biotic pollination links some quarter-million angiosperm species with a similar number of insect and other animal species. From this emerges another quantitative conclusion: that pollination by animals constitutes a dominant, largely (but not entirely; Renner, chap. 6 in this volume) mutualistic ecological interaction of terrestrial habitats, perhaps, joining the mycorrhizal mutualism between plants and fungi in its ubiquity.

Those pollination interactions that truly are mutualistic are beneficial to plants and animals, but this does not imply cooperation (Waser and Price 1983; Howe 1984; Westerkamp 1997). The mutualism also directly benefits humanity through crop productivity, and indirectly, through ecosystem health. Hence, pollination is an important (and gratis) ecosystem "service" (Costanza et al. 1997). As is true for other ecosystem services, pollination by animals is not replaceable to any appreciable degree by technology. Thus, it is of concern that the mutualism is under threat from habitat alteration, invasive species, climate change, and other factors (e.g., Nabhan and Buchmann 1997; Kearns et al. 1998; Kremen et al. 2002; Steffan-Dewenter et al. 2001, chap. 17 in this volume). Although uncertainty surrounds the final effect of these anthropogenic changes on pollination (e.g., Cane and Tepedino 2001), there is enough cause for alarm that conservation biologists—not only "pure" ecologists—seek a deeper understanding of plant–pollinator interactions.

A fundamental aspect of any ecological interaction, including any pollination mutualism, is the degree to which it is specialized or generalized. Specialization and generalization suggest a dichotomy, but this is incorrect and only reflects the limits of language (Waser et al. 1996). In fact, obligate specialization and extreme generalization represent two ends of a continuum in resource use or niche breadth. For plants, the resource is pollination services from one or more species of animals that visit flowers; for animals, it is food or other benefit gained from such visits to one or more species of flowers. For simplicity, this summary leaves aside cases of "cheating" (which in themselves demonstrate that the interaction is not cooperative and sometimes not even mutualistic), as exhibited, for example, by flowers that promise food or sex but provide none (Renner, chap. 6 in this volume) or by visitors that extract nectar from flowers without pollinating (Irwin et al. 2001). In short, a deeper understanding of pollination requires a clearer picture of the range of specialization and generalization of plants and pollinating animals, of temporal and spatial variation in these niche relationships, and of their deeper explanation in terms of factors that constrain and promote the evolution of niche breadth (Waser et al. 1996). This is a large endeavor!

The purpose of this chapter is to provide a brief sketch of this endeavor to date. I have argued that scientists may make indifferent historians of science (Waser 1997); therefore, I approach this task with some trepidation and with the warning that many historical threads will remain unexplored. I will attempt to trace the development of some major ideas about specialization and generalization in pollination interactions based on historical summaries in the literature and my own reading and interpretation of some of the classic works.

The Past

Kölreuter and Sprengel

Joseph Gottlieb Kölreuter (1733–1806) is largely responsible for inaugurating the scientific—including experimental—study of pollination by animals. While at the University of Tübingen, Kölreuter was exposed to the ideas of J. G. Gmelin on the possibility of hybridization of species and to the earlier work of R. J. Camerarius showing that plants reproduce sexually (Mayr 1986). This exposure led to Kölreuter's own work on plant sexuality, the most important of which was described in a "preliminary" report (*Vorläufige Nachricht*) in 1761 and in sequels appearing in 1763, 1764, and 1766. These reports are written in a fashion quite different from modern scientific writing: they explore pollination (including self-pollination, wind pollination, and pollination by insects), sexual characteristics of flowers, and hybridization of species in an almost stream-of-consciousness manner, with few landmarks to guide the reader. The result is that important insights often must be extracted from a welter of details, and a given topic repeatedly disappears and later reappears. The analysis by Mayr (1986) of parts of Kölreuter's *Voläufige Nachricht* is a valuable aid in reconstructing its major themes.

For the present purpose, what interests us most is the recognition of animals as pollinators (fig. 1.1). Earlier workers had hinted at this interaction, but Kölreuter was the first to fully recognize it. He observed that many of the insects seen flying around flowers are carriers of pollen, which adheres to their bodies and is transferred to stigmas as a fortuitous by-product of the animals' activities in extracting sweet droplets of nectar for their own nourishment. To support the premise that these activities cause pollination, Kölreuter turned to experiments in which he added pollen to stigmas by hand, and characterized the relationship between number of pollen grains and production of fruits and seeds, as well as experiments in which he excluded insects from flowers, showing that this caused failure of fruit production. These studies involved common plants of a northern European garden—in particular, cucurbits, irises, and mallows, but also snapdragons, elderberries, and others—and common visitors such as bumblebees, wasps, flies, and thrips. Kölreuter (1761, 36) ventured to extend his results from this context to flowering plants in general, concluding that "[insects] probably provide this uncommonly great service, if not to most plants, then at least to a very large portion of them" (und wahrscheinlich leisten [Insekten], wo nicht den allermeisten Pflanzen, doch wenigstens einem sehr großen Theil derselben, diesen ungemein großen Dienst). This conclusion seems prophetic in light of recent estimates that 90% or more of angiosperm species benefit from animal pollination (e.g., Renner 1998).

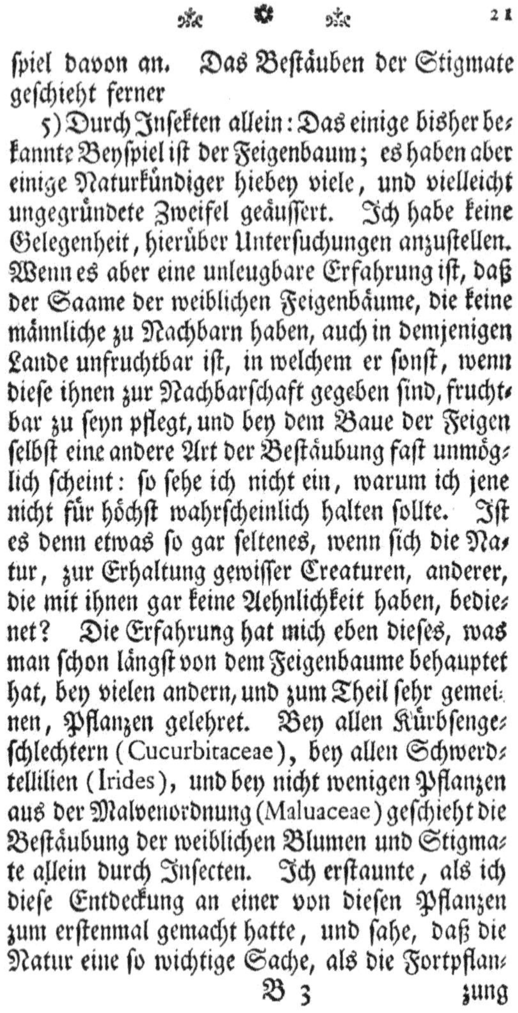

21

spiel davon an. Das Bestäuben der Stigmate geschieht ferner

5) Durch Insekten allein: Das einige bisher bekannte Beyspiel ist der Feigenbaum; es haben aber einige Naturkündiger hiebey viele, und vielleicht ungegründete Zweifel geäussert. Ich habe keine Gelegenheit, hierüber Untersuchungen anzustellen. Wenn es aber eine unleugbare Erfahrung ist, daß der Saame der weiblichen Feigenbäume, die keine männliche zu Nachbarn haben, auch in demjenigen Lande unfruchtbar ist, in welchem er sonst, wenn diese ihnen zur Nachbarschaft gegeben sind, fruchtbar zu seyn pflegt, und bey dem Baue der Feigen selbst eine andere Art der Bestäubung fast unmöglich scheint: so sehe ich nicht ein, warum ich jene nicht für höchst wahrscheinlich halten sollte. Ist es denn etwas so gar seltenes, wenn sich die Natur, zur Erhaltung gewisser Creaturen, anderer, die mit ihnen gar keine Aehnlichkeit haben, bedienet? Die Erfahrung hat mich eben dieses, was man schon längst von dem Feigenbaume behauptet hat, bey vielen andern, und zum Theil sehr gemeinen, Pflanzen gelehret. Bey allen Kürbsgeschlechtern (Cucurbitaceae), bey allen Schwerdtellilien (Irides), und bey nicht wenigen Pflanzen aus der Malvenordnung (Maluaceae) geschieht die Bestäubung der weiblichen Blumen und Stigmate allein durch Insecten. Ich erstaunte, als ich diese Entdeckung an einer von diesen Pflanzen zum erstenmal gemacht hatte, und sahe, daß die Natur eine so wichtige Sache, als die Fortpflan-

B 3 zung

Figure 1.1 The page from Kölreuter (1761) that introduces the discussion of pollination of flowers by insects alone

Kölreuter's painstaking observations in nature, including entire days spent watching single plants, gave him a remarkable insight into the use of insects by flowers and flowers by insects. In most cases, he described multiple types of insects visiting a given plant species and, although this was not explicitly discussed, the impression is that some of the insects were observed visiting several plant species. As the quotation at the beginning of this chapter suggests, Kölreuter recognized that such generalized niche relationships set the stage for hybridization between related plant species, a possibility of interest (and which he described as "unnatural") given his assumption that species were fixed entities with absolute barriers between them. Kölreuter's subsequent experimental crosses demonstrated to his surprise that hybridization, indeed, could occur.

Kölreuter (1761, 23) described his discoveries of plant-pollinator interaction as "secrets of nature" (*Geheimnisse der Natur*), a term that reappears in the title of Christian Konrad Sprengel's 1793 book on the structure and fertilization of flowers. This classic work describes in detail features of flowers and their functions, which are taken as evidence for intentional design for pollination by insects or wind. In introducing the thesis of intricate and beautiful design for pollination, Sprengel summarizes numerous patterns and deductions which are foundational to subsequent studies of pollination, such as the structure and function of nectaries (nectar-secreting tissues), the function of flower colors (including "nectar guides"; areas of contrasting color assumed to guide visitors to rewards) and odors, the existence of deceit flowers and of insects that rob for nectar, and many other themes (Vogel 1996). Sprengel (1793, 17) mentions Kölreuter's earlier recognition of dichogamy (different timing of the maturation of anthers and stigma within hermaphroditic flowers) and, therefore, had read the work of his predecessor, although there has been some confusion on this point (Faegri and van der Pijl 1966, 2) and some incorrect assignment of precedence of certain contributions (see Vogel 1996).

Sprengel (1793, 19–20) concluded that "It is certain that many flowers are fertilized by multiple species of insects . . . It also is certain that many flowers are fertilized solely by one species of insect, and this in a very distinct fashion" (Es ist gewiß, dass viele Blumen von mehrern Arten von Insekten befruchtet werden . . . Es ist aber auch gewiß, daß viele Blumen bloß von einer Art von Insekten, und zwar auf eine sehr bestimmte Art, befruchtet werden . . .) and furthermore (1793, 43) that "in the first case, the fertilization of the ovary and production of fruit must progress more easily" (so muß . . . im ersten Fall die Befruchtung des Fruchtknotens und die Erzielung der Frucht leichter vor sich gehen) since there is less danger that pollination will utterly fail. As Vogel (1996) points out, such conclusions are based on *comparative* studies of several hundred plant species observed in natural or nearly natural conditions, which contrasts with Kölreuter's *experimental* studies of pollination of a much shorter list of native plant species. There appears to be a trade-off here, however, because Sprengel's individual descriptions of plant species focus far more on floral features than on insect visitors, and the latter (when mentioned at all) sometimes are inferred from the former, all suggesting that Sprengel was unable to spend substantial time observing pollination species by species.

Sprengel's influential work helped to inaugurate two different themes in the study of pollination, both of which have been pursued until the present time. The first is the steady accumulation of empirical information on pollination relationships of individual plant species (and to a lesser extent on floral affinities of individual pollinator species), which at its best combines the comparative scope of Sprengel with the lengthy observations pioneered by Kölreuter of each species. The second theme is the attempt to find order in the diversity of floral phenotypes, ultimately in relationship to the diversity of pollination relationships. Thus, Sprengel categorized the hundreds of flowers he studied into "classes," based on Linné's system of number and arrangement of stamens, combined with reward for pollinators (with or without nectar) and sex expression (hermaphroditic, monoecious, or dioecious). This scheme of classification did not survive, but it does presage later attempts which culminate ultimately in later ideas of the *pollination syndromes.*

Encyclopedic Observations of Pollination

Charles Darwin was influenced by Sprengel in his studies of plant sexuality, just as he was influenced by Kölreuter in his thoughts on hybridization. Where Sprengel saw intentional design in floral features, Darwin (1859) argued convincingly for adaptation by natural selection and elaborated on Kölreuter's much earlier recognition of the differing interests of plants and pollinators (sexual reproduction vs. food acquisition) that drive the mutualism and explain many of its observable dynamics (e.g., Waser 2000). Except that it is recast as the study of floral adaptation, Darwin's (1862) famous examination of orchid flowers follows in the tradition of Sprengel, even to the extent that description of the pollinators constitutes a minor theme. Although Darwin (1869) later contributed more detailed observations of orchid pollination, and discussion of pollination and pollinator behavior are found in some of his other writings (e.g., Darwin 1876), his main purpose (other than a general argument for adaptation) was to explore the thesis that selection favors cross-fertilization.

Darwin, therefore, placed the scientific study of pollination in its modern evolutionary framework. But, whereas he did contribute to the growing empirical description of who pollinates whom, the major contributions fell to botanist Hermann Müller, a champion of Darwinism in Germany and continental Europe, and to Müller's successor, Paul Knuth. Müller's compendium, published in German in 1873 as *Die Befruchtung der Blumen durch Insekten* and a decade later in slightly truncated translation as *The fertilisation of flowers,* brings together lists of pollinators and details of flower and pollinator characteristics for some 400 plant species, and more cursory descriptions for several thousand more, and also presents details of morphology and behavior related to pollination for the major pollinating insect orders. The justification given for this effort was the study of adaptations for cross-pollination, echoing Darwin's (1876) focus. A similar structure is found in Paul Knuth's *Handbuch der Blütenbiologie,* which provides records of flower visitation and pollination for 4028 European species and 2357 species in other parts of the world. This massive effort was published in German between 1898 and 1905 in three volumes, the last of which appeared after Knuth's death;

the first two (but not the third!) were translated as *Handbook of flower pollination* between 1906 and 1909. (These were published as three volumes; nonetheless, they are missing some 1118 pages of Knuth's original.) Knuth (1898) introduced this work with a valuable historical overview of pollination research to the end of the 19th century, echoing the earlier historical summary of Müller (1883).

Attempts at Floral Classification

Baker (1983) refers to Knuth's *Handbook* as the "old Testament" of pollination biology, marking the close of an era of enthusiastic accumulation of information on pollination relationships. This accumulation continues, thereafter, at a slower rate up to the present. Over the same period, ideas were also circulating about how one might bring order to the welter of detail on who pollinates whom, by finding some natural scheme for classifying flowers.

As noted earlier, the scheme used by Sprengel (1793) did not persist, presumably because classification of flowers based on number of pistils and nectar reward is not particularly natural or useful. The efforts of Italian botanist Federico Delpino (1868–1875) were much more influential. As had his contemporaries, Delpino contributed to detailed records of pollination, including pollination by birds. But his more lasting contribution was in proposing not one, but two general schemes for the classification of flowers, with the explicit purpose of finding natural groupings (Vogel 1954).

Delpino's first scheme comprises no fewer than 47 "types" (*tipi*) of flowers, subsumed under 13 larger classes (Knuth 1898). Whereas proposals in the 1860s from Hildebrand and Axell arranged flowers in terms of sex expression and its presumed consequences for self-pollination (or lack thereof), Delpino's classes interpreted a much broader set of morphological and structural traits of flowers as convergent adaptations that suited flowers to certain behaviors or morphologies of visitors. For example, Class IX, Arrangements for hovering visitors (*Apparecchi circumvolatorii*), contains flowers of five types adapted for visits by hovering animals. Delpino's second scheme will seem more familiar to a modern reader: it arranges plants according to the agent of pollination in taxonomic terms. To pollination by water (hydrophilae), wind (anemophilae), small animals (zoidiophilae), birds (ornithophilae), and snails (malacophilae) is added pollination by insects (entomophilae), which is divided further according to insect order in a fashion that persists in some versions of the pollination syndrome idea. Whereas Delpino did describe some features of flowers within these last groups, he did not merge his two classification schemes more exactly (i.e., connect the 6 groups whose names end in the Latin "ae" with the 13 classes and their 47 subordinate flower types).

Delpino's efforts kindled an animated response in his contemporaries and inaugurated a debate on the ideal classification scheme that led 75 years later to the pollination syndromes. As Vogel (1954) describes, proposed schemes vacillated between emphasis on the type of reward gained by pollinators and features of floral morphology versus identity of major pollinator, never achieving a satisfactory fusion of the different perspectives. Using the flora of southern Africa as a basis (and as suggested by the quotation at the beginning of this chapter), Vogel (1954) proposed to finally effect this fusion with a detailed description of six floral "styles" encompassing details of reward, morphology, color, scent, and other phenotypic traits that represent convergent adaptations of unrelated flowers for particular types of pollinators (fig. 1.2). Van der Pijl (1961), Faegri and van der Pijl (1966), Baker and Hurd (1968), and others subsequently incorporated Vogel's efforts into a larger set of pollination syndromes, in the process simplifying and modifying Vogel's descriptions. Whereas most of the discussion of classification schemes had been published in German to this point, this last set of workers used English as the shared language to promulgate the syndromes, greatly increasing their impact.

The modern architects of the pollination syndromes were careful to retain a place for generalized flowers adapted to pollination by diverse insects. At the same time, the very enterprise of classifying flowers according to "the pollinator" furthers the idea that many species, particularly those with complex flowers, are pollinated by visitors of a single taxon or body plan. This conviction is manifest in the thinking of some workers who went beyond Darwinian adaptation in search of intrinsic organizing principles which ipso facto would justify a certain classification of flowers. One example is the argument by Vogel (1954, 10–23) for the reality of the floral "styles" because of intrinsic harmony within the flower and between flower and pollinator. That this was not unique explains the lengths that van der Pijl (1961) took to discredit such teleological approaches. Although those who subsequently elaborated the pollination syndromes cautioned against too narrow an interpretation (e.g., Baker and Hurd 1968, 15) and eschewed essentialism, orthogenesis, "autonomous correlated types," and the like, echoes remain, for example, in Faegri and van der Pijl's (1966) "harmonic relations between pollinators and blossoms."

In short, the history of floral classification is shot through with the conviction that a natural order *must* exist, reflecting Darwinian adaptation or more mysterious organizing forces that harmonize many flowers with a single pollinator type. The existence of generalized flowers is clearly stated, but such flowers often are referred to as "primitive" or "lower," indicating the additional conviction that evolutionary change has tended toward greater specialization of flowers for pollinators (e.g., Leppik 1957, 1964; van der Pijl 1961; Crepet 1983).

Modern Times

I have developed two historical themes that bear on our understanding of specialization and generalization. On the one hand, close observation of flowers in their natural setting, over long periods, sometimes will reveal a single species or

TABLE OF CHARACTERISTICS OF ZOOPHILOUS FLORAL STYLES

Style	Color and pattern (as perceived by humans)	Usual shape ("Gestalt")	Pecularities of shape and proportions	Nectar properties	Odor properties (as perceived by humans)	Periodicity of movement and scent	Anatomical peculiarities
Melittophily and micro-melittophily (large and small bees)	Blue, violet, purple, yellow, white; nectar guides usually present, subdivided	Papilionaceous flowers (flag type), lip- (gullet-) type, tubular type, "brush"-form (Nototribic and sternotribic pollination)	Underside of the flower expanded (landing platform), tube fairly narrow, flowers often designed for entry of an animal, filiform peduncles, entrance to nectar often hidden	Nectar hidden up to 15mm deep	Often strong and pleasant, more like honey	Blooms or opens during day, scent present during day	Silken or velvety sheen, ± robust
Psychophily (butterflies)	scarlet, purple, blue, yellow, white; nectar guides usually present, subdivided	Salverform type (with long tube or spur) (Pollination by wings, head, or proboscis)	± Disc-like display with simple margins, gullet and tube narrow, anthers often pendelous	Nectar hidden up to 40mm deep, or pollen-nectar distance up to 40mm	Pleasant, more like honey	Blooms or opens during day, scent present during day	Delicate
Sphingophily and phalenophily (hawkmoths and moths)	Sph.: white, cream, dull violet, underside washed with dull purple Phal.: yellow-green, green Both: Nectar guides always absent	Salverform type, "paintbrush" type (Pollination by wings, head or proboscis)	± Star-shaped display, often finely dissected wrinkled margins, gullet and tube very narrow, anthers often pendelous	Sph.: Nectar hidden up to 200mm and deeper, pollen-nectar distance up to 200mm Phal.: Nectar 4 to 20mm deep	Pleasant to the point of intoxicating, like perfume	Blooms or opens in evening and during night, scent present during night	Often with waxy surface, delicate
Myiophily and micromyiophily (flies)	brown-red, brown, flesh colors, dirty yellow, green-white Patterned with dots and stripes over entire petals	Basin- or saucer-shaped Kettle-shaped (Pollinated by proboscis or legs)	Flowers low to ground, more or less flat, entrance often like a camera aperture, formation of clear areas, wrinkles, motile appendages, flickering and shimmering bodies	Nectar exposed, easily accessible	Nauseating	Open during the day, usually without periodicity of movement or scent	Reflective or dull, warty surface, ciliated
Ornithophily (birds)	Scarlet, red-orange, carmine, yellow-green, pure blue, pure white, dark violet Nectar guides where present simple (not subdivided), black, yellow or green	Tubular, salverform type, "paintbrush" type (Pollination by throat, forehead, or beak)	Upper side of the flower expanded (entry open from underneath), that is, extension of the upper lip, sexual parts bent upward together, tube broadly sacklike, inflorescence lifted high up	Nectar at various depths, dilute, slimy, very plentiful	Lacking	Open during day, usually without periodicity	Strengthened mechanical elements, therefore robust, stiff filaments, capillary mechanism
Chiropterophily (bats) (after Porsch)	White, cream-colored, nectar guides absent	Wide open tube			Fruity and beet-like, unpleasant	Blooms or opens during night, smells during night	± Fleshy sepals and petals

Figure 1.2 A facsimile, in translation from the German, of the table from Vogel (1954, 38–39) that is the precursor of more recent classifications of flowers into "pollination syndromes"

type of visitor and often will reveal more than one species or type, sometimes many. I will characterize this as the view of the field ecologist, the empiricist who chronicles; it is a view that suggests frequent generalization. On the other hand, the human brain perceives patterns in the tremendous diversity of floral expression and perceives that these *clusters of similar phenotypes in phenotype space* are to be understood in relation to particular pollinators. I will tempt fate again and characterize this as the view of the orderly person, the synthesizer who seeks to find pattern; it is a view that suggests specialization as the norm.

By the early 1970s, revived interest in mutualisms (as compared to antagonistic ecological interactions), fascination with the idea of coevolution among species, and a focus on the measurement of niche relationships coincided with synthetic treatments by Faegri and van der Pijl (1966) and others to foster renewed study of pollination as a model interaction, of interest beyond the specific system. Many themes were pursued: the study of floral variation and adaptation; foraging behavior of pollinators in an economic context; pollination, gene flow, and genetic structure of plant populations; community-level patterns of pollination; experiments to determine which visitors were effective as pollinators; and others. (Real [1983] provides examples of these studies and further leads.) This effort ultimately served to juxtapose different views of specialization and generalization, including the two views just described. Ollerton (1996) clearly expressed the tension and questions raised by this juxtaposition: we see apparently specialized adaptations of flowers for particular pollinator types at the same time that careful observation often detects multiple types of visitors to the flowers (see also C. M. Herrera 1988, 1996; Waser et al. 1996; Waser 1998). How could such specialization arise in a generalized pollination environment? Indeed, is our perception of specialized adaptation an exaggeration of reality?

Some earlier workers foresaw Ollerton's paradox and its possible resolution. Stebbins (1970) proposed that floral adaptation is explained in reference to the "most effective" pollinator, with other pollinators safely ignored. Similarly, not all visitors to flowers are pollinators—a fact recognized by those who began to catalog pollination interactions 150 years ago, and all their successors. If most visitors are stealing resources, the generalized pollination environment might be an illusion (although floral adaptation still ought to be influenced by such interactions!). Whether these ideas or others that have been suggested (e.g., Waser 1998) can resolve Ollerton's paradox is largely an empirical issue—one needs to know which animals *do* pollinate and how many of them are surprises (e.g., bees as well as birds pollinating "bird flowers"). In short, the resolution depends on a more complete and quantitative understanding of the use of animals by plants as pollinators, and vice versa, and this across different ecosystems and geographic regions of the planet.

The Future

The two views of pollination I have developed here—that of the field ecologist and that of the orderly synthesizer—do not define all of pollination biology, and I do not juxtapose them to suggest that this is so or to generate a false dichotomy or disagreement. Instead, I contend that the tension between these views (and more generally among seemingly disparate views) shows us where some important unanswered questions lie and, in the specific case under discussion, defines growing points for our understanding of specialization and generalization. These questions and growing points are the focus of the book in your hands, and in reading it, I believe you will join me in concluding that the near future holds promise for rapid progress. I will end with some of my own guesses as to what is in store as we examine niche relationships in pollination systems more carefully and quantitatively. I will comment on network structure, floral classification, and floral adaptation.

Much of pollination biology over the past few centuries logically focused on a single plant or pollinator species and its mutualistic partners, whereas a focus at the level of entire communities was uncommon. Recently we see a revival of community studies, encouraged largely by new tools borrowed from the theory of food webs (and more broadly of interaction networks) that allow us to characterize and analyze the resulting patterns. For example, pollination networks show asymmetry—most specialist insects visit generalist plants, and most specialist plants are visited by generalist insects (e.g., Petanidou and Ellis 1996; Vázquez and Simberloff 2002; Dupont et al. 2003; Bascompte et al. 2003; Ollerton et al. 2003; Memmott et al. 2004; Vázquez and Aizen, chap. 9 in this volume; Petanidou and Potts, chap. 10 in this volume). This is a striking departure from the traditional implication of coevolved specialists! Pollination networks also seem to show scale invariance (i.e., the number of interactions per species, k, is distributed as a power law with negative exponent: $P(k) \propto k^{-\gamma}$), at least over most of the possible range of interactions per species (Jordano et al. 2003, chap. 8 in this volume). What do these discoveries suggest about the assembly of pollination communities in ecological and evolutionary time? Vázquez and Aizen (2003, chap. 9 in this volume) show that the number of interactions per species is strongly related to relative abundance of species. Therefore, might the succession of independent events in the life histories of organisms, which May (1975) posited to underlie species abundance distributions in communities, ultimately explain the topology of pollination networks? In this regard, can a lognormal distribution of relative abundance yield something close to the power-law (or truncated power-law) distributions that seem to describe pollination interactions? And, finally, if relative abundance of species explains X% of specialization and generalization in pollination networks (leaving aside the next deeper level of why this is so), what explains the remaining (100 − X)%, and how are we to explore this minority, perhaps, deceptively "statistically insignificant," set of influences?

Earlier, I characterized those who attempt to define pollination syndromes as synthesizers seeking a pattern. I hasten to add that I fully subscribe to MacArthur's (1972, 1) famous exhortation that "To do science is to search for repeated patterns, not simply to accumulate facts." At the same time, there is a danger that the endeavor of floral classification can *impose* pattern on nature, rather than discover it. Luckily we now have new tools at our disposal in the search for pattern that should minimize this risk. The pollination syndromes postulate the existence of clusters of similar floral phenotypes in a "phenotype space," and this space is necessarily multidimensional. The new tools are those of multivariate analysis, which allow us to search for clusters in such a space. What clusters will appear, and to what extent will they resemble those put forward by the architects of floral styles and pollination syndromes, who had available only the remarkable pattern recognition of the human brain? Steps toward answers have already been taken (J. Herrera 1988; Ollerton and Watts 2000; Alarcón 2004). Perhaps we will end up with classifications that apply only by ecoregion or by plant taxon, or that involve hitherto unappreciated phenotypic traits or trait combinations (e.g., Thomson et al. 2000; Kay and Schemske 2003; Wilson et al., chap. 3 in this volume; Corbet, chap. 14 in this volume), but the revised classifications surely will be a better match to our growing comprehension of pollination networks.

Thus, the future holds the certainty that we will reexamine the floral phenotype. How are we to understand floral adaptation? A subtheme in the development of pollination syndromes is that these suites of characters adapt unrelated flowers to specific types of pollinators because of perceptual or cognitive biases and constraints of the animals that limit them to certain flower colors, shapes, and so on (e.g., Leppik 1957, 1964; Faegri and van der Pijl 1966). But, a modern understanding of insect sensory and cognitive abilities suggests more subtlety and flexibility than imagined earlier (see Chittka and Thomson 2001), making this explanation unlikely and seeming to accord with the less-certain status of the classical syndromes. Nonetheless, floral complexity and diversity exist! The easiest explanation is that complexity and diversity reflect coevolution and cospecialization of flowers and pollinators, but this seems at odds with the multiple visitors to many complex flowers and the common association of generalists with specialists noted earlier. Stebbins (1970) acknowledged multiple pollinators of complex flowers, but sought to collapse this dimensionality back to a single pollinator of the flower. Is this resolution of adaptation sufficient? Careful studies of the pollination abilities of visitors to flowers, including the "wrong" visitors, suggest that it is not always sufficient or satisfying (e.g., Mayfield et al. 2001; Wolff et al. 2003). But, the possible explanations for floral complexity, diversity, and adaptation are far from exhausted. Radically different theoretical frameworks are beginning to appear (e.g., Aigner 2001, chap. 2 in this volume), and more are surely on the way.

Acknowledgments

For ideas, encouragement, and critical reading of drafts, I thank Paul Aigner, Ruben Alarcón, Scott Armbruster, Andreas Erhardt, Andrew Lack, Jeff Ollerton, and Mary Price. The insights of Andreas Erhardt into the contributions of earlier workers have greatly enriched my own understanding, and he and Susanne Renner kindly offered assistance with translation of more obscure parts of the table from Vogel (1954). Other than this, all translations from the German in this chapter are my own.

References

Aigner, P. A. 2001. Optimality modeling and fitness tradeoffs: when should plants become pollinator specialists? Oikos 95: 177–184.

Alarcón, R. 2004. The structure of plant-pollinator interactions in montane meadow environments. PhD dissertation, University of California, Riverside, CA.

Baker, H. G. 1983. An outline of the history of anthecology, or pollination biology. Pp. 7–28 *in* L. Real (ed.), Pollination biology. Academic Press, Orlando, FL.

Baker, H. G., and P. D. J. Hurd. 1968. Intrafloral ecology. Annual Review of Entomology 13: 385–414.

Bascompte, J., P. Jordano, C. J. Melián, and J. M. Olesen. 2003. The nested assembly of plant-animal mutualistic networks. Proceedings of the National Academy of Sciences (USA) 100: 9383–9387.

Cane, J. H., and V. J. Tepedino. 2001. Causes and extent of declines among native North American invertebrate pollinators: Detection, evidence, and consequences. Conservation Ecology 5. Online URL http://www.consecol.org/vol5/iss1/art1.

Chittka, L., and J. D. Thomson. 2001. Cognitive ecology of pollination. Cambridge University Press, Cambridge.

Costanza, R., R. d'Arge, R. de Groot, S. Farber, M. Grasso, B. Hannon, K. Limburg, S. Naeem, R. V. O'Neill, J. Paruelo, R. G. Raskin, P. Sutton, and M. van den Belt. 1997. The value of the world's ecosystem services and natural capital. Nature 387: 253–260.

Crepet, W. L. 1983. The role of insect pollination in the evolution of the angiosperms. Pp. 29–50 *in* L. Real (ed.), Pollination biology. Academic Press, Orlando, FL.

Darwin, C. 1859. On the origin of species by means of natural selection, or the preservation of favoured races in the struggle for life. Murray, London.

Darwin, C. 1862. The various contrivances by which orchids are fertilised by insects. Murray, London.

Darwin, C. 1869. Notes on the fertilization of orchids. Annals and Magazine of Natural History (Series IV) 4: 141–158.

Darwin, C. 1876. The effects of cross and self fertilisation in the vegetable kingdom. Murray, London.

Delpino, F. 1868–1875. Ulteriori osservazione e considerazioni sulla dicogamia nel regno vegetale. Atti della Societa Italiana di Scienze Naturale in Milano 11: 265–332; 12: 21–141, 179–233.

Dupont, Y. K., D. M. Hansen, and J. M. Olesen. 2003. Structure of a plant-animal flower-visitor network in the high altitude subalpine desert of Tenerife, Canary Islands. Ecography 26: 301–310.

Faegri, K., and L. van der Pijl. 1966. The principles of pollination ecology. Pergamon, Oxford.

Herrera, C. M. 1988. Variation in mutualisms: The spatiotemporal mosaic of a pollinator assemblage. Biological Journal of the Linnean Society 35: 95–125.

Herrera, C. M. 1996. Floral traits and plant adaptation to insect pollinators: A devil's advocate approach. Pp. 65–87 *in* D. G. Lloyd and S. C. H. Barrett (eds.), Floral biology: Studies on floral evolution in animal-pollinated plants. Chapman and Hall, New York.

Herrera, J. 1988. Pollination relationships in southern Spanish Mediterranean shrublands. Journal of Ecology 76: 274–287.

Howe, H. F. 1984. Constraints on the evolution of mutualism. American Naturalist 123: 764–777.
Irwin, R. E., A. K. Brody, and N. M. Waser. 2001. The impact of floral larceny on individuals, populations, and communities. Oecologia 129: 161–168.
Jordano, P., J. Bascompte, and J. M. Olesen. 2003. Invariant properties in coevolutionary networks of plant–animal interactions. Ecology Letters 6: 69–81.
Kay, K. M., and D. W. Schemske. 2003. Pollinator assemblages and visitation rates for 11 species of neotropical *Costus* (Costaceae). Biotropica 35: 198–207.
Kearns, C. A., D. W. Inouye, and N. M. Waser. 1998. Endangered mutualisms: The conservation of plant–pollinator interactions. Annual Review of Ecology and Systematics 29: 83–112.
Knuth, P. 1898. Handbuch der Blütenbiologie. I. Band: Einleitung und Literatur. Verlag Wilhelm Engelmann, Leipzig.
Kölreuter, J. G. 1761. Vorläufige Nachricht von einigen das Geschlecht der Pflanzen betreffenden Versuchen und Beobachtungen. Gleditschischen Handlung, Leipzig.
Kremen, C., N. M. Williams, and R. W. Thorp. 2002. Crop pollination from native bees at risk from agricultural intensification. Proceedings of the National Academy of Sciences (USA) 99: 16812–16816.
Labandeira, C. C., D. L. Dilcher, D. R. Davis, and D. L. Wagner. 1994. Ninety-seven million years of angiosperm–insect association: Paleobiological insights into the meaning of coevolution. Proceedings of the National Academy of Sciences (USA) 91: 12278–12282.
Leppik, E. E. 1957. A new system for classification of flower types. Taxon 6: 64–67.
Leppik, E. E. 1964. Floral evolution in the Ranunculaceae. Iowa State Journal of Science 39: 1–101.
MacArthur, R. H. 1972. Geographical ecology: Patterns in the distribution of species. Harper and Row, New York.
May, R. M. 1975. Patterns of species abundance and diversity. Pp. 81–120 *in* M. L. Cody and J. M. Diamond (eds.), Ecology and evolution of communities. Belknap Press of Harvard University Press, Cambridge, MA.
Mayfield, M. M., N. M. Waser, and M. V. Price. 2001. Exploring the "most effective pollinator principle" with complex flowers: Bumblebees and *Ipomopsis aggregata*. Annals of Botany 88: 591–596.
Mayr, E. 1986. Joseph Gottlieb Kölreuter's contributions to biology. Osiris (Series 2) 2: 135–176.
Memmott, J., N. M. Waser, and M. V. Price. 2004. Tolerance of pollination networks to species extinctions. Proceedings of the Royal Society of London B 271: 2605–2611.
Müller, H. 1883. The fertilisation of flowers, trans. and ed. D'Arcy Thompson. Macmillan, London.
Nabhan, G. P., and S. L. Buchmann. 1997. Services provided by pollinators. Pp. 133–150 *in* C. Daily (ed.), Nature's services: Societal dependence on natural ecosystems. Island Press, Washington, DC.
Ollerton, J. 1996. Reconciling ecological processes with phylogenetic patterns: The apparent paradox of plant–pollinator systems. Journal of Ecology 84: 767–769.
Ollerton, J., S. D. Johnson, L. Cranmer, and S. Kellie. 2003. The pollination ecology of an assemblage of grassland asclepiads in KwaZulu-Natal, South Africa. Annals of Botany 92: 807–834.
Ollerton, J., and S. Watts. 2000. Phenotype space and floral typology: Towards an objective assessment of pollination syndromes. Det Norske Videnskaps—Akademi. I. Matematisk Naturvidenskapelige Klasse, Skrifter, Ny Serie 39: 149–159.
Petanidou, T., and W. N. Ellis. 1996. Interdependence of native bee faunas and floras in changing Mediterranean communities. Pp. 201–226 *in* A. Matheson, S. L. Buchmann, C. O'Toole, P. Westrich, and I. H. Williams (eds.), The conservation of bees. Academic Press, London.
Proctor, M., P. Yeo, and A. Lack. 1996. The natural history of pollination. Timber Press, Portland, OR.
Real, L. A. 1983. Pollination biology. Academic Press, Orlando, FL.
Renner, S. S. 1998. Effects of habitat fragmentation on plant pollinator interactions in the tropics. Pp. 339–360 *in* D. M. Newbery, H. H. T. Prins, and N. D. Brown (eds.), Dynamics of tropical communities. Blackwell Science, Oxford.
Schultz, J. C. 2002. How plants fight dirty. Nature 416: 267.
Sprengel, C. K. 1793. Das entdeckte Geheimniss dem Natur im Bau und in der Befruchtung der Blumen. Friedrich Vieweg dem aeltern, Berlin.

Stebbins, G. L. 1970. Adaptive radiation of reproductive characteristics in angiosperms. I: Pollination mechanisms. Annual Review of Ecology and Systematics 1: 307–326.
Steffan-Dewenter, I., U. Münzenberg, and T. Tscharntke. 2001. Pollination, seed set, and seed predation on a landscape scale. Proceedings of the Royal Society of London B 268: 1685–1690.
Thomson, J. D., P. Wilson, M. Valenzuela, and M. Malzone. 2000. Pollen presentation and pollination syndromes, with special reference to *Penstemon*. Plant Species Biology 15: 11–29.
van der Pijl, L. 1961. Ecological aspects of flower evolution. II. Zoophilous flower classes. Evolution 15: 44–59.
Vázquez, D. P., and M. A. Aizen. 2003. Patterns of specialization in plant-pollinator interactions: Myth or reality? Ecology 84: 2493–2501.
Vázquez, D. P., and D. Simberloff. 2002. Ecological specialization and susceptibility to disturbance: Conjectures and refutations. American Naturalist 159: 606–623.
Vogel, S. 1954. Blütenbiologische Typen als Elemente der Sippengliederung, dargestellt anhand der Flora Südafrikas. Botanische Studien 1: 1–338.
Vogel, S. 1996. Christian Konrad Sprengel's theory of the flower: The cradle of floral ecology. Pp. 44–62 *in* D. G. Lloyd and S. C. H. Barrett (eds.), Floral ecology, studies on floral evolution in animal-pollinated plants. Chapman and Hall, New York.
Waser, N. M. 1997. Roots and new shoots of floral ecology. Trends in Ecology and Evolution 12: 40.
Waser, N. M. 1998. Pollination, angiosperm speciation, and the nature of species boundaries. Oikos 82: 198–201.
Waser, N. M. 2000. Pollination by animals. *In* Encyclopedia of life sciences, Macmillan References Ltd., London. Online URL http://www.els.net/elsonline/html/A0003163.html.
Waser, N. M., L. Chittka, M. V. Price, N. Williams, and J. Ollerton. 1996. Generalization in pollination systems, and why it matters. Ecology 77: 279–296.
Waser, N. M., and M. V. Price. 1983. Optimal and actual outcrossing in plants, and the nature of plant-pollinator interaction. Pp. 341–359 *in* C. E. Jones and R. J. Little (eds.), Handbook of experimental pollination biology, Van Nostrand Reinhold, New York.
Westerkamp, C. 1997. Flowers and bees are competitors—not partners: Towards a new understanding of complexity in specialised bee flowers. Acta Horticulturae 437: 71–74.
Wilson, E. O. 1992. The diversity of life. Belknap Press of Harvard University Press, Cambridge, MA.
Wolff, D., M. Braun, and S. Liede. 2003. Nocturnal versus diurnal pollination success in *Isertia laevis* (Rubiaceae): A sphingophilous plant visited by hummingbirds. Plant Biology 5: 71–78.

PART II

The Ecology and Evolution of Specialized and Generalized Pollination

Introductory Comments by Jeff Ollerton, W. Scott Armbruster, and Diego P. Vázquez

One must not be annoyed at having to spend a long time near a flowering plant, and at having often to repeat the same observations on any species of flower, for it is not always visited forthwith by the particular insect which is designed to fertilize it.
—Sprengel (1793) quoted in Knuth (1906)

Species interactions are a key component of biodiversity. Without the diversity of antagonistic, mutualistic, comensalistic, and amensalistic interactions, ecosystems as we know them would simply cease to function. Energy would not flow between trophic levels, primary productivity would not be turned into secondary productivity, and inorganic nutrients would not cycle. In addition, species interactions are thought to be at the heart of many speciation events, with tight and diffuse coevolution acting to generate taxonomic and genetic diversity. This central role of interactions, then, provides a unifying conceptual framework to part 2 of this book—that plant–pollinator interactions are vital to the continued functioning of most terrestrial ecosystems and may ultimately be responsible for much of angiosperm diversity and species diversity within major insect orders such as Hymenoptera and Lepidoptera.

The chapters in this part illustrate diverse approaches to the study of biotic pollination. This diversity emphasizes how far pollination biology has developed since the earliest systematic observations in the 18th and 19th centuries. In addition to careful natural history observations (as shown by the preceding quote, a cornerstone of the discipline from at least the time of Sprengel), pollination biologists can, at relatively low cost, utilize a vast variety of field and laboratory techniques, including analyses of nectar volume and concentration, flower color spectra, counts of pollen loads, indirect measures of pollen movement, direct measures of rates of interaction between plants and pollinators, indices of pollinator effectiveness, and even quantification of natural selection on phenotype. For rather more financial outlay, sophisticated techniques can be used to analyze genetic diversity within and between populations, and between species, backed up with a sometimes bewildering assortment of statistical and phylogenetic methods. Finally, formal mathematical models are becoming more and more sophisticated in their purview.

The authors of chapters in this part have used these techniques to answer a range of questions relating to the ecology and evolution of specialized and generalized plant–pollinator interactions. In chapter 2, Aigner reviews and develops ideas regarding trade-offs in floral traits that can affect a plant's adaptation to different pollinators via differential selection on flower characteristics such as shape or color. In contrast to theoretical speculation, there is surprisingly little evidence that trade-offs are important in generating floral diversity, although, as Aigner points out, studies that explicitly test for trade-offs remain rare. Results using the genus *Dudleya* (Crassulaceae) reinforce the conclusion that trade-offs are not always present, and different pollinator species are not necessarily equivalent in their ability to impose selection on floral traits and may in fact select for very different flower characteristics. These results also support Aigner's (2001) hypothesis that floral morphology may represent adaptations to pollinators that are neither most numerous nor most effective, but which provide a marginal fitness gain, in contrast to the commonly cited Most Effective Pollinator Principle.

Chapter 3, by Wilson et al., explores the evolution of hummingbird pollination from bee pollination in a diverse group of North American plants. The authors first map pollination systems onto a molecular phylogeny of the penstemons, showing that hummingbird pollination has multiple origins within this clade. Using information on floral characters associated with the different modes of pollination, and on the relative efficiencies of hummingbirds and bees as pollinators, the authors hypothesize that there may be a more or less fixed order to the evolution of floral traits, with nectar quantity and concentration preceding other characters. Penstemons attracting both bees and hummingbirds are interpreted as a transition toward hummingbird pollination. Interestingly, this conclusion contrasts with Aigner's hypothesis (chap. 2) that species with mixed pollinators may in fact benefit from the marginal pollination services offered by the minor pollinator and show appropriate floral adaptations.

Minckley and Roulston give an interesting overview of floral specialization among bees in chapter 4. As they point out, bees are particularly appropriate for the study of interaction specialization because they rely entirely on floral resources throughout their life cycle, something virtually unique among flower visitors. Minckley and Roulston base their discussion on two types of information: a survey of the literature on food sources of North American desert bees, and newly gathered field data in a Chihuahuan community. Taken together, this information suggests that extreme, reciprocal specialization is rare and that most oligolectic bees visit generalist plants that offer abundant resources. In turn, plants offering poor, nonfood floral resources are rather specialized, but their visitors are usually generalist pollinators. This study therefore leads to conclusions strikingly similar to those reached by Jordano et al. (chap. 8) and Vázquez and Aizen (chap. 9): a majority of flower visitors are rare and tend to be specialized on abundant, generalized plant species offering copious rewards, all of which results in plant-pollinator interactions being mostly asymmetrically specialized.

Continuing the theme of specialized bee-flower interactions, Cane and Sipes (chap. 5) detail the historical background and current confusion over the taxonomic spectrum of pollen collected by bees. They discuss the use of the term *oligolecty*, which in the past has meant slightly different things to different researchers. In addition, they show that a closer examination of the ecology of some bee species can lead to very different conclusions about their "lecty" than were originally proposed. The authors also review the pros and cons of different techniques for assessing the identity of pollen forage plants, and for presenting and interpreting the resulting data. This chapter therefore provides a useful reevaluation of pollen specialization by bees that is likely to stimulate lively discussion among apidologists.

In chapter 6, Renner tackles a long-standing problem in pollination biology—how plants with nonrewarding flowers can enjoy reproductive success in the face of the sophisticated learning abilities of many flower visitors. Rewardless pollination appears to have evolved repeatedly in many plant lineages, and Renner hypothesizes that transient rewardlessness (i.e., flowers that provide rewards only at times) has been key to this evolution. Transient rewardlessness would prevent the development of sophisticated discrimination mechanisms in pollinators. How does rewardlessness relate to specialization in plant-pollinator interactions? Although rewardlessness seems to be associated with "specialized" floral morphology (fusion of floral parts) in some angiosperm lineages otherwise offering nectar as a reward, there seems to be no general relationship to any aspect of pollinator or floral specialization where nectar is not the desired reward. Thus, available evidence suggests that both specialists and generalists can be deceived.

Finally, in chapter 7, Gómez and Zamora argue that evolutionary ecologists need to appreciate how generalization can be common in nature despite the frequently demonstrated ability of pollinators to select on plant phenotype. They suggest that the answer to this question will come from a reconciliation between theoretical predictions and empirical evidence, particularly through integrative studies that take a demographic approach to plant fitness throughout the whole life cycle. Gómez and Zamora review several factors that can prevent specialization, such as spatiotemporal variability and functional equivalence of different pollinator species. Under such circumstances, they argue, generalization can become an adaptive strategy. In their view, most previous studies have focused on nonadaptive generalization, which occurs when pollinators do not constitute real selective pressures, either because the selective regimes fluctuate in space or time or because the effect on the overall fitness of the plants is diluted by extrinsic factors operating throughout the plant life cycle. In contrast, adaptive generalization occurs in systems in which pollinators are strong agents of selection but have similar effectiveness, floral preferences, or mechanical fit, so that several pollinators can act in a concordant fashion to generate floral adaptations. The authors convincingly argue that both adaptive and nonadaptive generalization can occur in nature, and that both should be considered if we want to correctly understand the ecological and evolutionary consequences of plant–pollinator interactions.

As a whole, the studies included in this section provide excellent examples of the complexity of current ideas about how specialization and generalization can influence the ecology and evolution of plant–pollinator interactions. As our definitions of specialization and generalization are becoming more sophisticated and refined (as exemplified by chapters 4 and 5), we are coming to the realization that some of our old ideas—such as coadaptation resulting from tight reciprocal specialization or lack of adaptation resulting from extreme generalism—may have to be replaced by more elaborate concepts including adaptive generalism, asymmetric specialization and floral adaptation to uncommon or relatively ineffective floral visitors. These new concepts and elaborations lead to equally novel predictions, including the possible lack of importance of tradeoffs in generating floral diversity, the lack of relationship between pollinator specialization and the evolution of floral rewardlessness, and the strong selection experienced by generalized flowers pollinated by a diverse set of flower visitors. These six chapters are a vivid reminder that the study of the ecology and evolution of plant–pollinator interactions is a dynamic and exciting field of research.

References

Aigner, P. A. 2001. Optimality modeling and fitness tradeoffs: When should plants become pollinator specialists? Oikos 95: 177–184.

Knuth, P. 1906. Handbook of flower pollination I. Clarendon Press, Oxford.

Sprengel, C. K. 1793. Das entdeckte Geheimniss der Natur im Bau und in der Befruchtung der Blumen. Friedrich Vieweg dem aeltern, Berlin.

CHAPTER TWO

The Evolution of Specialized Floral Phenotypes in a Fine-grained Pollination Environment

Paul A. Aigner

The greatest paradox emerging from the study on intra- and interspecific differences in feeding repertoires of cichlid fishes is that the most specialized taxa are not only remarkable specialists in the narrow sense, but also jacks-of-all-trades.
—Liem (1980)

If a universally generalizable basis for trade-offs, such as energy conservation, does not exist, trade-offs cannot be assumed in the absence of functional analysis.
—Futuyma and Moreno (1988)

Introduction: The Trade-off Principle

The trade-off principle is a cornerstone of evolutionary biology. It holds that, in a heterogenous environment, a population may evolve increased adaptation to one environmental state, but that this will often entail a loss of adaptation to another—that is, a particular behavior, physiology, or morphology may function well (and hence increase fitness) in one situation or another, but not in both. Darwin (1859, chap. 4, "Divergence of Character") embraced the principle and attributed its first application to his contemporary, Henri Milne-Edwards, who discussed the advantage of physiological division of labor between organs of the same body. In the *Origin of Species,* Darwin extended the principle to populations and argued that specialization for "different habits of life" in the "general economy of any land" was the selective factor for the evolution of character divergence. Almost 150 years later, genetic trade-offs in adaptation to different environments are still viewed as essential to the evolution of many forms of phenotypic specialization—from habitat selection in desert rodents to host specialization in parasites and phytophagous insects (Levins 1962; Futuyma and Moreno 1988; Rosenzweig 1995; Fry 1996).

The aphorism "the jack-of-all-trades is master of none" is used so often in the ecological literature that the careless reader might assume trade-offs are universal. But whereas Darwin may have believed that natural selection always favors increased specialization and diversification of form, today we realize that not all characters are subject to trade-offs, nor is specialization always favored. We expect ubiquitous trade-offs among life-history traits (e.g., reproduction vs. growth) because of inherent constraints on the allocation of finite resources among competing demands within one individual, but constraints operating on other aspects of phenotype are not always so apparent. Trade-offs may be absent because a particular phenotype simultaneously optimizes several functions. Alternatively, some functions may be inherently less sensitive to phenotypic variation than others, so that phenotypic specialization for exacting tasks can evolve with little sacrifice in the efficiency of easier ones (Robinson and Wilson 1998). Ultimately, the key to understanding the evolutionary causes of and limits to biological diversity arguably lies in an understanding of trade-offs: what phenotypes are subject to them, and what environmental conditions foster them.

Flowers demand study of specialization and diversification, not only because angiosperms represent a significant fraction of biological diversity (Wilson 1988) but because much of the remarkable diversity seen in these organisms is thought to have evolved in response to a single and conspicuous element of the environment—pollination by animals (Baker 1963; Stebbins 1970; Crepet 1983; Grimaldi 1999; but see Gorelick 2001). Yet surprisingly, the trade-offs faced by plants in adapting to different pollinators are almost completely unexplored. Elegant experiments have revealed how and why a particular pollinator may act as an agent of selection on a particular floral trait, but few studies have examined how plant fitness is influenced by interactions between floral phenotype and different pollinator types (Wilson and Thomson 1996; Wilson et al., chap. 3 in this volume). In contrast, an exhaustive literature examines trade-offs in larval performance of phytophagous insects using different host plants (Jaenike 1990).

In this chapter, I suggest why pollination biologists have been slow to document trade-offs in floral adaptation to contrasting pollination environments. Much of the problem arises from a lack of explicit models of the evolution of specialization when environmental heterogeneity is fine-grained—a situation that I argue should frequently occur in flowering plants. In the following sections, I consider the distinction between coarse- and fine-grained environments and review a model of floral specialization in fine-grained environments. I argue that a common misconception about specialization is that those components of the environment that have the greatest effect on an organism's fitness are also the most important agents of selection. Finally, a review of the evidence for pollinator-mediated trade-offs in the function of floral phenotypes, including my own work with the genus *Dudleya,* suggests that strong trade-offs are not pervasive; therefore, I end with some speculations on how special adaptations might arise without functional trade-offs, particularly in the context of pollinator interactions and complex genetic architecture.

Fine- versus Coarse-grained Environments

Specialization may be viewed as adaptation to a particular environmental state in a heterogeneous environment that consists of multiple states. A specialized phenotype is optimal in one particular state, but suboptimal or one of multiple optima in another state (fig. 2.1A). In a given heterogeneous environment, natural selection should favor the phenotype that maximizes fitness, but identifying this phenotype requires several assumptions about how an organism experiences and responds to environmental heterogeneity. To begin, a useful dichotomy distinguishes between environments that are experienced as fine- versus coarse-grained (Levins 1968). A coarse-grained environment is one in which an organism experiences a single environmental state for all of its life (e.g., a larval insect that feeds exclusively on a single host plant). A fine-grained environment is one in which an organism experiences all environmental states within its lifetime (e.g., a fish encountering multiple prey types, a plant attacked by a variety of herbivores, or a flowering plant visited by a succession of animal pollinators).

In a coarse-grained environment, the expected fitness for a given phenotype, x, in an environment with $i = 1, \ldots, n$ states is the sum of its fitnesses in each environmental state, $w_i(x)$, weighted by the probabilities, p_i, of encountering those states: $E[W(x)] = \Sigma p_i w_i(x)$. Similarly, in a fine-grained environment, a given phenotype has fitness $W(x) = \Sigma p_i w_i(x)$, the sum of fitness values across environmental states, weighted by the frequencies, p_i, with which those states are experienced. The two equations are similar, but, in a fine-grained environment, the expected value notation is unnecessary because each individual experiences all environmental states with the same frequencies (whereas this is not so in the coarse-grained environment). A consequence of this difference is that some models of the evolution of specialization in coarse-grained environments do not apply to fine-grained environments. In particular, a number of recent models have demonstrated that even in the absence of trade-offs (i.e., when nonoverlapping sets of genetic loci are responsible for fitness in different environmental states) more specialized lineages (i.e., those that restrict their use of environmental states) have higher average fitness because they are more consistently exposed to selection and evolve more rapidly (Fry 1996; Kawecki 1996, 1997, 1998; Whitlock 1996). However, this logic does not apply in an environment where individuals cannot select the states to which they are exposed and all lineages experience the same environmental states.

A second major way in which coarse- and fine-grained environments differ is in the potential for interactions to occur among environmental states. In a coarse-grained environment, the fitness functions, $w_i(x)$, in each environmental state are independent. That is, the absolute fitness realized by an individual that experiences a particular state does not depend on which other states occur in the environment. For example, the performance of a phytophagous insect reared on

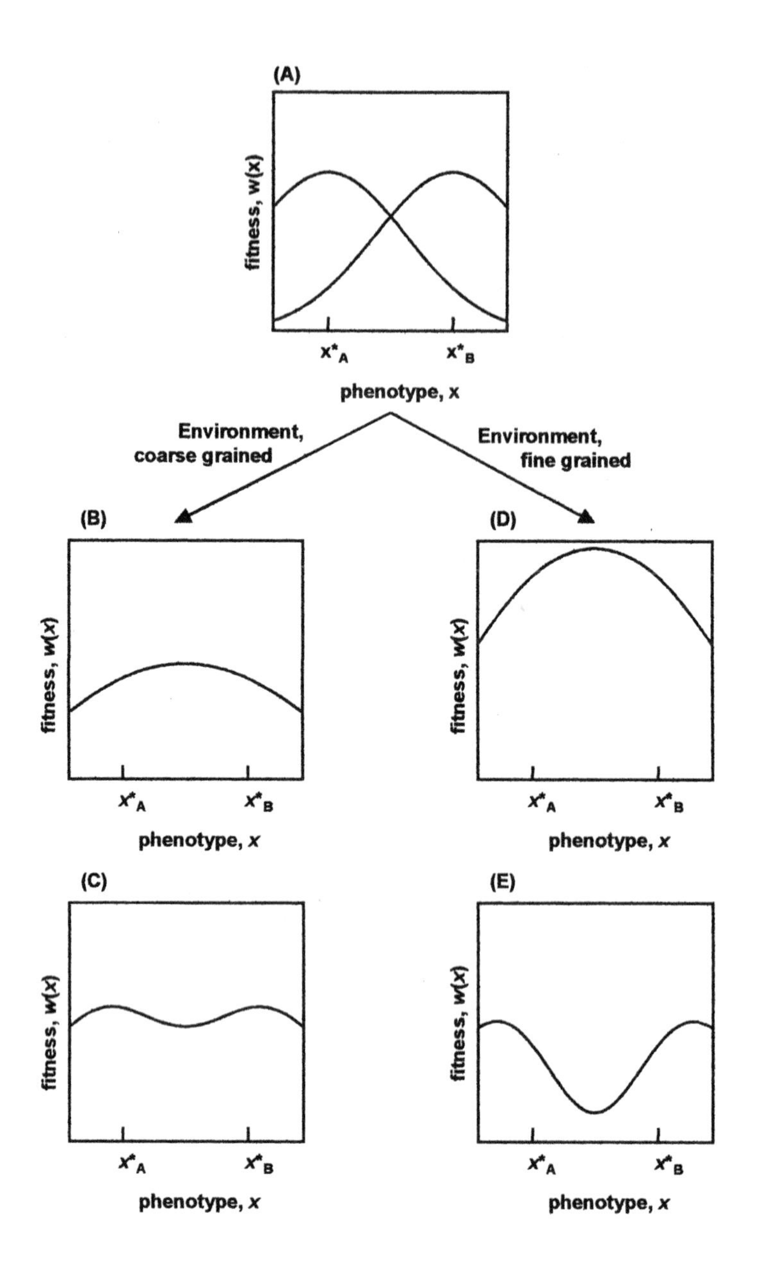

oak is independent of which other potential host species occur in the environment. In contrast, a fine-grained environment permits feedback from the individual to each environmental state so that performance in a particular state depends on what other states have been previously encountered. This opens up the possibility that $w_i(x)$ are not independent and that the net fitness function contains interactions. For example, $W(x) = \Sigma p_i w_i(x) + \Sigma\Sigma a_{ij} p_i p_j w_i(x) w_j(x)$, where $\Sigma\Sigma a_{ij} p_i p_j w_i(x) w_j(x)$ is an additional term reflecting interaction between fitnesses

Figure 2.1 Specialization in coarse- and fine-grained environments. (A) I begin by defining the underlying trade-off. Given two fitness functions, $w_A(x)$ and $w_B(x)$, for two environmental states, A and B, the phenotype x_A^* is specialized for state A because it is optimal in state A and suboptimal in B; likewise, x_B^* is specialized for state B. (B) In the simplest optimality model of a coarse-grained environment, organisms are randomly distributed among environmental states (e.g., seeds passively dispersed between two substrates) and the optimal phenotype is a generalist compromise that is not optimally adapted to either state: $E[W(x)] = p_A w_A(x) + p_B w_B(x)$. The location of the optimum depends on the frequency of each state in the environment. The net fitness function pictured is produced when both states occur with equal frequency ($p_A = p_B = 0.5$). (C) Many organisms can select the environmental state (e.g., habitat) they occupy. If an individual's probability of occupying a habitat is proportional to its relative fitness in that habitat, $[p_i(x) = w_i(x)/\Sigma w_i(x)]$, then selection is disruptive and there are two optimal phenotypes specialized for A or B: $E[W(x)] = \{w_A(x)/[w_A(x) + w_B(x)]\}w_A(x) + \{w_B(x)/[w_A(x) + w_B(x)]\}w_B(x)$. (D) In the simplest optimality model of a fine-grained environment, net fitness is the sum of fitness values across environmental states, weighted by the frequencies, p_i, in which those states are experienced: $W(x) = p_A w_A(x) + p_B w_B(x)$. This might be the case for a plant with two pollinators, A and B, which occur in the environment with absolute abundances of p_A and p_B, and visit plants at rates $r_A(x)$ and $r_B(x)$ per pollinator per plant, each visit resulting in the transfer of $c_A(x)$ and $c_B(x)$ pollen grains [$w_A(x) = c_A(x)r_A(x)$ and $w_B(x) = c_B(x)r_B(x)$]. Like the simple model for the coarse-grained environment, the optimal phenotype in this two-state environment is a generalist compromise; here $p_A = p_B = 1$. (E) In a fine-grained environment there may be an interaction between $w_A(x)$ and $w_B(x)$; that is, $W(x) = p_A w_A(x) + p_B w_B(x) - a[p_A w_A(x) p_B w_B(x)]$. For example, if pollinators A and B compete for resources and shift their visitation away from the plant phenotype for which there is the greatest overlap in preference, then selection becomes disruptive and the optimal phenotypes are specialists for A or B.

in any two environmental states *i* and *j*, and a_{ij} is a constant representing the direction and magnitude of the interaction between these two states. In environments with more than two states, higher-order interactions could be included in the model as well.

Pollination systems provide a simple hypothetical example of these potential interactions. Consider a flower that is visited by a mixture of nectar-feeding birds and bees. Let p_{bird} and p_{bee} be the relative abundances of birds and bees in the environment, independent of the rate at which they visit flowers. Thus, $w_{bird}(x)$ and $w_{bee}(x)$ represent the components of pollination performance (fitness) derived from birds and bees, respectively. In the absence of visitation by nectar-feeding birds, bees may preferentially visit and confer the highest fitness on flowers with the greatest nectar production. But, if birds also prefer flowers with more nectar and avoid those with the least nectar, then bees in the presence of birds may preferentially visit those flowers that were visited least by birds (i.e., there is a classic niche shift by bees under competition with birds). Thus, in a mixed bird- and bee-pollination environment, the fitness contribution of bees to plants with low nectar production differs from that which would be predicted for a pure bee-pollination environment.

The foregoing discussion suggests that predictions drawn from models of the evolution of phenotypic specialization depend critically on assumptions about how organisms experience environmental heterogeneity, how they respond to it, and whether feedback occurs between individuals and the environment. Figure 2.1 illustrates how changing these assumptions alters predictions about whether selection favors specialized or generalized phenotypes, given the same underlying fitness functions.

An Optimality Model for a Fine-grained Environment

Theoretical and empirical progress has been made toward understanding how specialization evolves in a coarse-grained environment, particularly in terms of host specialization by parasites and phytophagous insects (Futuyma and Moreno 1988; Jaenike 1990; Fry 1996; Kawecki 1998). Less progress has been made in understanding the process in a fine-grained environment.

The evolution of specialization by flowering plants for pollination by particular animals is likely to occur in a fine-grained environment. A specialized floral phenotype may evolve from a more generalized ancestor or from a phenotype that was specialized for a different type of pollinator, but in either case a transition phase is likely during which plants evolve adaptations to a particular type of pollinator while other types of pollinators also are present (Baker 1963).

Aigner (2001) used optimality modeling to explore how specialized floral phenotypes might evolve in such a fine-grained pollination environment containing two pollinator types (i.e., two environmental states from the perspective of plant reproduction). In the simplest model, the net function describing the relationship between floral phenotype and pollination success (here assumed equivalent to plant fitness) is the sum of the "fitness functions" for each pollinator type alone (fig. 2.2). The net fitness function is maximized at the phenotype where the marginal gain from increased adaptation to a particular pollinator exceeds the loss from becoming maladapted to other pollinators. Stated differently, specialization will primarily occur when there is an asymmetric trade-off in adapting to different pollinators. (A special case of specialization, as defined here, also occurs when one pollinator type is an agent of selection and the other is not, so that technically there is no trade-off.) If a single phenotype functions well with both pollinators or if trade-offs are broadly symmetric then the optimal phenotype can be considered generalized because it represents adaptation to both pollinator types (fig. 2.3). A key conclusion of this analysis is that specialization depends only on the marginal fitness effects of pollinators (i.e., the rate of change in fitness contributed by a given pollinator with change in floral phenotype); mean effects (i.e., the absolute fitness contributed by a pollinator) are irrelevant (cf. Schemske and Horvitz 1984).

Aigner extended this model to include the possibility that, when different pollinator types co-occur, they interact in their effects on plant fitness. These interactions can modify the trade-off structure of the additive model in complex and counterintuitive ways, but the basic conclusion remains unchanged: specialization is favored when the marginal fitness gains exceed losses.

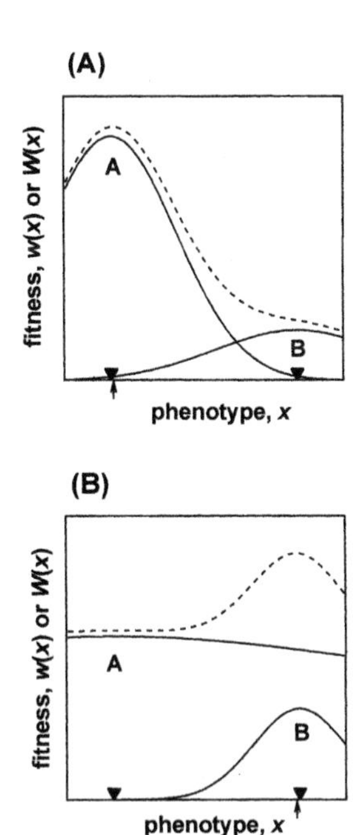

Figure 2.2 An optimality model for the evolution of floral phenotypes. Aigner (2001) presented an optimality model to predict when plants should evolve floral specializations for pollination. In the simplest scenario, total pollination performance, $W(x)$, in a mixed-pollinator environment is the sum of the performance gains from each pollinator type. With two pollinator types, A and B, $W(x) = w_A(x) + w_B(x)$, and $W(x)$ is maximized at the phenotype x^*, where $dW(x^*)/dx = 0$ and $d^2W(x^*)/dx^2 < 0$, which requires that $w_A'(x^*) = -w_B'(x^*)$. That is, when trade-offs are present, the optimal phenotype occurs where the marginal fitness gain from increased adaptation to one pollinator exactly cancels the loss from reduced adaptation to the other. As the graphical scenarios illustrate, conditions favoring the evolution of specialization are largely independent of differences in mean pollinator effectiveness. In each panel, total pollination performance as a function of floral phenotype (dashed line) is indicated as the sum of the performance gains made from each of two pollinator types, A and B. Triangles above the x axis indicate the two phenotypes that are specialized (optimal) for pollination by either A or B. The arrow below the axis indicates the optimal phenotype in the mixed-pollinator environment. (A) Stebbins (1970) predicted that floral evolution should proceed in response to selection from the most effective pollinators. This outcome is possible in the model but requires a particular form of trade-off in which the pollinator with the greatest average effectiveness also imposes the steeper selection differential. (B) If specialization evolves, it need not be in response to the most effective pollinator. When performance gains from adapting to the less-effective pollinator can be had with little loss in the performance contribution of the more effective pollinator (because less-effective pollinators impose steeper selection differentials), then the optimal phenotype in a mixed-pollinator environment is specialized for pollination by the less-effective pollinator.

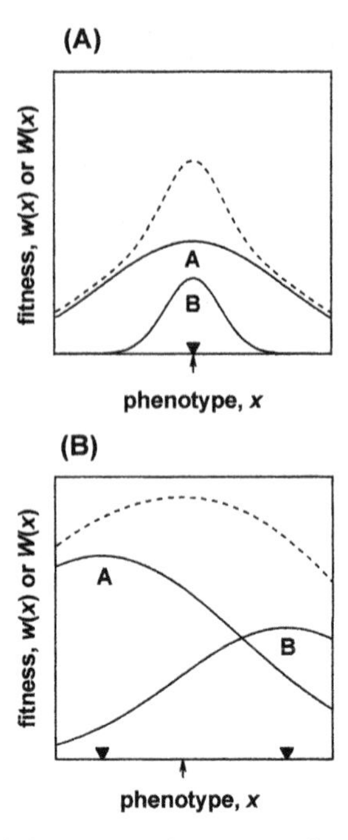

Figure 2.3 An optimality model of adaptive generalization. Not all pollinator types impose performance trade-offs on floral phenotype, and not all trade-offs result in specialization. Adaptive generalization occurs in a mixed-pollinator environment under either of two conditions. (A) When trade-offs are absent, a single floral phenotype may simultaneously maximize the performance contributions of different pollinator types. This generalist strategy should be resistant to evolutionary change regardless of how pollinator assemblages vary in space and time because the net phenotypic optimum remains constant despite the gain or loss of either pollinator type. (B) When trade-offs are present, but the ability of a plant to use each pollinator type is broad relative to the separation between the adaptive peaks for each (i.e., the resulting fitness set is convex; Levins 1968), then the optimal phenotype in a mixed pollinator environment is a generalist compromise that is not optimally adapted to either pollinator type. This generalist strategy should give rise to floral diversification under geographic or temporal shifts in the quantitative composition of the pollinator assemblage because a change in the relative abundance of either pollinator type will dramatically shift the net phenotypic optimum.

What Can Pollination Biologists Learn from Studies of Functional Morphology of Fishes? Special Adaptations Need Not Come at the Expense of General Utility

The foregoing section notwithstanding, a common misconception among evolutionary biologists is that the direction of adaptation for an organism influenced by multiple agents of selection depends on the mean fitness contributions of these agents. That is to say, organisms should adapt to those components of the environment that have the greatest effect on their economy (e.g., that move the most pollen grains). In pollination biology, this view has been formalized as the most effective pollinator principle (MEPP), which states that a plant should evolve specializations to its most effective pollinators at the expense of less effective ones (Stebbins 1970). Widespread acceptance of the MEPP by pollination biologists has led to a perceived paradox: whereas the evolution of floral diversity often seems to produce specialized adaptations to specific types of pollinators (e.g., birds vs. butterflies vs. bees), many species appear to be pollinated by multiple types of pollinators (Ollerton 1996). The paradox is that, although flowers appear phenotypically specialized, they tend to be ecologically generalized. Where can the selection for phenotypic specialization come from in a generalized environment?

Pollination biologists are not the only ones concerned with the evolution of specialized phenotypes in a fine-grained environment. A fish choosing between alternative food items or a larval frog attempting to survive multiple predators are also examples. Not surprisingly, analogs to the MEPP and the plant–pollinator paradox have been stated by evolutionary biologists working with other types of organisms.

The cichlid fish fauna of the great lakes of East Africa is remarkably diverse—the product of one of the most spectacular and rapid evolutionary radiations among vertebrates. Each of the largest lakes contains several hundred species, most of which are endemic to their particular lake (Meyer 1993). Many of the species show remarkable morphological specializations for feeding, and striking convergence of morphology has evolved among independent lineages (Liem 1980). Biologists long assumed that the rapid evolutionary radiation within each lake was driven largely by competition for food and subsequent specialization for particular trophic niches (McKaye and Marsh 1983). Yet, these fishes often do not obtain food in the manner for which they appear adapted. Many of the most specialized fishes forage in a generalized fashion, and many different specialist species share the same food resource. For example, *Petrotilapia tridentiger* and *Pseudotropheus zebra* are two cichlids with highly modified dentition for scraping algae from irregular rock surfaces; both are members of a larger group which could be said to constitute an algae-scraping syndrome. But, in nature both species forage extensively on plankton, detritus, fish fry, and fish eggs, and only sometimes on algae (Liem 1980; McKaye and Marsh 1983).

Such observations led Liem (1980) to question the efficacy of competition as a selective force for divergence: "If specialists are simultaneously jacks-of-all-trades, how could they have evolved according to the widely accepted hypothesis that broadening the range of usable resources prevents species from specializing on individual types?" Implicit in this question is the assumption that those food types that have been the strongest agents of selection on a species' morphology should be the most important component of that species' diet and, more specifically, that trade-offs are omnipresent to the extent that specializing on a relatively rare resource requires the loss of ability to use a more common one.

A similar view pervades the literature on predation and herbivory. A commonly cited prediction about induced defenses against multiple enemies is that the employment and magnitude of a specialized defense should be directly proportional to the cost of attack (Rhoades 1979; Sih 1987; Lima and Dill 1990; Van Buskirk 2001). As in the study of plant–pollinator relationships and foraging behavior of fish, the empirical data often fail to support the prediction that the specificity of adaptations is strongly correlated with the mean effects of different environmental states. Although evidence mounts that trade-offs are common in adaptation to multiple enemies (McIntosh and Peckarsky 1999; DeWitt et al. 2000; Decaestecker et al. 2002), victims do not necessarily exhibit the strongest or most specific response to enemies that impose the greatest fitness cost (Relyea 2001a, 2001b; Schmidt and Amézquita 2001). In some cases, highly specific responses to different enemies can evolve because there are no trade-offs in adopting the different strategies (Van Buskirk 2001).

In the preceding examples, confusion is generated when trade-offs are assumed to underlie the evolution of a particular behavioral or morphological trait. Models that predict that specialized adaptations should evolve for those elements of the environment that have the greatest impact on fitness have not explicitly included trade-offs. But, reconciling these models with others that do include trade-offs (Levins 1968; Aigner 2001) reveals that the prediction is based on two implicit assumptions about trade-offs: first, that trade-offs are ubiquitous, and second, that they take on a particular form. The form is such that the gain from becoming adapted to elements of the environment with the greatest mean effect on fitness exceeds the cost from becoming maladapted to environmental states with lesser mean effects (fig. 2.2A). To my knowledge, this latter assumption has never been validated in any system, and it is questionable whether a theoretical basis exists to assume its prevalence. Indeed, using the example of flowering plants and pollinators, one can construct a plausible argument for the likelihood of a very different form of trade-off. If one accepts the premise that an animal is a highly effective pollinator of a particular plant species because the plant comes from a lineage with a long evolutionary history of strong selection from and adaptation to that pollinator, then little genetically based phenotypic variance may remain that would influence the effectiveness of that pollinator. Instead, the strongest selection might come from more recent, but less effective, pollinators for which heritable variation in floral phenotype still has fitness consequences. Of course, at the extreme, one would not expect very rare pollinators to act as agents of selection, because an extremely low mean effect on plant fitness does not allow much room for covariation between fitness and floral phenotype. But, among pollinators that are more common, there is no reason to expect a strong relationship between mean effectiveness and strength as an agent of selection.

With respect to the foraging behavior of cichlid fishes, Robinson and Wilson (1998) pointed out that some food resources (environmental states) may be inherently easy to use. Thus, phenotypic specialization for obtaining relatively rare food items may evolve because it requires little sacrifice in the ability to obtain those that are easy to use. Such a view reconciles ecological and evolutionary theories of specialization, which at first appear at odds. Optimal foraging theory predicts that in ecological time organisms should expand their diet breadth as food becomes rare (MacArthur and Pianka 1966), whereas competition theory predicts that scarce resources should intensify interspecific competition, which should favor the evolution of more restricted niches. Robinson and Wilson (1998) suggested that, if trade-offs are absent, competition can favor the evolution of different phenotypic specializations for less-preferred food items in a community of fishes. The counterintuitive result of this model is that these highly specialized animals will usually reject the food for which they are adapted in favor of a common, but easy-to-use, resource.

Interspecific competition among plants for the services of pollinators has been repeatedly demonstrated (Free 1968; Waser 1978; Campbell and Motten 1985; Feinsinger and Tiebout 1991), but its importance as a selective force for floral divergence is only beginning to be explored (Caruso 2000, 2001). Nevertheless, it is conceivable that some pollinators are easier to use (i.e., more effective over a broad range of floral phenotypes) than others. In particular, although bumblebees sometimes act as agents of selection on floral phenotype (Galen and Newport 1987; Galen 1996), several careful studies have failed to demonstrate bumblebee-mediated selection where it might be expected (Wilson 1995; Aigner 2004). By constrast, studies of hummingbirds have consistently found them to be strong agents of selection (Waser and Price 1983; Campbell et al. 1991, 1994, 1996, 1997; Hurlbert et al. 1996; Schemske and Bradshaw 1999). Indeed, the hypothesis that some pollinators are inherently "easier to use" than others may explain why certain pollination syndromes appear more discrete than others (Ollerton and Watts 2000).

What *can* pollination biologists learn from the functional morphology of fishes? The important lesson from studies of organisms such as cichlid fishes that adapt to feeding on multiple food types or larval amphibians that adapt to escaping from multiple enemies is that the mean effect of an interaction between an organism and a component of its environment is an unreliable indicator of how important that component is as an agent of selection.

Evidence for Trade-offs in Adaptation to Multiple Pollinators

A growing number of studies provide evidence about trade-offs in floral adaptation to different pollinator types (Galen et al. 1987; Harder and Barrett 1993; Sutherland and Vickery 1993; Conner et al. 1995; Hurlbert et al. 1996; Wilson and Thomson 1996; Campbell et al. 1997; Goulson and Jerrim 1997; Goulson et al. 1998; Fulton and Hodges 1999; Schemske and Bradshaw 1999; Wesselingh and Arnold 2000; Thompson 2001; Sánchez-Lafuente 2002; Aigner 2003, 2004), but an understanding of the functional basis for trade-offs remains fragmentary in even the most well-studied systems. Most studies have focused on a specific component of pollination performance—either the frequency of visits by different pollinators (Sutherland and Vickery 1993; Campbell et al. 1997; Goulson and Jerrim 1997; Goulson et al. 1998; Schemske and Bradshaw 1999; Wesselingh and Arnold 2000) or their per-visit efficiency of pollen transfer (Harder and Barrett 1993; Wilson and Thomson 1996). Only a handful have simultaneously examined components of visitation frequency and per-visit effectiveness (Galen et al. 1987; Hurlbert et al. 1996; Fulton and Hodges 1999; Aigner 2004). This distinction is important because total pollination success is a product of visitation rate (pollination "quantity") and the per-visit efficiency with which pollen is transferred (pollination "quality"; Waser 1983; Herrera 1987, 1989; Campbell et al. 1996). Several studies of pollinator-mediated selection have demonstrated that the floral phenotype most preferred by pollinators is not necessarily the same one that maximizes the per-visit efficiency of pollen transfer (Galen and Newport 1987; Galen and Stanton 1989; Galen 1996; Hurlbert et al. 1996; Aigner 2004); conclusions about trade-offs that are based on only one component of pollination performance can be misleading.

Studies of trade-offs in attracting different pollinators have frequently been motivated by an interest in pollinators as agents of reproductive isolation rather than disruptive selection on particular aspects of floral form (Sutherland and Vickery 1993; Campbell et al. 1997; Goulson and Jerrim 1997; Fulton and Hodges 1999; Schemske and Bradshaw 1999; Wesselingh and Arnold 2000). As a consequence, many studies employ hybrid populations derived from plant species that use different types of pollinators. Usually the parental types have features that are interpreted as distinct specializations to different kinds of pollinators, so good reasons exist to predict a priori that trade-offs should be found; nevertheless, evidence for strong trade-offs has been mixed. A few floral traits do experience disruptive selection from different pollinator types (Schemske and Bradshaw 1999), but others experience selection from only one pollinator type and are selectively neutral with respect to another (Sutherland and Vickery 1993; Schemske and Bradshaw 1999). Still other traits have a single optimum that functions best in all pollination environments (Wesselingh and Arnold 2000).

The use of hybridizing plants in studying pollinator-mediated floral divergence and speciation is exemplified by work on *Aquilegia formosa* and *Aquilegia pubescens*. Grant (1949, 1952) featured these species in his initial argument that animal pollinators facilitate speciation in plants by functioning not only as agents of disruptive selection on floral phenotype but also as agents of reproductive isolation between incipient species. *Aquilegia formosa* has pendant flowers with short, red nectar spurs, red sepals, and bright yellow petals, whereas *A. pubescens* has upright flowers with long, pale yellow or white nectar spurs and pale yellow or white sepals and petals. The two species are largely segregated by elevation, but natural hybrid zones occur where they overlap. In one hybrid zone, Grant (1952) observed that *A. formosa* was visited primarily by hummingbirds and *A. pubescens* was visited primarily by hawkmoths; several trade-offs were suggested as the basis for this apparent isolation. First, he proposed that nectar in the long narrow spur of *A. pubescens* was accessible to long-tongued hawkmoths but not hummingbirds. Nectar in the short-spurred *A. formosa* is accessible by both pollinator types, but Grant suggested that the inverted flowers would be difficult for a hawkmoth to handle. Finally, Grant suggested a trade-off with respect to flower color, with pale flowers visible to nocturnal hawkmoths and red flowers preferred by diurnal hummingbirds. Grant suggested that these trade-offs would not operate with respect to pollen-collecting bees and proposed that bees bring about most occurrences of hybridization.

Although there is good evidence that hummingbirds and hawkmoths have strong preferences for *A. formosa* and *A. pubescens,* respectively (Grant 1952; Fulton and Hodges 1999; but see Chase and Raven 1975), functional trade-offs with respect to particular aspects of floral phenotype have yet to be established. Some components of Grant's hypothesized trade-offs were experimentally tested by Fulton and Hodges (1999). For example, by experimentally inverting flowers of *A. pubescens,* they demonstrated that hawkmoths strongly preferred upright flowers over pendant flowers that were otherwise identical. However, the adaptive value of pendant flowers for hummingbird pollination has yet to be established. Conventional wisdom holds that many hummingbird-pollinated flowers are pendant, but an understanding of the adaptive basis of the trait has been elusive (Tadey and Aizen 2001). It may be that the actual target of selection is a correlated trait such as pedicel flexibility rather than flower orientation per se (Hurlbert et al. 1996).

Closely related species that appear to have evolved distinct specializations for different pollinator types are a good starting point for investigating the evolution of specialization, because they should point to the floral features that are subject to the strongest trade-offs; however, there is also a pitfall in relying too much on studies that focus on pollinators as agents of reproductive isolation to achieve a broad understanding of which trade-offs have been important in the evolution of floral diversity. As Grant (1952) himself pointed out, the processes responsible for divergence of floral phenotype between populations and those responsible for reproductive barriers between incipient species may not be the same (Waser 1998). Absolute reproductive isolation, or even assortative mating strong enough to prevent gene flow from negating an evolutionary response to disruptive selection in sympatry, requires a more extreme trade-off than that necessary to cause divergent selection in allopatry. Many trade-offs in nature may be sufficient to cause adaptive floral divergence but may not meet the more stringent requirements for reproductive isolation (Waser 2001).

Floral biologists have long appreciated that floral divergence in allopatry may occur in response to quantitative changes in pollinator fauna (i.e., the relative proportions of different pollinator types) across geographic scales (Grant and Grant 1965, 162–163; Eckhart 1992; Dilley et al. 2000). A trade-off approach to studying adaptive floral divergence in this context is as equally valid as in the case where divergence seems to be driven by specialization to discrete pollinator types. If generalization (i.e., possession of a phenotype that is not singularly optimized for any one pollinator type) is the phenotypic optimum that results from a tension caused by complex trade-offs between adapting to different pollinator types, then decomposing the net fitness function into the contributions of pollinator types should reveal these trade-offs and allow one to predict how quantitative changes in the pollinator assemblage should shift the net phenotypic optimum. Few studies to date have taken such an approach (Harder and Barrett 1993; Conner et al. 1995; Wilson and Thomson 1996; Aigner 2003).

Of course, specialization and generalization (both in an evolutionary and ecological sense) represent extremes of a continuum, which depends in part on the taxonomic level at which one describes the pollination assemblage. A flower may be evolutionarily specialized for bee pollination, because it possesses features that are optimal for pollen transfer by bees, but suboptimal for pollen transfer by other insects or birds (although it may still be ecologically generalized if it continues to be visited by other animals in addition to bees). That same plant may be considered evolutionarily generalized with respect to an assemblage of bee pollinators if it possesses no traits that are particular adaptations to a single bee family, genus, or species.

In coastal southern California, some species within the genus *Dudleya* (Crassulaceae) are ecologically generalized at the coarsest taxonomic level, receiving visits from most major insect orders that act as pollinators as well as from hummingbirds. Aigner (2003) capitalized on this diversity of visitors to investigate trade-offs in adapting to broad categories of pollinators (hummingbirds vs. large bees vs. small bees and flies). *Dudleya* also appears to be evolutionarily generalized; that is, it lacks adaptive novelties that would allow it to use pollinators in a unique way. Flowers in the genus are radially symmetric and vary in basic features such as corolla flare (ranging from completely open to tubular), corolla length, color, and nectar production. Results from studies of *Dudleya* may reflect widespread trade-offs that have generally been important in angiosperm diversification. The activity of pollinators also shows substantial quantitative as well as qualitative variation. The primary floral visitors are hummingbirds, bumblebees, solitary bees, bee flies, hover flies, and butterflies. Some *Dudleya* species and populations are visited by all of these taxa, whereas others seem to be visited by only a subset.

Aigner (2003) used natural variation in pollinator assemblages to suggest where trade-offs might occur in adaptation to different pollination environments. Variation in floral phenotype was described using principal components analysis. Three components (PC1, PC2, and PC3) accounted for about 75% of the floral variation expressed in the original 10 metric traits, which were directly measured. PC1 was positively correlated with most floral measures, including corolla length, calyx width, and nectar production, suggesting that it reflected variation in overall flower size. PC2 was positively correlated with the degree of stigma and anther insertion and negatively correlated with corolla length and nectar production. Flowers with low scores on PC2 were longer and more rewarding, with more inserted reproductive parts. PC3 was correlated with corolla mouth width and the spatial separation of stigmas and anthers. The study used mixed-species arrays over two years at four sites in California. Pollinator assemblages differed among sites and between years but could be grouped depending on whether the primary floral visitors were hummingbirds, large-bodied nectar-collecting bees (*Bombus* and *Anthophora*), or a mixture of small-bodied solitary bees and flies. Covariance analysis revealed an overall negative relationship between pollen deposition and PC2, but the slope of this relationship varied in magnitude (although not in direction) among pollinator assemblages. Long flowers with more inserted reproductive parts had a performance advantage in all environments, but the advantage was greatest for hummingbird pollination, intermediate for large bee pollination, and very slight for pollination by small bees and flies. Although these results do not provide evidence for a trade-off in pollination performance per se, a trade-off may still exist with respect to total fitness if longer, more productive flowers are more costly to produce. In a small-bee- and fly-pollination environment, these costs may overwhelm the slight performance advantage of long flowers, so that the net form of selection favors shorter flowers.

An experimental approach yielded similar results. Aigner (2004) used manipulations of floral phenotype and controlled pollination environments to isolate targets and agents of selection. *Dudleya* flowers with three levels of corolla flare were exposed to pollination environments consisting exclusively of hummingbirds or bumblebees. There were strong interactions between pollinator type and corolla flare for both male (pollen-exported) and female (pollen-received) components of performance, but the interactions did not take the form of trade-offs. Birds and bees both deposited more pollen as flare decreased, but birds more strongly so, such that bees were more effective than birds at depositing pollen on wide flowers and birds were more effective than bees at narrow flowers. For pollen export, birds exported more dye particles (pollen analogs) as corolla flare decreased, but bees exported similar amounts of dye (pollen) across all phenotypes.

These results mirror those of other studies that have investigated trade-offs in generalized pollination systems. Trade-offs in the efficiency of pollination by different insect orders have rarely been documented (Conner et al. 1995; Conner and Rush 1996; Thompson 2001; but see Galen et al. 1987). For example, Conner et al. (1995) found that different pollinator types (syrphid flies, honeybees, butterflies, and small solitary bees) differed in the relationship between anther exsertion and the amount of pollen removed in single visits to flowers of wild radish (*Raphanus raphanistrum*), but this was because more pollen was removed as exsertion increased for some pollinator taxa, whereas for others removal was unrelated to exsertion. Similarly, in a number of plant species, large flowers have been shown to increase the efficiency of some pollinators, but rarely have large flowers been shown to entail a performance cost with respect to other pollinators (Galen and Newport 1987; Galen et al. 1987; Galen and Stanton 1989; Conner and Rush 1996; Thompson 2001).

We might expect trade-offs to be least apparent in adaptation to pollinator types that are most closely related or most similar in morphology and behavior. It is, perhaps, not surprising that, given the scarcity of evidence for trade-offs in adapting to pollination by different insect orders, there has also been little evidence for trade-offs in adapting to pollination by different species of bees (Eckhart 1992; Harder and Barrett 1993; Wilson and Thomson 1996). A conclusion common to these studies is that, although bees sometimes differed in their average contributions to plant fitness, and floral phenotype often had an effect on visitation rate or the amount of pollen transferred, there were no interactions between bee species and floral phenotype that would support a trade-off. That is, the same floral phenotype would be optimal in each bee environment. Only Eckhart (1992) found weak evidence for a trade-off: a specialized solitary wasp may have preferentially visited female plants of the gynodioecious *Phacelia linearis,* whereas four other groups of solitary bees preferred hermaphrodites.

Taken together, results from studies of "generalized" pollination systems provide no evidence for widespread trade-offs, let alone a particular trade-off structure that would support the MEPP. In fact, Aigner's (2004) results point out the limitation of the MEPP, particularly if selection were acting primarily through male reproductive success. In a hypothetical population of wide-flowered *Dudleya,* bumblebees would be more effective pollinators than hummingbirds; but, contrary to the prediction of the MEPP, evolution would occur entirely in response to selection imposed by hummingbirds because specialization for hummingbird pollination has no cost in reduced bumblebee pollination.

Adaptive Floral Divergence without Functional Trade-offs

What then is the state of our knowledge about the role of trade-offs in floral divergence? Although studies of closely related species that appear specialized for different pollinators sometimes reveal trade-offs between overall phenotypes in attracting visitors, in other cases trade-offs at the level of particular floral traits are absent. Studies of generalized systems are just beginning, but so far they have been able to document few trade-offs. A more common pattern seems to be that different pollinators act as agents of selection on different floral traits, so that few functional trade-offs exist with respect to individual traits. This may occur because different pollinators are attracted by different floral resources or cues and because pollinators differ in their relative ease of use by plants. For example, Aigner (2004) found that pollination success in *Dudleya* was insensitive to corolla flare for pollination by bumblebees. With respect to corolla flare, bumblebees were easy to use as pollinators. Even though corolla flare influenced where pollen was deposited on bumblebees, they were equally effective pollinators whether they carried pollen only on their face or distributed around their bodies. By contrast, pollination success was very sensitive to corolla flare when visitation was by hummingbirds. This was because pollen was deposited effectively on only one part of the hummingbird body (the bill) and this could only be achieved with a tubelike corolla, which restricts the angle at which hummingbirds enter the flower.

Studies that explicitly test for trade-offs in adaptation to different pollinator types are still rare, so it is too soon to dismiss the importance of trade-offs in floral divergence. But, given the paucity of evidence for trade-offs so far, particularly in systems where we would expect to find them, it seems necessary to consider what mechanisms could account for adaptive floral divergence in the absence of trade-offs.

If different pollinator types act as agents of selection on different floral traits, then the potential for the evolution of floral divergence will depend on the genetic architecture of floral phenotypes. Of particular importance is the degree to which morphological characters are correlated because of pleiotropy. If characters are determined by unlinked loci, then different characters should be able to adapt independently to different pollinators. The result should be the evolution of a single floral phenotype that is an overall generalist, with individual characters optimized for different pollinator types (e.g., mariposa lilies; Dilley et al. 2000). If, on the other hand, different floral characters are affected by closely linked loci or the same locus, then divergence can occur even when no functional trade-offs are apparent with respect to each individual character. Stated differently, genetically based fitness trade-offs (i.e., antagonistic pleiotropy) can still occur even if phenotype-based functional trade-offs are absent. This means that a complete understanding of floral specialization will require combined studies of phenotypic function and genetic architecture. Relatively few studies to date have taken this approach (Campbell et al. 1994; Campbell 1996; Conner 1996a, 1996b, 1997, 2002; Bradshaw et al. 1998; Schemske and Bradshaw 1999; Hodges et al. 2002), some by estimating the additive genetic variance-covariance matrix and others by mapping markers that are tightly linked to genes that have large effects on quantitative traits (QTLs).

An approach that combines tests for trade-offs in adaptation to different pollinators with mapping of QTLs for traits that affect pollinator preference has been completed most thoroughly for *Mimulus cardinalis* and *Mimulus lewisii*

(Bradshaw et al. 1998; Schemske and Bradshaw 1999), which are putative specialists for hummingbird and bee pollination, respectively. The results of these studies provide a good example of the importance of considering genetic architecture in understanding the potential for pollinator-mediated disruptive selection. Bradshaw et al. (1998) found a single locus that controlled carotenoid concentration in these two species and a major QTL which explained 41% of the variance in nectar volume between the species. Each of these QTLs strongly affected the visitation pattern of one pollinator type, but not the other; hummingbirds preferred flowers with high nectar volume, and bees preferred flowers low in carotenoids. Two other floral traits (with weaker QTLs) affected pollinator behavior: increased petal anthocyanin concentration had a positive effect on hummingbird visitation, but a negative effect on bee visitation, and increased corolla projected area had a positive effect on bee visitation, but no effect on bird visitation. Schemske and Bradshaw (1999) interpreted these results as demonstrating strong disruptive selection on floral form mediated by visitation preferences of the two pollinators. However, this interpretation must be treated with caution because only one of four floral traits (petal anthocyanin) affecting pollinator behavior generated a fitness trade-off. Furthermore, the two traits with QTLs of largest effect (carotenoid concentation and nectar volume) were under selection by only one pollinator, and, because both QTLs were in different linkage groups (probably on different chromosomes), each should respond independently to selection. In a mixed hummingbird/bee environment where the effects of bird and bee visits were additive, the optimal genotype would be predicted to have the *M. lewisii* allele at the carotenoid locus and the *M. cardinalis* allele at the nectar QTL.

To this point, the discussion has assumed that different pollinators have additive effects on plant fitness; however, interactions between pollinators in their effects on fitness may also affect their potential to act as agents of disruptive selection on floral form. Some studies of floral trade-offs in adaptation to different pollinators have included these interactions whereas others have not, but none has attempted to separately estimate the effect of the interaction on the form of selection or has explicitly considered how interactions might affect qualitative conclusions. Studies that have focused on pollinators as agents of ethological isolation (Grant 1949) implicitly include interactions between pollinator types. In these studies, pollinator preference is estimated in an "uncontrolled" pollination environment where multiple pollinators are present. If the preferences of particular pollinators depend on the presence of other pollinators, then these effects will be manifest in the mixed-pollinator environment.

When pollinators compete for floral resources it seems particularly likely that their behavior will be dependent on the presence of other pollinators. If interspecific competition between pollinators is intense, then pollinator behavior may show classic niche shifts: foraging behavior in the presence of competitors will change relative to that in the absence of competitors. Competitive interactions among pollinator types should increase the potential for disruptive selection on floral form, because each pollinator type should preferentially visit those phenotypes that are least preferred by its competitor. Growing evidence demonstrates that different pollinator species compete for floral resources (Brown et al. 1981; Roubik 1996; Thomson 2004) and that competition affects foraging behavior at flowers (Morse 1981, 1982; Feinsinger et al. 1985; Hingston and McQuillan 1999). If competitive interactions among pollinators are widespread, trade-offs may be apparent in mixed pollinator environments but absent if pollination environments are geographically segregated.

In contrast to studies of ethological isolation, studies that explicitly examine trade-offs in the mechanical fit of flowers to different pollinators typically exclude any potential interaction. This is because estimating the relationship between per-visit efficiency of a pollinator and floral phenotype requires isolated study of each pollinator. Interactions between pollinators can affect the mechanics of pollen transfer between, as well as frequency of visitation to, different floral phenotypes. For example, "ugly" pollinators are so named (Thomson and Thomson 1992) because they waste (from the plant's perspective) much of the pollen they remove from anthers because they either consume it immediately (in the case of many flies), collect it, or groom it off of body parts that are likely to come in contact with stigmas. Although ugly pollinators alone may have some positive effect on plant fitness because they do transfer some pollen, they may impose a fitness cost to a plant when other, less-wasteful pollinators are present. If this cost varies with floral phenotype, then the ugly pollinator interaction can quantitatively or qualitatively alter trade-offs that would be predicted by measuring the selective effects of pollinations in isolation. Aigner (2001) suggested that in some circumstances ugly pollinators could shift the balance of trade-offs to favor the evolution of specialized adaptations to these ugly pollinators, but much more theoretical and empirical investigation is needed before general predictions can be made about the outcomes of interactions that are caused by pollen wastage.

Many flower visitors are not pollinators, and trade-offs in utilizing pollinators alone may not explain all cases of adaptive floral divergence. Some animals remove nectar and pollen from flowers without providing any pollination service (Irwin et al. 2001), and others consume floral tissues directly (Breedlove and Ehrlich 1968). Like pollinators, these floral larcenists and floral herbivores are likely to impose selection on floral traits, the only difference being that their effects on fitness will usually be negative. If floral larcenists and herbivores select for floral traits in different directions than do pollinators, plants may face direct trade-offs in improving pollination service versus defending against enemies. Gómez and Zamora (chap. 7 in this volume) provide a good review of the evidence for such trade-offs. In addition to the direct effects of floral enemies, larcenists and herbivores may interact with one another and with pollinators in their effects on plant fitness. The potential mechanisms by which floral enemies might generate interactions are the same as for pollinators: floral enemies may change the visitation preferences of pollinators and other enemies (Krupnick et al. 1999) or may affect the relationship between the per-visit amount of pollen transferred by pollinators and floral phenotype. A complete view of adaptive floral divergence must necessarily consider all flower visitors, whether their individual effects on plant fitness are usually positive or negative.

Concluding Comments

An understanding of floral divergence that is simultaneously broad and mechanistic will be difficult to achieve, but investigating trade-offs in adaptation to different pollinators will be fundamental to this understanding. Whereas divergence may evolve in the absence of trade-offs, genetically based trade-offs in the function of particular floral traits seem the most likely mechanism that would generate an orderly and understandable pattern of floral diversity. That such order exists is beyond doubt; it is reflected in the long-standing propensity of pollination biologists to group flowers into pollination syndromes and to associate syndromes with patterns of pollinator visitation. Whereas these groupings and associations have been more or less qualitative, they are not illusory; their existence implies that particular types of functional trade-offs transcend taxonomic boundaries. Presumably, it is these ubiquitous trade-offs that generate nodes in phenotypic space which we recognize as syndromes. Traits that comprise these syndromes should be a fruitful starting point for identifying trade-offs that may be responsible for floral diversity at the coarsest scale.

In this chapter, I have argued for an approach that acknowledges that trade-offs for certain functions are not inevitable and may take on diverse forms. This should help make sense out of the observation that phenotypic novelty seems to evolve out of generalized ecological contexts, and should help in recognizing the potential for scenarios that fall outside the MEPP. We should be prepared to find adaptations to relatively uncommon or ineffective floral visitors when there is no sacrifice in the ability to use more common and effective ones. In short, sometimes a flower may be a jack-of-all-trades and still master one.

The debate among pollination biologists about the process by which plants evolve specialization for pollination parallels that in many other subdisciplines of biology concerned with the evolution of phenotypic specialization in fine-grained environments. Much insight could be gained by greater interchange between these scientists who study different manifestations of the same problem.

Acknowledgments

I thank Maria Clara Castellanos, Frances Knapczyk, Nick Waser, and Paul Wilson for comments and discussion that substantially improved the manuscript.

References

Aigner, P. A. 2001. Optimality modeling and fitness trade-offs: When should plants become pollinator specialists? Oikos 95: 177–184.

Aigner, P. A. 2003. The evolution of specialized floral phenotypes in a heterogeneous pollination environment: Fitness trade-offs in a *Dudleya* (Crassulaceae) species complex. PhD dissertation, University of California, Riverside, CA.

Aigner, P. A. 2004. Floral specialization without trade-offs: Optimal corolla flare in contrasting pollination environments. Ecology 85: 2560–2569.

Baker, H. G. 1963. Evolutionary mechanisms in pollination biology. Science 139: 877–883.

Bradshaw, H. D., Jr., K. G. Otto, B. E. Frewen, J. K. McKay, and D. W. Schemske. 1998. Quantitative trait loci affecting differences in floral morphology between two species of monkeyflower (*Mimulus*). Genetics 149: 367–382.

Breedlove, D. E., and P. R. Ehrlich. 1968. Plant-herbivore coevolution: Lupines and lycaenids. Science 162: 671–672.

Brown, J. H., A. Kodric-Brown, T. G. Whitham, and H. W. Bond. 1981. Competition between hummingbirds and insects for the nectar of two species of shrubs. Southwestern Naturalist 26: 133–145.

Campbell, D. R. 1996. Evolution of floral traits in a hermaphroditic plant: Field measurements of heritabilities and genetic correlations. Evolution 50: 1442–1453.

Campbell, D. R., and A. F. Motten. 1985. The mechanism of competition for pollination between two forest herbs. Ecology 66: 554–563.

Campbell, D. R., N. M. Waser, and E. J. Meléndez-Ackerman. 1997. Analyzing pollinator-mediated selection in a plant hybrid zone: Hummingbird visitation patterns on three spatial scales. American Naturalist 149: 295–315.

Campbell, D. R., N. M. Waser, and M. V. Price. 1994. Indirect selection of stigma position in *Ipomopsis aggregata* via a genetically correlated trait. Evolution 48: 55–68.

Campbell, D. R., N. M. Waser, and M. V. Price. 1996. Mechanisms of hummingbird-mediated selection for flower width in *Ipomopsis aggregata*. Ecology 77: 1463–1472.

Campbell, D. R., N. M. Waser, M. V. Price, E. A. Lynch, and R. J. Mitchell. 1991. Components of phenotypic selection: Pollen export and flower corolla width in *Ipomopsis aggregata*. Evolution 45: 1458–1467.

Caruso, C. M. 2000. Competition for pollination influences selection on floral traits of *Ipomopsis aggregata*. Evolution 54: 1546–1557.

Caruso, C. M. 2001. Differential selection on floral traits of *Ipomopsis aggregata* growing in contrasting environments. Oikos 94: 295–302.

Chase, V. C., and P. H. Raven. 1975. Evolutionary and ecological relationships between *Aquilegia formosa* and *A. pubescens* (Ranunculaceae), two perennial plants. Evolution 29: 474–486.

Conner, J. K. 1996a. Measurements of natural selection on floral traits in wild radish (*Raphanus raphanistrum*). I. Selection through lifetime female fitness. Evolution 50: 1127–1136.

Conner, J. K. 1996b. Measurements of natural selection on floral traits in wild radish (*Raphanus raphanistrum*). II. Selection through lifetime male and total fitness. Evolution 50: 1137–1146.

Conner, J. K. 1997. Floral evolution in wild radish: the roles of pollinators, natural selection, and genetic correlations among traits. International Journal of Plant Sciences 158: S108–S120.

Conner, J. K. 2002. Genetic mechanisms of floral trait correlations in a natural population. Nature 420: 407–410.

Conner, J. K., R. Davis, and S. Rush. 1995. The effect of wild radish floral morphology on pollination efficiency by four taxa of pollinators. Oecologia 104: 234–245.

Conner, J. K., and S. Rush. 1996. Effects of flower size and number on pollinator visitation to wild radish, *Raphanus raphanistrum*. Oecologia 105: 509–516.

Crepet, W. L. 1983. The role of insect pollination in the evolution of angiosperms. Pp. 29–50 *in* L. Real (ed.), Pollination biology. Academic Press, Orlando, FL.

Darwin, C. 1859. The origin of species by means of natural selection, or the preservation of favoured races in the struggle for life. Murray, London.

Decaestecker, E., L. De Meester, and D. Ebert. 2002. In deep trouble: Habitat selection constrained by

multiple enemies in zooplankton. Proceedings of the National Academy of Sciences (USA) 99: 5481–5485.
DeWitt, T. J., B. W. Robinson, and D. S. Wilson. 2000. Functional diversity among predators of a freshwater snail imposes an adaptive trade-off for shell morphology. Evolutionary Ecology Research 2: 129–148.
Dilley, J. D., P. Wilson, and M. R. Mesler. 2000. The radiation of *Calochortus:* Generalist flowers moving through a mosaic of potential pollinators. Oikos 89: 209–222.
Eckhart, V. M. 1992. Spatio-temporal variation in abundance and variation in foraging behavior of the pollinators of gynodioecious *Phacelia linearis* (Hydrophyllaceae). Oikos 64: 573–586.
Feinsinger, P., L. A. Swarm, and J. A. Wolfe. 1985. Nectar-feeding birds on Trinidad and Tobago: Comparison of diverse and depauperate guilds. Ecological Monographs 55: 1–28.
Feinsinger, P., and H. M. Tiebout III. 1991. Competition among plants sharing hummingbird pollinators: Laboratory experiments on a mechanism. Ecology 72: 1946–1952.
Free, J. B. 1968. Dandelion as a competitor to fruit trees for bee visits. Journal of Applied Ecology 5: 169–178.
Fry, J. D. 1996. The evolution of host specialization: Are trade-offs overrated? American Naturalist 148: S84–S107.
Fulton, M., and S. A. Hodges. 1999. Floral isolation between *Aquilegia formosa* and *Aquilegia pubescens.* Proceedings of the Royal Society of London B 266: 2246–2252.
Futuyma, D. J., and G. Moreno. 1988. The evolution of ecological specialization. Annual Review of Ecology and Systematics 19: 207–233.
Galen, C. 1996. Rates of floral evolution: Adaptation to bumblebee pollination in an alpine wildflower, *Polemonium viscosum.* Evolution 50: 120–125.
Galen, C., and M. E. A. Newport. 1987. Bumble bee behavior and selection on flower size in the sky-pilot, *Polemonium viscosum.* Oecologia 74: 20–23.
Galen, C., and M. L. Stanton. 1989. Bumble bee pollination and floral morphology: Factors influencing pollen dispersal in the alpine sky pilot, *Polemonium viscosum* (Polemoniaceae). American Journal of Botany 76: 419–426.
Galen, C., K. A. Zimmer, and M. E. Newport. 1987. Pollination in floral scent morphs of *Polemonium viscosum:* A mechanism for disruptive selection on flower size. Evolution 41: 599–606.
Gorelick, R. 2001. Did insect pollination cause increased seed plant diversity? Biological Journal of the Linnean Society 74: 407–427.
Goulson, D., and K. Jerrim. 1997. Maintenance of the species boundary between *Silene dioica* and *S. latifolia* (red and white campion). Oikos 79: 115–126.
Goulson, D., J. C. Stout, S. A. Hawson, and J. A. Allen. 1998. Floral display size in comfrey, *Symphytum officinale* L. (Boraginaceae): Relationships with visitation by three bumblebee species and subsequent seed set. Oecologia 113: 502–508.
Grant, V. 1949. Pollination systems as isolating mechanisms in angiosperms. Evolution 3: 82–97.
Grant, V. 1952. Isolation and hybridization between *Aquilegia formosa* and *A. pubescens.* Aliso 2: 341–360.
Grant, V., and K. A. Grant. 1965. Flower pollination in the phlox family. Columbia University Press, New York.
Grimaldi, D. 1999. The co-radiations of pollinating insects and angiosperms in the Cretaceous. Annals of the Missouri Botanical Garden 86: 373–406.
Harder, L. D., and S. C. H. Barrett. 1993. Pollen removal from tristylous *Pontederia cordata:* Effects of anther position and pollinator specialization. Ecology 74: 1059–1072.
Herrera, C. M. 1987. Components of pollinator "quality": Comparative analysis of a diverse insect assemblage. Oikos 50: 79–90.
Herrera, C. M. 1989. Pollinator abundance, morphology, and flower visitation rate: Analysis of the quantity component in a plant-pollinator system. Oecologia 80: 241–248.
Hingston, A. B., and P. B. McQuillan. 1999. Displacement of Tasmanian native megachilid bees by the recently introduced bumblebee *Bombus terrestris* (Linnaeus, 1758) (Hymenoptera: Apidae). Australian Journal of Zoology 47: 59–65.
Hodges, S. A., J. B. Whittall, M. Fulton, and J. Y. Yang. 2002. Genetics of floral traits influencing re-

productive isolation between *Aquilegia formosa* and *Aquilegia pubescens*. American Naturalist 159: S51–S60.
Hurlbert, A. H., S. A. Hosoi, E. J. Temeles, and P. W. Ewald. 1996. Mobility of *Impatiens capensis* flowers: Effect on pollen deposition and hummingbird foraging. Oecologia 105: 243–246.
Irwin, R. E., A. K. Brody, and N. M. Waser. 2001. The impact of floral larceny on individuals, populations, and communities. Oecologia 129: 161–168.
Jaenike, J. 1990. Host specialization in phytophagous insects. Annual Review of Ecology and Systematics 21: 243–273.
Kawecki, T. J. 1996. Sympatric speciation driven by beneficial mutations. Proceedings of the Royal Society of London B 263: 1515–1520.
Kawecki, T. J. 1997. Sympatric speciation via habitat specialization driven by deleterious mutations. Evolution 51: 1751–1763.
Kawecki, T. J. 1998. Red queen meets Santa Rosalia: Arms races and the evolution of host specialization in organisms with parasitic lifestyles. American Naturalist 152: 635–651.
Krupnick, G. A., A. E. Weis, and D. R. Campbell. 1999. The consequences of floral herbivory for pollinator service to *Isomeris arborea*. Ecology 80: 125–134.
Levins, R. 1962. Theory of fitness in a heterogeneous environment. I. The fitness set and adaptive function. American Naturalist 96: 361–373.
Levins, R. 1968. Evolution in changing environments. Princeton University Press, Princeton, NJ.
Liem, K. F. 1980. Adaptive significance of intraspecific and interspecific differences in the feeding repertoires of cichlid fishes. American Zoologist 20: 295–314.
Lima, S. L., and L. M. Dill. 1990. Behavioral decisions made under the risk of predation: A review and prospectus. Canadian Journal of Zoology 68: 619–640.
MacArthur, R. H., and E. R. Pianka. 1966. On optimal use of a patchy environment. American Naturalist 100: 603–609.
McIntosh, A. R., and B. L. Peckarsky. 1999. Criteria determining behavioural responses to multiple predators by a stream mayfly. Oikos 85: 554–564.
McKaye, K. R., and A. Marsh. 1983. Food switching by two specialized algae scraping cichlid fishes in Lake Malawi Africa. Oecologia 56: 245–248.
Meyer, A. 1993. Phylogenetic relationships and evolutionary processes in East African cichlid fishes. Trends in Ecology and Evolution 8: 279–284.
Morse, D. H. 1981. Interactions among syrphid flies and bumble bees on flowers. Ecology 62: 81–88.
Morse, D. H. 1982. Foraging relationships within a guild of bumble bees. Insectes Sociaux 29: 445–454.
Ollerton, J. 1996. Reconciling ecological processes with phylogenetic patterns: The apparent paradox of plant-pollinator systems. Journal of Ecology 84: 767–769.
Ollerton, J., and S. Watts. 2000. Phenotype space and floral typology: Towards an objective assessment of pollination syndromes. Det Norske Videnskaps—Akademi. I. Matematisk Naturvidenskapelige Klasse, Skrifter, Ny Serie 39: 149–159.
Relyea, R. A. 2001a. Morphological and behavioral plasticity of larval anurans in response to different predators. Ecology 82: 523–540.
Relyea, R. A. 2001b. The relationship between predation risk and antipredator responses in larval anurans. Ecology 82: 541–554.
Rhoades, D. F. 1979. Evolution of plant chemical defense against herbivores. Pp. 3–54 *in* G. A. Rosenthal and D. H. Jansen (eds.), Herbivores: Their interactions with secondary plant metabolites. Academic Press, New York.
Robinson, B. W., and D. S. Wilson. 1998. Optimal foraging, specialization, and a solution to Liem's paradox. American Naturalist 151: 223–235.
Rosenzweig, M. L. 1995. Species diversity in space and time. Cambridge University Press, Cambridge.
Roubik, D. W. 1996. African honey bees as exotic pollinators in French Guiana. Pp. 173–182 *in* A. Matheson, S. L. Buchmann, C. O'Toole, P. Westrich, and I. H. Williams (eds.), The conservation of bees. Linnean Society Symposium Series, no. 18. Academic Press, London.
Sánchez-Lafuente, A. M. 2002. Floral variation in the generalist perennial herb *Paeonia broteroi* (Paeoniaceae): Differences between regions with different pollinators and herbivores. American Journal of Botany 89: 1260–1269.

Schemske, D. W., and H. D. Bradshaw. 1999. Pollinator preference and the evolution of floral traits in monkeyflowers (*Mimulus*). Proceedings of the National Academy of Sciences (USA) 96: 11910–11915.
Schemske, D. W., and C. C. Horvitz. 1984. Variation among floral visitors in pollination ability: A precondition for mutualism specialization. Science 225: 519–521.
Schmidt, B. R., and A. Amézquita. 2001. Predator-induced behavioural responses: Tadpoles of the neotropical frog *Phyllomedusa tarsius* do not respond to all predators. Herpetological Journal 11: 9–15.
Sih, A. 1987. Predator and prey lifestyles: An evolutionary and ecological overview. Pp. 203–224 *in* W. C. Kerfoot and A. Sih (eds.), Predation: Direct and indirect impacts on aquatic communities. University Press of New England, Hanover, NH.
Stebbins, G. L. 1970. Adaptive radiation of reproductive characteristics in angiosperms, I: Pollination mechanisms. Annual Review of Ecology and Systematics 1: 307–326.
Sutherland, S. D., and R. K. Vickery. 1993. On the relative importance of floral color, shape, and nectar rewards in attracting pollinators to *Mimulus*. Great Basin Naturalist 53: 107–117.
Tadey, M., and M. A. Aizen. 2001. Why do flowers of a hummingbird-pollinated mistletoe face down? Functional Ecology 15: 782–790.
Thomson, D. 2004. Competitive interactions between the invasive European honey bee and native bumble bees. Ecology 85: 458–470.
Thomson, J. D. 2001. How do visitation patterns vary among pollinators in relation to floral display and floral design in a generalist pollination system? Oecologia 126: 386–394.
Thomson, J. D., and B. A. Thomson. 1992. Pollen presentation and viability schedules in animal-pollinated plants: Consequences for reproductive success. Pp. 1–24 *in* R. Wyatt (ed.), Ecology and evolution of plant reproduction. Chapman and Hall, New York.
Van Buskirk, J. 2001. Specific induced responses to different predator species in anuran larvae. Journal of Evolutionary Biology 14: 482–489.
Waser, N. M. 1978. Competition for hummingbird pollination and sequential flowering in two Colorado wildflowers. Ecology 59: 934–944.
Waser, N. M. 1983. The adaptive nature of floral traits: Ideas and evidence. Pp. 241–285 *in* L. Real (ed.), Pollination biology. Academic Press, Orlando, FL.
Waser, N. M. 1998. Pollination, angiosperm speciation, and the nature of species boundaries. Oikos 81: 198–201.
Waser, N. M. 2001. Pollinator behavior and plant speciation: Looking beyond the "ethological isolation" paradigm. Pp. 318–334 *in* L. Chittka and J. D. Thomson (eds.), Cognitive ecology of pollination. Cambridge University Press, Cambridge.
Waser, N. M., and M. V. Price. 1983. Pollinator behavior and natural selection for flower colour in *Delphinium nelsonii*. Nature 302: 422–424.
Wesselingh, R. A., and M. L. Arnold. 2000. Pollinator behaviour and the evolution of Louisiana iris hybrid zones. Journal of Evolutionary Biology 13: 171–180.
Whitlock, M. C. 1996. The red queen beats the jack-of-all-trades: The limitations on the evolution of phenotypic plasticity and niche breadth. American Naturalist 148: S65–S77.
Wilson, E. O. 1988. The current state of biological diversity. Pp. 3–18 *in* E. O. Wilson (ed.), Biodiversity. National Academy Press, Washington, DC.
Wilson, P. 1995. Selection for pollination success and the mechanical fit of *Impatiens* flowers around bumblebee bodies. Biological Journal of the Linnean Society 55: 355–383.
Wilson, P., and J. D. Thomson. 1996. How do flowers diverge? Pp. 88–111 *in* D. Lloyd and S. C. H. Barrett (eds.), Floral biology. Chapman and Hall, New York.

CHAPTER THREE

Shifts between Bee and Bird Pollination in Penstemons

Paul Wilson, Maria Clara Castellanos, Andrea D. Wolfe, and James D. Thomson

This is a story about change and lack of change. The subjects are penstemons. The flowers of certain closely related penstemons differ dramatically, whereas certain distantly related species are eerily similar. This is due to the repeated evolution of hummingbird pollination from bee pollination. Our chapter explores the biology surrounding these evolutionary shifts.

We start by defining some useful shorthand. We construe "penstemons" to include the genus *Penstemon* and closely related genera, particularly *Keckiella,* a genus that has both bee- and bird-pollinated species. By "bees" we mean the superfamily Apoidea plus (improperly) the wasp *Pseudomasaris vespoides,* which forages like a bee and like a bee rears its young on pollen. We use "bee-flowers," "bird-flowers," and "pollination syndromes" to refer to species on one side or another of a pollinator shift. The degree to which a shift from bees to birds has occurred varies from case to case, and, in general, the notion of pollination syndromes tends to have unfortunate typological connotations (Armbruster et al. 2000; Thomson et al. 2000; Fenster et al. 2004; Waser, chap. 1 in this volume).

By choosing penstemons, we focus on bilaterally symmetrical, nototribically pollinated flowers with deeply recessed nectaries capable of quickly replenishing nectar after it has been removed; on flowers that often bloom late in the season, when pollinators are generally abundant and hungry; and on plants that have diversified in patchy continental habitats. No doubt, the evolutionary dynamics are quite different for other kinds of flowers in other circumstances, and comparison of evolutionary dynamics among groups of plants holds great promise.

Our chapter shall be an overview of studies done by ourselves and others on penstemon floral evolution as it pertains to shifts from bee to bird pollination. Detailed accounts of the methods and analyses have been or will be published elsewhere. The research program that we outline is intended to not merely document the extent to which plants conform to lists of syndrome characteristics. We take the opportunity here to focus on the evolutionary mechanisms that cause syndrome shifts, just as others have done for evolutionary shifts from outcrossing to selfing, from homostyly to heterostyly, from blooming early to blooming late, and from semelparity to iteroparity.

We start by describing among-species patterns of floral characters and pollinator spectra. Next, we compare bees and birds at bee- and bird-adapted flowers in terms of the amounts of pollen they move. Then, we focus on differences in the nectar rewards for the two syndromes. And then, for a series of floral characters, we speculate on the form of natural selection that might make them change during pollinator shifts. Finally, we consider macroevolutionary dynamics that could explain why bee-pollinated species greatly outnumber hummingbird-pollinated species.

The Systematic Patterns

Penstemon is the largest and most derived genus in Cheloneae (Wolfe et al. 2002), with 270 species described. The closest relatives of *Penstemon* are *Nothochelone* (1 sp.), *Chionophila* (2 spp.), *Chelone* (4 spp.), and *Keckiella* (7 spp.). Two species, in *Penstemon* section *Ambigui,* may be adapted to pollination by long-tongued flies while retaining bees (Straw 1963), and several species in the section *Penstemon* seem to be adapted to pollination by Lepidoptera while retaining bees (Clinebell and Bernhardt 1998). The vast majority of penstemons are mostly (but not exclusively) pollinated by bees and *Pseudomasaris vespoides.* Thirty-nine species show noticeable adaptations for hummingbird pollination. The extent of the shift ranges from being mostly pollinated by bees while having also taken on hummingbirds, to being nearly exclusively pollinated by hummingbirds.

Ordinations

Using a standard protocol, we tabulated data on the floral morphology and pollinator spectra of 49 bee- and bird-pollinated species (Wilson et al. 2004). We scored floral characters that have been previously implicated in pollination syndromes, such as corolla color, anther exsertion, and the narrowness of the floral tube. Pollinator visits were recorded during many 30-minute field censuses. An ordination of the penstemons based on floral characters (fig. 3.1) produces two loose clusters, which correspond well to the two syndromes. The flowers on the left of the ordination are blue-violet or yellow or purple, colors associated with bee pollination. Those on the right are red or orange in the extreme and rose or magenta toward the middle of the ordination. Axis 1 correlates positively with attendance by hummingbirds and negatively with attendance by nectar-seeking bees. Interestingly, pollen-collecting bees tend to visit the bird-syndrome flowers more than the bee-syndrome flowers, possibly because the bird-syndrome flowers present their pollen more generously on anthers that are more accessible.

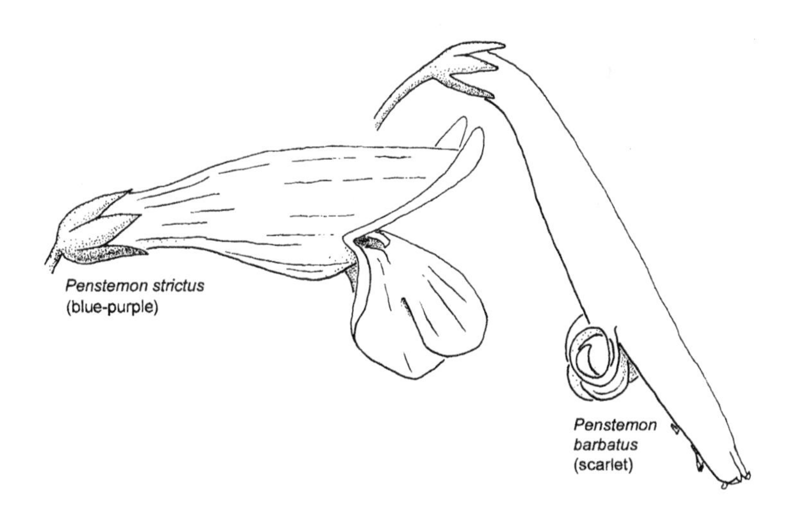

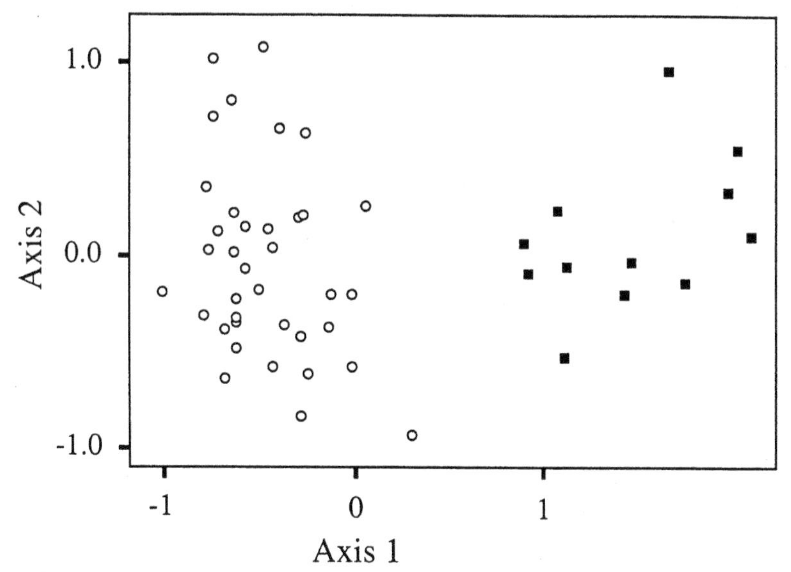

Correlations with Floral Characters

	Axis 1	Axis 2
Color	0.941	0.117
Lower lip reflexion	0.842	-0.148
Functional exsertion	0.918	0.107
Measured exsertion	0.519	0.498
Inclination	-0.640	0.128
Circumference	-0.044	0.712
Staminode length	-0.213	0.518
Tube length	0.419	0.315
Anther-floor distance	0.492	0.853

Correlations with Floral Visitors

	Axis 1	Axis 2
Hummingbirds	0.823	0.152
Pseudomasaris wasps	-0.225	-0.087
Osmia bees	-0.421	-0.482
Nectaring *Bombus*	-0.344	0.091
Pollen-collecting *Bombus*	0.114	0.054
Xylocopa bees	-0.205	0.526
Anthophora-sized bees	-0.293	0.091
Pollen-collecting *Lasioglossum*	0.358	0.107
Other pollen-collecting bees	0.179	0.042
Large hovering flies	-0.280	-0.195
Oligodranes flies	-0.140	-0.074
Small nectaring bees	-0.132	-0.124

Figure 3.1 Multidimensional scaling ordination of penstemon species based on floral characters. Circles represent species of the bee-pollination syndrome; squares represent the bird-pollination syndrome. These 49 species are those for which we have censused floral visitors. Axis 1 is positively correlated with hummingbird visitation at $r = 0.823$ and with many other types of animals. Above the ordination, a typical bee-pollinated species is shown on the left and a typical bird-pollinated species is shown on the right. Details are given by Wilson et al. (2004).

Figure 3.2 One of many parsimonious phylogenies of 194 species above the common ancestor of *Penstemon* and *Keckiella* (A. Wolfe, unpublished data), based on sequences of the internal transcribed spacer region. Outgroups have been pruned to save space. Open circles indicate species that are of the bee-pollination syndrome; solid circles, which are labeled, indicate those that are toward the hummingbird-pollination syndrome. There are 29 such hummingbird-pollinated species included in this phylogeny. Thickened bars represent hummingbird-pollinated lineages. As with other equally parsimonious phylogenies, this requires 23 shifts. Slightly longer trees allow for fewer shifts in pollination syndrome. Phylogenies based on chloroplast DNA sequences and traditional taxonomy also imply a large amount of homoplasy in pollination syndrome.

Phylogeny

The bird–bee contrasts become sharper if we consider a species' pollination system relative to that of its close relatives. For instance, *Penstemon newberryi* is visited by bees more than by hummingbirds, but its flowers show more hummingbird characteristics than those of its close relative *P. davidsonii,* which is nearly exclusively visited by bees. Given that *P. davidsonii* has the ancestral floral characters, we would say that there has been a shift toward hummingbird pollination in the lineage leading up to *P. newberryi* (Datwyler and Wolfe 2004). A phylogeny of *Keckiella* that shows two originations of hummingbird pollination is given by Freeman et al. (2003). A general phylogeny of penstemons is shown in figure 3.2 (A. Wolfe, unpublished data). The consensus trees from nuclear and plastid DNA data are not completely resolved, but all our analyses are robust to the conclusion that a remarkable amount of convergent evolution in the pollination system has occurred. Usually, if not always, the shifts appear to have been from bee to bird pollination. We have labeled the species on the phylogeny that have any tendency toward hummingbird-syndrome characters. Taking the phylogeny at face value, there might be as many as two dozen shifts between bee and hummingbird pollination, and, unless one is willing to accept a tree that is much longer than the shortest trees found in our analysis, one must believe there were at least 13 shifts. These shifts are not all equally extreme (Reid et al. 1988). Only about half of them stand out as excluding large nectaring bees.

Conservatism of Pollination System

Although the convergence is dramatic, the ordination and phylogeny also serve to highlight a subtle sort of "stasis" or maintenance of a clade's pollination system. There is nowhere near as much floral specialization as there could be. There are 576 branch segments above the common ancestor of *Penstemon* and *Keckiella* and 25 shifts between bee and bird pollination. Shifts between bees and birds are far rarer than they could be. Moreover, there is no reason to think that very many of the remaining branch segments are associated with shifts between different types of insects. We see one shift toward long-tongued-fly pollination, another toward butterfly pollination, and at most a half-dozen shifts toward specialization onto such large-bodied bees as *Xylocopa* or such small-bodied bees as *Osmia.*

K. corymbosa
K. ternata
K. cordifolia
rupicola
newberryi
labrosus
baccharifolius
cardinalis
bicolor
floridus
pseudospectabilis
murrayanus
barbatus
centranthifolius
eatonii
harvardii
wrightii
alamosensis
fasciculatus
hartwegii
isophyllus
kunthii
ramosus
parryi
superbus
lanceolatus
rostriflorus
utahensis
pinifolius

There are one or two additional shifts involving pollen placement on bees' heads rather than on their backs. Even the most liberal attribution of floral evolution to pollinator specialization would still leave (1) a great number of branch segments not associated with changes in pollinator type and (2) many species that differ in floral details but not in the principal mode of pollination. This is not taxon stasis, nor is it strictly character stasis; instead, it is stasis in the kind of pollinators used and the way in which they are used. Details change while the way of doing business remains the same. This brings up the question, "What has been the cause of the remaining evolutionary changes in the details of floral characters?"—possibly, fleeting adaptation to changing pollinator regimes that do not change the exclusivity of the flowers (Dilley et al. 2000); possibly, responses to antagonists, such as adaptations that deter floral parasites (Thomson et al. 2000); possibly, correlated responses to changes that are adaptive to other parts of the phenotype (Schemske and Bierzychudek 2001; Armbruster 2002).

Discussion

In the rest of the chapter, we envision lineages evolving from mostly bee pollination toward first including hummingbird pollination and then excluding nectar-feeding bees. In some instances, penstemons may have evolved in the opposite direction (at present, the phylogeny is not resolved sufficiently to be sure of the polarity of each shift), but it seems very likely that most if not all shifts were in the direction of bees to birds. For example, in figure 3.2, the most parsimonious tracing of character evolution would have 21 of the 23 changes be from bee to bird pollination, and postulating shifts as irreversible raises the amount of homoplasy by only two steps in details of the tree that are uncertain at this time.

Pollinator Effectiveness—the Impetus for Shifts

A possible explanation for alternative stable syndromes would be that bees have negative effects on flowers adapted to hummingbirds, while flowers adapted to bees are ill suited for hummingbird pollination. A trade-off of some sort could be involved (Aigner, chap. 2 in this volume). This was suggested by theorizing on optimal strategies for pollen presentation (Harder and Thomson 1989; Thomson and Thomson 1992; Thomson et al. 2000; Thomson 2003). When visitation rates are high, as is the case for most penstemons most of the time, flowers are thought to be under selection to place many small doses of pollen onto many individual pollinators. This is especially the case when the pollinators are bees that groom pollen into pollen-carrying structures or off of their bodies. With hummingbirds, it is reasonable to believe that pollen has greater carryover and should be presented by flowers more simultaneously and more generously. Anthers that open quickly and widely and are adapted to hummingbirds may still be visited by bees. However, in the presence of hummingbirds, the bees would be parasitic antagonists, since they would remove and waste large proportions of pollen that could otherwise be safely delivered to stigmas by the birds. Thus, flowers that are adapted to hummingbirds would be under selection to deter bees and avoid contact between anthers and bee bodies. For the other strategy, bee-syndrome flowers ought to have narrowly dehiscent anthers or pollen that is difficult to remove from anthers, making hummingbirds ineffective, we thought. These flowers would continue to be under selection to attract Hymenoptera. There would be selection against features that could make the flowers attractive to hummingbirds if it were at the expense of attracting bees. We have done a series of studies with the purple bee-syndrome *Penstemon strictus* and the red hummingbird-syndrome *Penstemon barbatus* to test some of the assumptions of the theory (Castellanos et al. 2003).

Bumblebees versus Hummingbirds on the Purple *Penstemon strictus*

We compared the numbers of pollen grains that bumblebees and hummingbirds remove and deposit per visit. A pollinator was given a male-phase donor flower followed by 15 emasculated female-phase recipient flowers. Our first comparison was between bees and birds visiting *P. strictus*. Hummingbirds were surprisingly good at removing pollen from the anthers of this bee-syndrome flower. On average, bumblebees removed 4507 grains, and hummingbirds removed 3148 grains (a nonsignificant difference). On the other hand, bumblebees deposited significantly more grains than did birds on the recipient flowers, 76 grains versus 32 on the 15 recipient stigmas. This seeming superiority of the bees at pollinating bee-adapted flowers was a result of better deposition in the first few recipients. The shape of the pollen carryover curves produced by birds and bees differed as predicted: the bumblebee curve quickly plummeted (presumably due to grooming), whereas the hummingbird curve showed essentially no sign of declining over the course of 15 recipient flowers. Extrapolating beyond 15 flowers, hummingbirds may deposit more of a donor's pollen to more recipients than bumblebees. Overall, both nectaring bumblebees and hummingbirds deliver onto the first 15 recipients about 1.6% of the grains they remove. Our experiments may overestimate deposition by bees because our use of emasculated recipients may have reduced grooming activity.

Bumblebees versus Hummingbirds on the Red *Penstemon barbatus*

In the field, nectar-seeking bumblebees do not visit *P. barbatus*, presumably because the floral tube is so narrow as to make it difficult for them to reach the nectar. We added nectar to flowers and trained small bumblebees to visit them. The visits were videotaped. No bee contacted any anther or stigma. As in other hummingbird-pollinated penstemons, *P. barbatus* has strongly exserted anthers and stigmas that are out of the way of a bumblebee entering the tubular corolla. Bumblebees foraging for nectar would be very ineffective pollinators of flowers with the morphology of *P. barbatus*. In the lineage leading up to *P. barbatus*, the flowers have in effect specialized on hummingbirds, encouraging them while excluding nectaring bees and evolving a morphology that is no longer mechanically harmonious with bees.

Hummingbirds Visiting *P. barbatus* versus *P. strictus*

The red *P. barbatus* has come to be pollinated with greater efficiency by hummingbirds than the purple *P. strictus*. The birds removed 9684 grains on average from *P. barbatus* compared to the 3148 from *P. strictus*. They deposited 182 grains on 15 *P. barbatus* stigmas compared to 32 on *P. strictus* stigmas. Moreover, they transferred a higher proportion of the grains that they removed (2.5 vs. 1.6%), and there was no significant difference in the shape of pollen carryover curves. All this is consistent with the suggestion that there has been selection on certain floral traits for better mechanical fit between the pollinator and the flower. When visiting *P. strictus,* for example, the birds came into contact with stigmas much less often than when visiting *P. barbatus*. We explain this as being due to the much broader corolla tube of *P. strictus,* which allows a bird to poke its beak into the flower from the side, often missing the anthers or stigma. Birds probably remove and deposit more pollen grains on *P. barbatus* than they do on *P. strictus* because the flowers of the former species fit the birds more snugly.

Hymenoptera Other Than Nectaring Bumblebees

The measurements mentioned thus far for pollen removal and deposition by bees are for bumblebees foraging only for nectar in a flight cage. We have also measured pollen removal by pollen-collecting queen bumblebees and by smaller *Osmia* bees, which combine nectar foraging with the deliberate rubbing of their backs against anthers. These bees removed much more pollen than nectaring bumblebees or hummingbirds. The majority of the animals that visit flowers like *P. strictus* are more interested in collecting pollen than our caged nectar-collecting bumblebees. Some, like the queen bumblebees, are primarily devoted to pollen collection and turn upside-down to manipulate the anthers. Others feed on nectar but also collect pollen. Frequent visitors in Colorado include *Osmia* bees, *Anthophora* bees, and *Pseudomasaris* wasps, all of which probe for nectar but also rub against anthers and groom pollen into pollen-carrying structures on their bodies from which, we presume, the pollen is very unlikely to ever reach stigmas. Many of these animals also seem to undervisit female-phase flowers. Indeed, there is every reason to imagine that the Hymenoptera, taken as a whole, deposit less of the pollen that they remove than hummingbirds, even though nectaring bumblebees are about equal to hummingbirds.

Discussion

Our theorizing was only partially confirmed by our studies with *P. strictus* and *P. barbatus*. Hummingbirds do not remove very much less pollen from anthers of *P.*

strictus than do bumblebees. In the first few recipients, birds are not superior at depositing it either, but, as predicted, they have pollen carryover curves that are flatter and more extended than those of bees. If we are correct about the pollen-transfer efficiencies of *Osmia, Anthophora,* and *Pseudomasaris,* and, if our data from emasculated flowers really do overestimate pollen deposition by bumblebees, hummingbirds are actually better on a per-visit basis at moving pollen of *P. strictus* than Hymenoptera considered collectively. In the bird-adapted *P. barbatus,* further adaptations have enhanced pollen-transfer efficiency by birds. In evaluating these results, it is worth mentioning that the *strictus/barbatus* species pair represents only one of many shifts between pollination systems, and this particular pair has *Habroanthus*-type anther dehiscence in which the anthers crack open incompletely and present pollen grudgingly. This is true of both species, and the amount of pollen moved from a single donor visit is probably much lower for both these species than it would be for species with widely opening anthers. Nevertheless, accepting our results as they stand, it would seem that (1) hymenopteran-adapted penstemon flowers are ready to be effectively pollinated by hummingbirds if only the birds would visit them; (2) when on occasion a lineage does acquire hummingbirds as regular visitors, further improvements in pollination efficiency by hummingbirds are likely; and (3) once this happens, hymenopteran visitors probably act as pollen-wasting antagonists when mutualistic hummingbirds are abundant.

The Primacy of Nectar Changes

All penstemons that have been tested replenish their nectar after it has been removed. They start replenishment soon after draining and complete it in 2 to 3 hours (Cruden et al. 1983; Castellanos et al. 2002). Consequently, a sizable patch of penstemon flowers is a rich resource that tends to attract heavy pollinator traffic. We postulate that penstemons cannot make hummingbirds ardently pollinate without providing high nectar rewards. The birds will investigate any penstemon, trying out a few flowers every now and then, but they will not work a patch unless the nectar is about as rewarding as that of other hummingbird flowers in the community. One circumstance that might make a bee-adapted penstemon attractive to birds would be the absence of other bird-adapted flowers or a drought that has made those bird-adapted flowers stop secreting nectar. For instance, in the Colorado Rockies during the drought of 2002, we observed more hummingbird visits at *P. strictus* than in previous years, possibly owing to a reduction in nectar offerings by *Ipomopsis aggregata.* At any rate, for a shift toward hummingbird pollination to proceed, we believe that there has to be evolution in the nectar characteristics of the plants. For birds, the economy of nectar is the bottom line.

Sugar Quantity

What hummingbirds presumably care about is how much energy can be consumed per unit time. The more nectar the better, assuming the concentration does not change. If the concentration does change—from, say, 40 to 20% sugar—then the quantity must also change in a more than compensatory way—from 2 µL to more than 4 µL. Comparing bee- and bird-syndrome flowers of *Mimulus,* which all secrete nectar at low concentrations of around 15% sugar, birds prefer species that produce more nectar, and they prefer F_2 hybrid individuals that produce more nectar when one holds other characters constant (Sutherland and Vickery 1993; Schemske and Bradshaw 1999). The nectar is hidden deep in the flower, so the birds cannot evaluate nectar rewards without visiting, but they quickly learn to associate high rewards with visible characters, such as the degree to which the corolla lobes are reflexed. We set up experimental arrays of the hummingbird-adapted *P. centranthifolius,* the hymenopteran-adapted *P. spectabilis,* F_1 hybrids, and backcrosses in both directions. Hummingbirds prefered *P. centranthifolius* and the backcrosses to *P. centranthifolius.* They visited plants of other parentages at a lower rate and probed fewer flowers on those plants before departing. After establishing the preferences, we added 5 µL of 19% sucrose solution hourly to the flowers of *P. spectabilis* and the backcross to *P. spectabilis.* Before nectar augmentation, the birds preferred the bird-adapted *P. centranthifolius* over *P. spectabilis* 3.7:1; after augmentation, the preference dropped to 1.7:1 (Jordan 2004).

Nectar Concentration

Hummingbird flowers tend to have nectar that is more watery than that of bee flowers (Pyke and Waser 1981). In penstemons bagged overnight, 21 bee-adapted species offered a median of 0.69 µL at 36%, whereas 14 bird-adapted species offered 7.76 µL at 26%. The bird-adapted *P. barbatus* brought its nectar level to about 5.4 µL at 25%, whereas the bee-adapted *P. strictus* brought its nectar level to about 0.4 µL at 42%. We believe that hummingbirds seldom visit *P. strictus* because it offers less nectar than other co-flowering plants. *Penstemon speciosus,* a species that is visually very similar to *P. strictus,* is aberrant for a bee-syndrome penstemon in bringing its nectar offerings to 2.7 µL at 13%, and it is attractive to hummingbirds. That hummingbird flowers have dilute nectar is most curious. At feeders, hummingbirds prefer concentrated nectar (Roberts 1996). The systematic pattern might have something to do with the capillary capacity of hummingbird tongues to take up less viscous, dilute nectar more quickly than concentrated nectar (Kingsolver and Daniel 1983; Roberts 1995; but see Roberts 1996). Hummingbirds may simply not mind dilute nectar as long as the total sugar content is high because they have extraordinarily good renal systems that allows them to excrete the unneeded water (McWhorter and Martínez del Rio 1999). It may also be that it is easier for the nectaries to produce copious nectar if it is watery. Bees probably prefer the concentrated nectars because dilute nectar is heavy to carry and inefficient to store in the nest. In the case of bees, it is not so costly for them to take their time drinking viscous syrup, since they do so while at rest inside the flower. At arrays of mechanical flowers when caloric return per unit time was held constant, bumblebees paid less than 10% of their visits to flowers with dilute nectar (13% sugar) and over 90% to flowers with concentrated nectar (40% sugar; J. Cnaani personal communication).

Sucrose-to-Hexose Ratio

There is a further difference in the nectar of bee- and hummingbird-pollinated penstemons; namely, the sucrose-to-hexose ratio (S:H) of bee nectars is lower than that of hummingbird nectars (Baker and Baker 1983). From our survey, bee-pollinated penstemons had a S:H of 0.315, on average, whereas bird-pollinated species had a S:H of 0.950. Hummingbirds do not mind sucrose because they have sucrase in the membranes of their intestines, and this allows them to break the sucrose down into hexoses (Martínez del Rio 1990a). It is not clear why hummingbirds prefer sucrose (it is energetically equivalent to hexoses), but, in fact, at feeders they do (Martínez del Rio 1990b; Martínez del Rio et al. 1992). One possibility is because a sucrose solution is less viscous than a hexose solution of the same caloric value. Bees do not seem to care whether their sugar is hexose or sucrose (Wells et al. 1992). Another fact worth considering is that, for the same amount of sugar, hexoses have almost double the osmotic potential of sucrose. Sucrose is the predominant sugar in phloem, so plants that offer hexose nectars must have nectaries that hydrolyze sucrose into hexoses. This could conceivably have something to do with getting the water to follow the sugar into the nectar or keeping the sugar from returning into the plant. By not hydrolyzing their sucrose, plants ought to be making their nectar more concentrated, not less, if it were only a matter of osmotic potentials (Nicolson 1998, 2002). The sucrose-rich composition of hummingbird-pollinated penstemons and their dilute sugar concentrations, therefore, are probably not merely different aspects of the physiology of secretion.

Discussion

We will consider nectar in comparison with other traits in the next section. For now, we emphasize that a penstemon must offer voluminous nectar, probably dilute and sucrose rich, to entrain hummingbird visitors. Without this evolutionary transition, hummingbirds will visit penstemons only casually unless ecological circumstances put the birds in dire need. Because of the primacy of nectar quantity and concentration in making hummingbirds frequent penstemon flowers, we believe it is the first character to change during a shift from bee to bird pollination. Evolution in other characters, including S:H, may or may not follow. This point of view arises from considering hummingbird preferences, not from a rigorous analysis of patterns of floral character states in our survey. Our conclusion here contrasts with the suggestion of Bradshaw and Schemske (2003), who were studying bee- and bird-pollinated *Mimulus*. Through back-crossing, they made *Mimulus lewisii* flowers yellow-orange instead of pink; this increased hummingbird visitation by a factor of 68 and decreased bumblebee visitation by a factor of 0.17 compared to control lines. Thus, perhaps in such a *Mimulus* a change in color could kick off a shift toward hummingbirds. However, the contrast in our conclusions is not as contradictory as it at first seems. Bradshaw and Schemske's control line of *M. lewisii* produced 2.3 μL on average, which, in our experience, is enough to get birds to visit penstemons (Castellanos et al. 2002). Their yellow-orange *M. lewisii* produced 5.1 μL, which is as much as many hummingbird-syndrome penstemons. And, although their yellow-orange *M. lewisii* attracted 68 times more hummingbird visits than the control *M. lewisii,* it was still only 0.008 times as much as *M. cardinalis* whether it had the yellow-orange allele or lacked it. This is presumably because *M. cardinalis* control lines offered 67 μL and lines without the yellow-orange allele offered 60 μL (T. Bradshaw, personal communication).

The Cascade of Changes

How might selection change a flower like that of *P. strictus* to a flower like *P. barbatus*? On the principle that every intermediate condition from one mode of life to another must function well, we are inclined to think that hymenopteran-pollinated lineages must have undergone "despecialization" in the sense of the flowers taking on hummingbirds as pollinators while still having the characters that allow for bee pollination (Baker 1963). Then, in some cases, there would have been subsequent "respecialization" in which Hymenoptera were no longer encouraged or were even excluded. It would not work to first evolve away from using bees and then evolve toward using birds. In imagining such shifts, there are several questions to consider. First, has a character state (red) been selected as a "positive" adaptation because it attracts hummingbirds or as a "negative" adaptation because it discourages bees (Faegri and van der Pijl 1979, 15)? Second, has a character state been selected because it attracts pollinators, because it fosters the mechanical interaction with the pollinators, or because it influences the pollinators' tendency to move on to another flower of the same species (Waser 1983; Wilson 1994, chap. 1)? Third, what is the level at which differential success is generated—among patches of kin, among individuals, or among flowers on different individuals (Goodnight et al. 1992)? We will take one syndrome character at a time and sketch out how selection most plausibly proceeded.

Nectar

Because hummingbirds seldom frequent penstemons that have bee-syndrome nectar, we expect that the evolution of copious nectar is the first stage by which despecialization comes to be genetically based rather than based on special external circumstances. We consider a volume increase to be a positive adaptation for hummingbird pollination. The dilution of nectar might also automatically reduce bee visitation, but we note that bees are happy to visit *P. speciosus* despite its unorthodox nectar of 13% sugar. The role of nectar in penstemons cannot be to advertise the flower that is producing the nectar, since pollinators cannot evaluate nectar rewards until after probing the flower. Pollinators are more likely to keep track of patches or parts of patches than flowers or individual penstemon plants in deciding on their foraging route. Increased nectar reward probably has little effect on pollen removal or deposition in a penstemon flower (cf. in *Erythronium;* Thomson 1986); instead, the effects it has are of two other sorts. First, nectar encourages the animal to continue foraging on that type of flower (i.e., to remain constant to the plant's species). This may be selected for because it ensures that pollen removed during a visit is thereafter transferred to appropriate stigmas. Second, nectar increases may be selected for because the other flowers in the immediate vicinity—on the same plant or nearby related plants—are visited by birds that have been well rewarded. Animals make choices about revisitation that are above the level of the flower and very likely above the level of the individual (Sutherland and Gass 1995). We suppose that copious nectar would have to spread by drift to the level of substantial parts of patches before hummingbirds would start discriminating among more- and less-rewarding items in a way that would cause the fixation of such alleles.

Color

Of the characters included in our ordination study, color was the one that best predicted hummingbird visitation (fig. 3.1; Wilson et al. 2004). Bee-syndrome penstemons are white, yellow, or most commonly blue-violet. Many have nectar guides that extend from the lower lip of the corolla to inside the vestibule. Some have staminodes of a contrasting color. Species of penstemon that have despecialized, acquiring hummingbirds as pollinators, are pink or magenta. This shift in color seems to have occurred in nearly all species that have begun a shift toward hummingbirds. Those species that have respecialized to deter bees are red or orange. We suppose from the distribution of characters that color evolves quickly at each stage in the transition. Evolution of magenta from blue-violet seems to be a positive adaptation for attracting hummingbirds. Further evolution to scarlet might be a positive attractant of birds, but we are inclined to think that it also discourages bees (Raven 1972). In Sutherland and Vickery's (1993) F_2 array of *Mimulus,* they found that bees responded to color but not to how reflexed the petal lobes were, whereas birds did not respond to color differences but did respond to petal reflexion. As already mentioned, Bradshaw and Schemske (2003) selectively bred *Mimulus* hybrids, producing plants with a flower pigment of one species in a genetic background near that of the other species. Both bumblebees and hummingbirds followed color by itself when the other characters of the pollination syndrome were nearly equal. Color is a character that is probably involved in initial attraction to the patch or the plant. It has also been widely implicated in affecting constancy (Gegear and Laverty 2001). We hypothesize that both Hymenoptera and hummingbirds have a statistical tendency to continue visiting penstemon flowers of precisely the same color as ones from which they have just received a satisfactory reward. This would, however, explain the maintenance of an established color more easily than the origin of a new one. The power of reds at predicting hummingbird attendance in penstemons and among western North American plants in general has mystified many biologists. Has red color been selected because it is superior at signaling to birds from a distance (Crosswhite and Crosswhite 1981)? Have associative learning and mimicry been important (Grant 1966; Brown and Kodric-Brown 1979; Bleiweiss 2001), and, if so, might the emergence of red as the dominant signal be a historical accident? Is it an outcome of physiological biases in color perception by birds versus bees? Bees have ultraviolet receptors. They can also discern red from other colors (Chittka and Waser 1997), although reds may be less strikingly distinct to bees than colors in the blue-violet range (Chittka et al. 2001). Birds have four eye pigments (maybe more) and oil droplets in their eyes that refract light—they are thought to be able to see a richer array of colors than either bees or humans (Bennett et al. 1994). Selection related to the role of red colors at promoting initial visitation seems most plausible among patches, not among individuals. By the time a bird is choosing among individuals in a patch, it can surely see all of them. As for ensuring constancy, selection could be among individuals, although the choices the animal would be making might be between penstemon flowers and co-flowering plants of no particular relation to *Penstemon.*

Floral Dimensions

Bee-pollinated and bird-pollinated species differ in many characters that seem to affect the mechanical fit of the flower around the pollinator, including exsertion of anthers and stigma, reflexion of the lower lip, narrowness of the flower tube, and inclination of the flower on its pedicel (Straw 1956). These characters may have evolved as positive adaptations to improve the efficiency of pollen transfer by hummingbirds by improving the mechanical fit or as negative adaptations to limit the impact of bees after they became antagonists. Using the bee-syndrome *P. strictus,* we altered the flower to have characteristics of the bird syndrome (Castellanos et al. 2004). We surgically extended anthers and stigmas, trimmed away lower lips, constricted the floral tube with rubber bands, and replaced the stiff horizontal pedicel with floppy fishing line. We then measured pollen removal and deposition by birds and bees. Extending the stigmas made them less likely to contact a bee and caused deposition by bees to be reduced. Trimming the lower lip caused removal by birds to be higher, although deposition was surprisingly reduced. Modestly constricting the tube increased removal by birds. Inclining the flowers surprisingly reduced pollen deposition by birds. We conclude that there is a potential for both positive and negative adaptation. We have evidence that narrow corolla tubes potentially increase pollen transfer by hummingbirds, whereas exserted organs reduce it for bees. The sequence in which these changes occur may contingently affect their value. For instance, pendant flowers may only be of value for bird-pollinated species after the flowers have a narrow tube and exserted organs; in the wide-mouthed *P. strictus,* floppy pedicels seem to let the stigma be missed by a bird entering laterally. In addition to affecting mechanical fit, floral morphology might affect attractiveness (Wilson 1995). We found no significant effect on hummingbird choices, although the flexible pedicel treatment caused the birds to take more time visiting the flowers. We have not tried to measure the effect on bee choices, although effects on bee handling time suggest that our morphological alterations might alter visitation rates. Trimming off the lip and making the pedicel flexible caused the bees to visit more slowly and might discourage visitation. Such effects would very likely be manifest among patches or parts of patches rather than among individuals in a patch. In contrast, the selection we envision via mechanical fit would be caused by differentials among individuals in their ability to disperse their pollen to stigmas.

Anther Dehiscence and Pollen Production

The anthers of bird-adapted penstemons tend to open more widely, more rapidly, and more synchronously than those of bee-adapted ones, at least if one compares pairs of related taxa (Thomson et al. 2000; Castellanos 2003, chap. 4). More widely or more simultaneously opening anthers in hummingbird-syndrome species would be a positive adaptation to hummingbirds. We cannot construe it as a negative adaptation for avoiding bees; rather, we suppose it comes at the cost of inefficient pollen transfer by bees, should they still be frequent visitors. We imagine that the generous pollen presentation of bird-adapted penstemons would arise only after flowers had started to respecialize away from bees. A character we once hypothesized to be related to the speed of pollen presentation is the amount of pollen produced in an anther. Because we believe birds to be more efficient, we predicted that bird-syndrome species should produce less pollen than bee-syndrome species after controlling for factors such as flower size and mode of dehiscence. We also predicted that, all other things being accounted for, anthers with narrower dehiscence should produce less pollen. In our survey of penstemon species, however, we failed to find any significant relationship between pollen production and syndrome or anther dehiscence (Castellanos 2003, chap. 4). Although pollen production is correlated with various aspects of flower size, it appears relatively invariant across shifts from bee to bird pollination. The wideness of anther dehiscence is thought to affect fitness via male–male competition because of the physical interaction of anthers with pollinators, not because of attraction. We suppose this differential success acts among flowers (on different individuals) that are in effect vying for the opportunity to fertilize the limited number of ovules that are available during their life.

The Reduction of Staminodes

In addition to the four fertile stamens, penstemon flowers have a sterile staminode. Staminodes vary in color, length, and hairiness—associated with the type of pollinator. Hummingbird-pollinated species tend to have short or flimsy hairless staminodes. Among insect-pollinated penstemons there is high variability in staminode characters, but in general staminodes are enlarged and often bearded at the end. Walker-Larsen and Harder (2001) suggested that the staminode is a vestigial organ in hummingbird-pollinated penstemons, while it functions to increase pollination success when the main pollinators are insects. By removing staminodes from flowers, they found in hummingbird-pollinated species (*P. centranthifolius* and *P. rostriflorus*) that the staminode had no detectable effect on hummingbird visitation or on pollen transfer. On the other hand, the large staminode of insect-pollinated *P. ellipticus* and *P. palmeri* increased pollen receipt, and, in the latter species, the rate of pollen removal. For the bee-pollinated *P. digitalis,* Dieringer and Cabrera (2002) found an effect on pollen deposition (albeit no effect on removal, which was always almost complete over the life of a flower), and the value of the staminode was greater when the pollinators were small. Staminodes in bee flowers seem to contribute to better contact of the insect body with the reproductive organs of the flowers, but the exact mechanism varies. In certain species, the staminode might act as a lever that pushes the reproductive organs down when a large insect puts pressure on it (Torchio 1974). If the visitor is a small bee, the hairy tip of the staminode might contribute to keeping the insect in contact with the stigma (Dieringer and Cabrera 2002). In other bee-pollinated species, even when the staminode is less enlarged, it might force a nectaring bee to probe one nectary at a time, pushing its body back and forth in contact with anthers and stigmas. We expect that reduction associated with hummingbird pollination ought to occur after the changes directly involved in attracting hummingbirds, and therefore only in a subset of the clades that have shifted away from bees.

Discussion

Our studies weakly support the existence of trade-offs, but of a multitrait sort and involving interactions between the effects of the two kinds of pollinators. It is rarely as simple as changing a single character that increases hummingbird visitation concomitantly with decreasing bee visitation by similar amounts. Rather, birds probably have steeper fitness functions for nectar (following the terminology of Aigner, chap. 2 in this volume), and bees are perhaps choosier about color, given that all penstemons make enough nectar to interest bees. For penstemons, we believe what prevents the flowers from being adapted to all pollinators is that better ones make inferior ones into conditional parasites (Thomson 2003). If we are correct about the sequence of stages in the evolution of hummingbird pollination, then there ought to be a nested statistical pattern among the many shifts. If we are absolutely correct, all shifts would involve a change in nectar rewards (as we see in the purple *P. speciosus*); in a subset of those, flowers would become pinkish or reddish (*P. newberryi*); in a subset of those, the anthers would open more widely (*P. hartwegii*) or be more strongly exserted (*P. kunthii*); in a subset of those, the floral tube would narrow to the extent that it would exclude bees, with a simultaneous evolution of orange or red rather than merely pink (*P. pinifolius*); and finally, in a subset of those, the staminode would be reduced (*P. centranthifolius*). At present, we do not have complete enough character data or sufficient phylogenetic resolution to statistically evaluate this prediction.

Possible Macroevolutionary Processes

Why are there about 245 bee-pollinated species of penstemons when there are only 39 hummingbird-pollinated species? For that matter, why aren't all of them generalized to be pollinated by both bees and birds? We only have data to show that hummingbirds are nearly as good as bumblebees on *P. strictus,* but we still suspect that hummingbird-pollinated flowers are across a saddle and up-slope on an adaptive landscape compared with bee-pollinated flowers (albeit for reasons of male–male competition that do not speak to species superiority or inferiority). So, why have all clades not specialized on birds?

Speciation Rates

Waser (1998) commented that it is hard to believe that pollinator specialization is so complete as to affect reproductive isolation in and of itself, but it is possible that pollination specialization affects speciation rates through assortative mating or in some other way (Jones 2001). Hypothetically, if pollination syndrome did affect net speciation rates, the difference in species numbers could be explained. All hummingbird species may cause nearly the same selection on hummingbird-adapted penstemons, so hummingbird pollination under this view is an innovation that would tend to disfavor subsequent radiation. In contrast, the clades that are hymenopteran-pollinated move through many different pollinator regimes with different types of Hymenoptera that, under this view, would be expected to cause local floral adaptation. This greater niche diversity in hymenopteran pollination than in hummingbird pollination could explain the greater species richness among hymenopteran-pollinated penstemons. One could statistically compare the number of species in hummingbird-pollinated clades to the number of species in sister clades that are bee-pollinated; this would tell of net differences in diversification affected by changes in speciation and/or extinction rates (Farrell 1998; Mitter et al. 1988). Our phylogeny does not yet support an analysis, but it is tantalizing to note individual cases in which a monotypic hummingbird-pollinated clade is sister to a clade of many bee-pollinated species (e.g., *P. rostriflorus* and *P. pinifolius* in fig. 3.2).

Are Hummingbird-Pollinated Lineages More Prone to Extinction?

Another possible explanation for the small number of hummingbird-pollinated penstemons would be a tendency of hummingbird-pollinated clades to have higher extinction rates. Perhaps specialization on hummingbirds repeatedly arises but tends to be a dead end. Under this explanation, one would expect that the extant hummingbird-pollinated species would be on short twigs of the phylogeny and that the shifts would be concentrated in branches that are high in the tree with a paucity of shifts inferred lower down. We wish to statistically evaluate this in future work. Many of the hummingbird-pollinated clades seem to be on short twigs, but certainly not all.

Limits to Genetic Variation and Ecological Circumstance

A final explanation (our null explanation) would be that evolutionary shifts away from hymenopteran pollination have been stymied by the response to selection being slower than the tempo of ecological change. Genetic variation would have to arise and be present in just the right ecological circumstance, such as when bees are rare or other hummingbird-pollinated plants are offering poor nectar rewards (Stebbins 1989). Cruden (1972) suggested that this might be the situation in mountains where the morning chill allows birds to visit before bees are active. The biogeography of the hummingbird-pollinated penstemons is not merely montane but includes many desert species. All we can say about the geography of hummingbird-syndrome penstemons is that they are in regions with abundant and diverse hummingbirds—the southwesterly subset of the geographic range of penstemons as a whole (Crosswhite and Crosswhite 1981). At any rate, we suppose there are few places where ecological circumstance is sustained for long enough to initiate a shift. Mitchell and Shaw (1993) and Mitchell et al. (1998) have reported heritability for nectar production in *P. centranthifolius* at $h^2 = 0.38$. With this heritability, how steep would the selection gradient have to be to change 1 μL to 5 μL in 100 generations? On average, the selection differential would have to be only $S = 0.11$ μL per generation. But, how commonly does a penstemon population live for 100 generations, or even a tenth of that period, with either a paucity of bees or a paucity of other flowers that reward hummingbirds? Shifts between modes of life are plausibly rare because over even tiny amounts of evolutionary time the small adaptive peak that a bee-syndrome species is on is on average slightly better than the intervening saddle that would have to be crossed to reach the hummingbird-syndrome peak. That selection, we think, would have to be above the level of individuals, which might make the bee-syndrome adaptive peak all the harder to escape from.

Discussion

Our phylogenetic analyses are premature, so we will not state any results regarding asymmetries in the phylogeny that might be consistent or inconsistent with hypotheses about speciation and extinction rates. Additional phylogenetic data promise to allow those patterns to be sought. As for the third hypothesis—macroevolutionary inequalities maintained by microevolutionary stabilizing selection—it is hard to gain hard data on the sustained press of ecological conditions favoring shifts toward bird pollination, but we can report that the traffic of hymenopteran visitors at bee-syndrome penstemons is consistently high in our experience.

Conclusion

Flowers of plants in the genus *Penstemon* and in related genera such as *Keckiella* can be arrayed along a "syndrome gradient," from having characters associated with bee and wasp pollination to having characters associated with hummingbird pollination. Evolution along this gradient (pollination shifts) seems to have occurred many times to varying degrees. *Penstemon barbatus,* a hummingbird-adapted flower, has its pollen more efficiently transferred by hummingbirds than *P. strictus,* a bee-adapted flower. On *P. strictus,* birds and bees move comparable amounts of pollen in a visit. If birds were to visit frequently enough, they would immediately be nearly as good at pollination and further adaptation would be possible by simple changes in floral dimensions. However, to secure such a shift requires evolution in nectar rewards. Once nectar has evolved to be more plentiful and possibly less viscous, other adaptations to hummingbirds generally ensue to various extents. In some cases, a further step is taken in which characters that exclude bees are favored. Curiously, there are still more bee-adapted than bird-adapted penstemons.

Acknowledgments

Many student assistants and technicians helped gather data on penstemons. We also thank J. Hogue for help identifying insects; J. Cnaani for sharing data on bees visiting mechanical flowers; and S. Kimball, S. Armbruster, and N. Waser for conversations that got us thinking. Funding was provided by the National Science Foundation (USA), and by the Natural Sciences and Engineering Research Council (Canada).

References

Armbruster, W. S. 2002. Can indirect selection and genetic context contribute to trait diversification? A transition-probability study of blossom-colour evolution in two genera. Journal of Evolutionary Biology 15: 468–486.

Armbruster, W. S., C. B. Fenster, and M. R. Dudash. 2000. Pollination "principles" revisited: Specialization, pollination syndromes, and the evolution of flowers. Det Norske Videnskaps—Akademi. I. Matematisk Naturvidenskapelige Klasse, Skrifter, Ny Serie 39: 179–200.

Baker, H. G. 1963. Evolutionary mechanisms in pollination biology. Science 139: 877–883.

Baker, H. G., and I. Baker. 1983. Floral nectar constituents in relation to pollinator type. Pp. 117–141 *in* C. E. Jones and R. J. Little (eds.), Handbook of experimental pollination biology. Van Nostrand Reinhold, New York.

Bennett, A. T. D., I. C. Cuthill, and K. J. Norris. 1994. Sexual selection and the mismeasure of color. American Naturalist 144: 848–860.

Bleiweiss R. 2001. Mimicry on the QT(L): Genetics of speciation in *Mimulus*. Evolution 55: 1706–1709.

Bradshaw, H. D., and D. W. Schemske. 2003. Allele substitution at a flower colour locus produces a pollinator shift in monkeyflowers. Nature 426: 176–178.

Brown, J. H., and A. Kodric-Brown. 1979. Convergence, competition, and mimicry in a temperate community of hummingbird-pollinated flowers. Ecology 60: 1022–1035.

Castellanos, M. C. 2003. The evolution of transitions between pollination modes in *Penstemon*. PhD dissertation, University of Toronto.

Castellanos, M. C., P. Wilson, and J. D. Thomson. 2002. Dynamic nectar replenishment in flowers of *Penstemon* (Scrophulariaceae). American Journal of Botany 89: 111–118.

Castellanos, M. C., P. Wilson, and J. D. Thomson. 2003. Pollen transfer by hummingbirds and bumblebees, and the divergence of pollination modes in *Penstemon*. Evolution 57: 2742–2752.

Castellanos, M. C., P. Wilson, and J. D. Thomson. 2004. "Anti-bee" and "pro-bird" changes during the evolution of hummingbird pollination in *Penstemon* flowers. Journal of Evolutionary Biology 17: 876–885.

Chittka, L., J. Spaethe, A. Schmidt, and A. Hackelsberger. 2001. Adaptation, constraint, and chance in the evolution of flower color and pollinator color vision. Pp. 106–126 *in* L. Chittka and J. D. Thomson (eds.), Cognitive ecology of pollination. Cambridge University Press, Cambridge.

Chittka, L., and N. M. Waser 1997. Why red flowers are not invisible to bees. Israel Journal of Plant Science 45: 169–183.

Clinebell, R. R., and P. Bernhardt. 1998. The pollination ecology of five species of *Penstemon* (Scrophulariaceae) in the tallgrass prairie. Annals of the Missouri Botanical Garden 85: 126–136.

Crosswhite, F. S., and C. D. Crosswhite. 1981. Hummingbirds as pollinators of flowers in the red-yellow segment of the color spectrum, with special reference to *Penstemon* and the "open habitat." Desert Plants 3: 156–170.

Cruden R. W. 1972. Pollinators in high-elevation ecosystems: Relative effectiveness of birds and bees. Science 176: 1439–1440.

Cruden, R. W., S. M. Hermann, and S. Peterson. 1983. Patterns of nectar production and plant-pollinator coevolution. Pp. 80–125 *in* B. Bentley and T. Elias (eds.), The biology of nectaries. Columbia University Press, New York.

Datwyler, S. L., and A. D. Wolfe. 2004. Phylogenetic relationships and morphological evolution in *Penstemon* subg. *Dasanthera* (Veronicaceae). Systematic Botany 29: 165–176.

Dieringer, G., and L. Cabrera R. 2002. The interaction between pollinator size and the bristle staminode of *Penstemon digitalis* (Scrophulariaceae). American Journal of Botany 89: 991–997.

Dilley, J., P. Wilson, and M. R. Mesler. 2000. The radiation of *Calochortus:* Generalist flowers moving through a mosaic of potential pollinators. Oikos 89: 209–222.

Faegri, K., and L. van der Pijl. 1979. The principles of pollination ecology, 3rd ed. Pergamon Press, Oxford.

Farrell B. D. 1998. "Inordinate fondness" explained: why are there so many beetles? Science 281: 555–559.

Fenster, C. B., W. S. Armbruster, P. Wilson, M. R. Dudash, and J. D. Thomson. 2004. Pollination syndromes and floral specialization. Annual Review of Ecology, Evolution, and Systematics 35: 375–403.

Freeman, C. E., J. S. Harrison, J. P. Janovec, and R. Scogin. 2003. Inferred phylogeny in *Keckiella* (Scrophulariaceae) based on noncoding chloroplast and nuclear ribosomal DNA sequences. Systematic Botany 28: 782–790.

Gegear, R. J., and T. M. Laverty. 2001. The effect of variation among floral traits on the flower constancy of pollinators. Pp. 1–20 *in* L. Chittka and J. D. Thomson (eds.), Cognitive ecology of pollination. Cambridge University Press, Cambridge.

Goodnight, C. J., J. M. Schwartz, and L. Stevens. 1992. Contextual analysis of models of group selection, soft selection, hard selection, and the evolution of altruism. American Naturalist 140: 743–761.

Grant, K. A. 1966. A hypothesis concerning the prevalence of red coloration in California hummingbird flowers. American Naturalist 100: 85–97.

Harder, L. D., and J. D. Thomson. 1989. Evolutionary options of maximizing pollen dispersal of animal-pollinated plants. American Naturalist 133: 323–244.

Jones, K. N. 2001. Pollinator-mediated assortative mating: Causes and consequences. Pp. 259–273 *in* L. Chittka and J. D. Thomson (eds.), Cognitive ecology of pollination. Cambridge University Press, Cambridge.

Jordan, E. 2004. Inheritance patterns in floral characters of *Penstemon* and pollinator preference. MS thesis, California State University, Northridge, CA.

Kingsolver, J. G., and T. L. Daniel. 1983. Mechanical determinants of nectar-feeding strategy in hummingbirds: Energetics, tongue morphology, and licking behavior. Oecologia 60: 214–226.

Martínez del Rio, C. 1990a. Dietary, phylogenetic, and ecological correlates of intestinal sucrase and maltase activity in birds. Physiological Zoology 63: 987–1011.

Martínez del Rio, C. 1990b. Sugar preferences in hummingbirds: The influence of subtle chemical differences on food choice. Condor 92: 1022–1030.

Martínez del Rio, C., H. G. Baker, and I. Baker. 1992. Ecological and evolutionary implications of digestive processes: Bird preferences and the sugar constituents of floral nectar and fruit pulp. Experientia 48: 544–551.

McWhorter, T. J., and C. Martínez del Rio. 1999. Food ingestion and water turnover in hummingbirds: How much dietary water is absorbed? Journal of Experimental Biology 202: 2851–2858.

Mitchell, R. J., and R. G. Shaw. 1993. Heritability of floral traits for the perennial wild flower *Penstemon centranthifolius* (Scrophulariaceae): Clones and crosses. Heredity 71: 185–192.

Mitchell, R. J., R. G. Shaw, and N. M. Waser. 1998. Pollinator selection, quantative genetics and predicted evolutionary responses of floral traits in *Penstemon centranthifolius* (Scrophulariaceae). International Journal of Plant Sciences 159: 331–337.

Mitter, C., B. Farrell, and B. Wiegmann. 1988. The phylogenetic study of adaptive zones: Has phytophagy promoted insect diversification? American Naturalist 132: 107–128.

Nicolson, S. W. 1998. The importance of osmosis in nectar secretion and its consumption by insects. American Zoologist 38: 418–425.

Nicolson, S. W. 2002. Pollination by passarine birds: Why are the nectars so dilute? Comparative Biochemistry and Physiology B 131: 645–652.

Pyke, G. H., and N. M. Waser. 1981. The production of dilute nectars by hummingbird and honeyeater flowers. Biotropica 13: 260–270.

Raven, P. H. 1972. Why are bird-visited flowers predominantly red? Evolution 26: 674.

Reid, W. H., P. Sensiba, and C. E. Freeman. 1988. A mixed pollination system in *Penstemon pseudospectabilis* M. E. Jones (Scrophulariaceae). Great Basin Naturalist 48: 489–494.

Roberts, W. M. 1995. Hummingbird licking behavior and the energetics of nectar feeding. Auk 112: 456–463.

Roberts, W. M. 1996. Humingbirds' nectar concentration preferences at low volume: The importance of time scale. Animal Behavior 52: 361–370.

Schemske, D. W., and P. Bierzychudek. 2001. Evolution of flower color in the desert annual *Linanthus parryae:* Wright revisited. Evolution 55: 1269–1282.

Schemske, D. W., and H. D. Bradshaw. 1999. Pollinator preference and the evolution of floral traits in monkey flowers (*Mimulus*). Proceedings of the National Academy of Sciences (USA) 96: 11910–11915.

Stebbins, G. L. 1989. Adaptive shifts toward hummingbird pollination. Pp. 39–60 *in* J. H. Bock and Y. B. Linhart (eds.), The evolutionary ecology of plants. Westview Press, Boulder, CO.

Straw, R. M. 1956. Adaptive morphology of the *Penstemon* flower. Phytomorphology 6: 112–118.

Straw, R. M. 1963. Bee-fly pollination of *Penstemon ambiguus*. Ecology 44: 818–819.

Sutherland, G. D., and C. L. Gass. 1995. Learning and remembering of spatial patterns by hummingbirds. Animal Behavior 50: 1273–1286.

Sutherland, S. D., and R. K. Vickery. 1993. On the relative importance of flower color, shape, and nectar rewards in attracting pollinators to *Mimulus*. Great Basin Naturalist 56: 282–282.

Thomson, J. D. 1986. Pollen transport and deposition by bumble bees in *Erythronium:* Influences of floral nectar and bee grooming. Journal of Ecology 74: 329–341.

Thomson, J. D. 2003. When is it mutualism? American Naturalist 162: S1–S9.

Thomson, J. D., and B. A. Thomson. 1992. Pollen presentation and viability schedules in animal-pollinated plants: Consequences for reproductive success. Pp. 1–24 *in* R. Wyatt (ed.), Ecology and evolution of plant reproduction. Chapman and Hall, New York.

Thomson, J. D., P. Wilson, M. Valenzuela, and M. Malzone. 2000. Pollen presentation and pollination syndromes, with special reference to *Penstemon*. Plant Species Biology 15: 11–29.

Torchio, P. F. 1974. Mechanisms involved in the pollination of *Penstemon* visited by the masarid wasp, *Pseudomasaris vespoides* (Cresson). Pan Pacific Entomologist 50: 226–234.

Walker-Larson, J., and L. D. Harder. 2001. Vestigial organs as opportunities for functional innovation: The example of the *Penstemon* staminode. Evolution 55: 477–487.

Waser, N. M. 1983. The adaptive nature of floral traits: Ideas and evidence. Pp. 241–285 *in* L. Real (ed.), Pollination biology. Academic Press, Orlando, FL.

Waser, N. M. 1998. Pollination, angiosperm speciation, and the nature of species boundaries. Oikos 81: 198–201.

Wells, H., P. S. Hill, and P. H. Wells. 1992. Nectarivore foraging ecology: Rewards differing in sugar type. Ecological Entomology 17: 280–288.

Wilson, P. 1994. The habits of selection for pollination success. PhD dissertation, State University of New York, Stony Brook, NY.

Wilson, P. 1995. Selection for pollination success and the mechanical fit of *Impatiens* flowers around bumble bee bodies. Biological Journal of the Linnean Society 55: 355–383.

Wilson, P., M. C. Castellanos, J. N. Hogue, J. D. Thomson, and W. S. Armbruster. 2004. A multivariate search for pollination syndromes among penstemons. Oikos 104: 345–361.

Wolfe, A. D., S. L. Datwyler, and C. P. Randle. 2002. A phylogenetic and biogeographic analysis of the Cheloneae (Scrophulariaceae) based on ITS and matK sequence data. Systematic Botany 27: 138–148.

CHAPTER FOUR

Incidental Mutualisms and Pollen Specialization among Bees

Robert L. Minckley and T'ai H. Roulston

Introduction: Pollination Mutualisms and Bees

Early studies of the spectacular, highly specific relationships between orchid bees and orchids, fig wasps and figs, and yucca moths and yuccas have generated a widespread notion that specialized pollination interactions are common in nature. However, it is now clear that pollination interactions with a high degree of reciprocal evolution between the plant and floral visitor are very rare or nonexistent in most ecosystems. This is not to say that all flowers lack extravagant features associated with pollination or that floral visitors lack modified behaviors, morphology, or life-history attributes closely tied to a preferred floral host; rather it is rare that both flowering plants and floral visitors participate in reciprocally specific pollination interactions (see Jordano et al., chap. 8 and Vázquez and Aizen, chap. 9 in this volume).

The change in perspective regarding the evolutionary history and ecological interactions among flowering plants and floral visitors stems from an increased interest in the perspective of floral visitors, further consideration of the conflict of interest on which plant–pollinator interactions are based, and community-level studies (Waser et al. 1996; Johnson and Steiner 2000; Pellmyr 2002). In most cases, the pollination interaction ultimately represents a meal to floral visitors, but it represents sexual reproduction to the plant. Natural selection should favor floral visitors that use host plants with abundant, spatially localized resources (when harvest rates are equal) to maximize resources acquired, minimize energy use, and minimize exposure to predators and parasites. Floral parasitism, including nectar robbery, represents a logical extension of this half of the plant–pollinator contest (Baker et al. 1971). The game for plants involves curbing wastage of their own gametes (pollen) and unnecessary expenditure on rewards. Some plant species take this to an extreme by investing in traits that attract bees but offering no reward at all (even the pollen is not viewed as a resource—e.g., orchids; see Renner, chap. 6 in this volume). These plants effectively parasitize both their pollinators and the local plant species that do sustain pollinator populations with their floral resources (Laverty 1992). Regardless of what outcome this plant–pollinator contest generates, a reasonable prediction based on the divergent interests of plants and pollinators is that instances when both parties converge on a relationship of mutual dependence should be rare.

We suggest here that before we can have a broad understanding of the ecological dynamics of pollination systems, and the conditions under which they evolved, the animal side of pollination must be more fully understood in the context of communities. We illustrate this with bees as examples, with particular emphasis on the patterns of floral host use by pollen specialist species from the North American deserts where we have mostly worked. These patterns show that (1) resource abundance is closely associated with evolution of pollen specificity in bees and the kinds of pollination interactions among bees and plants and (2) many floral traits often interpreted in the context of pollination have little effect on the species composition of the floral visitors. We also discuss how the same features that delimit the breadth of pollination interactions provide insight into the ecological advantages and evolutionary origins of trophic specificity by bees to one or a few pollen species. Exploration of community-level patterns of visitation and pollination will greatly enhance our understanding of those features of plants that promote pollinator specificity and that limit the evolution of specialized pollination mutualisms.

Definitions of Specialization and Generalization

There is inconsistency in the plant–pollinator literature in regard to the terms *specialization* and *generalization*. The terms are used to describe both the interactants (e.g., bees and flowering plants) and the nature of the interaction among them; this duplicity generates considerable confusion. For example, specialized flowers may be so described when there appears to be strong directional selection on traits believed to relate to pollination, such as long floral tubes and the production of unusual rewards. Generalized flowers are classified as those with an open floral design with typical floral rewards, which are both attractive and accessible to most floral visitors. This terminology relates to the pollination syndrome concept as codified by 20th century pollination researchers (van der Pijl 1961; Baker and Hurd 1968; Waser, chap. 1 in this volume) in which suites of floral traits are used to infer the identity of a plant's pollinators. A second meaning of specialized flower refers to the diversity of pollinators for each plant species. Armbruster and Baldwin (1998) define specialized pollination as "pollination by one or a few animal species" (see also Renner and Feil 1993; Fleming et al. 2001). Although this definition presumably distinguishes pollinators from non-pollinating visitors, it says nothing about which floral features influence the number and diversity of pollinators or visitors. In sum, pollination interactions of highly modified flowers are considered specialized a priori whereas pollination interactions of flowers with few visitors are discovered through observation

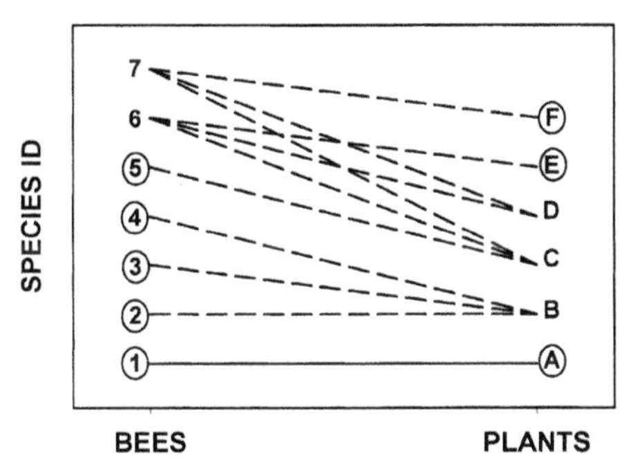

Figure 4.1 A range of pollination interactions among bees and plants. A circle indicates a specialized participant; a solid connecting line indicates a specialized interaction. In this example, all bee species are effective pollinators of the plants represented. Bee species 1 through 5 are all specialists of one plant species but only one bee specialist species is involved in a one-to-one interaction with a plant (bee species 1 with plant species A). All other bee specialist species are involved in generalized pollination interactions with either other bee specialist species (bee species 2–4), or generalist species (bee species 5). Bee species 6 and 7 are pollen generalists that interact with either generalist (C and D) or specialist (E and F) plant species.

and experimentation. For the purpose of this chapter, we confine ourselves to using the terms to refer to readily observable structures and phenomena. We follow Armbruster and Baldwin (1998) in denoting specialized flowers as those that attract few pollinator species. We use the term *simple flower* to describe an open floral morphology with readily accessible rewards, and *complex flower* for those that package their rewards in a way that excludes access for many potential visitors. For bees, we use the terms *specialized* and *generalized* based on the number and relatedness of each species' floral hosts, without regard to adaptive traits or to the presence of covisiting species on the same hosts. We reserve the use of the term *specialized interaction* for those cases in which flowers are pollinated by a narrow range of pollinators that harvest resources from very few plant species. A consequence of this terminology is that specialized flowers, complex flowers, and specialized bees commonly participate in generalized interactions (fig. 4.1).

A Brief Biology of Bees

The array of interactions among the 20,000 bee species and more than 250,000 species of flowering plants is complicated by the diverse biologies of both groups. Plant pollination strategies are the subject of numerous reviews, which we will not attempt to repeat. The literature is more diffuse on how differences in bee biology influence the interactions between bees and plants; here, we outline some important features of bee biology. Reviews by Barth (1991), Linsley (1958), Michener (2000), O'Toole and Raw (1991), Radchenko and Pesenko (1994), Roubik (1989), Wcislo and Cane (1996), and Westerkamp (1996) offer more information.

The association of bees and flowering plants is unusually close because both larvae and adults depend on floral resources. Only masarine wasps (Vespidae; Gess 1996) and one sphecid wasp (Krombein and Norden 1997) also collect pollen and and feed their larvae with it. Most other insects that consume pollen do so only as adults. Although bees as a group depend on pollen, bee species can be roughly divided into those that collect it from flowers (ca. 85% of described species) and parasitic species that lay their eggs in the brood cells of pollen-collecting species (ca. 15%).

Pollen-collecting bees can be further divided into social and solitary species. The honeybee (*Apis mellifera*), by far the most familiar bee species worldwide, is social and active year-round. Most bee species, however, are solitary and short-lived, either nesting in tunnels they excavate in the ground, or in wood, soft-pith, hollow plant stems, or other material above ground (Iwata 1966; Krombein 1967; Michener 2000). Differences among bee species in social behavior and longevity have implications for the evolution and ecology of floral host breadth. Social bee colonies can consist of thousands of individuals and remain active longer than the blooming period of most plant species (Visscher and Seeley 1982; O'Neal and Waller 1984); therefore, they use numerous floral hosts during their flight season. In contrast, many solitary species produce only 5–10 off-spring in their lifetime (Minckley et al. 1994), an output comparable to that of most primates. Because the activity period of short-lived, univoltine solitary species may fall within the flowering period of a single plant species, their foraging needs can be met by a single host species. Among solitary species, both floral specialists and floral generalists occur.

Host Breadth and Floral Resources

Bees visit flowering plants to collect resources, and although most plants provide floral visitors with pollen and nectar rewards, some offer other foods, scents, or nesting materials. Bees use these rewards differently, so to understand the breadth of the pollination interaction the rewards involved must be considered. Pollination interactions of plants that adopt very unusual rewards are specialized in the sense that they attract only a portion of the active bee fauna that would visit a host that offers nectar. For example, plants that offer oil or resin as the sole reward exclude males of most species because this sex visits plants solely for nectar, whereas plants that offer only chemical scents used by males exclude females. Although use of specialized rewards may reduce visitor diversity, it seldom leads to host-plant specialization among those visitors. To some extent, pollination interactions involving plants that offer only pollen as a reward (many Solanaceae, Australian *Acacia*), or that bloom at night, early in the morning, or very late in the day (Linsley 1978), work similarly. Temporal changes in blooming time, however, are sometimes associated with host specialization in bees. For example, two shifts from broad polylecty to oligolecty have occurred in the bee genus *Lasioglossum* (Danforth et al. 2003a). One species, *Lasioglossum*

(*Hemihalictus*) *lustrans,* restricts its foraging to *Pyrrhopappus carolinianus* (Asteraceae), which flowers from just after dawn until several hours later (Estes and Thorp 1975). The second involves eight species of *Sphecodagastra,* all of which collect pollen from night-blooming *Oenothera* (McGinley 2003).

Nonfood Resources

Some plant species attract bees with nonfood resources used for nesting (resins) and attracting mates (chemical scents; Zucchi et al. 1969; Wille and Michener 1973; Vogel 1974; Simpson and Neff 1987; Lokvam and Braddock 1999). The few studies of resin-producing plant species report visitation by multiple resin-collecting bee species, particularly euglossines and meliponines (Simpson and Neff 1983). Numerous Neotropical orchids produce volatile compounds collected by male euglossine bees, presumably as sex attractants or their precursors. These bees use their forelegs to scrape the compounds into a slit in their swollen, highly modified hind tibia. This interaction is commonly cited as specialized because these orchid species attract but a few euglossine bee species. However, it is little appreciated that the bees involved generally collect these compounds from various floral and nonfloral sources and are seldom, if ever, dependent on a single plant species (Ackerman 1983). Zucchi et al. (1969) list male euglossine bees that collect chemical compounds from plant species in the families Orchidaceae, Araceae, Gesneriaceae, Myrtaceae, and Fabaceae, as well as from decaying wood (see also Dodson 1966; Whitten et al. 1993). Although different plant species produce different composite fragrances, the most attractive components of those fragrances appear to be common natural compounds (Dodson et al. 1969). Overall, plants that offer nonfood floral resources are specialized to the extent that few bee species visit them, but their interactions are generalized to the extent that these bee visitors use other sources (floral or nonfloral) within their foraging range.

Food Resources: Oils, Nectar, and Pollen

Whereas plants that offer nonfood resources rely on bees that must visit other plant species for food resources, plants that offer food resources can potentially serve as the sole hosts of specialist bees. However, even among this latter group of floral hosts, reciprocally specific pollination interactions are exceedingly rare. Floral oils substitute for nectar in the larval provisions of oil-collecting bees and, thus, probably function primarily as an energy source (Vogel 1974; Simpson and Neff 1981; Buchmann 1987). Specialized scopal hairs, which facilitate collection and transport of floral oils, have evolved repeatedly among oil-collecting bees. Whether or not oil-collecting bees are pollen specialists or generalists is poorly documented. Nevertheless, some interactions of oil-collecting bees and their oil hosts are unusually specialized. Bee species in the genus *Rediviva* have elongated legs that they use to dab oils from the deep spurs of oil-secreting plants in the genera *Diascia* and *Ixianthes* (Scrophulariaceae), as well as of several species of orchids. This mutualism is close because these plants are visited by species of *Rediviva* and not bees from other genera. However, even the most extreme relationships in this system are decidedly asymmetrical. For example, *Ixianthes retzioides* is pollinated only by the large bee *R. gigas* (Steiner and Whitehead 1996), but the bee also visits other oil-secreting plants and does not occur in all populations of *I. retzioides*. Where the plant lacks a pollinator, its reproductive success declines considerably, but where the pollinator lacks a plant, it shifts to other floral hosts. The relationship of *Rediviva* and a few oil-secreting plant species may be typical among oil-producing flowers and oil specialist bees in the Eastern Hemisphere, where the few oil-producing flowers host the few oil-collecting bee species. These interactions are much broader in the Western Hemisphere, however, where plant species in 54 genera and 5 families offer oils (Vogel 1974; Vogel and Cocucci 1995). Females of some or all bee species of the eight Neotropical genera in the tribes Centridini, Tetrapediini, and Exomalopsini have modified structures and hairs on their legs for collecting and transporting oils and collect oil from multiple species of oil-producing plants (Simpson and Neff 1987; Schlindwein 1998). Greater generalization of the interactions in the Western Hemisphere may simply reflect the greater number of oil-secreting plant species and oil-collecting bees. These interactions are broadened further by the addition of bee species that do not collect oil but visit oil-secreting flowers for pollen (Vogel 1974).

Nectar and pollen are the two most common floral resources that bees collect. Nectar is a source of carbohydrates that adults consume directly to power flight and other activities or mix with pollen to feed developing larvae. Bees commonly use more plant species for nectar than they do for pollen (Thorp 1969; Klostermeyer et al. 1973; Danforth 1991; Neff and Danforth 1992; Rust et al. 2004). Evolutionary specialization to a nectar host has not been documented in bees, but many bees have morphological traits that influence their ability to extract nectar from different floral types. These include bizarrely elongate heads and/or mouthparts among bee species that visit deep flowers, or unusually shortened mouthparts in bee species that visit shallow flowers (Silveira 1993; Rozen and Ruz 1995; Alves-dos-Santos and Wittmann 1999). These unusual attributes, however, do not preclude bee species from using flowers with very different shapes and sizes. Bees with elongate mouthparts sometimes forage at shallow flowers (Dodson 1966; Inouye 1980; Bernhardt and Walker 1996; Minckley and Reyes 1996), and short-tongued species often crawl into or pierce deep flowers to extract nectar (Barrows 1980). Long tongues may allow bees to use a greater array of plant species, rather than fewer (Ranta and Lundberg 1980), as is true in hummingbirds (Bleiweiss 1999; Temeles and Kress 2003). Thus, selection on bee morphology apparently occurs in the absence of exclusive host-plant relationships.

Pollen is the primary source of protein for all bees, except for a few carrion-feeding species (Roubik 1982). Pollen proteins initiate and maintain oogenesis in adults and support development from egg to adult stage in offspring. All bee species limit their foraging to a subset of the available floral hosts (Ginsberg 1983), but species vary in the number and morphological diversity of hosts visited. At one extreme, broadly polylectic species collect pollen from numerous plant species from diverse lineages. Three colonies of the European honeybee (*A. mellifera*) in the Sonoran Desert collected pollen from 35–55 plant species per year—a quarter of all plant species in this area (Buchmann et al. 1990). At the species and colony level, all bumblebee species are broad generalists, except for *Bombus consobrinus* during bloom of *Aconitum* spp. (Löken 1961). Individual workers, however, often collect pollen from a single plant species during a foraging trip (Heinrich 1979; Thomson 1996) and entire colonies sometimes focus all foraging on one host (Visscher and Seeley 1982; O'Neal and Waller 1984). When this occurs, individual social bees and colonies exhibit host-plant fidelity; behavioral fidelity to a single host has been termed *floral constancy* (Waser 1986). This distinction between colony and individual foraging behaviors does not apply to solitary bee species. Only one forager provisions each nest, and the information gained on each foraging trip is the extent of her knowledge about the spatial and temporal distribution of floral resources. Like social bees, however, some polylectic solitary species visit one or several floral hosts during each foraging bout (Zavortink and LaBerge 1976; Bernhardt 1987), although observation of foraging females and of pollen composition in larval provisions that show floral constancy is frequent if not more typical (Westrich and Schmidt 1987; Cripps and Rust 1989; Bullock et al. 1991; Scott et al. 1993; but see Williams and Tepedino 2003).

Host specificity for pollen is the most common form of specialization by bees. Many bee species restrict pollen foraging throughout their geographic range to a single plant genus, tribe, subfamily, or family (Linsley and MacSwain 1958; Wcislo and Cane 1996). Pollen specialist species, or oligoleges, range from 60% of all pollen-collecting bee species in the warm deserts of California (Moldenke 1976), to 43% in the San Rafael Desert of Utah (Griswold et al. 1997), 28% in Carlinville, Illinois (Robertson 1926), 30% in central and northern Europe and subtropical Brazil (Westrich 1989; Monsevicius 1995; Schlindwein 1998), and 15–20% in Finland, Sweden (Pekkarinen 1997), and Nova Scotia (Sheffield et al. 2003). Monolecty, the use of a single plant species for pollen, is exceedingly rare among bees. Many cases of monolecty may reflect the isolation of a single plant species from congeners rather than a species-specific preference by the bee (Thorp 1969; Linsley 1978; Michener 2000), a condition referred to as enforced specialization in herbivorous insects (Fox and Morrow 1981). It is generally true that the majority of specialist bee species visit more than one plant species where closely related species co-occur and flower synchronously; this same pattern is documented for other herbivorous insects (Novotny et al. 2002). This does not mean, however, that specialist bees utilize all available congenerics of their floral host. For example, pollen-specialist bee species of the common sunflower (*Helianthus annuus*) are found rarely, if at all, on flowers of Jeruselum artichoke (*Helianthus tuberosa*) where both occur in eastern Kansas (C. Michener, personal communication). Although oligolecty is defined taxonomically, bee host-plant preference is expected to relate to floral characteristics that influence their ability to detect, recognize, and utilize host plants, and these characteristics may not always fall within simple taxonomic boundaries (Futuyma and Moreno 1988). Such a case in bees occurs in the genera *Perdita* and *Diadasia,* in which some lineages have shifted between distantly related, but morphologically similar, hosts in the plant families Cactaceae and Malvaceae (Danforth 1996; Sipes and Wolf 2001).

Oligolectic Bee Visitation in Floral Communities

To characterize a full range of bee–plant pollination interactions, we have begun a long-term community study in the upper Chihuahuan Desert along the U.S.–México border between Arizona and Sonora. The goals of the study are to examine the influence of plant traits on community-level interactions and on host specificity in bees, as well as the potential importance of oligoleges to host plant reproduction. Our study region contains a diverse flora of low-stature plants that can be readily observed from the ground and over 300 species of bees, 30–40% of which are oligolectic.

During April 2002 we established 1-ha plots at six sites within a 20 km^2 area of desert scrub. Within each plot we estimated, both through exhaustive counts and subsampling, the number of flowers of each plant species. On the following day, we netted insects on flowers, sampling each plant species three times a day, 15 minutes per survey. Overall, we sampled floral visitors on 34 plant species in 17 plant families. Twenty-seven of these species had an open floral morphology we classified as simple, whereas seven had complex flowers with deep corollas or required derived behavior or morphology on the part of the visitor to extract resources. This flora is primarily bee pollinated. Of the 748 insects collected at flowers, 83% were bees, representing 88 species. Twenty bee species were oligoleges, as defined by previous researchers, and the remaining 68 species were polyleges. Species accumulation curves based on data from these six sites estimated that more than 100 bee species were active during the three-week period of this study. The majority of other flower visitors were flies, represented by one to three individuals per species, and were usually only collected at one site. Plant species hosted from 0 to 13 species of floral visitor. Five important points could be drawn from our study.

1. Although most habitats were similar and a short distance apart (2–8 km), most species of flower visitors were rare and patchily distributed. Almost half of

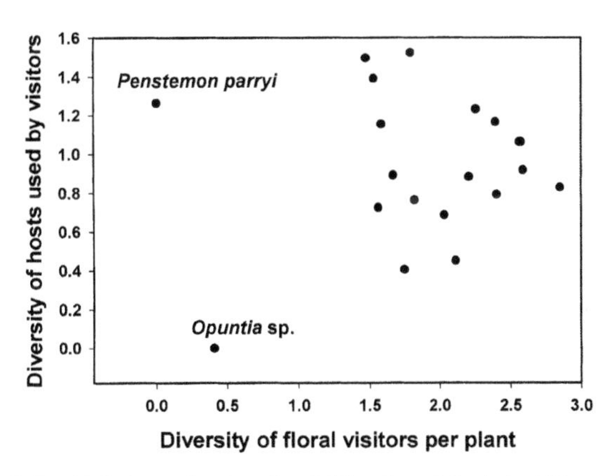

Figure 4.2 The relationship between the diversity of visitors to each host plant (Shannon–Wiener Index) and the relative specialization of those visitors.

the bee species caught were known only from a single specimen and only 3 of 120 insect species were found at all sites.

2. There was little evidence of cospecialization between plant and insect species. Within sites, there were 10 cases in which a plant species received visits from only a single species of insect. Of those 10, 8 of the visiting species were known generalists that were also visiting other plant species at the site. In two cases, the visitor was an oligolectic bee species; in these cases, the bee species was rare and probably visited relatively few of the open flowers of its host. Among the plant species for which at least five individual visitors were recorded, only *Opuntia* received visits from exclusively specialist visitors: the oligolectic bee species *Diadasia rinconis* and *Lithurge apicalis. Opuntia* spp., however, attract many generalist bee species at other sites (Simpson and Neff 1987). A low diversity of visitors to *Opuntia* in our study was probably due to it being only a minor element of the flora (8 stems, 16 open flowers), and most generalist bees were visiting more common plants. Figure 4.2 shows the relationship between the diversity of floral visitors to plant species across all six sites and the relative specialization of those visitors. Other than *Opuntia,* the only plant species to receive visits from a very narrow range of visitors was *Penstemon parryi,* the species with the second-longest floral tube of the plant species sampled. Visits to it were solely by generalist bees in the genus *Osmia.*

3. The number of floral visitor taxa to each plant species was strongly correlated with the total number of individual visitors (fig. 4.3). This provides evidence that the bee visitors seek similar resources (a common currency) that vary across plant species. As rewards increase, a greater proportion of the fauna utilizes the host if the rewards are readily available. If floral morphology or reward chemistry compartmentalized the fauna into isolated plant–pollinator relationships, there would be no expected overall relationship between total visitors and visitor diversity.

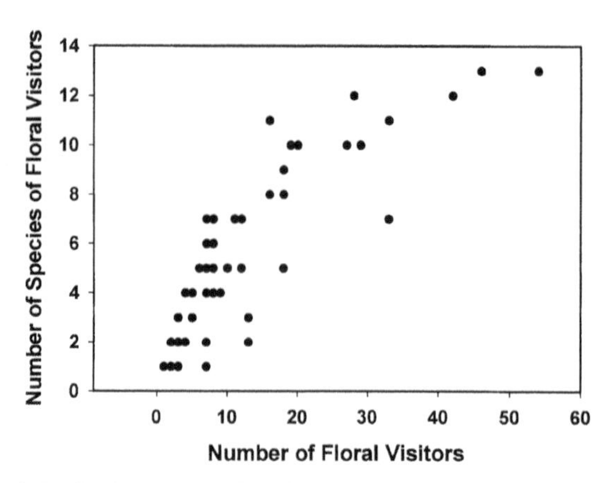

Figure 4.3 The relationship between number of insect visitors and number of insect species on host plants. Plants good for one visitor are good for most.

4. Oligolectic bee species generally visited plants that attracted numerous species of polylectic bees (fig. 4.3A, E; see also Vázquez and Simberloff 2002; Vázquez and Aizen, chap. 9 in this volume; Petanidou and Potts, chap. 10 in this volume).

5. Oligolectic bee species were primarily associated with simple flowers. If selection favors trait elaboration in specialized foragers to overcome restrictive floral morphology, then specialized foragers should be unusually common on complex flowers. We found no evidence of this. The plant species that hosted the greatest number of specialist species (five on creosote bush, five on mesquite) had notably simple, accessible flowers (illustrated by Simpson et al. 1977a). Simple flowers did not host significantly more visitor species than complex flowers (complex: $N = 7$, visitors $= 6 \pm 2.6$ S.E.; simple: $N = 23$, visitors $= 9.4 \pm 1.7$ S.E., $t = 1.10$, $P = .294$). These preliminary data show that most plant species at our study site are involved in generalized pollinator interactions. Furthermore, oligolectic bees are more likely to share host plants with generalists than they are to be involved in reciprocally specific plant–pollinator relationships with their hosts (see also Moldenke 1976; Petanidou and Potts, chap. 10 in this volume).

These findings, from a desert community in which oligolectic bee species are very common, suggest that specialization is associated with the utilization of highly attractive, readily accessible flowers, rather than underutilized flowers of plant species whose rewards are inaccesible to generalists. To see if this pattern occurs in nondesert communities, we reanalyzed three larger datasets of bee-host records: from Asteraceae, Malvaceae, Onagraceae, and Verbenaceae from Carlinville, Illinois, taken between 1884 and 1911 (Robertson 1929; Marlin and LaBerge 2001); from 451 bee-visited plant species in southern Germany (Westrich 1989); and from the bee-pollinated flora in subtropical Brazil (Schlind-

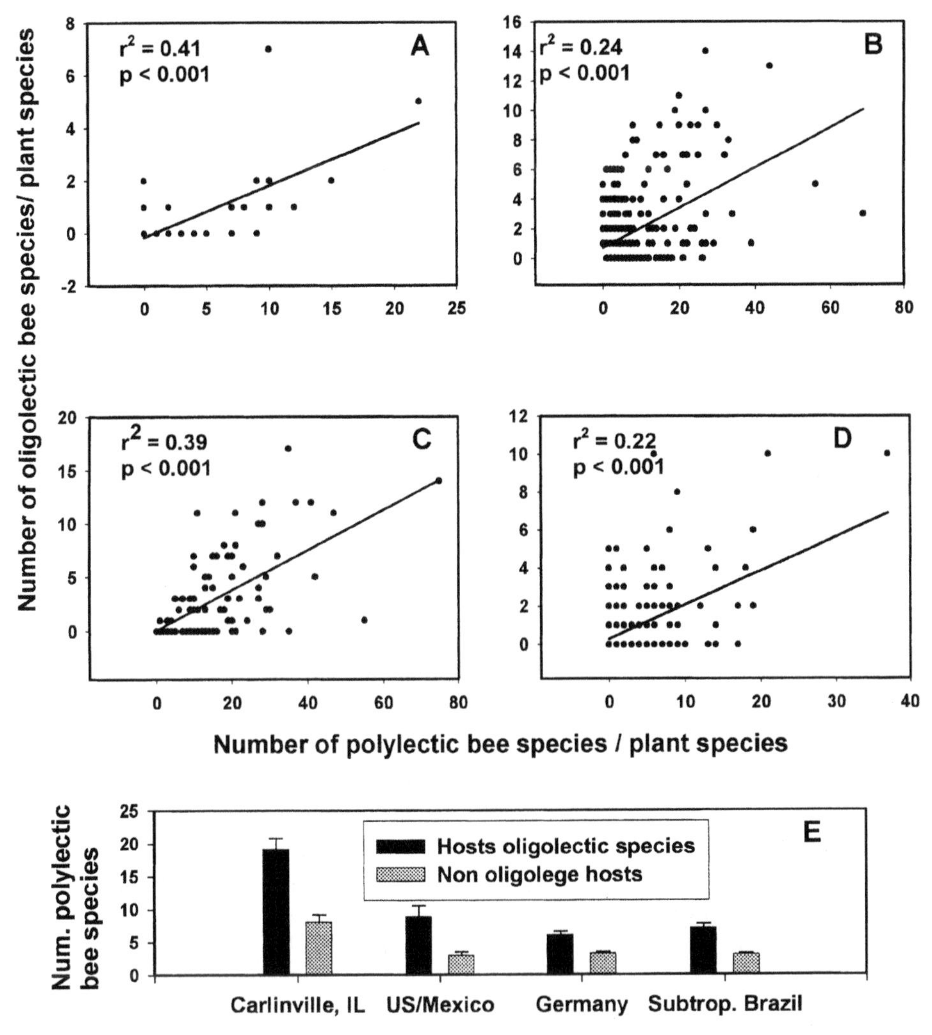

Figure 4.4 A comparison of the number of species of oligolectic and polylectic bee species on the same host plants. Data are from (A) México–Arizona border (Minckley and Roulston, unpublished data); (B) 451 plant species in southern Germany (Westrich 1990); (C) Carlinville, Illinois (Robertson 1929; plant families Asteraceae, Malvaceae, Onagraceae, and Verbenaceae); and (D) subtropical Brazil (Schlindwein 1998). Panel (E) compares the number of polylectic bees hosted by plant species that also host oligolectic bees (black) or do not also host oligolectic bees (gray). Differences in the total number of species between sites partially reflect sampling intensity.

wein 1998; fig. 4.4B–E). The data were not collected in the same manner across studies, yet all indicate that oligoleges and polyleges converge on the same hosts. The pattern is similar for each individual plant family in the Carlinville data, even though these plant groups are fundamentally different with respect to the resources and flower types they present to bees. Among species of Verbenaceae and Convolvulaceae are floral features that have generated structural changes in the specialist bees they host, which might be expected to promote specificity in pollination interactions. *Verbena* offer pollen at the base of long floral tubes, and bees in *Calliopsis* subgenus *Verbenapis* have modified tarsi and stout hairs on their proboscises that allow them to gather pollen (Robertson 1914; Shinn 1967). Plants of the genus *Convolvulus* have exceptionally large pollen, and their bee oligoleges consistently have broadly spaced hairs that enable females to carry this pollen back to the nest (Thorp 1979). In contrast, Asteraceae flowers have an open structure with easily accessible resources—attributes that might be expected to promote strongly generalized relationships. However, this group attracts the greatest number of oligolectic bee species, many of which lack overt morphological structures or unusual behaviors that suggest adaptation to their primary host plants.

Costs and Benefits of Specialization and Generalization to Plants

It has long been recognized that floral constancy among bees promotes successful pollination (Jones 1977; Waser 1986). Because oligolecty is functionally similar to floral constancy and independent of relative resource abundance, it is tempting to hypothesize that host-plant specialization by bees would yield extraordinary reproductive success among their floral hosts. Although some specialists are good pollinators of their hosts (Tepedino 1981), many are not (Barrows 1976; Zavortink 1992; Bernhardt and Weston 1996; Olsen 1997). Pollination efficacy can be diminished because of both morphological incongruence between the plant and bees and inappropriate bee behavior such as floral robbing or chewing into unopened flowers (Hurd and Linsley 1963). Different ecological requirements of floral hosts and specialist bees also limit their utility as pollinators. Here we discuss the geographic and temporal disparities among floral hosts and their specialist bees, and the propensity of specialist bees to use plants that attract numerous other pollinators. Any or all of these may limit the potential for specialized pollination interactions to develop over evolutionary time.

Herrera (1988) and many others have shown that floral visitors occur irregularly at their host across sites and through time (but see Cane et al. 2005). Oligolectic bee species are comparably variable. Steiner and Whitehead (1990) found the oil-specialist bees *R. neliana* at 21 of 22 populations of its host plant, *Diascia* spp., but the other 11 *Rediviva* species were collected at an average of only 2.2 ± 0.5 (mean ± S.E., range 1–7) sites. The xerophytic shrub *Larrea tridentata* (creosote bush) occurs throughout the Chihuahuan, Mojave, and Sonoran Deserts (an area covering approximately one-seventh of the North American continent), but only 2 of the 22 oligolectic bee species it hosts have distributions that overlap more than half of the host plant's range (Hurd and Linsley 1975; Simpson and Neff 1987; Minckley et al. 1999). Fifteen of the bee species occur only in the Lower Sonoran Desert and adjoining areas of the Mojave Desert (Minckley et al. 2000). Dispersal capability is not obviously related to this pattern. The two cosmopolitan species are above and below the mean body size of all 22 creosote bush oligoleges. A study at finer spatial scales found that species turnover of oligolectic bees on creosote bush was more than 65% at sites from 0.5 to more than 1400 km apart and as variable as polylectic bee species on the same host (Minckley et al. 1999). These values may overestimate species co-occurrence, because sites were matched for their similarity and not chosen randomly. Studies of other oligolectic bee species indicate that this pattern of spatial unreliability at their host plant is common (Linsley et al. 1973; Rozen 1977; Michener 1979; Pesenko and Radchenko 1993). Sometimes the absence of oligolectic bee species can be attributed to a lack of preferred nesting substrates, such as sand dunes (*Colletes stepheni, Calliopsis larrea;* Hurd and Powell 1958; Rust 1988), vertical cliffs protected from rain (*Melitoma, Ancyloscelis;* Linsley et al. 1980), or narrow cavities (*Hoplitis biscutellae;* Cane et al. 2006). However, in most cases, the absence of a specialist at its host cannot be attributed to obvious ecological parameters.

In addition to spatial variability, there is considerable variation through the season and among years for local bee faunas worldwide (reviewed by Williams et al. 2001). The variation is no less for plants that host specialist bees and further limits their potential value as exclusive pollinators. It is not unusual to observe specialist bees active well before (Michener and Rettenmeyer 1956) or well after (Thorp 1969) host bloom. In the absence of their usual host, females of some species shift to new pollen sources that may or may not be closely related to the normal host (Michener and Rettenmeyer 1956; Linsley 1958; Thorp 1969; see also Cane and Sipes, chap. 5 in this volume). Another form of temporal unpredictability occurs among plants that bloom twice per year but host oligolectic bees that are active during only one of the bloom events. For example, only 2 of the 22 species of creosote bush specialists are active in the late-summer bloom (Hurd and Linsley 1975). Five oligolectic species of *Perdita* on mesquite (*Prosopis glandulosa*) were active at a site in southern Arizona during the spring bloom, yet only one of these species emerged for the late-summer bloom (Simpson and Neff 1987). Diverse species of oligolectic and polylectic bees visit *Euphorbia* in the Upper Chihuahuan Desert when it blooms in the late summer, but no oligoleges and few polyleges can be found there in the spring (R. Minckley and T. Roulston, unpublished data). Bees at globe mallow show a more complex pattern, wherein one species of *Diadasia* visits its flowers in the spring and three species of *Perdita* visit its flowers in the fall. The factors that favor specialist activity in one season over the other presumably relate to resource abundance, temporal reliability of the bloom, adequate emergence cues to trigger synchrony, or other phylogenetic and ecological constraints.

The high spatial and temporal unpredictability of oligolectic bee species at their host plants is consistent with the hypothesis that the reproductive success of host plants is not often dependent on the presence of a specialist bee. It also suggests that obligately outcrossed plants that were successful at limiting the number of pollinator species might frequently experience pollination deficit. Exclusive pollinator relationships may be rare because high pollinator unpredictability renders them unstable over evolutionary time.

Ecological Basis of Evolutionary Change

Considerable work has focused on how the reproductive success of bee-pollinated plants depends on visitation rate, relative abundance of the different species of visitors, and the effectiveness and efficiency of different bees as pollinators. Fewer studies have examined how community dynamics (e.g., pollinator visitation patterns across different plant species assemblages and variation in pollinator abundance) over time and space influence the evolutionary potential of plant–pollinator interactions. Studies of whole communities over multiple years are badly needed to assess the potential evolutionary significance of the results of short-term, single-species studies. For example, Thompson and Pellmyr (1992) showed that, even when a highly effective specialist pollinator is present, it may exert only modest selection pressure. In their relatively simple system, infrequent visitation by a poorly suited generalist pollinator effectively counteracted selection on floral traits by a specialist pollinator. How ecological traits influence the evolutionary trajectory of pollination interactions and the level of specificity or generality of pollination mutualisms remains another more difficult, yet crucial, level to understand.

Evolutionary Patterns in Floral Host Breadth

Most bee clades contain all oligolectic or all polylectic bee species. Among those groups where the phylogeny has been studied, speciation events in oligolectic lineages most commonly involve no host shift or shifts to a closely related host, and seldom to distantly related hosts. Within lineages containing both specialists and generalists, shifts from specialization to generalization may be as common as shifts from generalization to specialization. Among 72 species of Old World Anthidiini, only four or five shifts in host breadth occurred, and in all cases the change was from oligolecty to polylecty (Müller 1996). In an unrelated group—the *Hemihalictus* series of *Lasioglossum*—two shifts occurred and both were in the opposite direction, from polylecty to oligolecty, in *L.* (*Hemihalictus*) *lustrans,* and *Lasioglossum* (*Sphecodagastra*) spp. (Danforth et al. 2003a). Two hypotheses are consistent with this pattern of evolutionary stasis for specialization and generalization in bee lineages. One is that plants have similar floral chemistry and/or morphology to bees which shift between them (this being especially common when the plants are closely related), and the other is that bees are cognitively limited to detecting many or few hosts, rather than constrained by intrinsic host features (floral chemistry, nutrient quality, morphology; see Menken 1996 for an example of this last phenomenon in herbivores). Both these hypotheses deserve scrutiny in future studies of the physiological mechanisms underlying host detection and attraction among specialist bees. Also critical to future progress toward understanding host specificity in bees will be the differentiation between the origin and maintenance of host specificity. Studies of specialists that inherited narrow host breadth from a common ancestor but use different hosts (e.g., sequential specialists) may tell us more about conditions that maintain specialization than about conditions that lead to specialization. Comparative studies of closely related specialist and generalist species should reveal much about conditions favoring the origin of host specificity.

Why Should Bees Specialize on Pollen Sources?

Here we shift from a focus on why plant–pollinator interactions are usually generalized, even when oligolectic bees are involved, to ask what advantages pollen specialization may offer bees. We contend that the common pattern of specialists converging on the same hosts as generalists provides a clue to plant features that promote the evolution of pollen specialization in bees. At the same time, this leads to questions concerning the factors that promote specialized foraging and feeding behavior. Together, these pieces of information may help reveal the conditions under which specialized mutualisms evolve in bees and plants.

Competition and Foraging

The oldest and most widely discussed hypothesis for the advantage of floral specialization among bees is increased foraging efficiency by specialists (Robertson 1914; Lovell 1918; Michener 1954; Linsley and MacSwain 1958; Thorp 1969). Some experiments have compared the foraging attributes of generalist and specialist bees. Strickler (1979) compared foraging rate and pollen harvest of one oligolectic and four polylectic bee species on one floral host and found that the oligolege was more efficient at producing offspring per unit foraging time. In this case, the oligolege *Hoplitis anthocopoides* gained its advantage by flying rapidly between the plants of its host, *Echium vulgare* (Boraginaceae), and quickly harvesting the pollen, rather than by utilizing derived morphological structures. Similarly, *Habropoda laboriosa,* a specialist bee of *Vaccinium* spp., gathered more resources per unit time from its host than did generalists on the same host (Cane and Payne 1988). Naïve foragers of the only known oligolectic social bee species, *Bombus consobrinus,* handled the flowers of its primary host, *Aconitum* spp., more efficiently than did naïve foragers of generalist species (Laverty and Plowright 1988). These findings support the hypothesis that evolutionary specialization can lead to traits that improve host-plant utilization, thereby giving specialists an advantage. Many plant species, however, host multiple species of specialists and generalists, and it seems unlikely that specialists will always cluster as the

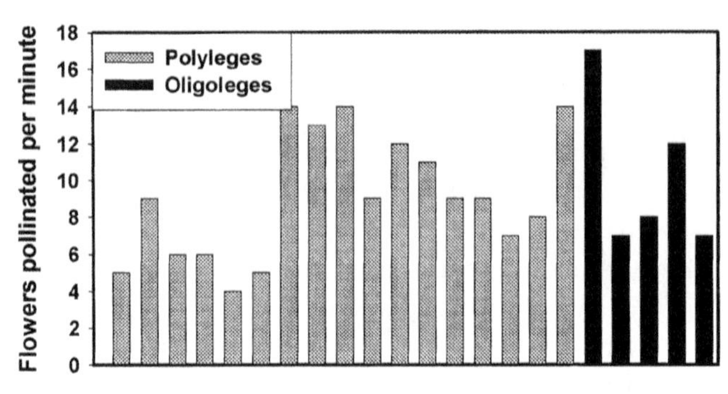

Figure 4.5 A comparison of foraging rates of oligolectic and polylectic bees on alfalfa (data from Pesenko and Radchenko 1993)

most efficient foragers. Data from Pesenko and Radchenko (1993) suggest that specialists as a whole are not always faster than generalists at utilizing shared hosts (fig. 4.5), although the critical test would be comparisons of pollen removal, deposition, and efficiency rates among species.

Even when specialists do forage more efficiently on a particular host, there are some difficulties in extrapolating a highly manipulative captive experiment or basic field observations to an evolutionary argument. For one, these experiments assume that reproductive success of wild bees is pollen-limited. Data show that nectar resources are drained to depletion during the day (Heinrich 1976), as are the pollen resources of some plants (Minckley et al. 1994; Willis and Kevan 1995). Tepedino and Stanton (1980), however, tracked floral resources between years and concluded that changes were so variable and unpredictable that the solitary bee community could only rarely experience density-dependent selection due to pollen limitation (see also Ranta 1984; Roubik 1992). Available pollen resources from the creosote bush appear to be much greater than the biomass of bees at their flowers (Minckley et al. 2003). Because bee populations of univoltine species lag one year out of phase with floral resources (i.e., bee populations represent the floral resources available to the previous bee generation), competition for floral resources will vary greatly between seasons or years, depending on the sequence of good and poor flowering periods.

Second, even when specialists do forage more efficiently on a particular host, they suffer the potential cost of restricting their foraging to an inferior host in years of poor bloom or at sites in which their host is rare. This cost remains, regardless of the efficiency of the specialist on its preferred host. Experimental tests of the trade-off between increased foraging efficiency on a given host and decreased resource availability on fewer potential hosts have not been done. They are necessary, however, to discern if increased efficiency can drive shifts to specialization or merely improve resource harvest after specialization has already arisen.

Factors other than resource availability are known to limit reproduction of solitary bees. Although the queens of social insects including bees can be highly fecund, most solitary bees produce only one or two eggs over one or a few days (Danforth 1989; Neff and Simpson 1991; Neff and Danforth 1992; Minckley et al. 1994; Field 1996). Slow egg maturation or other physiological constraints may reduce the competitive interactions among solitary bees for total utilizable resources, at least in resource-rich (relative to number of floral visitors) environments (Minckley et al. 1994; Field 1996; Rosenheim 1999). Selection could still strongly favor rapid foraging if it reduces exposure of an unguarded nest to parasites, decreases predation probability at flowers, or lessens harassment by mate-seeking males (Dukas 2001; Goodell 2003).

Competition and Character Displacement

Examples of closely related oligolectic bee species on different floral hosts have been cited as evidence that host-plant specialization reduces interspecific competition (Robertson 1925; Thorp 1969; Linsley 1978). Several observations indicate that this is unlikely to be a primary mechanism promoting specialization.

Host breadth is evolutionarily conservative (generalists giving rise to generalists, and specialists to specialists) among bee lineages, and most speciation events do not include host shifts (Danforth 1996; Müller 1996; Sipes and Wolf 2001). In most cases, sister species in oligolectic lineages use the same food sources and should compete maximally when they occur in sympatry. Therefore, competition for resources appears to be a less-important cause of diversification than allopatry or other speciation mechanisms (Linsley 1961; Thorp 1969; Hurd and Linsley 1970; Linsley et al. 1973; Sipes and Wolf 2001).

A pattern widely recognized by students of bees, but less recognized by others, is that those plant species that host oligoleges are often the most bee-attractive plants in a particular community (see fig. 4.3E). Plants in North America with diverse bee fauna include sunflowers (*Helianthus* spp.), with 39 oligolectic and more than 280 polylectic species (Hurd et al. 1980); creosote bush, with 22 oligolectic and more than 100 polylectic species (Hurd and Linsley 1975); prickly pear cactus (*Opuntia* spp.), with about 16 oligolectic and 85 polylectic species (Grant et al. 1979); and mesquite (*Prosopis*), with 30 oligolectic and 130 polylectic species (Simpson and Neff 1987). Species of willow (*Salix*) in the Holarctic host many bee species and also appear to follow this pattern (Westrich 1989; Pekkarinen 1997), as does *Ludwigia* in the Neotropics (Martins and Borges 1999). These oligolectic bee faunas of one plant species often comprise several distant lineages, indicating that repeated independent host shifts have occurred. For the creosote bush, 12–15 host shifts appear to be represented among the 22 oligolectic bee species it supports. Rather than restricting foraging to host plants not used by other bees, and where competition is likely to be minimal, oligolectic bee species specialize on bee-attractive hosts, where competition for resources appears most likely. Frequent evolutionary transitions to the same plant group suggest that specialization is selected for or maintained by specific plant traits rather than by the avoidance of interspecific competition among bees.

Physiological Trade-offs

Host specificity among bees could be favored when resource quality differs greatly among potential hosts. Some herbivores that must overcome host defenses perform best on their preferred host, but this pattern is not universal and may not even be particularly common (Futuyma and Moreno 1988). Herbivores and bees both consume plant products, but pollen may serve as a reward to mutualistic pollinators, thereby reducing or nullifying selection for defensive compounds in pollen. Nevertheless, some defensive compounds that discourage foliar herbivory (Coley et al. 1985) also occur in floral structural tissues, pollen, and nectar (Detzel and Wink 1993; Carisey and Bauce 1997), albeit usually at lower concentrations than in leaves. The potential significance of toxic compounds in nectar has been discussed at some length (Adler 2000; Adler et al. 2001; Irwin et al. 2004), but the occurrence and importance of such compounds in pollen is poorly known (Roulston and Cane 2000).

Nutrient concentration of pollens differs greatly among plant species (Roulston and Cane 2000), which could favor bee specialization if differences are difficult to detect by foragers but strongly influence performance. Pollen protein concentration varies widely among plant species and is the most studied nutrient (Roulston et al. 2000). In one study, a generalist bee species produced large offspring from protein-rich pollen and small offspring from protein-poor pollen (Roulston and Cane 2002); thus, one benefit of host specificity could be specialization on plants with high-quality pollen. However, Roulston et al. (2000) found no association between pollen protein concentration and oligolecty in bees. Pollens also differ greatly in starch content, which has been implicated as an indicator of pollen nutrient quality (Baker and Baker 1979). However, there is little experimental evidence supporting its nutritional significance and, as with protein pollen, little evidence that it has evolved under selection by pollinators (Roulston and Buchmann 2000). Although there is little evidence that bees assess pollen nutrients, there is evidence that foragers distinguish pollen based on nonnutritive qualities such as scent (Dobson 1985). An obstacle to direct detection of pollen nutrients is that most nutrients are located inside the pollen wall and are only released during a slow digestive process.

More studies are needed to determine if oligolectic bee larvae generally perform better on their usual host. Those tested thus far can survive on some alternative hosts (Tepedino 1997), though in some cases they do better on preferred hosts (Guirguis and Brindley 1974; Bohart and Youssef 1976). At present, larvae are not known to refuse provided pollens, but adults may refuse to collect pollen from alternative sources even when their host is unavailable (Strickler 1979;

Table 4.1 Pollen production among several plants that host oligolectic bee species

Plant species	Flowers/plant	Pollen/plant	Reference
Helianthus annuus	108	27,941 mg	Minckley et al. 1994
Prosopis velutina	126.5	3161 mg	Simpson 1977
Larrea tridentata	2766	5200 mg	Simpson et al. 1977b
Encelia farinosa	545	2.4×10^9 grains	Visscher and Danforth 1993

McIntosh 2001; Williams 2003). Williams (2003) transplanted eggs of oligolectic and polylectic bee species onto preferred and nonpreferred hosts and found that, contrary to expectations, the specialist performed better than the generalist on novel hosts. In some pollen mixtures that included both the usual and a novel host, performance of the oligolege was *better* than on its usual specialized diet.

Cost of Choosing

Bernays (1999 and references therein) has proposed that the costs of evaluating and switching to different host plants may be great among short-lived insects, and selective attention to one host explains why most insects are specialists. There is evidence that bees have limited learning and memory capacity and that there is a cost imposed by switching hosts (Bernays and Wcislo 1994; Chittka et al. 2001; Dukas and Kamil 2001); however, it is not known if these costs outweigh the cost of remaining on a host plant that becomes depleted.

Synchronization with a Dominant Host

To the extent that specialists rely on a given host, their reproduction is limited to the resources of that host. Calculations of the relative contribution to total floral resources of all plant species in any plant community have never been published, but calculations of the floral resources produced by several highly bee-attractive plant species are impressive (table 4.1). The pollen of a single wild sunflower plant could produce 360 bees of the sunflower specialist *Dieunomia triangulifera* (adult female weight is 25 mg; Minckley et al. 1994), whereas the pollen of one brittlebush *Encelia farinosa* could produce 1600 offspring of the Asteraceae specialist *Calliopsis pugionis* (adult female weight is 7 mg; Visscher and Danforth 1993). When these plant species remain abundant across years they may promote, or at least sustain, specialization among bees. However, not all bee-attractive, dominant, dependable plant species host many species of oligolectic bees. Paloverde (*Cercidium* spp.), for example, blooms profusely and attracts many bees and bee species during spring in the Sonoran Desert, but few of these are specialists. Examining the traits of productive plant species that do not host specialists may be very helpful in discovering both the key plant traits and the advantages associated with pollen specialization.

In xeric habitats, bloom is often restricted to short periods of time and may be unreliable across years. A study of the bee fauna of creosote bush flowers throughout the southwestern United States revealed the seemingly contradictory pattern that species richness of the 22 specialist bee species was greatest in those areas where bloom was least dependable between years (Minckley et al. 2000). In desert areas that received sufficient rainfall for creosote bush flowering in fewer than 50 of the preceding 100 years, oligolectic species predominated, but polyleges predominated when rainfall was sufficient for bloom in at least 95% of years. Our hypothesis is that this odd pattern is the result of generalists being extirpated because they are poorly synchronized with the dominant plant species in that community. Oligoleges persist because they facultatively remain in diapause in years when their host plant does not flower, and they emerge synchronously in years when the host plant does flower.

The cost to a bee of emerging at the wrong time is low in mesic areas because it is very rare that other plant species are not in flower. However, the cost to being poorly synchronized with a host plant in xeric areas can be great in years when either no other plants flower or those that do flower provide relatively meager resources. Selection on bees for synchronization with a dominant host plant should be particularly strong under these conditions. Further evidence for the importance of bloom predictability to specialist bees in these environments comes from seasonal activity of the oligolectic bee species of creosote bush. Twenty of the 22 oligolectic bee species of creosote bush emerge during the spring bloom, 2 emerge during both the spring and late-summer blooms, and none emerge only for the late-summer bloom. In years with winter rain, creosote bush predictably initiates flowering during a two- to three-week period in the spring as soil and air temperatures increase (Abe 1982). In contrast, bloom later in the year is triggered by each precipitation event of more than 12 mm (Bowers and Dimmitt 1994). In years when summer rains occur in the Sonoran Desert, bloom initiation varies from late June to September, which may be difficult for specialist bees to track.

The creosote bush–bee study highlights several conclusions: not only must plant hosts of oligolectic bee species be sufficiently abundant and rewarding to support the bee population, but they must also flower predictably, either through adherence to a regular annual cycle or in response to environmental cues (e.g., rainfall) that also trigger specialist emergence. Bee species in deserts synchronize activity with host-plant bloom by facultatively remaining in diapause in dry years with no host bloom and emerging in years when rains trigger their host to bloom (reviewed by Danforth 1999). Droughts lasting for more than five years are not uncommon (Schmidt 1989); some bees are known to have remained in diapause for 7 (Rozen 1990) and 10 years (Houston 1991), and many bee species skip one or a few years. Diapause constantly depletes body fats, however, increasing risk of mortality and increasing sensitivity to emergence cues (Danforth 1999). In areas where flowering is sporadic or rare among plant species, selection should synchronize bee emergence with the flowering of productive plants that bloom with minimal precipitation (*L. tridentata*) or that have roots tapped into deep subsurface moisture (e.g., mesquite). Creosote bush blooms at the lowest rainfall amount of any plant species studied in the Sonoran Desert (Bowers and Dimmitt 1994). Bees that synchronize adult activity with its bloom therefore take advantage of an unusually predictable source of floral resources in a floristically depauperate environment where no plants bloom in some, if not most, years (Davidowitz 2002). Oligolecty among bees is most common in xeric environments (Ayala et al. 1993; Griswold et al. 1997; Michener 2000). Where flowering is episodic, the match of bee emergence and host-plant bloom should be crucial to bee population survival (Danforth et al. 2003b).

Selection for synchronization of emergence with the bloom of productive plant species guarantees that food resources are available when adults are active but does not guarantee that bees will visit the flowers of the plant with which it is synchronized. When many plant species are in bloom, other more productive plant species may be available that should attract most generalist bees. Where several oligolege-hosting plant species bloom simultaneously, it is unlikely they all represent the best-available host choice, as determined by resource availability, to their oligolectic fauna. Synchronization of bee activity with host-plant bloom is therefore probably most typically a modification associated with specialization.

Conclusions

It has become apparent that reciprocal specialization in plant–pollinator interactions is rare, although specialized plants and specialized pollinators are common in some environments (Robertson 1925; Johnson and Steiner 2000; Ollerton et al., chap. 13 in this volume). For plants to draw few species of visitors, they must produce little or no reward (thereby relying on naïve, generalist foragers; though see Renner, chap. 6 in this volume), produce an unusual reward/attractant, or restrict access to rewards through complex floral morphology. Of these mechanisms, only the last two involve food that would contribute to maintaining a pollinator population. Unusual rewards, such as floral oils and volatile compounds used as probable sex attractants, clearly reduce the diversity of visitors to flowers, but the visitors that seek these resources commonly extract them from various floral and nonfloral sources and seldom rely on any one plant species. Complex floral morphology may diminish visitor diversity somewhat, but most often the visitors that remain are generalists.

Pollinators visit flowers for the resources they offer. Much theoretical and experimental work describes how selection operates on generalist foraging strategies to maximize return (Pyke 1980) or minimize exposure time by initially choosing rewarding host plants and then switching hosts as resources diminish. These cost-benefit views of foraging provide a model for examining generalist behavior, but it is not obvious how they can explain the origin and maintenance of specialist behavior, which is common among solitary bee species. If selection favors floral sampling and host shifts among generalists, how can it favor fixed host-plant preference in the same group of organisms? This conundrum is not unique to bees and flowers and deserves more attention.

Among taxonomically rich pollinator groups, specialization is common only in solitary bees. We considered five ideas for the origin of host-plant specialization in this group: (1) efficient resource extraction, (2) efficient nutrient processing, (3) competition avoidance, (4) limited time span/selective attention, and (5) synchronization with dominant resource presentation. Some evidence supports increased foraging efficiency of specialists over that of generalists (Strickler 1979; Cane and Payne 1988; Laverty and Plowright 1988), but it is not clear that any advantage associated with increased harvesting efficiency on one host outweighs the loss of alternative hosts. Little evidence demonstrates that specialist bee larvae require their specialized diet (Williams 2003), and no evidence indicates that specialists have evolved to overcome host-plant defensive compounds in pollen. Evidence is strongly against specialization as a form of niche partitioning to reduce competition: most speciation events within specialist lineages do not involve host shifts (Müller 1996; Sipes and Wolf 2001), and oligolectic bees commonly converge on bee-attractive hosts. At present, it is difficult to evaluate the effect of selective attention on specialization. Generalist bees often confine foraging bouts to one plant species and there is no evidence yet that the cost associated with finding and switching hosts is greater than the benefit of discovering more productive hosts. One pattern that is fairly clear among oligolectic bees is that most specialize on abundant, productive, dependable hosts that often represent a major pollen resource in their habitat (Minckley et al. 1999). Thus, synchronization of bee emergence with the flowering of these plant species may be crucial to bee survival in marginal habitats. Suites of ecological factors such as resource abundance, predation pressure, foraging efficiency, and synchronization may combine to promote shifts in specialization and generalization among bees. These shifts are tempered by history; stasis in lineages suggests that ancestral conditions strongly influence the host breadth of descendant species.

Why is cospecialization rare in plant–pollinator interactions? Plant species that produce enough resources to support populations of specialist bees rarely also have features that dissuade generalist bee visitors, and plant species that attract bees to non-food resources, via deception or with morphologically elaborate flowers, attract only generalists. Even occasional pollination by generalists can prevent selection on plant traits toward an exclusive relationship with a specialist pollinator (Thompson and Pellmyr 1992). Given the conflicting goals of the plants and floral visitors involved in pollination interactions, it should not surprise anyone to find that the specialization of one partner sometimes represents maximal exploitation of the other partner, and very rarely leads to coadaptation. In addition, specialized pollination interactions may be rare if they often result in extinction due to pollinator unpredictability while more generalized interactions persist. We do not doubt that plants and pollinators have coevolved in some diffuse fashion (Janzen 1980; Grimaldi 1999). Mainland–island comparisons and many other studies clearly show that plant reproductive strategies can change when the pollinator community changes (Carlquist 1974; Linhart and Feinsinger 1980). The question that remains is how such a wealth of relationships and interesting structures of bees and flowers has evolved, given the chaotic conditions found in extant plant–pollinator communities.

Acknowledgments

Bryan Danforth, Paul Fine, Sedonia Sipes, and Neal Williams provided needed comments on early drafts of this chapter. We are grateful to Jose Arturo Romero Gutierrez, Noelia de la Torre, and Baruk Maldonado for long hours of able assistance in the field. Financial support to RLM from the World Wildlife Fund and U.S. Fish and Wildlife Service helped considerably. Finally, we thank William Radke (San Bernardino National Wildlife Refuge) and Josiah and Valer Austin for much logistic support, encouragement, and permission to work on property under their purview.

References

Abe, Y. 1982. Phenology of tetraploid creosotebush, *Larrea tridentata* (DC) Cov., at the northeastern edge of the Sonoran Desert. PhD dissertation, University of Arizona, Tucson, AZ.

Ackerman, J. D. 1983. Specificity and mutual dependency of the orchid–euglossine bee interaction. Biological Journal of the Linnean Society, London 20: 301–314.

Adler, L. S. 2000. The ecological significance of toxic nectar. Oikos 91: 409–420.

Adler, L. S., R. Karban, and S. Y. Strauss. 2001. Direct and indirect effects of alkaloids on plant fitness via herbivory and pollination. Ecology 82: 2032–2044.

Alves-dos-Santos, I., and D. Wittmann. 1999. The proboscis of the long-tongue *Ancyloscelis* bees (Anthophoridae/Apoidea), with remarks on flower visits and pollen collecting with the mouthparts. Journal of the Kansas Entomological Society 72: 277–288.

Armbruster, W. S., and B. C. Baldwin. 1998. Switch from specialized to generalized pollination. Nature 394: 632.

Ayala, R., T. L. Griswold, and S. H. Bullock. 1993. The native bees of Mexico. Pp. 179–228 *in* T. P. Ramamoorthy, R. Bye, A. Lot, and J. Fa (eds.), Biological diversity of Mexico: Origins and distribution. Oxford University Press, Oxford.

Baker, H. G., and I. Baker. 1979. Starch in angiosperm pollen grains and its evolutionary significance. American Journal of Botany 66: 591–600.

Baker, H. G., R. W. Cruden, and I. Baker. 1971. Minor parasitism in pollination biology and its community functions: The case of *Ceiba acuminata*. Bioscience 21: 1127–1129.

Baker, H. G., and P. D. Hurd Jr. 1968. Intrafloral ecology. Annual Review of Entomology 13: 385–414.

Barrows, E. M. 1976. Nectar robbing and pollination of *Lantana camara* (Verbenaceae). Biotropica 8: 132–135.

Barrows, E. M. 1980. Robbing of exotic plants by introduced carpenter and honey bees in Hawaii, with comparative notes. Biotropica 12: 23–29.

Barth, F. G. 1991. Insects and flowers: The biology of a partnership. Princeton University Press, Princeton, NJ.

Bernays, E. A. 1999. Plasticity and the problem of choice in food selection. Annals of the Entomological Society of America 92: 944–951.

Bernays, E. A., and W. T. Wcislo. 1994. Sensory capabilities, information processing, and resource specialization. Quarterly Review of Biology 69: 187–204.

Bernhardt, P. 1987. A comparison of the diversity, density, and foraging behavior of bees and wasps on Australian *Acacia*. Annals of the Missouri Botanical Garden 74: 42–50.

Bernhardt, P., and K. Walker. 1996. Observations on the foraging preferences of *Leioproctus* (*Filiglossa*) Rayment (Hymenoptera: Colletidae) in eastern Australia. Pan-Pacific Entomologist 72: 130–137.

Bernhardt, P., and P. Weston. 1996. The pollination ecology of *Persoonia* in eastern Australia. Telopea 6: 775–804.

Bleiweiss, R. 1999. Joint effects of feeding and breeding behaviour on trophic dimorphism in hummingbirds. Proceedings of the Royal Society of London B 266: 2491–2497.

Bohart, G. E., and N. N. Youssef. 1976. The biology and behavior of *Evylaeus galpinsiae* Cockerell (Hymenoptera: Halictidae). Wasmann Journal of Biology 34: 185–234.

Bowers, J. E., and M. A. Dimmitt. 1994. Flowering phenology of 6 woody plants in the northern Sonoran desert. Bulletin of the Torrey Botanical Club 121: 215–229.

Buchmann, S. L. 1987. The ecology of oil flowers and their bees. Annual Review of Ecology and Systematics 18: 343–369.

Buchmann, S. L., M. K. O'Rourke, and C. W. Shipman. 1990. Pollen preferences and dietary breadth of managed and feral Sonoran honey bees. American Bee Journal 130: 797–798.

Bullock, S. H., R. Ayala, D. Rodriguez-Zamora, D. L. Quiroz-Garcia, and M. D. Luz Arreguin-Sanchez. 1991. Nest provision and pollen foraging in three Mexican species of solitary bees (Hymenoptera: Apoidea). Pan-Pacific Entomologist 67: 171–176.

Cane, J. H., R. L. Minckley, L. Kervin, and T. H. Roulston. 2005. Temporally persistent patterns of incidence and abundance in a pollinator guild at annual and decadal scales: The bees of *Larrea tridentata*. Biological Journal of the Linnean Society 85: 319–329.

Cane, J. H., R. L. Minckley, T. H. Roulston, L. J. Kervin, and N. M. Williams. 2006. Complex responses within a desert bee guild (Hymenoptera: Apiformes) to urban habitat fragmentation. Ecological Applications. In press.

Cane, J. H., and J. A. Payne. 1988. Foraging ecology of the bee *Habropoda laboriosa* (Hymenoptera: Anthophoridae), an oligolege of blueberries (Ericaceae: *Vaccinium*) in the southeastern United States. Annals of the Entomological Society of America 81: 419–427.

Carisey, N., and E. Bauce. 1997. Impact of balsam fir flowering on pollen and foliage biochemistry in relation to spruce budworm growth, development and food utilization. Entomologia Experimentalis et Applicata 85: 17–31.

Carlquist, S. 1974. Island biology. Columbia University Press, New York.

Chittka, L., J. Spaethe, A. Schmidt, and A. Hickelsberger. 2001. Adaptation, constraint, and chance in the evolution of flower color and pollinator color vision. Pp. 106–126 *in* L. Chittka and J. D. Thomson (eds.), Cognitive ecology of pollination. Cambridge University Press, Cambridge.

Coley, P. D., J. P. Bryant, and F. S. Chapin III. 1985. Resource availability and plant antiherbivore defense. Science 230: 895–899.

Cripps, C., and R. W. Rust. 1989. Pollen preferences of seven *Osmia* species (Hymenoptera: Megachilidae). Environmental Entomology 18: 133–138.

Danforth, B. N. 1989. Nesting behavior of four species of *Perdita* (Hymenoptera: Andrenidae). Journal of the Kansas Entomological Society 62: 59–79.

Danforth, B. N. 1991. Female foraging and intranest behavior of a communal bee, *Perdita portalis* (Hymenoptera: Andrenidae). Annals of the Entomological Society of America 84: 537–548.

Danforth, B. N. 1996. Phylogenetic analysis and taxonomic revision of the *Perdita* subgenera *Macrotera, Macroteropsis, Macroterella,* and *Cockerellula* (Hymenoptera: Andrenidae). University of Kansas Science Bulletin 55: 635–692.

Danforth, B. N. 1999. Emergence dynamics and bet hedging in a desert bee, *Perdita portalis*. Proceedings of the Royal Society of London B 266: 1985–1994.

Danforth, B. N., L. Conway, and S. Q. Ji. 2003a. Phylogeny of eusocial *Lasioglossum* reveals multiple

losses of eusociality within a primitively eusocial clade of bees (Hymenoptera: Halictidae). Systematic Biology 52: 23–36.
Danforth, B. N., J. Shuqing, and L. J. Ballard. 2003b. Gene flow and population structure in an oligolectic desert bee, *Macrotera (Macroteropsis) portalis* (Hymenoptera: Andrenidae). Journal of the Kansas Entomological Society 72: 221–235.
Davidowitz, G. 2002. Does precipitation variability increase from mesic to xeric biomes? Global Ecology and Biogeography 11: 143–154.
Detzel, A., and M. Wink. 1993. Attraction, deterrence or intoxication of bees (*Apis mellifera*) by plant allelochemicals. Chemoecology 4: 8–18.
Dobson, H. E. M. 1985. Role of volatile pollen coat oils in host plant recognition by solitary bees. American Journal of Botany 6: 850.
Dodson, C. H. 1966. Ethology of some bees of the tribe Euglossini (Hymenoptera: Apidae). Journal of the Kansas Entomological Society 39: 607–629.
Dodson, C. H., R. L. Dressler, H. G. Hills, R. M. Adams, and N. H. Williams. 1969. Biologically active compounds in orchid fragrances. Science 164: 1243–1249.
Dukas, R. 2001. Effects of predation risk on pollinators and plants. Pp. 214–236 *in* L. Chittka and J. D. Thomson (eds.), Cognitive ecology of pollination. Cambridge University Press, Cambridge.
Dukas, R., and A. C. Kamil. 2001. Limited attention: The constraint underlying search image. Behavioral Ecology 12: 192–199.
Estes, J. R., and R. W. Thorp. 1975. Pollination ecology of *Pyrrhopappus carolinianus* (Compositae). American Journal of Botany 62: 148–159.
Field, J. 1996. Patterns of provisioning and iteroparity in a solitary halictine bee, *Lasioglossum (Evylaeus) fratellum* (Perez), with notes on *L. (E.) calceatum* (Scop) and *L. (E.) villosulum* (K). Insectes Sociaux 43: 167–182.
Fleming, T. H., C. T. Sahley, J. N. Holland, J. D. Nason, and J. L. Hamrick. 2001. Sonoran Desert columnar cacti and the evolution of generalized pollination systems. Ecological Monographs 71: 511–530.
Fox, L. R., and P. A. Morrow. 1981. Specialization: Species property or local phenomenon? Science 211: 887–893.
Futuyma, D. J., and G. Moreno. 1988. The evolution of ecological specialization. Annual Review of Ecology and Systematics 19: 207–233.
Gess, S. K. 1996. The pollen wasps: Ecology and natural history of the Masarinae. Harvard University Press, Cambridge, MA.
Ginsberg, H. S. 1983. Foraging ecology of bees in an old field. Ecology 64: 165–175.
Goodell, K. 2003. Food availability affects *Osmia pumila* (Hymenoptera: Megachilidae) foraging, reproduction, and brood parasitism. Oecologia 134: 518–527.
Grant, V., K. A. Grant, and P. D. Hurd. 1979. Pollination of *Opuntia lindheimeri* and related species. Plant Systematics and Evolution 132: 313–320.
Grimaldi, D. 1999. The co-radiations of pollinating insects and angiosperms in the Cretaceous. Annals of the Missouri Botanical Garden 86: 373–406.
Griswold, T., F. D. Parker, and V. J. Tepedino. 1997. The bees of the San Rafael Desert: Implications for the bee fauna of the Grand Staircase-Escalante National Monument. Pp. 175–186 *in* L. M. Hill (ed.), Proceedings of the Grande Staircase–Escalante National Monument science symposium. U.S. Department of the Interior, Bureau of Land Management, Salt Lake City, UT.
Guirguis, G. N., and W. A. Brindley. 1974. Insecticide susceptibility and response to selected pollens of larval alfalfa leafcutting bees, *Megacile pacifica* (Panzer) (Hymenoptera: Megachilidae). Environmental Entomology 3: 691–694.
Heinrich, B. 1976. Resource partitioning among some eusocial insects: Bumblebees. Ecology 57: 874–889.
Heinrich, B. 1979. "Majoring" and "minoring" by foraging bumblebees, *Bombus vagans:* An experimental analysis. Ecology 60: 245–255.
Herrera, C. M. 1988. Variation in mutualisms: The spatio-temporal mosaic of a pollinator assemblage. Biological Journal of the Linnean Society, London 35: 95–125.
Houston, T. F. 1991. Ecology and behaviour of the bee *Amegilla (Asarapoda) dawsoni* (Rayment) with

notes on a related species (Hymenoptera: Anthophoridae). Records of the Western Australian Museum 15: 591–609.

Hurd, P. D., Jr., W. E. LaBerge, and E. G. Linsley. 1980. Principal sunflower bees of North America with emphasis on the southwestern United States (Hymenoptera: Apoidea). Smithsonian Contributions to Zoology 310: 1–158.

Hurd, P. D., Jr., and E. G. Linsley. 1963. Pollination of the unicorn plant (Martyniaceae) by an oligolectic, corolla-cutting bee (Hymenoptera: Apoidea). Journal of the Kansas Entomological Society 36: 248–252.

Hurd, P. D., Jr., and E. G. Linsley. 1970. A classification of the squash and gourd bees *Peponapis* and *Xenoglossa* (Hymenoptera: Apoidea). University of California Publications in Entomology 62: 1–39.

Hurd, P. D., Jr., and E. G. Linsley. 1975. The principal *Larrea* bees of the southwestern United States (Hymenoptera: Apoidea). Smithsonian Contributions to Zoology 193: 1–74.

Hurd, P. D., Jr., and J. A. Powell. 1958. Observations on the nesting habits of *Colletes stepheni* Timberlake. Pan-Pacific Entomologist 34: 147–153.

Inouye, D. W. 1980. The effect of proboscis and corolla tube lengths on patterns and rates of flower visitation by bumblebees. Oecologia 45: 197–201.

Irwin, R. E., L. S. Adler, and A. K. Brody. 2004. The dual role of floral traits: Pollinator attraction and plant defense. Ecology 85: 1503–1511.

Iwata, K. 1966. Evolution of instinct: Comparative ethology of Hymenoptera. Amerind Publishing Company, New Delhi.

Janzen, D. H. 1980. When is it coevolution? Evolution 34: 611–612.

Johnson, S. D., and K. E. Steiner. 2000. Generalization versus specialization in plant pollination systems. Trends in Ecology and Evolution 15: 140–143.

Jones, C. E. 1977. Pollinator constancy as a pre-pollination isolating mechanism between sympatric species of *Cercidium*. Evolution 32: 189–198.

Klostermeyer, E. C., S. J. Mech Jr., and W. B. Rasmussen. 1973. Sex and weight of *Megachile rotundata* (Hymenoptera: Megachilidae) progeny associated with provision weights. Journal of the Kansas Entomological Society 46: 536–548.

Krombein, K. V. 1967. Trap-nesting wasps and bees: Life histories, nests, and associates. Smithsonian Press, Washington, DC.

Krombein, K. V., and B. B. Norden. 1997. Nesting behavior of *Krombeinictus nordenae* Leclercq, a sphecid wasp with vegetarian larvae (Hymenoptera, Sphecidae, Crabroninae). Proceedings of the Entomological Society of Washington 99: 42–49.

Laverty, T. M. 1992. Plant interactions for pollinator visits: A test of the magnet species effect. Oecologia 89: 502–508.

Laverty, T. M., and R. C. Plowright. 1988. Flower handling by bumblebees: A comparison of specialists and generalists. Animal Behaviour 36: 733–740.

Linhart, Y. B., and P. Feinsinger. 1980. Plant–hummingbird interactions: Effects of island size and degree of specialization on pollination. Journal of Ecology 68: 745–760.

Linsley, E. G. 1958. The ecology of solitary bees. Hilgardia 27: 543–599.

Linsley, E. G. 1961. The role of flower specificity in the evolution of solitary bees. Internationaler Kongress für Entomologie Wien XI: 593–596.

Linsley, E. G. 1978. Temporal patterns of flower visitation by solitary bees, with particular reference to the southwestern United States. Journal of the Kansas Entomological Society 51: 531–546.

Linsley, E. G., and J. W. MacSwain. 1958. The significance of floral constancy among bees of the genus *Diadasia* (Hymenoptera: Anthophoridae). Evolution 12: 219–223.

Linsley, E. G., J. W. Macswain, and C. D. Michener. 1980. Nesting biology and associates of *Melitoma* (Hymenoptera, Anthophoridae). University of California Publications in Entomology 90: 1–45.

Linsley, E. G., J. W. MacSwain, P. H. Raven, and R. W. Thorp. 1973. Comparative behavior of bees and Onagraceae V. *Camissonia* and *Oenothera* bees of cismontane California and Baja California. University of California Publications in Entomology 71: 1–76.

Löken, A. 1961. *Bombus consobrinus* Dahlb., an oligolectic bumble bee (Hymenoptera, Apidae). XIth International Congress of Entomology 1960. 1: 598–603.

Lokvam, J., and J. F. Braddock. 1999. Anti-bacterial function in the sexually dimorphic pollinator rewards of *Clusia grandiflora* (Clusiaceae). Oecologia 119: 534–540.

Lovell, J. H. 1918. The flower and the bee. C. Scribner's and Sons, New York.

Marlin, J. C., and W. E. LaBerge. 2001. The native bee fauna of Carlinville, Illinois, revisited after 75 years: A case for persistence. Conservation Ecology 5. Online URL http://www.consecol.org/vol5/iss1/art9.

Martins, R. P., and J. C. Borges. 1999. Use of *Ludwigia* (Onagraceae) pollen by a specialist bee, *Diadasina distincta* (Hymenoptera: Apidae), at a nesting site in southeastern Brazil. Biotropica 31: 530–534.

McGinley, R. J. 2003. Studies of Halictinae (Apoidea: Halictidae), II: Revision of *Sphecodogastra* Ashmead, floral specialists of Onagraceae. Smithsonian Contributions to Zoology 610: 1–55.

McIntosh, M. 2001. Pollen preferences of four solitary, cactus specialist bees foraging for larval provisions. PhD dissertation, University of Arizona, Tucson, AZ.

Menken, S. B. J. 1996. Pattern and process in the evolution of insect-plant associations: *Yponomeuta* as an example. Entomologia Experimentalis et Applicata 80: 297–305.

Michener, C. D. 1954. Bees of Panama. Bulletin of the American Museum of Natural History 104: 1–176.

Michener, C. D. 1979. Biogeography of the bees. Annals of the Missouri Botanical Garden 66: 277–347.

Michener, C. D. 2000. The bees of the world. John Hopkins University Press, Baltimore, MD.

Michener, C. D., and C. W. Rettenmeyer. 1956. The ethology of *Andrena erythronii* with comparative data on other species (Hymenoptera, Andrenidae). University of Kansas Science Bulletin 37: 645–684.

Minckley, R. L., J. H. Cane, and L. J. Kervin. 2000. Origins and ecological consequences of pollen specialization among desert bees. Proceedings of the Royal Society of London B 267: 265–271.

Minckley, R. L., J. H. Cane, L. Kervin, and T. H. Roulston. 1999. Spatial predictability and resource specialization of bees (Hymenoptera: Apoidea) at a superabundant, widespread resource. Biological Journal of the Linnean Society 67: 119–147.

Minckley, R. L., J. H. Cane, L. Kervin, and D. Yanega. 2003. Biological impediments to measures of competition among introduced honey bees and desert bees (Hymenoptera: Apiformes). Journal of the Kansas Entomological Society 76: 306–319.

Minckley, R. L., and S. G. Reyes. 1996. Capture of the orchid bee, *Eulaema polychroma* (Friese) (Apidae: Euglossini) in Arizona, with notes on northern distributions of other mesoamerican bees. Journal of the Kansas Entomological Society 69: 102–104.

Minckley, R. L., W. T. Wcislo, D. Yanega, and S. L. Buchmann. 1994. Behavior and phenology of a specialist bee (*Dieunomia*) and sunflower (*Helianthus*) pollen availability. Ecology 75: 1406–1419.

Moldenke, A. R. 1976. California pollination ecology and vegetation types. Phytologia 34: 305–361.

Monsevicius, V. 1995. A check list of wild bee species (Hymenoptera, Apoidea) of Lithuania with data to their distributions and bionomics. New and rare for Lithuania insect species. Records and descriptions of 1994–1995. Institute of Ecology and Lithuanian Entomological Society, Vilnius.

Müller, A. 1996. Host-plant specialization in western palearctic anthidiine bees (Hymenoptera: Apoidea: Megachilidae). Ecological Monographs 66: 235–257.

Neff, J. L., and B. N. Danforth. 1992. The nesting and foraging behavior of *Perdita texana* (Cresson) (Hymenoptera: Andrenidae). Journal of the Kansas Entomological Society 64: 394–405.

Neff, J. L., and B. B. Simpson. 1991. Nest biology and mating behavior of *Megachile fortis* in central Texas (Hymenoptera: Megachilidae). Journal of the Kansas Entomological Society 64: 324–336.

Novotny, V., Y. Basset, S. E. Miller, G. D. Weiblen, B. Bremer, L. Cizek, and P. Drozd. 2002. Low host specificity of herbivorous insects in a tropical forest. Nature 416: 841–844.

Olsen, K. M. 1997. Pollination effectiveness and pollinator importance in a population of *Heterotheca subaxillaris* (Asteraceae). Oecologia 109: 114–121.

O'Neal, R. J., and G. W. Waller. 1984. On the pollen harvest by the honey bee (*Apis mellifera* L.) near Tucson, Arizona (1976–1981). Desert Plants 6: 81–109.

O'Toole, C., and A. Raw. 1991. Bees of the world. Blandford Publishing, London.

Pekkarinen, A. 1997. Oligolectic bee species in northern Europe (Hymenoptera, Apoidea). Entomologica Fennica 8: 205–214.

Pellmyr, O. 2002. Pollination by animals. Pp. 157–184 *in* C. M. Herrera and O. Pellmyr (eds.), Plant-animal interactions: An evolutionary approach. Blackwell Science, Oxford.
Pesenko, Y. A., and V. D. Radchenko. 1993. The use of bees (Hymenoptera, Apoidea) for alfalfa pollination: The main directions and modes, with methods of evaluation of populations of wild bees and pollinator efficiency. Entomological Review 72: 101–119.
Pyke, G. H. 1980. Optimal foraging in bumblebees: Calculation of net rate of energy intake and optimal patch choice. Theoretical Population Biology 17: 232–246.
Radchenko, V. G., and Y. A. Pesenko. 1994. Biology of bees. Russian Academy of Sciences, St. Petersburg.
Ranta, E. 1984. Proboscis length and the coexistence of bumblebee species. Oikos 43: 189–196.
Ranta, E., and H. Lundberg. 1980. Resource partitioning in bumblebees: The significance of differences in proboscis length. Oikos 35: 398–302.
Renner, S. S., and J. P. Feil. 1993. Pollinators of tropical dioecious angiosperms. American Journal of Botany 80: 1100–1107.
Robertson, C. 1914. Origin of oligotropy of bees (Hym.). Entomological News 25: 67–73.
Robertson, C. 1925. Heteroptropic bees. Ecology 6: 412–436.
Robertson, C. 1926. Revised list of oligolectic bees. Ecology 7: 378–380.
Robertson, C. 1929. Flowers and insects. Lists of visitors to four hundred and fifty-three flowers. C. Robertson, Carlinville, IL.
Rosenheim, J. A. 1999. The relative contributions of time and eggs to the cost of reproduction. Evolution 53: 376–385.
Roubik, D. W. 1982. Obligate necrophagy in a social bee. Science 217: 1059–1060.
Roubik, D. W. 1989. Ecology and natural history of tropical bees. Cambridge University Press, Cambridge.
Roubik, D. W. 1992. Loose niches in tropical communities: Why are there so few bees and so many trees? Pp. 327–354 *in* M. D. Hunter, T. Ohgushi, and P. W. Price (eds.), Effects of resource distribution on animal-plant interactions. Academic Press, San Diego, CA.
Roulston, T. H., and S. L. Buchmann. 2000. A phylogenetic reconsideration of the pollen starch–pollination correlation. Evolutionary Ecology Research 2: 627–643.
Roulston, T. H., and J. H. Cane. 2000. Pollen nutritional content and digestibility for animals. Plant Systematics and Evolution 222: 187–209.
Roulston, T. H., and J. H. Cane. 2002. The effect of pollen protein concentration on body size in the sweat bee *Lasioglossum zephyrum* (Hymenoptera: Apiformes). Evolutionary Ecology 16: 49–65.
Roulston, T. H., J. H. Cane, and S. L. Buchmann. 2000. What governs the protein content of pollen: Pollinator preferences, pollen–pistil interactions, or phylogeny? Ecological Monographs 70: 617–643.
Rozen, J. G. 1977. Biology and immature stages of the bee genus *Meganomia* (Hymenoptera, Melittidae). American Museum Novitates 2630: 1–14.
Rozen, J. G. 1990. Pupa of the bee *Pararhophites orobinus* (Hymenoptera: Apoidea: Megachilidae). Journal of the New York Entomological Society 98: 379–382.
Rozen, J. G., and L. Ruz. 1995. South American panurgine bees (Andrenidae: Panurginae), Part II. Adults, immature stages, and biology of *Neffapis longilingua,* a new genus and species with an elongate glossa. American Museum Novitates 3136: 1–15.
Rust, R. W. 1988. Biology of *Nomadopsis larreae* (Hymenoptera: Andrenidae), with an analysis of yearly appearance. Annals of the Entomological Society of America 81: 99–104.
Rust, R. W., G. Cambon, J.-T. Torre Grossa, and B. E. Vassiere. 2004. Nesting biology and foraging ecology in the weed-boring bee *Lithurgus chrysurus* (Hymenoptera: Megachilidae). Journal of the Kansas Entomological Society 77: 269–279.
Schlindwein, C. 1998. Frequent oligolecty characterizing a diverse bee–plant community in a xerophytic bushland of subtropical Brazil. Studies on Neotropical Fauna and Environment 33: 46–59.
Schmidt, R. H. J. 1989. The arid zones of Mexico: Climatic extremes and conceptualization of the Sonoran Desert. Journal of Arid Environments 16: 241–256.
Scott, P. E., S. L. Buchmann, and M. K. O'Rourke. 1993. Evidence for mutualism between a flower-

piercing carpenter bee and ocotillo: Use of pollen and nectar by nesting bees. Ecological Entomology 18: 234–240.
Sheffield, C. S., P. G. Kevan, R. F. Smith, S. M. Rigby, and R. E. L. Rogers. 2003. Bee species of Nova Scotia, Canada, with new records and notes on bionomics and floral relations (Hymenoptera: Apoidea). Journal of the Kansas Entomological Society 76: 357–384.
Shinn, A. F. 1967. A revision of the bee genus *Calliopsis* and the biology and ecology of *C. andreniformis* (Hymenoptera: Andrenidae). University of Kansas Science Bulletin 46: 753–936.
Silveira, F. A. 1993. The mouthparts of *Ancyla* and the reduction of the labiomaxillary complex among long-tongued bees (Hymenoptera: Apoidea). Entomologica Scandinavica 24: 293–300.
Simpson, B. B. 1977. Breeding systems of dominant perennial plants of two disjunct warm desert ecosystems. Oecologia 27: 203–226.
Simpson, B. B., and J. L. Neff. 1981. Floral rewards: Alternatives to pollen and nectar. Annals of the Missouri Botanical Garden 68: 301–322.
Simpson, B. B., and J. L. Neff. 1983. Evolution and diversity of floral rewards. Pp. 142–159 *in* C. E. Jones and R. J. Little (eds.), Handbook of experimental pollination biology. Van Nostrand Reinhold Company, New York.
Simpson, B. B., and J. L. Neff. 1987. Pollination ecology in the arid southwest. Aliso 11: 417–440.
Simpson, B. B., J. L. Neff, and A. R. Moldenke. 1977a. *Prosopis* flowers as a resource. Pp. 85–107 *in* B. B. Simpson (ed.), Mesquite: Its biology in two desert ecosystems. Dowden, Hutchinson, and Ross, Stroudsburg, PA.
Simpson, B. B., J. L. Neff, and A. R. Moldenke. 1977b. Reproductive systems of *Larrea*. Pp. 92–114 *in* T. J. Mabry, J. H. Hunziker, and D. R. J. Difeo (eds.), Creosote bush: Biology and chemistry of *Larrea* in new world deserts. Dowden, Hutchinson, and Ross, Stroudsburg, PA.
Sipes, S. D., and P. G. Wolf. 2001. Phylogenetic relationships within *Diadasia*, a group of specialist bees. Molecular Phylogenetics and Evolution 19: 144–156.
Steiner, K. E., and V. B. Whitehead. 1990. Pollinator adaptation to oil-secreting flowers: *Rediviva* and *Diascia*. Evolution 44: 1701–1707.
Steiner, K. E., and V. B. Whitehead. 1996. The consequences of specialization for pollination in a rare South African shrub, *Ixianthes retzioides* (Scrophulariaceae). Plant Systematics and Evolution 201: 131–138.
Strickler, K. 1979. Specialization and foraging efficiency of solitary bees. Ecology 60: 998–1009.
Temeles, E. J., and W. J. Kress. 2003. Adaptation in a plant–hummingbird association. Science 300: 630–633.
Tepedino, V. J. 1981. The pollination efficiency of the squash bee (*Peponapis pruinosa*) and the honey bee (*Apis mellifera*) on summer squash (*Cucurbita pepo*). Journal of the Kansas Entomological Society 54: 359–377.
Tepedino, V. J. 1997. A comparison of the alfalfa leafcutting bee (*Megachile rotundata*) and the honey bee (*Apis mellifera*) as pollinators for hybrid carrot seed in field cages. Pp. 457–461 *in* K. W. Richards (ed.), Proceedings of the 7th international symposium on pollination, Lethbridge, Alberta, Canada, 23–28 June 1996. International Society for Horticultural Science, Leiden.
Tepedino, V. J., and N. L. Stanton. 1980. Spatiotemporal variation in phenology and abundance of floral resources on shortgrass prairie. Great Basin Naturalist 40: 197–215.
Thompson, J. N., and O. Pellmyr. 1992. Mutualism with pollinating seed parasites amid copollinators: Constraints on specialization. Ecology 73: 1780–1791.
Thomson, J. D. 1996. Trapline foraging by bumblebees: I. Persistence of flight-path geometry. Behavioral Ecology 7: 158–164.
Thorp, R. W. 1969. Systematics and ecology of bees of the subgenus *Diandrena* (Hymenoptera: Andrenidae). University of California Publications in Entomology 52: 1–146.
Thorp, R. W. 1979. Structural, behavioral, and physiological adaptations of bees (Apoidea) for collecting pollen. Annals of the Missouri Botanical Garden 66: 788–812.
van der Pijl, L. 1961. Ecological aspects of flower evolution. II. Zoophilous flower classes. Evolution 15: 44–59.
Vázquez, D. P., and D. Simberloff. 2002. Ecological specialization and susceptibility to disturbance: Conjectures and refutations. American Naturalist 159: 606–623.

Visscher, P. K., and B. N. Danforth. 1993. Biology of *Calliopsis pugionis* (Hymenoptera: Andrenidae): Nesting, foraging, and investment sex ratio. Annals of the Entomological Society of America 86: 822–832.

Visscher, P. K., and T. D. Seeley. 1982. Foraging strategy of honeybee colonies in a temperate deciduous forest. Ecology 63: 1790–1801.

Vogel, S. 1974. Ölblumen und ölsammelnde Bienen. Franz Steiner Verlag, Wiesbaden.

Vogel, S., and A. Cocucci. 1995. Pollination of *Basistemon* (Scrophulariaceae) by oil-collecting bees in Argentina. Flora 190: 353–363.

Waser, N. M. 1986. Flower constancy: Definition, cause, and measurement. American Naturalist 127: 593–603.

Waser, N. M., L. Chittka, M. V. Price, N. M. Williams, and J. Ollerton. 1996. Generalization in pollination systems, and why it matters. Ecology 77: 1043–1060.

Wcislo, W. T., and J. H. Cane. 1996. Floral resource utilization by solitary bees (Hymenoptera: Apoidea) and exploitation of their stored foods by natural enemies. Annual Review of Entomology 41: 195–224.

Westerkamp, C. 1996. Pollen in bee–flower relations: Some considerations on melittophily. Botanica Acta 109: 325–332.

Westrich, P. 1989. Die Wildbienen Baden-Württembergs. Eugen Ulmer, Stuttgart.

Westrich, P., and K. Schmidt. 1987. Pollen analysis, an auxiliary tool to study the collecting behaviour of solitary bees. Apidologie 18: 199–213.

Whitten, W. M., A. M. Long, and D. L. Stern. 1993. Non-floral sources of chemicals that attract male euglossine bees. Journal of Chemical Ecology 15: 1285–1295.

Wille, A., and C. D. Michener. 1973. The nest architecture of stingless bees with special reference to those of Costa Rica (Hymenoptera: Apidae). Revista de Biologia Tropical, Universidad de Costa Rica 21: 1–278.

Williams, N. M. 2003. Use of novel pollen species by specialist and generalist solitary bees (Hymenoptera: Megachilidae). Oecologia 134: 228–237.

Williams, N. M., R. L. Minckley, and F. A. Silveira. 2001. Variation in native bee faunas and its implications for detecting community changes. Conservation Ecology 5. Online URL http://www.consecol.org/vol5/iss1/art7.

Williams, N. M., and V. J. Tepedino. 2003. Consistent mixing of near and distant resources in foraging bouts by the solitary mason bee *Osmia lignaria*. Behavioral Ecology 14: 141–149.

Willis, D. S., and P. G. Kevan. 1995. Foraging dynamics of *Peponapis pruinosa* (Hymenoptera: Anthophoridae) on pumpkin (*Cucurbita pepo*) in southern Ontario. Canadian Entomologist 127: 167–175.

Zavortink, T. J. 1992. A new subgenus and species of *Megandrena* from Nevada, with notes on its foraging and mating behavior. Proceedings of the Entomological Society of Washington 74: 61–75.

Zavortink, T. J., and W. E. LaBerge. 1976. Bees of the genus *Martinapis* Cockerell in North America. Wasmann Journal of Biology 34: 119–145.

Zucchi, R., S. F. Sakagami, and J. M. F. Camargo. 1969. Biological observations on a neotropical parasocial bee, *Eulaema nigrita*, with a review on the biology of Euglossinae. Journal of the Faculty of Science, Hokkaido University, Series VI, Zoology 17: 271–380.

CHAPTER FIVE

Characterizing Floral Specialization by Bees: Analytical Methods and a Revised Lexicon for Oligolecty

James H. Cane and Sedonia Sipes

Introduction

Pollinators do not visit flowers haphazardly, whether in a meadow or forest canopy. Rather, their foraging bouts consist of strings of visits to conspecific flowers, sometimes interrupted by switches between species. Flowering plants depend on conspecific visitation for pollination. The taxonomic foraging fidelities of bees represent two very different phenomena: individual constancy and inherent host specialization. Floral constancy is a plastic attribute of individual foragers. In contrast, bee species vary in the taxonomic range of floral hosts from which they inherently collect pollen. This specialization is a species-specific trait.

Traditionally, bee species have been classified as monolectic, oligolectic, or polylectic, depending on whether they use one, few, or many pollen host taxa, respectively. However, these terms have been applied inconsistently for at least three reasons: (1) the taxonomic spectrum of pollen use by bees cannot be jammed into only three categories, (2) our knowledge of bee host use is incomplete, and (3) some of the analytical methods used defy comparison. In this chapter, we first review methods for characterizing ranges of taxonomic pollen fidelities of bees and then refine and expand upon the traditional classifications of pollen host breadth specialization in bees.

Two Forms of Floral Fidelity: Constancy versus Oligolecty

Foraging bees express taxonomic floral fidelity at two different scales. *Floral constancy* is the temporary tendency of individual foragers to sequentially visit conspecific flowers. In contrast, oligolecty is a largely fixed specialization shared by all conspecific bees for the same few pollen hosts. Floral constancy can manifest itself in response to any floral resource (nectar, pollen, oil, etc.); oligolecty manifests itself in pollen foraging. We first briefly discuss constancy, which must be understood to appreciate its profound difference from oligolecty.

Aristotle, and much later, Arthur Dobbs (in 1750), described what we now call floral constancy (quoted in Proctor et al. 1996, 16–17). During a foraging bout, a flower-constant bee switches between different host species much less frequently than would be expected from the local mix of flowering species in a patch or artificial floral array (reviewed by Grant 1950; Wilson and Stine 1996). To demonstrate constancy, foragers can be observed visiting sequential flowers. This simple method is widely used in studies of social bees (honey bees, bumblebees, stingless bees; e.g., Wilson and Stine 1996) and some studies of nonsocial (hereafter "solitary") bees (e.g., Jones 1978). Tracked floral visits can reveal constancy to nectar, oil, or pollen hosts. A less common approach involves sampling pollen loads from foragers to evaluate their taxonomic purity (e.g., Grant 1950; Chambers 1968; O'Neal and Waller 1984; Macior 1994; Ramalho et al. 1994). Foraging constancy of solitary species also can be revealed by the taxonomic constitution of pollen amassed by mother bees for each of their progeny. On sequential days, even polylectic individuals often assemble provision masses that are taxonomically pure, but different individuals often choose to focus on different floral hosts for their pollen (e.g., Westrich and Schmidt 1987; Rust 1990).

Under rare circumstances, wherein pollen of congeneric flowering species differs in appearance, floral constancy can even be detected in provisions of oligolectic bees; thus, the two forms of fidelity are not mutually exclusive. For some *Dufourea* and *Proteriades* bees that are *Phacelia* specialists, exhumed provision masses of individual nests were either blue or yellow, depending on the pollen color of the *Phacelia* species foraged on a given day (Torchio et al. 1967; Rust et al. 1974).

Floral constancy represents ephemeral sequences of host choices made by individual foragers. Conspecific social bees often choose to work different flowering species in the same patch. Furthermore, individual foragers often switch hosts from one foraging bout or day to another. The thrust of our discussion will focus instead on a range of more predictable taxonomic floral fidelities, such as oligolecty, that typify entire species of bees.

Discovery of Oligolecty

We owe our concepts of oligolecty and polylecty to Charles Robertson, amateur naturalist and the son of an American country doctor. From 1884 to 1916, he guided his horse-drawn buggy about the environs of Carlinville, Illinois USA, scouring the prairies, wood lots, and farms for native bees found visiting 441 flowering species (Robertson 1925 and references therein). Unlike his contemporaries, who were content with mere species lists or anecdotal accounts (e.g., Lovell 1918), Robertson also meticulously distinguished nectaring from pollen foraging for individuals of many of the 296 bee species that he collected. (Sample pages from his ledger book can be viewed online at http://www.consecol.org/vol5/iss1/art9/index.html.) Robertson's foraging distinction proved to be insightful, revealing to him numerous Midwest bee species that obtained nectar from many plants but pollen from only a few genera (Robertson 1926). Indiscriminate nectaring by many species belied their much narrower pollen diets.

Robertson coined the terms *monolecty, oligolecty,* and *polylecty* to categorize each bee species for taxonomic breadth of its pollen foraging habits; these terms refer to bees with a single host species, a few related hosts, and many hosts, respectively (Robertson 1925, tables corrected in Robertson 1926). Linsley and MacSwain (1958) later refined Robertson's definition of oligolecty to emphasize invariant taxonomic specificity for a few related floral genera, musing on its evolutionary origins and adaptive advantages. They largely dismissed Robertson's term *monolecty.* Some investigators have added "broad" and "narrow" as informal qualifiers (e.g., broad oligolege), but, in essence, the spectrum of pollen–host associations of bees has been fit to this same binary classification scheme since the days of Robertson's buggy rides. We propose a new, more encompassing and precise lexicon for categorizing bees on the basis of pollen dietary breadth.

Practical Evidence for Classifying Pollen Usage

Successful application of our lexicon will depend on the data on which we base the classifications. Robertson distinguished pollen hosts from nectar hosts by directly observing bees foraging at flowers; later studies have often used pollen analysis or manipulative experiments. Unfortunately, different studies use summary statistics and criteria that are incommensurate, which prevents meaningful compilation and comparison. We first review and critique the various data and criteria used to classify bees' diet breadths and then suggest ways to focus and standardize future investigations.

Floral Visitation as Evidence of Diet Breadth

Prior to Charles Robertson, observers of bees did not discern foraging for pollen versus foraging for nectar. Most floral host lists found in taxonomic revisions or catalogs of bees merely compile label data on pinned specimens, which rarely specify pollen hosts. For example, museum specimens of *Andrena dunningi* often report willow or woody genera of Rosaceae on host labels, but only pollen of the latter prevailed in pollen loads of 131 females intercepted at their nests (Johnson 1984). The author regularly saw females just nectaring at willow or maple. Thus, uncritical compilation and analysis of museum host label data can lead to very flawed interpretations of pollen host use by bees.

Observed floral visits will continue to be our primary clue for pollen-feeding habits of most bees. Despite Robertson's formidable evidence that bees regularly have more nectar than pollen hosts, many researchers still confound nectar and pollen hosts when reporting floral visitation patterns of bees. Some fail to distinguish males from female bees, and males do not even collect pollen! Moreover, some floral records for bees come from pollination studies, where the emphasis is often on pollen transfer between particular conspecific plants rather than the pollinators' diet. Compiled floral records can indicate oligolecty if, throughout a bee's range, the same taxonomic specificity is recorded (especially in the presence of alternative pollen sources). But floral records alone may be insufficient to refute polylecty because nectar and pollen hosts may not be distinguished. Robertson (1929) disparaged mere floral visitation records, calling them "worthless . . . because the collector is looking out for particular kinds of insects and not for all of the kinds which occur on the flowers." He intimated that collectors may overlook alternative pollen hosts when studying presumed oligoleges. These factors must temper our use of floral visitation observations to infer bee diet breadth.

Floral records will constitute stronger evidence of pollen-feeding habits when investigators distinguish nectar feeding and pollen feeding via behavioral observations or other means. For example, for flowers whose nectaries are spatially separated from anthers, bees may be seen either working the anthers for pollen or bypassing the anthers altogether while drinking nectar. Grooming behavior to sweep bodily pollen to the pollen "baskets" (corbiculae) and brushes (scopae) for transport also indicates pollen collection, though bees often visit many flowers before pausing to groom (Jander 1976).

In some cases, the host plants themselves aid in the distinction between nectar foraging and pollen foraging. For example, many species of Solanaceae and Melastomataceae are nectarless; bees must sonicate the species with poricidally dehiscent anthers to harvest their pollen (Buchmann 1983; Renner 1989). A visual and auditory record of bees buzzing such flowers constitutes a clear record of pollen, and not nectar, collection. Similarly, bees visiting other nectarless taxa, such as Papaveraceae (Vogel 1978) and lupine, gather only pollen. Conversely, pollen from the orchids and some Apocynaceae never feeds larval bees. Pollen of these taxa comes packaged in pollinia, which are transported only incidentally by bees seeking nectar, oil, or (in the case of deceptive orchids) the empty promise of a reward. Intelligent interpretation of floral records can help characterize bee pollen hosts.

Observation of floral visits facilitates quantitative and taxonomic assessment of a flowering community, thereby placing the pollen choices of bees within the context of alternative floral hosts. We, therefore, encourage at least qualitative assessment of available floral resources. Comparisons with pollen foraging by local, broadly polylectic species, such as honey bees or bumblebees, will nicely characterize host choices available to putative oligoleges. Even a simple survey list—tallying local plant genera used for pollen by any other co-occurring bee species that are concurrently active—can be useful (e.g., Cane and Payne 1993). Oligolecty is only demonstrable if alternative pollen hosts are flowering. For instance, the alfalfa pollinators *Megachile rotundata* and *Nomia melanderi* consistently gather pure provisions of alfalfa pollen from large farm fields, but both species are clearly polylectic when given floral choices (Packer 1970; Small et al. 1997).

A potential drawback to using floral records lies in the typical lack of vouchered herbarium specimens of host plants. Thus, there is usually no way to confirm or reconsider field identifications (a contrast with pollen load studies, where the pollen vouchers itself). This lack of vouchers can be vexing when taxonomies change. For example, floral records indicate that some *Diadasia* species visited "*Sida*" (Malvaceae). One of these, previously known as *S. hederacea* (Hooker) A. Gray, is now known as *Malvella leprosa* (Ortega) Krapov. Yet pollen analyses show other *Diadasia* species collecting pollen from taxa still classified as *Sida* (Sipes 2001; Sipes and Tepedino, unpublished data). Hence, a floral record of *Sida,* without a pollen or plant voucher, narrows the observed host to two genera at best.

Floral visitation by bees has also been studied in artificial or manipulated environments, wherein host choices of captive bees are evaluated using "cafeteria" experiments (more than one host offered) or no-choice experiments (e.g., Raw 1974; Small et al. 1997; Williams 2003). Such manipulations have appeal, because females can be denied access to strongly preferred pollen hosts, revealing grudging acceptance or outright rejection of less preferred pollen hosts. Rejection of an offered pollen host, especially without choice, clearly signifies that the given flower is not part of the bee species' pollen menu (e.g., Raw 1974). But, what if they are accepted? We exemplify this conundrum with polylectic *Megachile rotundata.* Free-flying, nesting populations of this bee take nectar but no pollen from cranberry (*Vaccinium macrocarpon*), even amid hectares of blooming commercial monoculture (Mackenzie and Javorek 1997). However, mated females caged solely with blooming cranberry promptly commence to forage and provision their nest cells with its pollen (Cane and Schiffhauer 2003). Is cranberry, therefore, a pollen host or not? And, more broadly, when can such results be used to confirm or extend a bee species' pollen host menu? For now, we advise cautious interpretation of no-choice experiments for evaluating bee diet breadth, because we need greater behavioral and physiological insights into maternal foraging decisions and their consequences for larval performance.

Pollen Analysis

Taxonomic identification and numerical summary of pollen gathered by bees for nest provisioning offers a more direct and quantitative method for assessing bee diet breadth. Specifically, pollen analysis offers two advantages over field observations discussed earlier: unambiguous and unbiased discrimination of pollen versus nectar hosts, and unbiased representation of inaccessible or unanticipated pollen hosts (e.g., forest canopy species or alternative hosts of presumed oligolectic species, respectively). The obvious disadvantage of pollen analysis is logistic feasibility, because pollens of related species and even genera are often indistinguishable by gross morphology; even familial discrimination often requires considerable training and practice. Thus, analyses of pollen loads alone may not be sufficient to delineate bee diets. In cases of indistinguishable pollen types, field observations will also be necessary.

Sampling Strategies

Researchers have analyzed pollen spectra using diverse sampling methods and sources of pollen. First, pollen loads collected by foraging females can be taken for analysis (Chambers 1946; Cane and Payne 1988; Cripps and Rust 1989a, 1989b). An advantage of this method is that the researcher can also identify all flowering plants in the area and collect pollen reference samples from them to assist in identifying the scopal pollen. Such a reference collection may greatly refine the hierarchical level to which the pollen of very species-rich taxa, such as Asteraceae, can be identified. Additionally, the relative abundances of the flowering taxa can be noted or measured; this could help explain the foraging patterns of individuals in different places and times. For example, bees may visit alternative hosts only when preferred hosts are scarce; knowing which plants are abundant and which are not will aid in recognizing such patterns.

A limitation of field interception is that it samples pollen collected by individual females during a single foraging visit, and the sample may underestimate the dietary breadth of both the individual and the species (e.g., a single pollen load may reflect constancy and/or host availability, not inherent host preferences). Moreover, collection of females at flowers may inject a bias, because the collector may unevenly sample different taxa based on their showiness, convenience, or the collector's expectations of which plants are "correct" host plants (Robertson 1929; Chambers 1946). An alternative to analyzing pollen from foraging females is either to collect females returning to nesting sites (Chambers 1946, 1968) or to analyze the pollen provisions found within nests of individual females (e.g., Cane and Payne 1988; Rust 1990; Bullock et al. 1991; Martins and Borges 1999; Archibald et al. 2001; Thiele 2002; Williams and Tepedino 2003) or colonies (e.g., Edwards-Anderka and Pengelly 1970; Ramalho et al. 1985). This method has several advantages: first, lifetime foraging patterns can be assessed for individual females (if all provisioned cells can be analyzed); second, it avoids unintended sampling biases during surveys of flowering communities; and third, it can better represent floral host choices throughout a bee species' flight season. Besides sampling freshly assembled pollen provisions, pollen diets can be evaluated after consumption (Brian 1951; Visscher and Danforth 1993). Most bee larvae defecate following consumption of their entire pollen provision; this fecal material remains in the nest cell after emergence of the adult bee. Exines (exterior shells) of pollen grains often remain largely intact in these larval feces. Thus, old nests hold clues to bee pollen diets. The impediment to sampling nest provisions is its substantial difficulty and time invested in locating and excavating nests, particularly for solitary ground-nesting bees.

Because sampling nest provisions or foraging females in the field for pollen analysis is labor intensive, spatiotemporal variation is difficult to characterize. Therefore, individual constancy, populational differences in a species' host preferences, and spatiotemporal variation in available host plants may obscure species-level dietary patterns. An instructive example comes from a study of pollen hosts foraged by 31 species of British *Andrena* (Chambers 1946, 1968). Chambers evaluated individual pollen loads of females returning to their nests in several locations of mixed land use. One species, *A. varians,* was sampled over eight years at three locations (102 pollen loads). At one site, Flitwick, the nesting site bordered an orchard. During bloom, 44 of 51 pollen loads consisted of more than 90% fruit tree pollen. This single site's data led him to designate the species as oligolectic, but the species is in fact polylectic: 102 pollen loads from throughout the flight season contained significant amounts of pollen from 19 genera in 10 plant families. Broader sampling of pollen-foraging efforts of bees may be needed before extrapolating from populations and seasons to species-specific generalizations (e.g., Cane and Payne 1993).

A valuable way to sample broadly across years and geographic ranges is to sample museum specimens (Cruden 1972; Dyer and Shinn 1978; Müller 1996; Sipes 2001; Sipes and Tepedino, unpublished data). The pollen loads of foraging females neatly withstand collecting, pinning, and dry storage if captured with nets or dry malaise traps (as opposed to pan traps). A drawback of sampling from museum specimens is that a bee's available pollen host choices remain largely unknown. What was in bloom at the time? Pollen identification also becomes problematic. Some pollen types can be ascribed to a genus or tribe, but others only to a family when surveying pollen loads of pinned museum specimens (e.g., Müller 1996; Sipes 2001; Sipes and Tepedino, unpublished data). Plant distributional data can help narrow the options for such indistinguishable pollen types.

Pollen Numbers versus Volume

Once pollen samples are obtained, sundry methodologies facilitate analysis. Methods for preparing, staining, identifying, and counting are detailed elsewhere (e.g., O'Rourke and Buchmann 1991). Unlike pollination, for which the count of grains transferred is a key reproductive measure, grain volume has relevance for quantifying bee diets (Buchmann and O'Rourke 1991; da Silveira 1991). Nutritional gain is better defined by the overall volume consumed and not the count of individual units. Calculating diet using pollen volume more realistically portrays a bee's dietary attributes. For example, *Diadasia diminuta* prefers *Sphaeralcea* pollen, but it also takes some from *Opuntia* (Cactaceae) and a few Asteraceae (Sipes 2001; Sipes and Tepedino, unpublished data). Grain size

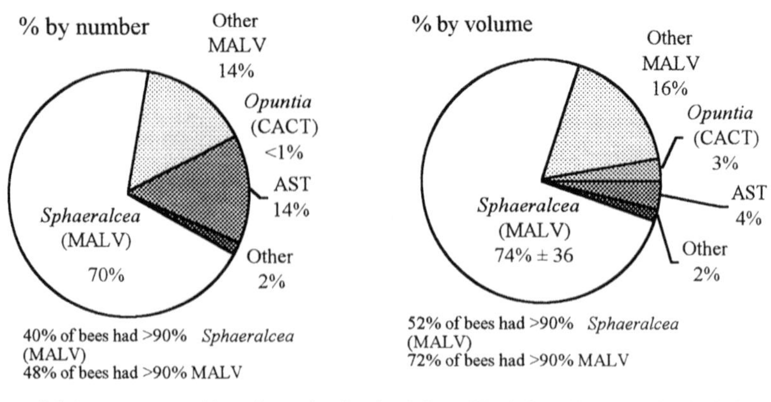

Figure 5.1 Average composition of scopal pollen loads from 25 *Diadasia diminuta* individuals, from 18 sites, calculated by grain number (*left*) and grain volume (*right;* from Sipes 2001). Because of the large differences in pollen sizes, the composition by grain number underestimates the importance of large pollen grains (e.g., Cactaceae) and overestimates the importance of small pollen grains (e.g., Asteraceae). AST, Asteraceae; CACT, Cactaceae; MALV, Malvaceae.

differs markedly among these pollen hosts: *Opuntia* (diameter ca. 100 μm) has a volume larger than 10 *Sphaeralcea* grains (ca. 45 μm) or larger than 60 small asteraceous grains (ca. 25 μm). Use of volumes dramatically changes dietary characterization of this bee species (fig. 5.1).

Another important issue in pollen analyses of bee diets is the choice of statistics for summarizing dietary data. Here we critique three major methods of summary: (1) averaging host use across all individuals, (2) reporting the incidence of pure versus mixed loads, and (3) using diversity indices. Some researchers average percentages of pollen hosts used across all individuals or nest cells sampled (e.g., Visscher and Danforth 1993; Müller 1996). However, Sipes and Tepedino (unpublished data) argued that such averages can mislead, because they fail to indicate the incidence of pure pollen loads. For example, if a bee species collects, on average, 95% host A and 5% host B, does host B represent a common contaminant from a nectar plant, or did a few individuals collect pure loads of host B? Thus, if the purpose of a pollen analysis study is to characterize the pollen host preferences of bee species, and to apply labels such as polylecty, oligolecty, and so forth, statistics reflecting the incidence of pure and mixed loads of pollen are more informative (e.g., Cruden 1972; Dyer and Shinn 1978; Martins and Borges 1999; Sipes 2001; Sipes and Tepedino, unpublished data). The difference between these statistics can be biologically important. Turning again to the example of *D. diminuta* (Sipes 2001; Sipes and Tepedino, unpublished data), 25 pollen loads from 18 sites across the species' range averaged 72% *Sphaeralcea* pollen, yet only 52% of individuals carried "pure" loads (defined as greater than or equal to 90%, to allow for contamination) of *Sphaeralcea* pollen.

Another method of summarizing data from pollen analyses is by diversity or niche-breadth indices. These have been used to characterize the diets of various insects (reviewed by Jaenike 1990; Symons and Beccaloni 1999), including bees (Cane and Payne 1988; Cripps and Rust 1989a; Rust 1990; Martins and Borges 1999; Quiroz-Garcia et al. 2001). Diversity indices, such as the Levins index (Levins 1968) or the Shannon–Wiener index (see Colwell and Futuyma 1971), account for abundance as well as richness of dietary host taxa, thus downplaying the importance of "contaminant" or minor host pollens. Niche-breadth indices offer advantages over simple counts or lists of host taxa, which may not differentiate between preferred and less-preferred hosts. Such indices are perhaps best used to compare the diets of different bee taxa foraging within the same plant community (Cripps and Rust 1989a; Quiroz-Garcia et al. 2001) or for a single bee species throughout its season (Martins and Borges 1999).

Problems generally arise, however, when comparing bee dietary niche breadths of different studies, because incommensurate indices may be used and methods by which pollen is sampled and identified may differ. Usually, some host plants represented in a pollen load are not identifiable at the species level; rather, mixtures of higher taxonomic levels and/or morphotypes are used when unknown or indistinguishable taxa must be pooled. For example, indistinguishable pollen had to be pooled for two host genera (*Balsamorhiza, Wyethia*) constituting the preferred hosts of two oligolectic *Osmia* spp. (Cripps and Rust 1989a), whereas the preferred host of *Diadasina distincta* (*Ludwigia laruotteana*) was identifiable to species (Martins and Borges 1999). Thus, the Levins niche-breadth indices reported for the oligolectic bees in these two studies are hardly comparable. Moreover, because each study used a single locality, extrapolation of diet breadth from one population to the species' ranges warrants caution. Such inconsistencies result in diversity indices being largely incomparable between studies by different researchers or encompassing different geographical areas. They are useful but should not constitute the sole summary of floral host use.

An additional method of summarizing dietary data, one that has yet to be applied to bees, is a phylogeny-based diversity index, such as the phylogenetic diversity index (Faith 1992) or the root phylogenetic diversity index of Symons and Beccaloni (1999). Proponents of phylogeny-based diversity indices note that simple counts of host taxa, as well as traditional niche-breadth indices, may implicitly assign biological importance to higher taxonomic levels. But hierarchical levels of the taxonomic system are arbitrary and cannot be compared across taxa. For example, stating that all bee species that use only one host genus are equally specialized implies that all plant genera are equivalent and comparable units, which is not true in either a phylogenetic or a phenotypic sense (see also Ollerton, chap. 18 in this volume).

Phylogeny-based diversity indices, as applicable to dietary breadths, require robust host phylogenies and are especially difficult to implement when comparing nonoverlapping diet breadths of animal species using distantly related hosts. Thus, although the potential advantages of such indices are obvious, they do not represent realistically feasible options to be adopted as a common method of describing bee diets, at least at the present time. Once enough phylogenetic data for angiosperms accrue, such indices will become feasible.

Quantifying Oligolecty

Given a dataset for the pollen hosts of a particular bee species, the species can be fit to one of the categories outlined in our lexicon. But first, one must deal with several sources of pollen contamination. Traces of pollen from nectar hosts and wind-pollinated flowers may get into the scopal pollen and nest provisions of bees. Also, when accumulated in a killing jar, bee bodies may pick up pollen from each other, a problem of little consequence if just the scopal pollen is sampled. How much should be allowed for such contaminants? Minimally, it should at least cover the percentage of known wind-dispersed pollen, such as that from conifers and grasses that appear in pollen loads (Rust 1987b). Many investigators allow for other sources of contaminants. The threshold for discarded taxa has ranged from 4% (Müller 1996) to 10% (Sipes 2001; Sipes and Tepedino, unpublished data) to a liberal 15% (Dyer and Shinn 1978) constitution of a load or provision. We suggest that a 5–10% allowance be used per load for trace contamination, in accord with amounts of wind-blown pine pollen reported for bee loads (Rust 1987b).

A second cutoff that must be delineated in designating monoleges and oligoleges involves pollen loads containing alternative hosts that are minor and/or infrequent but that exceed the cutoff amount for contamination. Müller (1996) classified oligoleges as having pollen loads comprising, on average, 95% as one genus, tribe, or family. However, as discussed previously, averages of pooled loads can confound interpretations. Do all individuals bear the same trace of pollen, or do rare individuals collect it in purity? Sipes (2001) classified oligoleges as those species in which 90% or more of females collect pure loads of one genus or family. We suggest that classifications be made based on the incidence of pure loads of individuals rather than pooled averages across loads, but a more focused study is needed to justify specific quantitative cutoffs. For now, we advise investigators to consider their sample sizes and objectives. If one has few samples (less than 20 individual loads or provisions), then one may want to score all pollen taxa present as pollen hosts, even those carried by only one or a few individuals. In other words, we suggest a null hypothesis that pollens above contamination levels serve as biologically important pollen hosts to all or some individuals in the bee species.

A Revised Lexicon of Pollen Specialization

Monolecty: Singular Pollen Hosts

Robertson's characterization of *monolecty* was in our opinion lamentable. He invoked it for bee species that collect pollen from either monotypic floral genera or floral genera represented by a single species within the bee's geographic range; the few examples seem to be mostly idiosyncratic (e.g., Austin 1978; Houston 1992; Neff and Rozen 1995; Cane et al. 1996; and perhaps Eickwort 1973). Other cases reflect evolutionary accumulation of variously related bee species on the sole regional representative of a plant genus, such as the 21 species of vernal bees reliant on *Larrea tridentata* for pollen (Rust 1987a; Minckley et al. 2000). However, monolecty appears most often to be "oligolecty without choice," reflecting an absence of sympatric, synchronic host congeners (Linsley 1958; Thorp 1969). In our opinion, monolecty should have designated those bee species restricted to a single floral host *genus* rather than a species. However, we acknowledge precedence in Robertson's definition; adjusting his definition now would engender confusion.

Degrees of Oligolecty: Using Few Pollen Hosts

Narrow Oligolecty: A Singular Generic Focus

Rather than redefine monolecty, we consider bee species to be *narrowly oligolectic* if they gather pollen solely from species comprising a single small host clade (typically a genus). This term has precedent (e.g., Pekkarinen 1997, 1999); we simply propose to standardize its application. Plant genera that are pollen hosts to narrow oligoleges vary in both species richness and morphological diversity. Some richly speciose plant genera host numerous oligoleges (e.g., *Eucalyptus, Opuntia, Prosopis*); others are monotypic or depauperate in species (e.g., *Larrea*). Species of some genera popular with oligoleges vary widely in floral phenotype (e.g., *Penstemon, Clarkia*); others are quite uniform (e.g., *Helianthus, Sphaeralcea, Salix*).

Examples of narrow oligolecty are scattered among bee taxa, bee and plant communities, and biomes worldwide. Narrow oligolecty is frequent in some bee genera of different families (e.g., *Dufourea* of Halictidae, *Perdita* and *Andrena* of Andrenidae, *Melitta* of Melittidae, *Proteriades* of Megachilidae, and *Diadasia* of Apidae); the various species are oligoleges for sundry host genera and families (see table 1 of Wcislo and Cane 1996). Many pollen wasps (Masarinae: Vespidae) are also oligolectic (reviewed by Gess 1996) or even narrowly oligolectic (e.g., Houston 1984; Mauss et al. 2003). Narrow oligoleges are associated with plant families small and large. For instance, *Andrena hattorfiana* uses only *Knautia* for pollen; the host family, Dipsacaceae, has only 290 species (Westrich 1989; Müller 1996). In contrast, *Hoplitis* (= *Osmia*) *anthocopoides* is a narrow oligolege of *Echium,* which belongs to a much larger plant family (Boraginaceae, with more than 2300 species). An extensive and intensive sampling of provision masses of one narrow oligolege, *Proteriades bullifacies,* found them to contain 97% *Phacelia* pollen (Hydrophyllaceae; 11 sites, 606 nests, and 2400 nest cells sampled; Parker 1978). Narrow oligolecty is no mere illusory result of undersampling! (See also Minckley and Roulston, chap. 4 in this volume.)

Some narrowly oligolectic bee taxa also depend on their host plants for an unusual floral resource—oil—which they substitute for nectar in larval provisions (Buchmann 1987). Narrow oligolecty results when these oil bees only take pollen from their limited suite of floral oil hosts (e.g., Steiner and Whitehead 1990; Houston et al. 1993). The Holarctic bee genus *Macropis* (Melittidae) is a prominent example, all species of which are tightly associated with one or more species of *Lysimachia* (Primulaceae) for all of their oil and pollen needs (Popov 1958; Vogel 1976; Cane et al. 1983). *Macropis* and *Lysimachia* represent one of the few bee–host associations in which cospeciation is a distinct possibility.

For many cases, oligolectic bees handle diverse congeneric flowers and their pollen with equal proficiency. Morphologies and sizes of congeneric flowers and especially pollen are often very similar. Pollen chemistries, such as protein concentration, are also highly taxonomically conserved (Roulston et al. 2000). Thus, it is reasonable to assume, in the absence of contrary evidence, that bee adaptations (behavioral, physiological, and morphological) for finding, working, or eating pollen from one plant species should generally suffice when using other members of that plant genus (Thorp 1979).

Oligolecty: The Limited Menu

That oligolecty exists is indisputable, but its application to particular bee species engenders confusion and debate. The pollen host menu of an oligolege may be regarded as being exclusive of many potential pollen hosts rather than inclusive of a few. Thus, oligolectic bee species are those that pass up opportunities to take pollen from diverse available plant genera and families used for pollen by broadly polylectic species.

Choice of definition can shade one's interpretation of oligolecty. Robertson was somewhat vague (or else broad-minded) on the issue, applying his new term *oligolecty* to "a bee [species] collecting pollen from a species, genus or family [of plant]." Linsley and MacSwain (1958) declared oligolectic species to be those bees that "collect pollen . . . from a group of similar or related plant species." But how distantly related? Most bee ecologists would agree (Robertson 1925; Cripps and Rust 1989b; Westrich 1989; Wcislo and Cane 1996) that a bee species is oligolectic if its pollen foraging is restricted to one or a few host genera of the same plant family. Having concisely defined narrow oligolecty, and seeing that this "middle ground" of oligolecty is also little disputed, our problem lies in the boundary separating oligolecty from broader categories.

It should be evident that a simple, straightforward, consistent, and universal method for precisely setting the bounds of oligolecty will remain elusive. We propose a definition of oligolecty that is sufficiently inclusive without diluting the intended meaning for undisputed cases: we define oligolecty as consistent and predictable reliance of a species on a few plant genera for their pollen needs. Except for cases of eclectic oligolecty (see later discussion), these plant genera will be taxonomically related, either as members of the same clade or plant family (smaller families) or as members of the same tribe or subfamily (large families).

If a species' oligolectic habits are, to a degree, facultative, we can expect foragers to occasionally turn to additional pollen taxa, often of unrelated clades, but only rarely when their regular pollen hosts are *temporarily* absent, as was emphasized by Linsley and MacSwain (1958). On the other hand, predictable and regular use of a pollen host by some but not most populations of a bee species merits recognition as a bona fide pollen host (as we discuss for eclectic oligoleges).

Plant families vary widely in their content and diversity of species, with practical implications for classifying a bee's pollen host breadth, at least until phylogenetic indices become practical (Symons and Beccaloni 1999). For example, *Osmia spinulosa* was declared oligolectic although it gathers pollen from genera in more than five tribes of Asteraceae (Müller 1994), a family with more than 25,000 species. We contend that this bee is much less of a specialist than, for instance, *Andrena hattorfiana,* the oligolege of *Knautia* (Dipsacaceae, 290 species). To improve the equivalence of such comparisons across disparate taxa, we contend that, for plant families rich in species (e.g., Asteraceae, Fabaceae), only those bee species restricted to genera of a single tribe or subfamily should be considered oligolectic. If a bee species gathers pollen from multiple tribes of a big plant family, we suggest designating it as *mesolectic* (see discussion to follow).

How does an oligolege respond if its pollen hosts fail to bloom? Are there reproductive consequences? Most oligoleges have one generation per year (univoltine), so adult emergence can and must synchronize with host bloom (e.g., Cane and Payne 1993). Interestingly, a few oligoleges such as *Calliopsis hesperia equina* and *C. helianthi* are reportedly bivoltine (Rozen 1958). Some oligoleges seem to be *obligate* oligoleges, refusing to provision or even nest in the absence of their floral host. They simply await host bloom, and in so doing they minimize risks of predation, nest usurpation, and general wear. Some examples of oligoleges that forgo nesting and foraging until host bloom is available include *Evylaeus galpinsiae* (host is *Oenothera;* Bohart and Youssef 1976), *Hoplitis anthocopoides* (host is *Echium;* Strickler 1979), and *Osmia californica* (hosts are vernal Heliantheae; Williams 2003). More species are probably *facultative* oligoleges, turning to alternative pollen sources when their local hosts are unavailable. Some cactus oligoleges in the genera *Diadasia, Lithurge,* and *Idiomelissodes* take pollen from alternative hosts when denied cactus (McIntosh 2001). Conceivably, some oligolectic species may consist of both facultative and obligate individuals. Some vernal forest oligoleges of the eastern United States and Europe turn to oak catkins for pollen (Chambers 1945; Michener and Rettenmeyer 1956; Cane and Payne 1988) when their pollen hosts are unavailable (Michener and Rettenmeyer 1956; Cane and Payne 1988). Few studies evaluate larval performance on substitute pollen, but those that have done so show that larvae of specialists can perform well on nutritious nonhost pollen (e.g., Levin and Haydak 1957; Williams 2003). Why these emergency hosts do not feature in the bees' normal diet remains puzzling.

Eclectic Specialization

We introduce the new term *eclectic oligolecty* to separately group those bee species that also are restricted to just a few genera of floral hosts. Unlike traditional oligoleges, these pollen host genera belong to widely disparate clades (different families, even different orders). "Disjunctive oligophagy," a term tersely introduced by Jolivet (1992), applies to other phytophagous insects. At this time, we are familiar with cases from only a few bee genera. Other published examples may lurk either under the heading of "polylecty," because the species gather pollen from multiple plant families, or under "oligolecty," wherein one host family has been considered to be primary and the other(s) dismissed as accidental (e.g., Torchio 1990). Both groupings overlook or deny apparent attributes of the phenomenon.

A case study illustrates the distinctive nature of eclectic oligolecty. The megachilid bee *Osmia* (*Osmia*) *ribifloris* is widely but patchily distributed from the Texas Plains westward into California (Rust 1974). Wild populations of this vernal bee gather pollen from just *Cercis* (redbud—Fabaceae; J. L. Neff, personal communication), *Berberis* (mahonia—Berberidaceae; Rust 1986), and *Arctostaphylos* (manzanita—Ericaceae; Torchio 1990). The host clades are unrelated and have radically different floral morphologies. *Osmia ribifloris* will pollinate and provision with pollen from some *Vaccinium* spp. (Ericaceae; Torchio 1990; Sampson and Cane 2000). Concurrently flowering, sympatric species of Rosaceae, Salicaceae, and Hydrophyllaceae are used by its close relative, *O. lignaria,* but *O. ribifloris* remains tenaciously faithful to one (Torchio 1990) or another (Rust 1986) of its trio of floral hosts. Could *O. ribifloris* be an amalgam of three narrowly oligolectic, cryptic sibling species? Assortative mating in not evident in a six-year-old captive composite population (Blair Sampson and J. H. Cane personal observation). No single host genus of *O. ribifloris* is primary; the host trio is nowhere broadly sympatric, and quite often, only one of the host genera is present in a region with the bee. Hence, we have a case of a bee species fixated on a few genera of distantly related pollen hosts.

Other indisputable eclectic oligoleges exist within *Osmia.* For example, the common vernal European bee, *Osmia rufa,* specializes on just oak and buttercup for pollen. From six populations in France, pollen stores of 52 nest cells of this bee averaged 56% oak, 30% buttercup, and 10% pear and hawthorn (Tasei 1973). At four sites in the United Kingdom, 239 provision masses from 48 nests averaged 52% buttercup and 38% oak (Raw 1974), not unlike an earlier British study that sampled 60 nest provisions (Free and Williams 1970). This bee's peculiar predilection for pollen from two genera in two unrelated clades is thus spatiotemporally consistent. As with *O. ribifloris,* floral morphology and display of these two host genera are utterly different (drab catkins in trees vs. bold yellow bowl-shaped flowers of an herb); *O. rufa* is thus also predictably eclectic in its pollen-foraging habits.

Some *Diadasia* bees appear to be eclectic oligoleges, though somewhat different from the preceding examples. Sipes (2001) and Sipes and Tepedino (unpublished data) analyzed scopal pollen loads taken from museum specimens of 25 North American *Diadasia* species. Individuals of eight species had pollen from more than one plant family in significant amounts. Among these eight, pollen of one or two genera of a single plant family usually predominated, but 10–26% of individuals sampled had either pure loads of an alternative host family or mixed loads ranging from 10 to greater than 90% of the alternative host. Some eclectic species using *Sphaeralcea* (Malvaceae) as their primary host also collect Heliantheae (Asteraceae) and *Opuntia* pollen, whereas those emphasizing *Opuntia* also collected Heliantheae and *Sphaeralcea* pollen. In all but one case, the alternative hosts are the primary hosts of other *Diadasia* species. We do not know if alternative-host use reflects shortages of primary hosts, but their predictable selectivity when choosing alternative hosts is noteworthy. Eclectic oligoleges seem to be ideal candidates for experimental manipulation and evaluation of potential larval imprinting on chemical cues of host pollen (Hopkin's host selection model). If eclectic oligolecty proves to be illusory, then, by our choice of terminology, its cases will remain retrievable with the search word "oligolecty."

Mesolecty: The Middling Menu

We propose a new term, *mesolecty,* to group those bee species that collect pollen from various species and genera of flowering plants drawn from the same few plant families (or tribes of large plant families). Meso is a combining form from the Greek μεσος (mesos), meaning "intermediate" or "middle." In our lexicon, mesolecty is an intermediate between oligolecty and polylecty. Mesolecty absorbs many of the cases ascribed to "broad oligolecty" or "narrow polylecty" (e.g., Small et al. 1997)—terms that are oxymora.

Contrasted with eclectic oligoleges, mesolectic bees are less predictable for their choices of pollen hosts from among the genera available in their pollen host families; the species and genus of the floral host is more predictable by familial association in a mixed-plant community. In contrast, eclectic oligoleges take pollen from the same few host genera of their several host families, ignoring the available alternatives in those families. Pollen host choices by eclectic oligoleges are, therefore, predictable at both the generic and familial ranks; those of mesoleges are only predictable at the familial level.

We provide a few examples of mesolectic bee species. European *Anthidium oblangatum* gathers its pollen from numerous floral species of six or more genera in four families (55 pollen loads, 55 localities; Westrich 1989; Müller 1996), which is comparable to European *A. manicatum* (Westrich 1989; Müller 1996; Banaszak and Romasenko 1998). Where these two species escaped in the United States, they remain faithful to these same pollen hosts (Hoebeke and Wheeler 1999). Both *Calliopsis andreniformis* (Dyer and Shinn 1978) and *Megachile brevis* (Michener 1953) are well-studied mesoleges emphasizing six and eight genera of Fabaceae in their pollen diets, respectively, but each regularly provisions with pollen from other plant families as well. The contrast of mesolecty with polylecty is one that emphasizes the magnitude of host diversity.

Degrees of Polylecty: Using Diverse Pollen Hosts

Polylecty: Many Pollen Hosts

We assign the term *polylecty* to species that gather significant amounts of pollen from more than three plant families, but not as many as are used by a co-occurring broad polylege (table 5.1). As discussed under mesolecty, we consider tribes of big plant families to count as families for this purpose. Exclusion of rarely used hosts and traces of pollen was discussed earlier. Polylecty becomes a noticeably more homogeneous category once we differentiate it from its former extremes, mesolecty and broad polylecty.

Most bees are of necessity polylectic if they are eusocial, multivoltine, or have prolonged active adult lifespans, because their periods of adult foraging exceed blooming periods of any one floral host. Polylecty may be ubiquitous among the large and small carpenter bees (Xylocopinae; Hurd and Moure 1963; Michener 1971; Daly 1973) and the Oxaeidae (Hurd and Linsley 1976), although surprisingly few species have been analyzed; this also seems common among Halictinae (Moure and Hurd 1987), Megachilinae, and many tribes of Apidae sensu Michener (Michener 2000). Orchid bees may be polylectic or even broadly polylectic, but few species have been studied (Roubik 1989). Much of Hylaeinae (at least outside of Australia) may also be polylectic. However, hylaeines and related euryglossines transport pollen internally, thereby hiding their pollen preferences (Jander 1976).

Broad Polylecty: Extreme Host Diversity

In our lexicon, bee species that are *broadly polylectic* can collect and use pollen from most genera and species of numerous plant families. Taxonomic versatility in pollen foraging is the hallmark of broad polyleges, although even the honey bee takes pollen from only a minority of the angiosperms available within its global range (e.g., Percival 1947) and ignores pollen from nearly all anemophilous monocots (e.g., sedges, grasses) and some dicots (e.g., temperate-zone trees); inevitable exceptions exist, however, such as maize and oak (e.g., Raw 1974). Nonetheless, broad polyleges utilize more than 10% of the pollen hosts of

Table 5.1 Tabular lexicon for taxonomic pollen specialization by bee species throughout their geographic range

	Number of plant taxa		
Class of specialization	**Species**	**Genera**	**Tribes or families**
Broad polylecty	many	many	>25% of available families
Polylecty	many	many	4 to <25% of available families
Mesolecty	many	>4	1–3 families or big tribes
Oligolecty	>1	1–4	1
Eclectic	>1	2–4	2–3
Narrow	>1	1	1
Monolecty	1	1	1

the entire melittophilous flora at a locale, an assertion that can be appreciated by watching such bees collect pollen in a botanical garden.

The European honeybee, *Apis mellifera,* is a familiar broad polylege. In Cardiff, United Kingdom, foragers of one hive took pollen from 86 of 225 (28%) available flowering species (Percival 1947). At the University Botanical Garden in São Paulo, Brazil, Cortopassi-Laurino (cited by Ramalho et al. 1985) reported honey bees gathering pollen from 44 families (104 of 190 flowering species). In the desert around Tucson, Arizona, honey bees gathered pollen from more than 90 species of 41 families (O'Neal and Waller 1984). It appears that *A. mellifera* generally uses 20–40% or more of the "showy" flowering species in an area. Presumably, other *Apis* (Apini) species are likewise broadly polylectic.

Other groups of broad polyleges include the stingless bees (Meliponini), bumblebees (*Bombus*), and some sweat bees (Halictini). Stingless bees are tropical and, like honey bees, have perennial social colonies, which necessitate foraging versatility. In a single year and location, three colonies of *Plebia remota* collected pollen from 64 species (29 families; Ramalho et al. 1985). At a São Paulo botanical garden, *Trigona spinipes* took pollen from 34 species and as many families as *A. mellifera* (cited in Ramalho et al. 1985). Colonies of nearly all bumblebee species are annual, but they too are versatile pollen foragers. Seven colonies of *Bombus lucorum* in Scotland collected pollen from 23 species in 17 families (Brian 1951). Ten families were found represented in just 34 pollen loads from *Bombus huntii* netted in southeastern Wyoming (Tepedino 1982). Pollen-foraging habits of social halictine bees are poorly detailed, but some appear to also be broadly polylectic. In eastern Kansas, the sweat bee *Lasioglossum (Dialictus) versatum* took pollen from 31 species in 16 families (Michener 1966). From pollen loads of 28 female *Halictus confusus* netted in southeastern Wyoming, 9 families (17 genera) were present (trace pollen not scored; Tepedino 1982). Many social bees, owing to their long-lived colonies, are broadly polylectic.

There is less evidence for broad oligolecty among solitary bees. The pollen-foraging habits of orchid bees (Euglossini—solitary relatives of honeybees and bumblebees) are poorly known; they inhabit Neotropical forests, often foraging at canopy flowers inaccessible to us. Their free-standing nests are generally not amenable to standard trap-nesting methods either. Males can be attracted to scent baits, reflecting their specialized pollination relationships with orchids, but this has no relation to female pollen foraging. Nonetheless, pollen-foraging habits of a few species show broad polylecty. From a single year and site in Chiapas, Mexico, 51 nests of *Euglossa atroveneta* had pollen from 74 species in 41 families (Arriaga and Hernandez 1998). Adults have unusually prolonged lives of active foraging, which underscores their need for foraging versatility.

Large carpenter bees (Xylocopini) are more amenable to analyses of pollen collection, being taxonomically well-understood cavity nesters found worldwide, but they have been largely overlooked for detailed pollen studies. They are certainly polylectic, and some may be broadly polylectic. From a single Argentinean site, 131 nest cells of *Xylocopa splendidula* sampled during three one-month periods yielded pollen representing 11 families (Tellería 2000). A thorough pollen study of any *Xylocopa* would be manageable and revealing.

Broad polylecty merits consideration for several reasons. First, these bees represent extreme taxonomic versatility for pollen foraging among diverse, distantly related clades. Their broad polylecty may stem from life-history attributes different from other polyleges (sociality and/or long-lived active adults), and so its evolutionary origins and adaptive qualities might be associated with these life-history traits. Broadly polylectic bees are ideal subjects for studying learned behavior in pollen foraging. They are also of interest for the versatile digestive physiologies that should accompany such a mixed diet either fed directly to larvae (for mass-provisioning species) or transformed into dietary glandular secretions by nurse bees (highly social species).

As a practical matter, a list of local pollen hosts used by a broad polylege best represents the subset of flowering species available to all other local bee species when we evaluate degrees of taxonomic pollen specialization. Obviously, every flowering community has species whose pollen is eschewed by all bees, but ad hoc human judgments invariably include mistaken decisions about usable flowering species. For instance, many trees of the North Temperate Zone are wind pollinated. Among these, we noted that some bee species use oak and willow pollen, but what about catkins of birches, hickories, and alders? Dietary pollen spectra for broad polyleges are imperfect indicators of the subset of local flowers usable by bees but are superior to a simple checklist of available plant species. Pollen spectra for honey bees are available for many locations worldwide (e.g., table 1 of O'Neal and Waller 1984), facilitating such summaries without recourse to a separate pollen study.

Conclusions

We have seen that, at one extreme, social bees and some solitary species are *polylectic* or even *broadly polylectic*. Pollen that they gather to feed their progeny comes from many host species, representing some to many families of flowering plants. At the other extreme, a few species of bees are *monolectic*, relying on a single floral host species to fulfill their pollen needs.

Oligolectic species of solitary bees satisfy their pollen needs using only a restricted subset of the host-plant menu available to them, year after year and throughout their geographic range. Traditionally defined oligolecty encompasses several discernable taxonomic patterns in dietary breadth: *narrow oligoleges* are restricted to a single plant genus for pollen; *oligoleges* are restricted to several genera, traditionally all of the same clade. We introduce two new categories here: *eclectic oligoleges* are those species whose few fixed genera of pollen hosts belong to different clades (families); *mesolectic* bee species gather pollen from more than four genera in two to three different clades (families or tribes of large families). Evidence for these pollen host classifications comes from observations of floral visitation, analysis of pollen gathered or provisioned by females, or (rarely) experimental feeding trials. For many of the approximately 16,000 known species of bees, host associations are little known, often only from floral visitation records that typically do not distinguish between nectar and pollen hosts. Analyses of pollen collected by females or stored in nest provisions offer several advantages over floral visitation alone: pollen and nectar hosts are better discerned, and inaccessible or unanticipated pollen hosts will be represented (e.g., forest canopy species). Analysis of pollen foraging patterns can be most confidently interpreted and compared when (1) pollen grain volumes of different hosts are used to estimate proportional composition; (2) classification is based on the incidence of pure- versus mixed-host loads/provision, rather than host-taxon averages pooled across samples; and (3) pollens from bees or nests are sampled from multiple years and locations. Our methodological suggestions, if adopted, should enhance synthesis and comparison with other studies. There are many evolutionary, ecological, behavioral, and physiological insights to be gained, including bee–plant cospeciation, innate and learned floral perception, digestive specialization, mismatching of maternal choice and larval performance, optimal foraging decisions, mating strategies, and pollination consequences of specialization in both wild and agricultural landscapes (for example, Minckley and Roulston, chap. 4 in this volume).

Acknowledgments

We thank Dr. Vincent Tepedino for numerous insightful discussions; he was the impetus for our ideas about mesolecty. Karen Strickler and an anonymous reviewer provided helpful comments that improved the manuscript.

References

Archibald, J. K., P. G. Wolf, V. J. Tepedino, and J. Bair. 2001. Genetic relationships and population structure of the endangered Steamboat buckwheat, *Eriogonum ovalifolium* var. *williamsiae* (Polygonaceae). American Journal of Botany 88: 608–615.

Arriaga, E. R., and E. M. Hernandez. 1998. Resources foraged by *Euglossa atroveneta* (Apidae: Euglossinae) at Unión Juárez, Chiapas, Mexico. A palynological study of larval feeding. Apidologie 29: 347–359.

Austin, D. F. 1978. Morning glory bees and the *Ipomoea pandurata* complex (Hymenoptera: Anthophoridae). Proceedings of the Entomological Society of Washington 80: 397–402.

Banaszak, J., and L. Romasenko. 1998. Megachilid bees of Europe. Pedagogical University Bydgoszcz, Bydgoszcz, Poland.

Bohart, G. E., and N. N. Youssef. 1976. The biology and behavior of *Evylaeus galpinsiae* Cockerell (Hymenoptera: Halictidae). Wassmann Journal of Biology 34: 185–234.

Brian, A. D. 1951. The pollen collected by bumblebees. Journal of Animal Ecology 20: 191–194.

Buchmann, S. L. 1983. Buzz pollination in angiosperms. Pp. 73–113 *in* C. E. Jones and R. J. Little (eds.), Handbook of experimental pollination biology. Van Nostrand Reinhold, New York.

Buchmann, S. L. 1987. The ecology of oil flowers and their bees. Annual Review of Ecology and Systematics 18: 343–369.

Buchmann, S. L., and M. K. O'Rourke. 1991. Importance of pollen grain volumes for calculating bee diets. Grana 30: 591–595.

Bullock, S. H., R. Ayala, D. Rodriguez-Zamora, D. L. Quiroz-Garcia, and M. D. de la Luz Arreguin-Sanchez. 1991. Nest provision and pollen foraging in three Mexican species of solitary bees (Hymenoptera: Apoidea). Pan-Pacific Entomologist 67: 171–176.

Cane, J. H., G. C. Eickwort, F. R. Wesley, and J. Spielholz. 1983. Foraging, grooming, and mate-seeking behaviors of *Macropis nuda* (Hymenoptera, Melittidae) and use of *Lysimachia ciliata* (Primulaceae) oils in larval provisions and cell linings. The American Midland Naturalist 110: 257–264.

Cane, J. H., and J. A. Payne. 1988. Foraging ecology of the bee *Habropoda laboriosa* (Hymenoptera: Anthophoridae), an oligolege of blueberries (Ericaceae: *Vaccinium*) in the southeastern United States. Annals of the Entomological Society of America 81: 419–427.

Cane, J. H., and J. A. Payne. 1993. Regional, annual, and seasonal variation in pollinator guilds: Intrinsic traits of bees (Hymenoptera: Apoidea) underlie their patterns of abundance at *Vaccinium ashei* (Ericaceae). Annals of the Entomological Society of America 86: 577–588.

Cane, J. H., and D. Schiffhauer. 2003. Dose-response relationships between pollination and fruiting refine pollinator comparisons for cranberry (*Vaccinium macrocarpon* Ait.). American Journal of Botany 90: 1425–1432.

Cane, J. H., R. R. Snelling, L. J. Kervin, and G. C. Eickwort. 1996. A new monolectic coastal bee, *Hesperapis oraria* Snelling and Stage (Hymenoptera: Melittidae), with a review of desert and neotropical disjunctives in the southeastern U.S. Journal of the Kansas Entomological Society 69: 238–247.

Chambers, V. H. 1945. British bees and wind-borne pollen. Nature 155: 145.

Chambers, V. H. 1946. An examination of the pollen loads of *Andrena:* The species that visit fruit trees. Journal of Animal Ecology 15: 9–21.

Chambers, V. H. 1968. Pollens collected by species of *Andrena* (Hymenoptera: Apidae). Proceedings of the Royal Entomological Society of London A 43: 155–160.

Colwell, R. K., and D. J. Futuyma. 1971. On the measurement of niche breadth and overlap. Ecology 52: 567–576.

Cripps, C., and R. W. Rust. 1989a. Pollen foraging in a community of *Osmia* bees (Hymenoptera: Megachilidae). Environmental Entomology 18: 582–589.

Cripps, C., and R. W. Rust. 1989b. Pollen preferences of seven *Osmia* species (Hymenoptera: Megachilidae). Environmental Entomology 18: 133–138.

Cruden, R. W. 1972. Pollination biology of *Nemophila menziesii* (Hydrophyllaceae) with comments on the evolution of oligolectic bees. Evolution 26: 373–389.

Daly, H. V. 1973. Bees of the genus *Ceratina* in America north of Mexico (Hymenoptera: Apoidea). University of California Publications in Entomology 74: 1–114.

da Silveira, F. A. 1991. Influence of pollen grain volume on the estimation of the relative importance of its source to bees. Apidologie 22: 495–502.
Dyer, J. G., and A. F. Shinn. 1978. Pollen collected by *Calliopsis andreniformis* Smith in North America (Hymneoptera: Andrenidae). Journal of the Kansas Entomological Society 51: 787–795.
Edwards-Anderka, C. J., and D. H. Pengelly. 1970. Pollen analysis in the ecology of bees of the genus *Bombus* Latr. (Hymenoptera: Apidae) in southern Ontario. Proceedings of the Entomological Society of Ontario 100: 170–176.
Eickwort, G. C. 1973. Biology of the European mason bee, *Hoplitis anthocopoides* (Hymenoptera: Megachilidae), in New York State. Search Agriculture (Cornell University Agricultural Experiment Station) 3: 1–31.
Faith, D. P. 1992. Conservation evaluation and phylogenetic diversity. Biological Conservation 61: 1–10.
Free, J. B., and I. H. Williams. 1970. Preliminary investigations on the occupation of artificial nests by *Osmia rufa* L. (Hymn. Megachilidae). Journal of Applied Ecology 7: 559–566.
Gess, S. K. 1996. The pollen wasps: Ecology and natural history of the Masarinae. Harvard University Press, Cambridge, MA.
Grant, V. 1950. The flower constancy of bees. Botanical Review 16: 379–398.
Hoebeke, E. R., and A. G. J. Wheeler. 1999. *Anthidium oblangatum* (Illiger): An Old World bee (Hymenoptera: Megachilidae) new to North America, and new North American records for another adventive species, *A. manicatum* (L.). University of Kansas Publications, Museum of Natural History 24: 21–24.
Houston, T. F. 1984. Bionomics of a pollen wasp, *Paragia tricolor* (Hymenoptera: Vespidae: Masarinae), in Western Australia. Records of the Western Australia Museum 11: 141–151.
Houston, T. F. 1992. Three new, monolectic species of *Euryglossa* (*Euhesma*) from Western Australia (Hymenoptera: Colletidae). Records of the Western Australia Museum 15: 719–728.
Houston, T. F., B. B. Lamont, S. Radford, and S. G. Errington. 1993. Apparent mutualism between *Verticordia nitens* and *V. aurea* (Myrtaceae) and their oil-ingesting bee pollinators (Hymenoptera, Colletidae). Australian Journal of Botany 41: 369–380.
Hurd, P. D., Jr., and E. G. Linsley. 1976. The bee family Oxaeidae with a revision of the North American species (Hymneoptera: Apoidea). Smithsonian Contributions to Zoology 220: 1–75.
Hurd, P. D., Jr., and J. S. Moure. 1963. A classification of the large carpenter bees (Xylocopini). University of California Publications in Entomology 24: 1–365.
Jaenike, J. 1990. Host specialization in phytophagous insects. Annual Review of Ecology and Systematics 21: 243–273.
Jander, R. 1976. Grooming and pollen manipulation in bees (Apoidea): The nature and evolution of movements involving the foreleg. Physiological Entomology 1: 179–194.
Johnson, M. D. 1984. The pollen preferences of *Andrena dunningi* (Hymenoptera: Andrenidae). Journal of the Kansas Entomological Society 57: 34–43.
Jolivet, P. 1992. Insects and plants: Parallel evolution and adaptations, 2nd ed. Sand Hill Press, Gainesville, FL.
Jones, C. E. 1978. Pollinator constancy as a pre-pollination isolating mechanism between sympatric species of *Cercidium*. Evolution 32: 189–198.
Levin, M. D., and M. H. Haydak. 1957. Comparative value of different pollens in the nutrition of *Osmia lignaria*. Bee World 38: 221–226.
Levins, R. 1968. Evolution in changing environments. Princeton University Press, Princeton, NJ.
Linsley, E. G. 1958. The ecology of solitary bees. Hilgardia 27: 543–599.
Linsley, E. G., and J. W. MacSwain. 1958. The significance of floral constancy among bees of the genus *Diadasia* (Hymenoptera: Anthophoridae). Evolution 12: 219–223.
Lovell, J. H. 1918. The flower and the bee. C. Scribner's and Sons, New York.
Macior, L. W. 1994. Pollen-foraging dynamics of subalpine bumblebees (*Bombus* Latr.). Plant Species Biology 9: 99–106.
Mackenzie, K., and S. Javorek. 1997. The potential of alfalfa leafcutter bees (*Megachile rotundata* L.) as pollintors of cranberry (*Vaccinium macrocarpon* Aiton). Acta Horticulturae 437: 345–351.

Martins, R. P., and J. C. Borges. 1999. Use of *Ludwigia* (Onagraceae) pollen by a specialist bee, *Diadasina distincta* (Hymenoptera: Apidae), at a nesting site in southeastern Brazil. Biotropica 31: 530–534.

Mauss, V., A. Muller, and E. Yildirim. 2003. Nesting and flower associations of the pollen wasp *Ceramius fonscolombei* Latreille, 1810 (Hymenoptera: Vespidae: Masarinae) in Spain. Journal of the Kansas Entomological Society 76: 1–15.

McIntosh, M. E. 2001. Interactions between cactus-specialist solitary bees and their host cacti. PhD dissertation, University of Arizona, Tucson, AZ.

Michener, C. D. 1953. The biology of the leafcutter bee *Megachile brevis* and its associates. University of Kansas Science Bulletin 35: 1659–1748.

Michener, C. D. 1966. The bionomics of a primitively social bee, *Lasioglossum versatum* (Hymenoptera: Halictidae). Journal of the Kansas Entomological Society 39: 193–217.

Michener, C. D. 1971. Biologies of African allodapine bees (Hymenoptera, Xylocopinae). Bulletin of the American Museum of Natural History 145: 219–302.

Michener, C. D. 2000. The bees of the world. Johns Hopkins University Press, New York.

Michener, C. D., and C. W. Rettenmeyer. 1956. The ethology of *Andrena erythronii* with comparative data on other species (Hymenoptera, Andrenidae). University of Kansas Science Bulletin 37: 645–684.

Minckley, R. L., J. H. Cane, and L. Kervin. 2000. Origins and ecological consequences of pollen specialization among desert bees. Proceedings of the Royal Society of London B 267: 265–271.

Moure, J. S., and P. D. Hurd Jr. 1987. An annotated catalog of the halictid bees of the Western Hemisphere (Hymenoptera: Halictidae). Smithsonian Institution Press, Washington, DC.

Müller, A. 1994. Die Bionomie der in leeren Schneckengehäusen nistenden Biene *Osmia spinulosa* (Kirby 1802). Veröffentlichungen Naturschutz Landschaftspflege Baden-Württemberg 68/69: 291–334.

Müller, A. 1996. Host-plant specialization in Western Palearctic anthidiine bees (Hymenoptera: Apoidea: Megachilidae). Ecological Monographs 66: 235–257.

Neff, J. L., and J. G. J. Rozen. 1995. Foraging and nesting biology of the bee *Anthemurgus passiflorae* (Hymenoptera: Apoidea), descriptions of its immature stages, and observations on its floral host (Passifloriaceae). American Museum Novitates 3138: 1–19.

O'Neal, R. J., and G. W. Waller. 1984. On the pollen harvest by the honey bee (*Apis mellifera* L.) near Tucson, Arizona (1976–1981). Desert Plants 6: 81–109.

O'Rourke, M. K., and S. L. Buchmann. 1991. Standardized analytical techniques for bee-collected pollen. Annals of the Entomological Society of America 20: 507–513.

Packer, J. S. 1970. The flight and foraging behavior of the alkali bee (*Nomia melanderi*) (Ckll.) and the alfalfa leaf cutter bee (*Megachile rotundata*). PhD dissertation, Utah State University, Logan, UT.

Parker, F. D. 1978. Biology of the bee genus *Proteriades* Titus (Hymenoptera: Megachilidae). Journal of the Kansas Entomological Society 51: 145–173.

Pekkarinen, A. 1997. Oligolectic bee species in Northern Europe (Hymenoptera, Apoidea). Entomologica Fennica 8: 205–214.

Pekkarinen, A. 1999. Oligolectic bee species and their decline in Finland (Hymenoptera: Apoidea). Proceedings of the XXIV Nordic Congress of Entomology 24: 151–156.

Percival, M. 1947. Pollen collection by *Apis melifera*. New Phytologist 46: 142–173.

Popov, V. V. 1958. Special features of the correlated evolution of *Macropis, Epeoloides* (Hymenoptera, Apoidea), and *Lysimachia* (Primulaceae). Entomological Review (Entomologicheskoe Obozrenie) 37: 433–451.

Proctor, M., P. Yeo, and A. Lack. 1996. The natural history of pollination. Timber Press, Portland, OR.

Quiroz-Garcia, D. L., E. Martinez-Hernandez, R. Palacios-Chavez, and N. E. Galindo-Miranda. 2001. Nest provisions and pollen foraging in three species of solitary bees (Hymenoptera: Apidae) from Jalisco, Mexico. Journal of the Kansas Entomological Society 74: 61–69.

Ramalho, M., T. C. Giannini, K. S. Malago-Dibraga, and V. L. Imperatriz-Fonseca. 1994. Pollen harvest by stingless bee foragers (Hymenoptera, Apidae, Meliponinae). Grana 33: 239–244.

Ramalho, M., V. L. Imperatriz-Fonseca, A. Kleinert-Giovannini, and M. Cortopassi-Laurino. 1985. Ex-

ploitation of floral resources by *Plebia remota* Holmberg (Apidae, Meliponinae). Apidologie 16: 307–330.
Raw, A. 1974. Pollen preferences of three *Osmia* species (Hymenoptera). Oikos 25: 54–60.
Renner, S. S. 1989. A survey of reproductive biology in neotropical Melasomataceae and Memecylaceae. Annals of the Missouri Botanical Garden 76: 496–518.
Robertson, C. 1925. Heteroptropic bees. Ecology 6: 412–436.
Robertson, C. 1926. Revised list of oligolectic bees. Ecology 7: 378–380.
Robertson, C. 1929. Flowers and insects. Lists of visitors to four hundred and fifty-three flowers. C. Robertson, Carlinville, IL.
Roubik, D. W. 1989. Ecology and natural history of tropical bees. Cambridge University Press, Cambridge.
Roulston, T. H., J. H. Cane, and S. L. Buchmann. 2000. What governs protein content of pollen: Pollinator preferences, pollen-pistil interactions, or phylogeny? Ecological Monographs 70: 617–643.
Rozen, J. G., Jr. 1958. Monographic study of the genus *Nomadopsis* Ashmead (Hymenoptera: Andrenidae). University of California Publications in Entomology 15: 1–202.
Rust, R. W. 1974. The systematics and biology of the genus *Osmia*, subgenera *Osmia*, *Chalcosmia*, and *Cephalosmia* (Hymenoptera: Megachilidae). Wassmann Journal of Biology 32: 1–93.
Rust, R. W. 1986. Biology of *Osmia* (*Osmia*) *ribifloris* Cockerell (Hymenoptera: Megachilidae). Journal of the Kansas Entomological Society 59: 89–94.
Rust, R. W. 1987a. Biology of *Nomadopsis larreae* (Hymenoptera: Andrenidae), with an analysis of yearly appearance. Annals of the Entomological Society of America 81: 99–104.
Rust, R. W. 1987b. Collecting of *Pinus* (Pinaceae) pollen by *Osmia* bees (Hymenoptera: Megachilidae). Environmental Entomology 16: 668–671.
Rust, R. W. 1990. Spatial and temporal heterogeneity of pollen foraging in *Osmia lignaria propinqua* (Hymenoptera: Megachilidae). Environmental Entomology 19: 332–338.
Rust, R. W., R. W. Thorp, and P. F. Torchio. 1974. The ecology of *Osmia nigrifrons* with a comparison to other *Acanthosmoides*. Journal of Natural History 8: 29–47.
Sampson, B. J., and J. H. Cane. 2000. Pollination efficiencies of three bee (Hymenoptera: Apoidea) species visiting rabbiteye blueberry. Journal of Economic Entomology 93: 1726–1731.
Sipes, S. D. 2001. Phylogenetic relationships, taxonomy, and evolution of host choice in *Diadasia* (Hymenoptera: Apoidea). PhD dissertation, Utah State University, Logan, UT.
Small, E., B. Brookes, L. P. Lefkovitch, and D. T. Fairey. 1997. A preliminary analysis of the floral preferences of the Alfalfa Leafcutting Bee, *Megachile rotundata*. Canadian Field Naturalist 111: 445–453.
Steiner, K. E., and V. B. Whitehead. 1990. Pollinator adaptation to oil-secreting flowers: *Rediviva* and *Diascia*. Evolution 44: 1701–1707.
Strickler, K. 1979. Specialization and foraging efficiency of solitary bees. Ecology 60: 998–1009.
Symons, F. B., and G. W. Beccaloni. 1999. Phylogenetic indices for measuring the diet breadths of phytophagous insects. Oecologia 119: 427–434.
Tasei, J.-N. 1973. Le comportement de nidification chez *Osmia* (*Osmia*) *cornuta* Latr. et *Osmia* (*Osmia*) *rufa* L. (Hymenoptera Megachilidae). Apidologie 4: 195–225.
Tellería, M. C. 2000. Exploitation of pollen resources by *Xylocopa splendidula* in the Argentine pampas. Journal of Apicultural Research 39: 55–60.
Tepedino, V. J. 1982. Flower visitation and pollen collection records for bees of high altitude shortgrass prairie in southeastern Wyoming, USA. Southwestern Entomologist 7: 16–25.
Thiele, R. 2002. Nesting biology and seasonality of *Duckeanthidium thielei* Michener (Hymenoptera: Megachilidae), an oligolectic rainforest bee. Journal of the Kansas Entomological Society 75: 274–282.
Thorp, R. W. 1969. Systematics and ecology of bees of the subgenus *Diandrena* (Hymenoptera: Andrenidae). University of California Publications in Entomology 52: 1–146.
Thorpe, R. W. 1979. Structural, behavioral, and physiological adaptations of bees (Apoidea) for collecting pollen. Annals of the Missouri Botanical Garden 66: 788–812.

Torchio, P. F. 1990. *Osmia ribifloris,* a native bee species developed as a commercially managed pollinator of highbush blueberry (Hymenoptera: Megachilidae). Journal of the Kansas Entomological Society 63: 427–436.

Torchio, P. F., J. G. Rozen Jr., G. E. Bohart, and M. S. Favreau. 1967. Biology of *Dufourea* and of its cleptoparasite, *Neopasites* (Hymenoptera: Apoidea). Journal of the New York Entomological Society 75: 132–146.

Visscher, P. K., and B. N. Danforth. 1993. Biology of *Calliopsis pugionis* (Hymenoptera: Andrenidae): Nesting, foraging, and investment sex ratio. Annals of the Entomological Society of America 86: 822–832.

Vogel, S. 1976. *Lysimachia:* Ölblumen der Holarktis. Naturwissenschaften 63: 44–45.

Vogel, S. 1978. Evolutionary shifts from reward to deception in pollen flowers. Pp. 89–96 *in* A. J. Richards (ed.), The pollination of flowers by insects. Academic Press, London.

Wcislo, W. T., and J. H. Cane. 1996. Floral resource utilization by solitary bees (Hymenoptera: Apoidea) and exploitation of their stored foods by natural enemies. Annual Review of Entomology 41: 195–224.

Westrich, P. 1989. Die Wildbienen Baden-Württembergs allgemeiner Teil: Lebensräume, Verhalten, Ökologie und Schutz. Verlag Eugen Ulmer, Stuttgart.

Westrich, P., and K. Schmidt. 1987. Pollenanalyse, ein Hilfsmittel beim Studium des Sammelverhaltens von Wildbienen (Hymenoptera, Apoidea). Apidologie 18: 199–213.

Williams, N. M. 2003. Use of novel pollen species by specialist and generalist solitary bees (Hymenoptera: Megachilidae). Oecologia 134: 228–237.

Williams, N. M., and V. J. Tepedino. 2003. Consistent mixing of near and distant resources in foraging bouts by the solitary mason bee *Osmia lignaria.* Behavioral Ecology 14: 141–149.

Wilson, P., and M. Stine. 1996. Floral constancy in bumble bees: Handling efficiency or perceptual conditioning? Oecologia 106: 493–499.

CHAPTER SIX

Rewardless Flowers in the Angiosperms and the Role of Insect Cognition in Their Evolution

Susanne S. Renner

The long-term stability of mutualisms depends on each partner's ability to limit deception and cheating. Even so, antagonistic elements are widespread in mutualisms. This chapter reviews antagonistic pollinator–flower interactions in which flowers do not provide a reward to their pollinators. The other side of the coin—antagonistic interactions in which visitors exploit flower resources without delivering pollination—lies outside the focus of this review. After defining pollinator deception and summarizing the data on its phylogenetic distribution in the angiosperms, recent insights into insect, especially bee, cognition (e.g., Chittka and Thomson 2001) are introduced to answer the question of why exploitative systems persist. A key point is that pollinators constantly encounter temporarily or permanently rewardless flowers, and it is against this background that the evolution of exploitative systems has to be seen.

Insect behavior and cognition are especially relevant because insects are the predominant pollinators of flowering plants (Regal 1977, 1982; Crepet 1984); of the 13,500 genera of angiosperms, about 500 contain bird-pollinated species, 250 contain bat-pollinated species, 874 contain solely wind or water pollinated species, and the remainder contain mostly insect-pollinated species (Renner and Ricklefs 1995). Understanding insect vision, olfaction, and learning is therefore key to understanding pollinator deception. This is taken up at the end of this chapter, together with the question of how deception may relate to ecological generalization or specialization. Generalization and specialization in ecology usually refer to niche breadth and are quantified as the number of different resource items used (e.g., the number of pollinator species used by a plant or the number of plant species visited by a pollinator). As pointed out by Herrera (2005), using pollinator species numbers when comparing levels of generalization between plants or communities does not adequately account for differential pollinator abundances, and when abundances are accounted for (which is rarely done), generalization may turn out not to be an invariant species-level property. The term *generalization* for behaviorists also has another meaning relevant in this chapter, namely, a behavior evoked by a stimulus other than the one an animal was conditioned to, which results in the transfer of a learned response from one situation to another similar to the original situation. This kind of generalization is the precondition for the deception of pollinators and is discussed under the section Insect Cognition and the Evolution and Persistence of Rewardlessness.

There is no consistent relationship between the degree of specialization of a flower and the degree of specialization of its interaction partners, an asymmetry that precludes extrapolation from flower specialization to visitor specialization (e.g., Renner 1998; Armbruster et al. 2000; Vázquez and Simberloff 2002; several chapters in this volume). Fieldwork-based lists of pollinators used by phylogenetically related deceptive and nondeceptive plant species (occurring in similar habitats and studied over similar lengths of time) are one way to arrive at solid data on generalization or specialization in deceptive flowers; another way is to experimentally add a reward to normally rewardless flowers to assay effects on pollinator spectra. Results from such studies are discussed at the end of this review.

Deceptive flowers exploit pollinators by signaling the presence of a reward without providing the reward (Sprengel 1793; Knuth 1898–1908; Gilbert 1975; Vogel 1975, 1993; Baker 1976; Wiens 1978; Little 1983; Dafni 1984; Schemske et al. 1996; Smithson and Gigord 2003). Flowers that are *permanently* deceptive have evolved in numerous lineages of flowering plants, including species with different mating systems and pollinators (transiently deceptive flowers are discussed in the next section). Deception is particularly well known in orchids (Sprengel 1793; Ackerman 1986; Paulus 1988; Nilsson 1992; van der Cingel 1995; Johnson and Nilsson 1999). The possibility of deception often arises from an insect's inability to distinguish between rewarding and nonrewarding flowers until after a visit has begun (e.g., Johnson 2000), which in many flowers is sufficient to result in pollen deposition and/or uptake.

In plants whose male and female functions are performed by separate flowers or floral phases, additional exploitative interactions become possible, including starvation or drowning of pollinators (after they have delivered male gametes) or starvation of pollinator offspring (once parents have transferred gametes). Pollinator "termination" occurs only in female flowers or female-stage flowers because male flowers or male-stage flowers need to export gametes.

For deceptive systems to be stable, the energetic or fitness costs incurred by pollinators when they visit rewardless flowers must be balanced by rewards obtained from other living or dead organisms in the vicinity—usually other flowers, but also fungi, carrion, or other insects. The fossil record of angiosperms flowers, together with molecular phylogenetic trees, indicates that angiosperms started out as insect pollinated, with wind and water pollination independently arising numerous times (Crane et al. 1995; Friis et al. 2001). Most evolutionary transitions from animal pollination to wind or water pollination appear accompanied by loss of floral rewards. Rewardlessness, however, not only became fixed in lineages that left animal pollination, but it is also a transient stage (during anthesis) in many animal-pollinated, normally rewarding flowers.

Difficulties in Recognizing Rewardlessness

Recognizing when one is dealing with permanent rewardlessness is often problematic. First, as already mentioned, flowers can be transiently rewardless because they have been visited and rewards are depleted, or because nectar may be produced only during the female phase (e.g., Delph and Lively 1992), pollen only during the male phase, or because of local conditions such as water stress. In addition, differences in nectar production among flowers, plants, and populations are common (Frankie and Haber 1983; Gilbert et al. 1991; Mitchell and Shaw 1993; López-Portillo et al. 1993; Wilson et al., chap. 3 in this volume), and flowers may or may not replenish nectar rewards that have been depleted (Percival 1965). Such reversible and unpredictable rewardlessness—at least when it comes to nectar—forms the backdrop of the evolutionary theater in which exploitative pollination systems evolve.

A second problem in deciding whether a flower is rewardless is that it depends on the reward being sought; the matter is complicated because of the many kinds of animals and rewards involved. As a reward, pollinators get to (1) lay their eggs into developing seeds or surrounding structures that provide larval hatching and/or feeding sites; (2) feed on pollen, nectar, or tissues; (3) gather nest-building material; (4) obtain pollen or fatty oil to feed their brood; (5) gather liquid fragrances used in mating displays; or (6) find shelter, warmth, or rendezvous places (see Ollerton, chap. 18 in this volume). The mere absence of nectar or pollen (in female flowers) is, thus, insufficient for deciding whether one is dealing with deceptive pollination. Even knowing the species of pollinator is sometimes insufficient information for deciding between a rewarding and a nonrewarding pollination system; a pollinator's sex can influence whether a flower is rewarding or not. Flowers that pretend to be egg-laying sites may deceive female flies, whereas they may reward male flies that seek mating opportunities with females physiologically ready to lay eggs (*Rafflesia, Arum*). Similarly, droplets of nectar or stigmatic secretions offered in trap flowers of Araceae or Aristolochiaceae sustain imprisoned flies, but it was the deceptive odor of seeming oviposition sites that lured the female flies into the kettles (see table 6.1 for references). Thus, in spite of offering floral secretions, the system is exploitative because fly larvae that hatch inside trap flowers die. Yet another example comes from sexually deceptive orchids that differently "harm" their wasp pollinators, depending on the pollinators' sex (Wong and Schiestl 2002).

A third complication is that flowers may exploit their primary pollinator but reward minor or secondary visitors. This, of course, is the expected situation in

Table 6.1 Angiosperms with rewardless flowers

Taxon: sexual system	Reward[a]	Pollinator	Reference
Annonaceae			
Stelechocarpus burahol: monoecious	Pollen (m); no reward (f)	Pollen-feeding beetles	Arroyo and Ganders in Baker 1976
Apocynaceae			
Plumeria rubra: hermaphroditic	All flowers rewardless	Nectar-feeding Sphingidae	Haber 1984
Nerium oleander: hermaphroditic	All flowers rewardless	Nectar-feeding moths and/or bees	Herrera 1991
Aspidosperma quebracho-blanco: hermaphroditic	All flowers rewardless	Nectar-feeding Noctuidae	Lin and Bernardello 1999
Araceae			
Arum maculatum: monoecious	All inflorescences rewardless traps[b]	Ovipositing *Psychoda* and carrion flies	Vogel 1965
Arisaema spp.: monoecious, but all c. 150 species with environmental sex determination	All inflorescences rewardless traps; female inflorescences lethal	Ovipositing mycophagous flies	Vogel 1965, 1978, 1993
Arisarum proboscideum: monoecious	All inflorescences rewardless traps	Ovipositing mycophagous flies	Vogel 1978
Cryptocoryne ciliata: monoecious	All inflorescences rewardless traps	Foraging and ovipositing flies	Vogel 1965
Helicodiceros muscivorum: monoecious	All inflorescences rewardless traps	Foraging and ovipositing carrion flies	Vogel 1965, 1993
Pinellia ternata: monoecious	All inflorescences rewardless traps; female-stage inflorescences lethal	Foraging and ovipositing flies, mainly Ceratopogonidae	Vogel 1965, 1993
Stylochaeton spp.: monoecious	All inflorescences rewardless traps	Foraging and ovipositing carrion beetles	Vogel 1965
Typhonium spp.: monoecious	All inflorescences rewardless traps	Foraging and ovipositing carrion beetles	Van der Pijl 1953; Vogel 1965
Arecaceae			
Attalea colenda: monoecious	Pollen (m); no reward (f)	Pollen-foraging bees	J. P. Feil, personal communication 1995
Attalea funifera: monoecious	Pollen, petals (m); no reward (f)	Feeding Curculionidae, Nitidulidae	Voeks 1988
Geonoma acaulis: monoecious	Pollen (m); no reward (f)	Feeding Curculionidae: Derelomini	Listabarth 1993
Orbignya spectabilis: monoecious	Pollen, brood site (m); no reward (f)	Feeding and ovipositing Curculionidae, Nitidulidae, Staphylinidae	Küchmeister et al. 1993
Mauritia flexuosa: dioecious	Pollen (m); no reward (f)	Feeding Chrysomelidae	Ervik 1993
Phytelephas seemannii: dioecious	Pollen, brood site (m); no reward (f)	Feeding and ovipositing Staphilinidae	Bernal and Ervik 1996
Aristolochiaceae			
Asarum caudatum, A. hartwegii: monoecious	All flowers rewardless	Ovipositing mycophagous flies	Vogel 1978; Mesler and Lu 1993
Aristolochia spp.: hermaphroditic	All flowers rewardless	Feeding and ovipositing flies	Vogel 1965, 1978, 1993
Heterotropa tamaensis: hermaphroditic	All flowers rewardless	Ovipositing mycophagous flies	Sugawara 1988
Asclepiadaceae			
Ceropegia spp.: hermaphroditic	All flowers rewardless[b]	Mate-seeking flies: Milichiidae, also Chloropidae, Ceratopogonidae	Vogel 1961, 1993
Hoodia triebneri: hermaphroditic	All flowers rewardless[b]	Foraging carrion (*Calliphora*) flies	Meve and Liede 1994
Stapelia spp.: hermaphroditic	All flowers rewardless[b]	Foraging and ovipositing carrion flies, also Muscidae	Meve and Liede 1994
Asteraceae			
Antennaria parvifolia: dioecious	Pollen (m); no reward (f)	Pollen-foraging bees	Bierzychudek 1987
Petasites hybridus: often dioecious	Pollen (m); No reward except in one central nectariferous flower (f)	Pollen- or nectar-foraging bees	Knuth 1898–1908, 576

Table 6.1 *(continued)*

Taxon: sexual system	Reward[a]	Pollinator	Reference
Tussilago farfara: monoecious	Pollen, nectar (m); no reward (f)	Pollen- or nectar-foraging bees	Knuth 1898–1908, 576
Begoniaceae			
Begonia boliviensis: monoecious	All flowers nectarless	Nectar-seeking hummingbirds	Vogel 1975, 1993
Begonia sp. (Colombia): monoecious	All flowers nectarless	Nectar-seeking hummingbirds	Kraemer and Schmitt 1991
Begonia ferruginea: monoecious	No nectar (m); nectar (f)	Nectar-seeking hummingbirds	Vogel 1975
Begonia involucrata: monoecious	Pollen (m); no reward (f)	Pollen-foraging *Trigona grandipennis*	Ågren and Schemske 1991
Begonia oaxacana: monoecious	Pollen (m); no reward (f)	Pollen-foraging *Bombus ephippiatus*	Schemske et al. 1996
Berberidaceae			
Podophyllum peltatum: hermaphroditic	All(?) flowers nectarless	Nectar-feeding *Bombus* spp.	Laverty and Plowright 1988
Bignoniaceae			
Lundia obliqua: hermaphroditic	All(?) flowers nectarless	Nectar-feeding bees	Gentry 1974
Tabebuia sp.: hermaphroditic	All(?) flowers nectarless	Nectar-feeding bees	Vogel 1993
Burmanniaceae			
Burmannia sp.: hermaphroditic	All flowers rewardless traps	Foraging and ovipositing carrion beetles	Vogel 1965
Buxaceae			
Sarcococca ruscifolia: monoecious	Pollen, nectar (m); no reward (f)	Pollen- or nectar-foraging bees	Baker 1976; Vogel, personal communication 1993
Campanulaceae			
Lobelia: hermaphroditic	Only some populations nectarless	Nectar-feeding hummingbirds	Brown and Kodric-Brown 1979
Caricaceae			
Carica papaya: monoecious	Pollen, nectar (m); no reward (f)	Nectar-feeding Sphingidae: *Hyles* sp.	Baker 1976
Jacaratia dolichaula: dioecious	Pollen, nectar (m); no reward? (f)	Nectar-feeding Sphingidae	Baker 1976; Bawa 1980
Clusiaceae			
Clusia spp.: dioecious	Pollen (m); no reward (f)	Pollen-foraging *Euglossa, Xylocopa*	Bittrich and Amaral 1996
Clusia criuva: dioecious	Pollen (m); no reward (f)	Pollen-feeding beetles	Rodrigues Correia et al. 1993
Havetiopsis laurifolia: dioecious	Pollen (m); no reward (f)	Pollen-foraging bees	V. Bittrich, personal communication 1993
Tovomita sp.: dioecious	Pollen (m); no reward (f)	Pollen-foraging bees	J. P. Feil, personal communication 1994
Cucurbitaceae			
Ecballium elaterium: monoecious	Pollen, nectar (m); no reward (f)	*Apis mellifera, Andrena florea*	Vogel 1981; Dukas 1987
Gurania and *Psiguria* spp.: monoecious and with environmental sex determination	Pollen, nectar (m); no reward (f)	Pollen-feeding *Heliconius*	Gilbert 1975
Lagenaria sp., *Peponium* sp.: both monoecious, rarely dioecious	Pollen, nectar (m); no reward (f)	Nectar- or pollen-foraging bees	Vogel 1981
Telfairia sp.: dioecious			

(continued)

Table 6.1 *(continued)*

Taxon: sexual system	Reward[a]	Pollinator	Reference
Euphorbiaceae			
Dalechampia subternata: gynodioecious	Pollen (hermaphrodite); no reward (f)	Pollen-foraging bees and/or beetles	Armbruster et al. 1994
Hura crepitans: monoecious	Edible flower (m); no reward (f)	Flower-eating bats	Steiner 1992
Plukenetia volubilis: monoecious	Pollen (m); apparently no reward (f)	Pollen-foraging bees	S. Armbruster personal communication, 1994
Tragia spp.: monoecious	Pollen (m); no reward (f)	Pollen-foraging bees	S. Armbruster personal communication, 1994
Fabaceae			
Prosopis glandulosa: hermaphroditic	Some populations with 54% nectarless individuals	Nectar- and pollen-foraging bees	López-Portillo et al. 1993
Fagaceae			
Castanea vesca: monoecious	Pollen (m); no reward (f)	Pollen-foraging bees	Knuth 1899–1908, 389
Hydnoraceae			
Hydnora africana: hermaphroditic	All flowers rewardless traps	Foraging and ovipositing carrion beetles	Vogel 1965
Iridaceae			
Iris pumila: hermaphroditic	All (?) flowers nectarless	Nectar-feeding bumble bees	Vogel 1993
Malpighiaceae			
Banisteriopsis muricata: hermaphroditic	Some individuals rewardless	Oil-foraging bees: *Epicharis, Centris*	Sazima and Sazima 1989
Banisteriopsis lutea: hermaphroditic	All flowers rewardless	Oil-foraging bees: *Epicharis, Centris*	Sazima and Sazima 1989
Heteropterys aceroides: hermaphroditic	Some individuals rewardless	Oil-foraging bees: *Epicharis, Centris,* four spp. of smaller Exomalopsini	Sazima and Sazima 1989
Malvaceae (including Sterculiaceae)			
Ambroma augusta: hermaphroditic	All flowers rewardless traps	Mate-seeking flies: Milichiidae	Van der Pijl 1953
Moraceae: examples			
Ficus asperifolia: fct. dioecious	Brood site (m syconia/plants); no reward (f syconia/plants)	Ovipositing wasps: *Kradibia gestroi*	Verkerke 1987
Ficus carica: fct. gynodioecious	Brood site (hermaphroditic syconia); no reward (female syconia/plants)	Ovipositing wasps: *Blastophaga psenes*	Valdeyron and Lloyd 1979; Beck and Lord 1988a, 1988b
Ficus fistulosa: dioecious	Brood site (m syconia/plant); no reward (f syconia/plant)	Ovipositing wasps: *Ceratosolen hewittii*	Galil 1973
Myristicaceae			
Myristica fragrans: dioecious	Pollen (m); no reward (f)	Pollen-feeding beetles: Anthicidae	Armstrong and Drummond 1986
Myristica insipida: dioecious	Pollen (m); no reward (f)	Pollen-feeding Curculionidae, Nitidulidae, and Staphylinidae	Armstrong and Irvine 1989
Nymphaeaceae			
Nymphaea capensis: hermaphroditic	Pollen (m); female-stage inflorescences lethal	Pollen-feeding flies, bees, and beetles	Vogel, personal communication 1994
Orchidaceae: reviews and examples			
10,000 spp.	All flowers rewardless	Diverse foraging animals; flies, ants, wasps, bees seeking mates; insects seeking egg-laying sites	Ackerman 1986; Nilsson 1992

Table 6.1 *(continued)*

Taxon: sexual system	Reward[a]	Pollinator	Reference
4800 spp. (1/4 of 19,500)	All flowers rewardless	Foraging animals, mainly bees	Dressler 1993, 222–223
Calopogon spp.: hermaphroditic	All flowers rewardless	Pollen-foraging *Bombus* bees	Firmage and Cole 1988
Cephalanthera spp.: hermaphroditic	All flowers rewardless	Nectar-feeding bees	Dafni and Ivri 1981a; Nilsson 1983b
Calypso bulbosa: hermaphroditic	All flowers rewardless	Nectar-feeding bees	Ackerman 1981; Alexandersson and Ågren 1996
Dactylorhiza spp.: hermaphroditic	All flowers rewardless	Nectar-feeding bees	Nilsson 1980; Lammi and Kuitunen 1995; Kropf and Renner, in press; Gigord et al. 2001
Ophrys spp.: hermaphroditic	All flowers rewardless	Mate-seeking bees or wasps	Paulus and Gack 1990
Orchis spp.: hermaphroditic	All flowers rewardless	Usually nectar-feeding bees	Vogel 1972; Dafni and Ivri 1981b; Nilsson 1983a, 1984; Fritz 1990; Johnson et al. 2003
Thelymitra spp.: hermaphroditic	All flowers rewardless	Pollen-foraging bees	Bernhardt and Burns-Balogh 1986; Dafni and Calder 1987
Rafflesiaceae			
Rafflesia pricei: dioecious	All flowers rewardless	Foraging carrion flies	Beaman et al. 1988
R. kerrii: dioecious	All flowers rewardless[b]	Foraging carrion flies	Bänzinger 1991
Santalaceae			
Osyris alba: dioecious	Pollen, nectar, staminal hairs (m); no reward (f)	Flies: Mycetophilidae, Muscidae; beetles: *Anaspis* sp.	Aronne et al. 1993
Scrophulariaceae			
Mohavea confertiflora: hermaphroditic	All flowers nectarless	Nectar-feeding bees: Halictidae and *Hesperapis* (Melittidae)	Little 1983, 298–299
Siparunaceae			
Siparuna spp.: dioecious	Brood site (m); probably no reward (f)	Ovipositing gall midges: Cecidomyiidae	Feil 1992
Solanaceae			
Solanum polygamum: dioecious	Pollen (m); no reward (f)	Pollen-foraging bees	Knapp et al. 1998

Note: Families are those of the Angiosperm Phylogeny Group (APGII 2003). For Orchidaceae, only a few exemplars are listed.

[a]Male: m; female: f.

[b]Some trap flowers in the Araceae and Rafflesiaceae provide stigmatic secretions that sustain trapped flies and, in the Asclepiadaceae, serve to correctly position the mouthparts of the insects for pollen removal and deposition (see especially Bänzinger 1991; Ollerton, personal communication, 2005).

species undergoing a pollinator shift from, for instance, a nectar-seeking to an ovipositing species, which occurred during evolution of the *Yucca* pollination system (Pellmyr 2003). Mosaics, with some populations being rewarded and others not or less so, have been described from nectariferous *Lobelia* (Brown and Kodric-Brown 1979), nectariferous *Prosopis* (López-Portillo et al. 1993; Golubov et al. 1999), and normally oil-producing *Byrsonima* (Malpighiaceae), in which oil glandless populations exploit oil-seeking anthophorid bees but reward pollen-collecting *Trigona* bees (Sazima and Sazima 1989). In these cases, it is impossible to designate the respective plant species as rewardless or rewarding; instead one has to specify the particular plant morph, plant population, and pollinator species that is exploited. Based on the presently known cases of exploitative pollination (table 6.1), it appears that both flowers with relatively broad pollinator spectra and flowers relying on a single pollinator species can be deceptive (see discussion following). Clearly, however, for floral rewardlessness to become an evolutionary strategy, it must involve pollinators that affect gene flow for the plant in question.

The fourth and final problem in recognizing rewardlessness is that species with unisexual flowers may have one rewarding sex and one more-or-less rewardless one (Willson and Ågren 1989). For example, in *Musa paradisiaca* and *M. velutina,* male flowers produce four times more nectar than do perfect flowers (Percival 1965); in *Oreopanax confusum* (Araliaceae), male flowers produce 12 times more nectar than female flowers (A. Gumbert, personal communication). In other cases, functionally female flowers produce sterile pollen (e.g., *Actinidia chinensis, Decaspermum parviflorum, Rosa setigera, Saurauia veraguensis, Solanum* spp.; Haber and Bawa 1984; Kevan and Lack 1985; Anderson and Symon 1988; Corbet et al. 1988; Kevan et al. 1990). Some workers (Vogel 1975) consider species with sterile pollen deceptive, whereas others (Cane 1993) find that bees actively harvest sterile pollen, probably for its fat-rich pollenkitt coating. Cane found sterile pollen produced by female *Saurauia* flowers to be equivalent in amino acid constitution to fertile pollen of that species.

Phylogenetic Distribution, Evolution, and Kinds of Rewardlessness in the Angiosperms

Table 6.1 lists known cases of rewardlessness in the angiosperms (that is, cases of permanent, not transient, rewardlessness). Dates were taken from individual studies and the reviews by Wiens (1978), Little (1983), Willson and Ågren (1989), Renner and Feil (1993), and Vogel (1975, 1993). Rewardless flowers were defined as flowers that offer neither food nor viable brood sites, construction materials, floral oils, mating partners, or any of the rewards listed at the beginning of this chapter to a flower's primary pollinators (see later discussion for rewards that can or cannot be faked). The table includes rewardless species in 63 genera (not counting orchid genera) from 32 families, together with data on sexual system, kind of rewardlessness, and pollinators.

Rewardlessness has evolved in almost all major groups of angiosperms (APGII 2003 classification), including basal lineages such as Nympheaceae, monocots, Laurales, and Magnoliales. Any estimate of the number of genotypically rewardless species in the angiosperms depends on the assumed number of rewardless orchids. Ackerman (1986) estimated that one-third of the Orchidaceae—that is, between 6500 and 10,000 species—are deceptive in some way, and Dressler

(1993, 223) suggests that there are 4800 deceptive generalized "food flower" mimics (flowers that pretend to offer pollen or nectar but do not resemble a specific model) among the orchids. Assuming that 6500 orchid species and 1000 non-orchid species have genotypically rewardless flowers, the percentage of angiosperms regularly deceiving their pollinators may lie in the vicinity of 3.7%. (Table 6.1 contains a few non-orchid genera that have hundreds of species, only some of them with rewardless populations or morphs, such as *Ficus* and *Begonia.*) This percentage is based on an absolute figure of 222,500 species of flowering plants (Scotland and Wortley 2003) minus 18,000 species that are wind pollinated (Ricklefs and Renner 1994) and 150 that are water pollinated (Cox 1988), leaving 204,350 animal-pollinated angiosperms of which 7500 may be deceptive. Vogel (1993), based on a lifetime of study of different floral rewards, estimated that 13,000, or 6%, of angiosperms have deceptive flowers.

Most of the cases listed in table 6.1 represent independent origins of rewardlessness because they involve phylogenetically distant taxa that are embedded in clades that are otherwise rewarding. A few lineages that are rewardless are species rich. For example, *Aristolochia, Arisaema,* and the orchid genera *Eulophus* and *Epidendrum* each have over 100 species, most or all of which are probably frauds. Thus, deceptive strategies may promote diversification within reward-labile lineages or may at least enable shifts away from resource-intensive food-based pollination systems (comparable to the shift away from a nectar reward system that occurred during the evolution of *Yucca;* see preceding discussion).

Genera in which there is evidence of repeated parallel evolution of rewardlessness are *Begonia, Banisteriopsis, Clusia,* and *Lobelia* (table 6.1). *Begonia,* with some 1475 species (W. Goodall-Copestake, personal communication), comprises numerous species in which all flowers are nectarless and rewardless (table 6.1; pollen is not a reward for the hummingbirds that pollinate the respective species) and at least one species with rewardless male flowers, but rewarding female flowers (Vogel 1975). The Malpighiaceae *Banisteriopsis muricata* has single populations that contain empty and rewarding individuals (at a ratio of 1:2), whereas *B. lutea* is entirely rewardless (Sazima and Sazima 1989). *Lobelia cardinalis* (Campanulaceae) a hummingbird-pollinated species, has rewardless and nectariferous populations (Brown and Kodric-Brown 1979). Yet another species with labile rewardlessness is *Himantoglossum hircinum* (Orchidaceae), contradicting Smithson and Gigord's (2003) assertion that polymorphism for the presence or absence of nectar production is unknown in orchids. Its floral spurs are usually completely dry, but nectar-producing flowers have been reported from at least two populations in southern Germany and Austria (Teschner 1980; Vöth 1999; M. Nickol, personal communication).

Reversion from being rewardless to being rewarding may have occurred in *Disa* (Johnson et al. 1998), *Arisaema* (Renner et al. 2004), and *Arum* (A. Diaz, personal communication). *Arum* is normally pollinated by psychodid midges temporarily trapped in its foul-smelling kettles, but *Arum creticum* is sweet smelling, provides abundant pollen, and is pollinated by pollen-collecting female bees. The converse is likely the case in *Nymphaea,* in which one species has lethal (female-stage) trap flowers, whereas the remaining species apparently all provide pollen as a reward for beetles or flies. In *Arisaema,* female plants are lethal traps for fly pollinators and their offspring (e.g., Vogel 1965; Vogel and Martens 2000), but a recently discovered subspecies of *A. flavum* that is nested high in a molecular phylogenetic tree of *Arisaema* offers nectar in its kettles (Renner et al. 2004).

Based on table 6.1, two kinds of exploitative pollination systems may be distinguished: those that involve deception by one sexual phase or morph and may rely mainly on the other phase or morph to reward the pollinator(s), and those that involve complete rewardlessness and, therefore, rely on heterospecific flowers or other biological or nonbiological entities to keep the pollinator in the general vicinity. All antagonistic pollination systems in which flowers kill pollinators or pollinator offspring belong to the second kind. Polymorphic exploitative systems come in many shades; the relative frequencies of rewarding and nonrewarding flowers probably differ among species with different sexual systems (perfect-flowered, monoecious, dioecious, or gynodioecious, all of which include rewardless species; table 6.1). Although the distinction between monomorphic and polymorphic exploitative flower populations at first may appear of interest only to the botanist, the two kinds of systems pose different challenges for the pollinator. A monomorphic situation (all plant individuals equally rewardless) makes it possible to learn the location of deceptive patches and to avoid them. In a polymorphic situation, where an unpredictable fraction of conspecific (and, thus, morphologically similar) individuals or flowers is deceptive, the cost of learning the reward status of flowers may exceed the cost of mistake visits (e.g., Schemske et al. 1996).

For rewardless orchids, population structure has been seen as causally linked to the evolution of rewardlessness (Ackerman 1986). With plants few and far between, as is typical of many orchids, visitation frequencies may be low under the best of circumstances and may not differ much between rewarding plants and rewardless mutants. A second way in which low population density and the evolution of deceptive pollination may be linked is that pollinators cannot afford to depend on rare flowers for food (but they probably can for other needs, such as for obtaining fragrances). Visits to rare flowers will always be interspersed with visits to more abundant flowers. This poses a problem for plants with loose pollen that may be lost on heterospecific stigmas during such visits, and there should be strong selection in such species to increase pollinator faithfulness and thereby conspecific pollen deposition—for example, by offering a reward. In orchids, the danger of pollen loss is reduced via pollen packaging and precise emplacement, and orchids' pollinia or pollinaria remain available for deposition on the "correct" stigma even after numerous heterospecific flowers have been visited for food. As long as pollinaria are fairly long lived, orchids may suffer little from "sending" disappointed pollinators off to find food elsewhere; they may gain genetic benefits from reduced pollination of neighboring flowers on the same plant (Johnson and Nilsson 1999).

An orchid for which the effect of rewardlessness might be studied under natural conditions is *Himantoglossum hircinum,* in which some populations offer nectar while the remainder may be more or less rewardless (e.g., slight nectar production along one spur furrow—see Teschner 1980; nectar for short-tongued visitors on papillae at the spur entrance, but deception of long-tongued visitors—see Vöth 1999; Pridgeon et al. 2001). The effects of experimental nectar supplementation on the reproductive success of rewardless orchids are often negative (e.g., Ackerman 1981; Smithson and Gigord 2001; Smithson 2002; but see Johnson and Nilsson 1999 and Johnson et al. 2004), throwing doubt on the strength of the positive correlation between male fitness and a flower's investment in nectar, at least in orchids.

Unisexuality and Rewardlessness

Perfect-flowered species predominate among rewardless angiosperms because orchids are the single most important rewardless group, and Orchidaceae are predominantly perfect-flowered (The few species of orchids that have unisexual flowers all offer floral fragrances in both sexes.) Other perfect-flowered rewardless species occur in Apocynaceae, Aristolochiaceae, Bignoniaceae, Campanulaceae (or Lobeliaceae), Iridaceae, Fabaceae, Malpighiaceae, Scrophulariaceae, and Sterculiaceae. However, all remaining rewardless angiosperms have unisexual flowers and are monoecious, gynodioecious, or dioecious (table 6.1). In dioecious species, rewardless female flowers are many times more frequent than rewardless males (in a ratio of 29 to 1, counting only one species per genus and not counting figs and *Siparuna* to avoid problems of phylogenetic independence). Only one species, *Begonia ferruginea,* has rewarding female and rewardless male flowers (Vogel 1975).

These results agree with the assessment of Willson and Ågren (1989) that, among species with unisexual flowers, rewardlessness is more frequent in female flowers. Willson and Ågren suggested that gender-specific limitations on reproductive success and risk of damage to the (more) rewarding morph might determine which floral morph should be rewarding. Thus, plants that offer floral egg-laying sites or entire flowers as food may need to protect their ovaries from destruction by making female flowers less attractive (i.e., unrewarding); this seems to apply to *Siparuna* and figs (table 6.1). In dioecious species, female plants may also be more resource limited and thus under stronger selection to economize resources by not offering nectar. In contrast, male individuals are selected to maximize pollen dispersal by repeatedly rewarding mutualists, thereby attracting more visits, and so dispense more pollen (Bell 1985; Harder and Thomson 1989; Selten and Shmida 1991; Mitchell 1993). Conversely, female fitness may sometimes be saturated with a few visits.

Whether rewardlessness evolves more readily in species with unisexual flowers than in perfect-flowered species is difficult to assess because available data are biased by the ±6500 rewardless perfect-flowered orchids whose precise phylogenetic relationships are unknown, which prevents comparisons of minimal numbers of independent acquisition in orchids versus non-orchids. Species with unisexual flowers, which predominantly are tropical trees and climbers (Ricklefs and Renner 1994), are underrepresented among plants whose rewards and pollinators are known, whereas (rewardless) terrestrial European orchids have received comparatively greater attention.

Kinds of Rewards and Pollinators Involved in Deceptive Pollination Systems

Of the major kinds of rewards known in the angiosperms—mating and oviposition sites, pollen, nectar, nest-building materials, fatty oils, fragrant oils, shelter, and warmth—a few do not lend themselves to faking. Thus, none of the rewardless flowers (table 6.1) pretends to offer resins, shelter, or liquid fragrances (which would be collected by perfume-seeking male Euglossine bees).

In terms of the kinds of pollinators that are being deceived, flies, moths, beetles, wasps, and solitary or hemisocial bees predominate. Hummingbirds are involved in two cases (*Begonia* spp., *Lobelia* sp.) and bats in one (*Hura crepitans*). It is not clear, however, whether bees, wasps, and flies are more easily deceived than birds and bats or whether the numbers simply reflect the scarcity of bird and bat pollination in the angiosperms in general; fewer than 1% of flowering plants are adapted for bird pollination. Honeybees, which maximize food intake by concentrating on the food source with the best cost-to-reward ratio (Greggers and Menzel 1993), may also be rare pollinators of deceptive flowers (table 6.1), yet *Ecballium elaterium* (Cucurbitaceae) appears to be regularly pollinated by *Apis mellifera.* However, the solitary bee *Andrena florea* is a more effective pollinator of *Ecballium* (Dukas 1987).

Insect Cognition and the Evolution and Persistence of Rewardlessness

Theoretical biologists have decried the lack of data on the occurrence and frequency of rewardless flowers (e.g., Bell 1986; Gilbert et al. 1991; Selten and Shmida 1991) or have pointed out that, besides orchids, "complete rewardlessness is rarely recorded in other plant families" (Smithson and Gigord 2003), where the intended stress is possibly not quite correct; it should be on *recorded.* Theory indicates that nonrewarding plant individuals should constantly threaten to invade pollinator–flower mutualisms but be held in check by (1) discrimination by pollinators against rewardless flowers and (2) strong positive correlations between pollen export (male fitness) and a flower's investment in pollinator rewards (Bell 1986; Pleasants and Chaplin 1983; Ågren et al. 1986; Mitchell 1993). (As noted earlier, the effects of experimental nectar supplementation on the reproductive success of rewardless orchids so far are equivocal, but sample sizes are small.)

In the face of these two factors limiting the spread of rewardlessness, how can we explain its repeated evolution and persistence? One approach is to focus on pollinator behavior and cognition—specifically on the question of why stronger discrimination against rewardlessness does not appear to have evolved. A key to answering this question may be the continuum, already stressed at the beginning of this chapter, between transiently and permanently rewardless populations, individuals, and flowers. Insect pollinators are likely to encounter rewardless flowers during most foraging bouts. As an evolutionary result of this, bumblebees and honeybees, for example, learn negative stimuli (nonrewarding visits) more slowly than they do positive stimuli (Menzel and Greggers 1992). (Little seems to be known about effects of flower rewardlessness on the behavior of flies, moths, or beetles.) In addition, when bumblebees must divide attention among an increasing number of rewarding and nonrewarding flower types, their ability to discriminate decreases rapidly (Dukas and Real 1993). Bees usually do not learn the position of individual flowers (most of which fade after a day) but instead memorize the location of rewarding and nonrewarding plants (Menzel 1985). As discussed earlier (Phylogenetic Distribution, Evolution, and Kinds of Rewardlessness in the Angiosperms), polymorphic and monomorphic deceptive systems, therefore, pose different learning challenges.

Naturalists and theoretical ecologists alike often assume that rewardless plants rely on naïve pollinators, such as recent emergents or immigrants that have not yet learned to avoid empty flowers (Vogel 1975, 1993; Ackerman 1981; Little 1983; Dafni 1984; Herrera 1991; Dressler 1993, 222–223; Ferdy et al. 1998; Smithson 2002; Smithson and Gigord 2003). An almost paradigmatic idea is that "most rewardless orchids are pollinated by naïve pollinators searching for nectar" (Smithson and Gigord 2003). This, however, involves two problems: first, naïve pollinators quickly become experienced pollinators, in the case of some bee species after a single rewarding visit; second, behavioral studies show that it is precisely a bee's individual foraging experience that determines which rewardless flowers it is likely to visit.

The naïve pollinator idea, therefore, needs modification. Although it may hold for northern European spring orchids, it is unlikely to apply to the thousands of rewardless tropical orchids pollinated by long-lived bees (such as euglossines) or the 1000 or so rewardless other taxa that do not all flower right when bees emerge. Even in the spring-blooming rewardless orchid *Dactylorhiza sambucina,* pollinia export proportions of flowers higher up on the inflorescence, which open late in the season, can be significantly higher than those of flowers further down, which opened early in the spring (Kropf and Renner

2005). However, visitation rates to this species vary widely, depending on locality and year, and some individual bumblebees may indeed learn to avoid flowers (Nilsson 1980; Vogel 1993; Kropf and Renner 2005). In *Orchis dinsmorei,* too, the pollination rate of 70% in its first week of flowering decreased to 10% in the final week (Dafni 1986). These contrasting results are probably due to interacting factors at work at different times, most important among them the density of other, rewarding flowers; yearly variation in the emergence of bees; and for species with color morphs, such as *D. sambucina,* the frequencies of yellow and purple morphs, with the more common morph being avoided more rapidly (Gigord et al. 2001). A study of *Anacamptis* (= *Orchis*) *morio* found that even experienced *Bombus lapidarius* queens continued to probe flowers as long as these empty flowers were rare (Johnson et al. 2003). Strikingly, queens carrying pollinia and thus obvious return visitors to *A. morio* were more likely to probe flowers than queens not carrying pollinia!

Other experimental setups, using potted plants in the field, found visitation rates in rewardless orchids to depend on not only relative but also absolute abundances of empty and rewarding flowers, which in turn affected pollinator density in a patch and movement patterns between patches (see Johnson et al. 2003 and references therein). An experiment that directly addresses the influence of an individual bee's experience on visitation rates of food-deceptive orchids is that of Gumbert and Kunze (2001). They let initially flower-naïve bumblebees forage on either of two rewarding species that differentially matched two simultaneously presented food-deceptive species. Experiments were repeated with different species ratios, and resemblance of the four species to each other was assessed in terms of shape, size, smell, and color according to a model of bumblebee vision. In each experiment, bumblebees approached and visited predominantly that orchid that more closely resembled the nectar-providing species it had visited just before. If being naïve were an important factor, one would have expected the bees to visit deceptive flowers in proportion to their abundance in the experimental populations; instead, visitation rates depended on previous experience. Moreover, the experiment showed that size, shape, and smell were irrelevant compared to color.

As has often been pointed out, flower similarity and flower mimicry need to be assessed in the context of the sensory abilities of the relevant pollinators (Nilsson 1980, 1983a, 1983b, 1984; Ackerman 1981; Dafni and Ivri 1981a, 1981b; Dafni 1984, 1986; Bernhardt and Burns-Balogh 1986; Fritz 1990; Alexandersson and Ågren 1996). However, measuring color and scent involves expensive and sophisticated technical setups, and available data on flower parameters relevant to deception are few. In the aforementioned experiment of Gumbert and Kunze (2001), pollination success of the food-deceptive orchid species *Orchis boryi* and *O. provincialis* correlated strongly with degree of similarity in color to particular rewarding species known to individual pollinators. This implies a more generalized form of mimicry than the traditionally envisioned imitation by a rewardless mimic of key characteristics of a rewarding model. Instead, "purple color" in the mimic may be sufficient to deceive many experienced bumblebees, which in Europe, at least, are bound to have handled numerous purple, richly nectariferous flowers. Choice among similarly colored flowers, once bees are experienced, is controlled by generalization from previously learned colors (Gumbert 2000; Gigord et al. 2002; Johnson et al. 2003). In another experiment that used artificial rewarding and nonrewarding flowers, experimentally adding scent cues to the rewarding flowers improved the speed with which bumblebees (*Bombus terrestris*) learned to discriminate against the rewardless mimic, regardless of whether the mimic had a different scent or was scentless (Kunze and Gumbert 2001). Discrimination of the mimic was poorest if it had the same scent as the model, demonstrating a potential for scent mimicry. Taken together, all these results imply that the best strategy for a rewardless flower growing together with several potential models would be to have a common flower color and no scent.

To summarize this section, bees need experience with rewarding flowers to develop expectations about signal-reward relationships (Greggers and Menzel 1993); the firmer the expectations and entrained behavioral sequences, the easier they may be exploited. Part of each bee's foraging experience will have involved visits to empty flowers, and this background of transient rewardlessness—to which pollinators must respond by flexible foraging (not by complete avoidance of flowers, plants, or populations)—probably explains why pollinators do not discriminate more strongly against rewardless species. They may not be able to afford strong discrimination, being constantly faced with a temporally and spatially varying kaleidoscope of rewarding and almost empty flowers. Moreover, flower discrimination itself involves costs such as extended decision times, reduced flight speed, or longer flights (as demonstrated by Smithson and Gigord 2003).

Deceptive Pollination and Ecological and Evolutionary Specialization

Deciding whether a pollination interaction is ecologically specialized (i.e., involves only a few species of pollinators; Armbruster et al. 2000) requires long-term observations and, importantly, experiments (because visitor lists can be misleading about the effectiveness of visitors as pollinators). To my knowledge, such data are unavailable for most rewardless species, and there are no sister taxon studies where one sister would be rewarding, the other rewardless (see Phylogenetic Distribution, Evolution, and Kinds of Rewardlessness in the Angiosperms for genera in which such pairs could be found). The dearth of knowledge about most pollinators' foraging breaths so far also precludes an answer to the question whether oligolectic or polylectic foragers are more likely to be deceived. One might argue, however, that successful deception requires that a flower closely match the sensory abilities and foraging behaviors of the deceived animal species. This might place an upper limit on the number of pollinator types that can be deceived by any one species of plant; alternatively, selection might favor rewardless morphs capable of pollination by a broader range of visitors/pollinators. This may be the case in some species with unisexual flowers in which the female morph is rewardless (lacking pollen) but apparently visited by a relatively broad range of animals; an example may be *Osyris* (Santalaceae), which is pollinated by beetles and species from two families of flies.

What about evolutionary specialization—that is, the process of evolving in the direction of increasing specialization (Armbruster et al. 2000)? From molecular phylogenies it is clear that lineages with fused flower parts, which are the ones most likely to offer nectar in spurs or similar deep containers, are among the most derived angiosperms. These may be the very lineages in which flowers can falsely advertise nectar because pollinators often cannot gauge the content of deep nectaries without actually probing them (some nectars, however, are scented, and some pollinators are able to use the scent of nectar to detect its presence or absence; Marden 1984; Raguso 2004 and references therein). In this broad sense, evolutionary flower specialization (i.e., fusion of parts) may facilitate deception where the reward is nectar; however, where nectar is not the sought reward, no phylogenetic clustering of rewardlessness among derived angiosperms is apparent (see Phylogenetic Distribution, Evolution, and Kinds of Rewardlessness in the Angiosperms). Earlier conclusions that "floral deception should be regarded as a recent development, which appears mainly in highly evolved families" (Dafni 1984) are possibly too orchid-centered. Obviously, deception can be an ancient or recent development in basal angiosperm families such as Nympheaceae, monocots, Laurales, and Magnoliales (table 6.1).

Conclusions and Outlook

This review suggests that rewardlessness has evolved in at least 7500 species in 63 genera and 32 families of angiosperms. Results of experiments as well as natural history data suggest that pollinators need not be naïve to be deceived. Instead insects, especially bees, develop expectations about floral syndromes and their typical rewards (i.e., they generalize in the behavioral sense of the word), leading them to judge by appearance and to rely on a few simple cues. Close resemblance between species is thus not necessary to understand the persistence of rewardlessness. Scenarios that paint the evolution of rewardlessness in a world of reliably rewarding flowers (Ferdy et al. 1998; Smithson and Gigord 2003) may posit a false contrast that makes it unnecessarily difficult to understand why pollinators have failed to evolve stronger remote discrimination against cheaters. Not only is it unclear how pollinators might reliably achieve remote discrimination against cheating flowers, but it is also unclear that they would benefit from strong discrimination, given that so many flowers are transiently rewardless. Transiently rewardless flowers indeed may have facilitated the evolution of rewardlessness as an evolutionary stable strategy because their presence lowers the strength of selection on pollinators to consistently discriminate against lack of reward.

Acknowledgments

I thank Steven Johnson, Nick Waser, Irene Terry, and three anonymous reviewers for their comments.

References

Ackerman, J. D. 1981. Pollination biology of *Calypso bulbosa* var. *occidentalis* (Orchidaceae): A food-deception system. Madroño 28: 101–110.

Ackerman, J. D. 1986. Mechanisms and evolution of food-deceptive pollination systems in orchids. Lindleyana 1: 108–113.

Ågren, J., T. A. Elmquist, and A. Tunlid. 1986. Pollination by deceit, floral sex ratios and seed set in dioecious *Rubus chamaemorus* L. Oecologia 70: 332–338.

Ågren, J., and D. W. Schemske. 1991. Pollination by deceit in a neotropical monoecious herb, *Begonia involucrata.* Biotropica 23: 235–241.

Alexandersson, R., and J. Ågren. 1996. Population size, pollinator visitation and fruit production in the deceptive orchid *Calypso bulbosa.* Oecologia 107: 533–540.

Anderson, G. J., and D. E. Symon. 1988. Insect foragers on *Solanum* flowers in Australia. Annals of the Missouri Botanical Garden 75: 842–852.

APGII 2003. An update of the Angiosperm Phylogeny Group classification for the orders and families of flowering plants. Botanical Journal of the Linnean Society 141: 399–436.

Armbruster, W. S., M. E. Edwards, J. F. Hines, R. L. A. Manhunnah, and P. Munyenyembe. 1994. Pollination ecology. National Geographic Research and Exploration 9: 460–474.

Armbruster, W. S., C. B. Fenster, and M. R. Dudash. 2000. Pollination "principles" revisited: Specialization, pollination syndromes, and the evolution of flowers. Det Norske Videnskaps—Akademi. I. Matematisk Naturvidenskapelige Klasse, Skrifter, Ny Serie 39: 179–200.

Armstrong, J. E., and B. A. Drummond III. 1986. Floral biology of *Myristica fragrans* Houtt. (Myristicaceae), the nutmeg of commerce. Biotropica 18: 32–38.

Armstrong, J. E., and A. K. Irvine. 1989. Floral biology *of Myristica insipida* (Myristicaceae), a distinctive beetle pollination syndrome. American Journal of Botany 76: 86–94.

Aronne, G., C. C. Wilcock, and P. Pizzolongo. 1993. Pollination biology and sexual differentiation of *Osyris alba* (Santalaceae) in the Mediterranean region. Plant Systematics and Evolution 188: 1–16.

Baker, H. G. 1976. "Mistake" pollination as a reproductive system with special reference to the Caricaceae. Pp. 161–169 *in* J. Burley and B. T. Styles (eds.), Tropical trees: Variation, breeding system, and conservation. Academic Press, London.

Bänzinger, H. 1991. Stench and fragrance: Unique pollination lure of Thailand's largest flower, *Rafflesia kerrii* Meijer. Natural History Bulletin of the Siam Society 39: 19–52.

Bawa, K. S. 1980. Mimicry of male by female flowers and intrasexual competition for pollinators in *Jacaratia dolichaula* (D. Smith) Woodson (Caricaceae). Evolution 34: 467–474.

Beaman, R. S., P. J. Decker, and J. H. Beaman. 1988. Pollination of *Rafflesia* (Rafflesiaceae). American Journal of Botany 75: 1148–1162.

Beck, N. G., and E. M. Lord. 1988a. Breeding system in *Ficus carica,* the common fig. I. Floral diversity. American Journal of Botany 75: 1904–1912.

Beck, N. G., and E. M. Lord. 1988b. Breeding system in *Ficus carica,* the common fig. II. Pollination events. American Journal of Botany 75: 1913–1922.

Bell, G. 1985. On the function of flowers. Proceedings of the Royal Society of London B 224: 223–265.

Bell, G. 1986. The evolution of empty flowers. Journal of Theoretical Biology 118: 253–258.

Bernal, R., and F. Ervik. 1996. Floral biology and pollination of the dioecious palm *Phytelephas seemannii* in Colombia: An adaptation to staphylinid beetles. Biotropica 28: 682–696.

Bernhardt, P., and P. Burns-Balogh. 1986. Floral mimesis in *Thelymitra nuda* (Orchidaceae). Plant Systematics and Evolution 151: 187–202.

Bierzychudek, P. 1987. Pollinators increase the cost of sex by avoiding female flowers. Ecology 68: 444–447.

Bittrich, V., and M. C. E. Amaral. 1996. Flower morphology and pollination biology of *Clusia* species from the Gran Sabana (Venezuela). Kew Bulletin 51: 681–694.

Brown, J. H., and A. Kodric-Brown. 1979. Convergence, competition, and mimicry in a temperate community of hummingbird-pollinated flowers. Ecology 60: 1022–1035.

Cane, J. H. 1993. Reproductive role of sterile pollen in *Saurauia* (Actinidiaceae), a cryptically dioecious neotropical tree. Biotropica 25: 493–495.

Chittka, L., and J. D. Thomson. 2001. Cognitive ecology of pollination. Cambridge University Press, Cambridge.

Corbet, S. A., H. Chapman, and N. Saville. 1988. Vibratory pollen collection and flower form: Bumble-bees on *Actinidia, Symphytum, Borago,* and *Polygonatum.* Functional Ecology 2: 147–155.

Cox, P. A. 1988. Hydrophilous pollination. Annual Review of Ecology and Systematics 19: 261–280.

Crane, P. R., E. M. Friis, and K. R. Pedersen. 1995. The origin and early diversification of angiosperms. Nature 374: 27–33.

Crepet, W. L. 1984. Advanced (constant) insect pollination mechanisms: Pattern of evolution and implication vis-a-vis angiosperm diversity. Annals of the Missouri Botanical Garden 71: 607–630.

Dafni, A. 1984. Mimicry and deception in pollination. Annual Review of Ecology and Systematics 15: 259–278.

Dafni, A. 1986. Pollination in *Orchis* and related genera: Evolution from reward to deception. Pp. 81–104 *in* J. Arditti (ed.), Orchid biology. Cornell University Press, Ithaca, NY.

Dafni, A., and D. M. Calder. 1987. Pollination by deceit and floral mimesis in *Thelymitra antennifera* (Orchidaceae). Plant Systematics and Evolution 158: 11–22.

Dafni, A., and Y. Ivri. 1981a. The flower biology of *Cephalanthera longifolia* (Orchidaceae): Pollen imitation and facultative floral mimicry. Plant Systematics and Evolution 137: 229–240.

Dafni, A., and Y. Ivri. 1981b. Floral mimicry between *Orchis israelitica* Baumann and Dafni (Orchidaceae) and *Bellevalia flexuosa* Boiss (Liliaceae). Oecologia 49: 229–232.

Delph, L. F., and C. M. Lively. 1992. Pollinator visitation, floral display, and nectar production of the sexual morphs of a gynodioecious shrub. Oikos 63: 161–170.

Dressler, R. L. 1993. Phylogeny and classification of the orchid family. Dioscorides Press, Portland, Oregon.

Dukas, R. 1987. Foraging behavior of three bee species in a natural mimicry system: Female flowers which mimic male flowers in *Ecballium elaterium* (Cucurbitaceae). Oecologia 74: 256–263.

Dukas, R., and L. A. Real. 1993. Cognition in bees: From stimulus reception to behavioral change. Pp. 343–373 *in* D. R. Papaj and A. C. Lewis (eds.), Insect learning. Chapman and Hall, London.

Ervik, F. 1993. Notes on the phenology and pollination of the dioecious palms *Mauritia flexuosa* and *Aphandra natalia* in Ecuador. Pp. 7–12 *in* W. Barthlott, C. M. Naumann, K. Schmidt-Loske, and K.-L. Schuchmann (eds.), Animal–plant interactions in tropical environments, 7–12. Zoologisches Forschungsinstitut und Museum Alexander König, Bonn, Germany.

Feil, J. P. 1992. Reproductive ecology of dioecious *Siparuna* (Monimiaceae) in Ecuador: A case of gall midge pollination. Botanical Journal of the Linnean Society 110: 171–203.

Ferdy, J.-B., P.-H. Gouyon, J. Moret, and B. Godelle. 1998. Pollinator behavior and deceptive pollination: Learning process and floral evolution. American Naturalist 152: 696–705.

Firmage, D. H., and F. R. Cole. 1988. Reproductive success and inflorescence size of *Calopogon tuberosus* (Orchidaceae). American Journal of Botany 75: 1371–1377.

Frankie, G. W., and W. A. Haber. 1983. Why bees move among mass-flowering Neotropical trees. Pp. 360–372 *in* C. E. Jones and R. J. Little (eds.), Handbook of experimental pollination biology. Van Nostrand Reinhold, New York.

Friis, E. M., K. R. Pedersen, and P. R. Crane. 2001. Origin and radiation of angiosperms. Pp. 97–102 *in* D. E. G. Briggs and P. R. Crowther (eds.), Palaeobiology II. Blackwell Science, Oxford.

Fritz, A. L. 1990. Deceit pollination of *Orchis spitzelii* (Orchidaceae) on the Island of Gotland in the Baltic: A suboptimal system. Nordic Journal of Botany 9: 577–587.

Galil, J. 1973. Pollination in dioecious figs: Pollination of *Ficus fistulosa* by *Ceratosolen hewitti*. Garden Bulletin (Singapore) 26: 303–311.

Gentry, A. H. 1974. Coevolutionary patterns in Central American Bignoniaceae. Annals of the Missouri Botanical Garden 61: 728–759.

Gigord, L. D. B., M. R. Macnair, and A. Smithson. 2001. Negative frequency-dependent selection maintains a dramatic flower color polymorphism in the rewardless orchid *Dactylorhiza sambucina* (L.) Soò. Proceedings of the National Academy of Sciences (USA) 98: 6253–6255.

Gigord, L. D. B., M. R. Macnair, M. Stritesky, and A. Smithson. 2002. The potential for floral mimicry in rewardless orchids: An experimental study. Proceedings of the Royal Society of London B 269: 1389–1395.

Gilbert, F. S., N. Haines, and K. Dickson. 1991. Empty flowers. Functional Ecology 5: 29–39.

Gilbert, L. E. 1975. Ecological consequences of a coevolved mutalism between butterflies and plants. Pp. 210–240 *in* L. E. Gilbert and P. H. Raven (eds.), Coevolution in animals and plants. University of Texas Press, Austin, TX.

Golubov, J., L. E. Eguiarte, M. C. Mandujano, J. Lopez-Portillo, and C. Montaña. 1999. Why be a honeyless honey mesquite? Reproduction and mating system of nectarful and nectarless individuals. American Journal of Botany 86: 955–963.

Greggers, U., and R. Menzel. 1993. Memory dynamics and foraging strategies of honeybees. Behavioral Ecology and Sociobiology 32: 17–29.

Gumbert, A. 2000. Color choices by bumble bees (*Bombus terrestris*): Innate preferences and generalization after learning. Behavioral Ecology and Sociobiology 48: 36–43.

Gumbert, A., and J. Kunze. 2001. Colour similarity to rewarding model plants affects pollination in a food deceptive orchid, *Orchis boryi*. Biological Journal of the Linnean Society 72: 419–433.

Haber, W. A. 1984. Pollination by deceit in a mass-flowering tropical tree, *Plumeria rubra* L. (Apocynaceae). Biotropica 16: 269–275.

Haber, W. A., and K. S. Bawa. 1984. Evolution of dioecy in *Saurauia* (Dilleniaceae). Annals of the Missouri Botanical Garden 71: 289–293.

Harder, L. D., and J. D. Thomson. 1989. Evolutionary options of maximizing pollen dispersal of animal-pollinated plants. American Naturalist 133: 323–244.

Herrera, J. 1991. The reproductive biology of a riparian Mediterranean shrub, *Nerium oleander* L. (Apocynaceae). Botanical Journal of the Linnean Society 106: 147–172.

Herrera, J. 2005. Plant generalization on pollinators: species property or local phenomenon. American Journal of Botany 92: 13–20.

Johnson, S. D. 2000. Batesian mimicry in the non-rewarding orchid *Disa pulchra*, and its consequences for pollinator behaviour. Biological Journal of the Linnean Society 71: 119–132.

Johnson, S. D., H. P. Linder, and K. E. Steiner. 1998. Phylogeny and adaptive radiation of pollination systems in *Disa* (Orchidaceae). American Journal of Botany 85: 402–411.

Johnson, S. D., and L. A. Nilsson. 1999. Pollen carryover, geitonogamy, and the evolution of deceptive pollination systems in orchids. Ecology 80: 2607–2619.

Johnson, S. D., C. I. Peter, and J. Ågren. 2004. The effects of nectar addition on pollen removal and geitonogamy in the non-rewarding orchid *Anacamptis morio*. Proceedings of the Royal Society of London B 271: 803–809.

Johnson, S. D., C. I. Peter, L. A. Nilsson, and J. Ågren. 2003. Pollination success in a deceptive orchid is enhanced by co-occurring rewarding magnet species. Ecology 84: 00–00.

Kevan, P. G., D. Eisikowitch, J. D. Ambrose, and J. R. Kemp. 1990. Cryptic dioecy and insect pollination in *Rosa setigera* Michx. (Rosaceae), a rare plant of Carolinian Canada. Biological Journal of the Linnean Society 40: 229–243.

Kevan, P. G., and A. J. Lack. 1985. Pollination in a cryptically dioecious plant *Decaspermum parviflorum* (Lam.) A. J. Scott (Myrtaceae) by pollen-collecting bees in Sulawesi, Indonesia. Biological Journal of the Linnean Society 25: 319–330.

Knapp, S., V. Persson, and S. Blackmore. 1998. Pollen morphology and functional dioecy in *Solanum* (Solanaceae). Plant Systematics and Evolution 210: 113–139.

Knuth, P. 1898–1908. Handbuch der Blütenbiologie, vols. I–III. W. Engelmann, Leipzig, Germany.

Kraemer, M., and U. Schmitt. 1991. Interaktionen zwischen andinen Kolibris (Trochilidae) und ihren

Nahrungspflanzen im Paramó de los Dominguez, Cordillera Central, Colombia. MS thesis, University of Bonn, Bonn, Germany.
Kropf, M., and S. S. Renner. 2005. Pollination success in monochromic yellow populations of the rewardless orchid *Dactylorhiza sambucina*. Plant Systematics and Evolution 254 (3): 185–198.
Küchmeister, H., G. Gottsberger, and I. Silberbauer-Gottsberger. 1993. Pollination biology of *Orbignya spectabilis*, a "monoecious" Amazonian palm. Pp. 67–76 *in* W. Barthlott, C. M. Naumann, K. Schmidt-Loske, and K.-L. Schuchmann (eds.), Animal–plant interactions in tropical environments. Zoologisches Forschungsinstitut und Museum Alexander König, Bonn, Germany.
Kunze, J., and A. Gumbert. 2001. The combined effect of color and odor on flower choice behavior of bumble bees in flower mimicry systems. Behavioral Ecology 12: 447–456.
Lammi, A., and M. Kuitunen. 1995. Deceptive pollination of *Dactylorhiza incarnata:* An experimental test of the magnet species hypothesis. Oecologia 101: 500–503.
Laverty, T. M., and R. C. Plowright. 1988. Fruit and seed set in mayapple (*Podophyllum peltatum*): Influence of intraspecific factors and local enhancement near *Pedicularis canadensis*. Canadian Journal of Botany 66: 173–178.
Lin, S., and G. Bernardello. 1999. Flower structure and reproductive biology in *Aspidosperma quebracho-blanco* (Apocynaceae), a tree pollinated by deceit. International Journal of Plant Sciences 160: 869–878.
Listabarth, C. 1993. Pollination in *Geonoma macrostachys* and three congeners, *G. acaulis, G. gracilis,* and *G. interrupta*. Botanica Acta 106: 496–506.
Little, R. J. 1983. A review of floral food deception mimicries with comments on floral mutualism. Pp. 294–309 *in* C. E. Jones and R. J. Little (eds.), Handbook of experimental pollination biology. Van Nostrand Reinhold, New York.
López-Portillo, J., L. E. Eguiarte, and C. Montaña. 1993. Nectarless honey-mesquites. Functional Ecology 7: 452–461.
Marden, J. H. 1984. Remote perception of floral nectar by bumblebees. Oecologia 64: 232–240.
Menzel, R. 1985. Learning in honey bees in an ecological and behavioural context. Pp. 55–74 *in* B. Hölldobler and M. Lindauer (eds.), Experimental behavioral ecology and sociobiology. Fortschritte der Zoologie 31. Gustav Fischer, Stuttgart, Germany
Menzel, R., and U. Greggers. 1992. Temporal dynamics and foraging behaviour in honeybees. Pp. 303–318 *in* J. Billen (ed.), Biology and evolution of social insects. Leuven University Press, Leuven, Belgium.
Mesler, M. R., and K. L. Lu. 1993. Pollination biology of *Asarum hartwegii* (Aristolochiaceae): An evaluation of Vogel's mushroom-fly hypothesis. Madroño 40: 117–125.
Meve, U., and S. Liede. 1994. Floral biology and pollination in stapeliads: New results and a literature review. Plant Systematics and Evolution 192: 99–116.
Mitchell, R. J. 1993. Adaptive significance of *Ipomopsis aggregata* nectar production: Observation and experiment in the field. Evolution 47: 25–35.
Mitchell, R. J., and R. G. Shaw. 1993. Heritability of floral traits for the perennial wild flower *Penstemon centranthifolius* (Scrophulariaceae): Clones and crosses. Heredity 71: 185–192.
Nilsson, A. 1980. The pollination of *Dactylorhiza sambucina* (Orchidaceae). Botaniske Notiser 133: 367–385.
Nilsson, A. 1983a. Anthecology of *Orchis mascula* (Orchidaceae). Nordic Journal of Botany 3: 157–179.
Nilsson, A. 1983b. Mimesis of bellflower (*Campanula*) by the red helleborine orchid *Cephalanthera rubra*. Nature 305: 799–800.
Nilsson, A. 1984. Anthecology of *Orchis morio* (Orchidaceae) at its outpost in the north. Nova Acta Regiae Societatis Scientiarum Upsaliensis, Series 5, C 3: 166–179.
Nilsson, A. 1992. Orchid pollination biology. Trends in Ecology and Evolution 7: 255–259.
Paulus, H. F. 1988. Co-Evolution und einseitige Anpassungen in Blüten-Bestäuber-Systemen: Bestäuber als Schrittmacher in der Blütenevolution. Verhandlungen der Deutschen Zoologischen Gesellschaft 81: 25–46.
Paulus, H. F., and C. Gack. 1990. Pollinators as prepollinating isolation factors: Evolution and speciation in *Ophrys* (Orchidaceae). Israel Journal of Botany 39: 43–79.

Pellmyr, O. 2003. Yuccas, yucca moths and coevolution: A review. Annals of the Missouri Botanical Garden 90: 35–55.
Percival, M. S. 1965. Floral biology. Pergamon Press, Oxford.
Pleasants, J. M., and S. J. Chaplin. 1983. Nectar production rates of *Asclepias quadrifolia:* causes and consequences of individual variation. Oecologia. 59: 232–238.
Pridgeon, A., M. Chase, P. Cribb, and F. P. N. Rasmussen (eds.). 2001. Genera orchidacearum, vol. 2, part 1. Oxford University Press, Oxford.
Raguso, R. A. 2004. Why are some floral nectars scented? Ecology 85: 1486–1494.
Regal, P. J. 1977. Ecology and evolution of flowering plant dominance. Science 196: 622–629.
Regal, P. J. 1982. Pollination by wind and animals: Ecology of geographic patterns. Annual Review of Ecology and Systematics 13: 497–524.
Renner, S. S. 1998. Effects of habitat fragmentation on plant pollinator interactions in the tropics. Pp. 339–360 *in* D. M. Newbery, H. H. T. Prins, and N. D. Brown (eds.), Dynamics of tropical communities. Blackwell Science Publishers, Oxford.
Renner, S. S., and J. P. Feil. 1993. Pollinators of tropical dioecious angiosperms. American Journal of Botany 80: 1100–1107.
Renner, S. S., and R. E. Ricklefs. 1995. Dioecy and its correlates in the flowering plants. American Journal of Botany 82: 596–606.
Renner, S. S., L.-B. Zhang, and J. Murata. 2004. A chloroplast phylogeny of *Arisaema* (Araceae) illustrates tertiary floristic links between Asia, North America, and East Africa. American Journal of Botany 91: 881–888.
Ricklefs, R. E., and S. S. Renner. 1994. Species richness within families of flowering plants. Evolution 48: 1619–1636.
Rodrigues-Correira, M. C., W. T. Ormond, M. C. B. Pinheiro, and H. A. de Lima. 1993. Estudo de biología floral de *Clusia criuva* Camb.: Um caso de mimetismo. Bradea 6: 209–219.
Sazima, M., and I. Sazima. 1989. Oil-gathering bees visit flowers of eglandular morphs of the oil-producing Malpighiaceae. Botanica Acta 102: 106–111.
Schemske, D. W., J. Ågren, and J. Le Corff. 1996. Deceit pollination in the monoecious Neotropical herb *Begonia oaxacana* (Begoniaceae). Pp. 292–318 *in* D. G. Lloyd and S. C. H. Barrett (eds.), Floral biology. Chapman and Hall, New York.
Scotland, R. W., and A. H. Wortley. 2003. How many species of seed plants are there? Taxon 52: 101–104.
Selten, R., and A. Shmida. 1991. Pollinator foraging and flower competition in a game equilibrium model. Pp. 195–256 *in* R. Selten (ed.), Game equilibrium models, vol. 1, Evolution and game dynamics. Springer-Verlag, Berlin.
Smithson, A. 2002. The consequences of rewardlessness in orchids: Reward-supplementation experiments with *Anacamptis morio* (Orchidaceae). American Journal of Botany 89: 1579–1587.
Smithson, A., and L. D. B. Gigord. 2001. Are there fitness advantages in being a rewardless orchid? Reward supplementation experiments with *Barlia robertiana.* Proceedings of the Royal Society of London B 268: 1435–1441.
Smithson, A., and L. D. B. Gigord. 2003. Evolution of empty flowers. American Naturalist 161: 537–552.
Sprengel, C. K. 1793. Das entdeckte Geheimnis der Natur im Bau und in der Befruchtung der Blumen. Friedrich Vieweg dem aeltern, Berlin.
Steiner, K. E. 1992. Mistake pollination of *Hura crepitans* (Euphorbiaceae) by frugivorous bats. PhD dissertation, University of California, Davis, CA.
Sugawara, T. 1988. Floral biology of *Heterotropa tamaensis* (Aristolochiaceae) in Japan. Plant Species Biology 2: 133–136.
Teschner, W. 1980. Sippendifferenzierung und Bestäubung bei *Himantoglossum* Koch. Die Orchidee, Special Issue: 104–115.
Valdeyron, G., and D. G. Lloyd. 1979. Sex differences and flowering phenology in the common fig, *Ficus carica* L. Evolution 33: 673–685.
van der Cingel, N. A. 1995. An atlas of orchid pollination: European orchids. Balkema, Rotterdam.

van der Pijl, L. 1953. On the flower biology of some plants from Java, with general remarks on fly-traps. Annales Bogoriensis 1: 77–99.
Vázquez, D. P., and D. Simberloff. 2002. Ecological specialization and susceptibility to disturbance: Conjectures and refutations. American Naturalist 159:
Verkerke, W. 1987. Syconial anatomy of *Ficus asperifolia* (Moraceae), a gynodioecious tropical fig. Kongelige Nederlands Akademi van Wetenschappen C 90: 461–492.
Voeks, R. A. 1988. Changing sexual expression of a Brazilian rain forest palm (*Attalea funifera* Mart.). Biotropica 20: 107–113.
Vogel, S. 1961. Die Bestäubung der Kesselfallen-Blüten von *Ceropegia*. Beiträge zur Biologie der Pflanzen 36: 159–237.
Vogel, S. 1965. Kesselfallen-Blumen. Umschau 1965: 12–18.
Vogel, S. 1972. Pollination von *Orchis papilionacea* L. in den Schwarmbahnen von *Eucera tuberculata* F. Jahresbericht des Naturwissenschaftlichen Vereins Wuppertal 25: 67–74.
Vogel, S. 1975. Mutualismus und Parasitismus in der Nutzung von Pollenträgern. Verhandlungen der Deutschen Zoologischen Gesellschaft: 102–110.
Vogel, S. 1978. Pilzmückenblumen als Pilzmimeten. Flora 167: 329–398.
Vogel, S. 1981. Trichomatische Blütennektarien bei Cucurbitaceen. Beiträge zur Biologie der Pflanzen 55: 325–353.
Vogel, S. 1993. Betrug bei Pflanzen: Die Täuschblumen. Akademie der Wissenschaften (Mainz), Franz Steiner Verlag, Stuttgart, Germany.
Vogel, S., and J. Martens. 2000. A survey of the function of the lethal kettle traps of *Arisaema* (Araceae), with records of pollinating fungus gnats from Nepal. Botanical Journal of the Linnean Society 133: 61–100.
Vöth, W. 1999 Lebensgeschichte und Bestäuber der Orchideen am Beispiel von Niederösterreich. Stapfia 65: 1–257.
Wiens, D. 1978. Mimicry in plants. Pp. 365–403 *in* M. K. Hecht, W. C. Steere, and B. Wallace (eds.), Evolutionary biology 11. Plenum, New York.
Willson, M. F., and J. Ågren. 1989. Differential floral rewards and pollination by deceit in unisexual flowers. Oikos 55: 23–29
Wong, B. B. M., and F. P. Schiestl. 2002. How an orchid harms its pollinator. Proceedings of the Royal Society of London B 269: 1529–1532.

CHAPTER SEVEN

Ecological Factors That Promote the Evolution of Generalization in Pollination Systems

José M. Gómez and Regino Zamora

Specialization is a key concept in the discipline of pollination ecology. Discovering the causes and consequences of the specialization of plants on certain pollinators is an empirical and theoretical challenge that has pervaded pollination ecology since its foundation (Darwin 1862, 1877). The conventional wisdom shared by most evolutionary ecologists has been that specialization is very advantageous for plants and, thus, is a crucial feature of many pollination systems. For example, most pollinator–plant coevolutionary models require the occurrence of specialization (Thompson 1994). In addition, many theoretical approaches still accept an evolutionary trend toward specialization in plants, where "advanced" plants are more specialized than "primitive" plants (see Ollerton 1999). Furthermore, it is assumed that specialization promotes pollinator-mediated speciation, floral divergence, and reproductive isolation, thereby explaining the current diversity of flowering plants (Waser 1998; Levin 2000).

In pollination ecology, the study of the evolution of specialization is driven by the widely known and amply cited "most effective pollinator principle" (Stebbins 1970), which states that natural selection should modify plant phenotypes to increase the frequency of interaction of the plants with the pollinators that confer the best services. As a consequence of the selection exerted by the pollinators, we would expect the flowers of most plants to be visited predominantly by a reduced group of highly effective pollinators (also called *adaptive specialization;* Thompson 1994; Herrera 1996; Armbruster et al. 2000; Johnson and Steiner 2000). According to this line of thought, pollinators are considered strong selective agents, and many plant traits are considered to be adaptations to specific pollinators (Johnson and Steiner 2000). It is possible to predict the most important pollinators of a plant simply by looking at the suite of floral traits displayed by the plants (called *pollination syndromes;* Faegri and van der Pijl 1979; Waser, chap. 1 in this volume; Corbet, chap. 14 in this volume). Support for these ideas is provided by a wide spectrum of studies that have demonstrated that pollinators can exert significant phenotypic selection on many morphological and phenological plant traits, many of which show moderate-to-high heritability (Johnson and Steiner 2000). Indirect support comes from the observation of many plant species with very complex and restrictive flowers, from which many evolutionary ecologists infer specialization on specific pollinators.

Nevertheless, generalization is quite frequent in natural pollination systems (Herrera 1996; Waser et al. 1996; Armbruster et al. 2000; Olesen 2000). Indeed, extreme specialization is only observed among plants that provide neither pollen nor nectar as rewards, but some unusual products like seeds, resins, nonvolatile oils, or fragrances (Buchman 1987; Armbruster 1997; Pellmyr 1997; Fleming and Holland 1998; Steiner and Whitehead 2002) or no reward is offered at all and pollination occurs by deceit (Dafni and Berhardt 1990; see also Renner, chap. 6 in this volume). In these systems, reduction in common rewards (pollen and nectar) facilitates the evolution of specialization by discouraging the visits of copollinators (Fenster and Dudash 2001). In contrast to these odd examples of specialization, the flowers of many plant species are visited by a diverse assemblage of pollinators (e.g., Herrera 1988; Horvitz and Schemske 1990; Gómez et al. 1996; Gómez and Zamora 1999; Dilley et al. 2000; Lippok et al. 2000; Olesen 2000; Thompson 2001; and references therein). Studies that draw into question the predictive power of pollination syndromes pose an additional challenge to the traditional view of specialization. Community surveys demonstrate the difficulty of inferring the type of pollinator that visits a plant based exclusively on the plant's floral traits (Ollerton and Watts 2000).

A dilemma exists, therefore, about the extent and importance of generalization versus specialization in pollination systems. Specifically, evolutionary ecologists need to resolve an intriguing question: how is it possible for generalization to be common in nature despite the frequently demonstrated ability of pollinators to select on plant phenotype? The reconciliation between theoretical predictions and empirical evidence, although difficult and far from complete, is crucial if we want to build a more general theory about the ecology and evolution of the plant–pollination interactions.

Necessary Conditions for Ecological Specialization

Specialization and generalization are comparative terms (species A is less or more specialized than species B in a given environment) that do not represent exclusive, dichotomous solutions but a gradient along the resource-use niche axis (Futuyma 2001). In practice, specialization is simply defined as the use of a narrow spectrum of resources (Futuyma 2001). In pollination biology, an operational definition of specialization proposed by Armbruster et al. (1999, 2000) is *successful pollination by a small number of animal species.* We begin with this definition but propose two refinements. First, we consider the phyletic diversity of the pollinators, because a plant pollinated by few species of disparate phylogenetic affiliations is probably more generalist than a plant pollinated by a moderate

Table 7.1 Main requirements for specialization under the framework of the most effective pollinator principle

Requirements	Mechanisms
Pollinators exert actual selective pressure	
Pollinators must benefit fitness	The effect of a pollinator species on fitness is called the *importance* of that pollinator for a plant species and is a consequence of its abundance at flowers (namely the *quantity component* of the interaction) **and** its per-visit effectiveness (the *quality component* of the interaction).
and	
Pollinators must produce a significant covariance between fitness value and a given trait value	This pollinator-mediated fitness-trait covariation is considered a *pollinator-mediated phenotypic selection* and is a consequence of the pollinator ability to discriminate between different plant phenotypes (*pollinator preference*) **or** the pollinator ability to match to specific plant phenotypes (*pollinator mechanical fit*).
Each pollinator species represents a distinctive selective agent	
Variation among floral visitors in their effect on fitness	This among-pollinator variation permits the establishment of a ranking of different pollinators according to their service to the plants, and it occurs because pollinators consistently differ in per-visit effectiveness **or** abundance.
and	
Among-pollinator variation in the phenotypic selection that they originate	This pollinator-mediated trade-off in phenotypic selection is crucial for successful specialization and is produced by among-pollinator difference in preference **or** mechanical fit.

Source: Based on Gómez 2002.
Note: Note the "and" and "or" operations, which integrate the several mechanisms under each requirement.

number of pollinators that belong to the same genus or family. Second, we consider the number of pollinators available in a region, because a plant visited by all the pollinators inhabiting a depauperate ecosystem could be considered more generalist than a plant visited only by a small fraction of the pollinators that occur in a species-rich ecosystem, even although the absolute number of pollinators is higher in the latter case.

Pollinators as Selective Agents

A necessary (although not sufficient) condition for a plant to develop an adaptive trait that attracts the most efficient pollinator is the occurrence of natural selection mediated by that pollinator.

For natural selection to occur, pollinators must first benefit plant fitness (table 7.1). Pollinators enhance plant fitness due to both their abundance at flowers and their per-visit effectiveness, which provide the quantity and quality com-

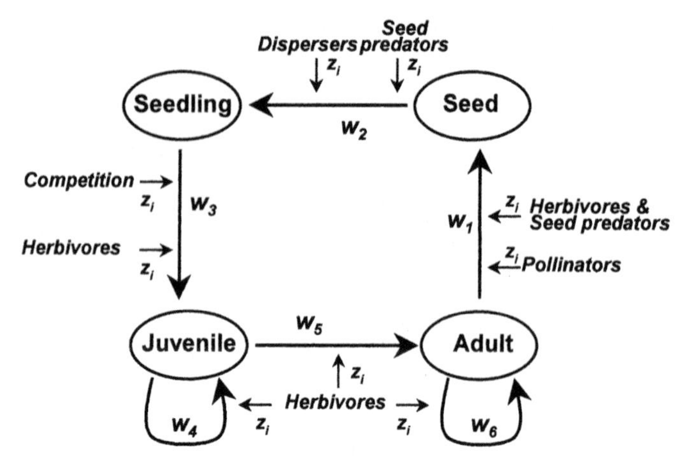

Figure 7.1 An extremely simplified representation of the life cycle of a hypothetical plant, where W_k are the sequential multiplicative fitness components (≈ transition rates; $W_t = \prod W_k$), and z_i are the phenotypic traits covarying with each fitness component due to the effect of different ecological factors that act as selection agents. We have not assigned specific subindices to z_i to point out that they can covary with more than one fitness component. Each $W_k(z)$ can itself be decomposed into several to many fitness components. For example, most pollination biologists divided $W_1(z)$ into pollen deposition, pollination success, seed ripening, predispersal seed predation probability, and so forth.

ponents of the interaction, respectively (Waser 1983; Herrera 1987, 1989; Gómez and Zamora 1992, 1999; Gómez 2000a). The *importance* of a given pollinator for a plant depends on these two components (table 7.1). However, many studies on pollinator importance may overestimate pollinator benefits to plant fitness when no appropriate definition of fitness is used (Mitchell et al. 1998; Herrera et al. 2002). Plants, like other organisms, have complex life cycles (fig. 7.1). In a life-cycle context, fitness is the rate of increase of a given individual genotype, and it is composed of many sequential fitness components such as germination rate, seedling survival, or fecundity, all of them multiplicatively contributing to the total fitness: $W_t(z) = \prod W_k(z)$, where $W_k(z)$ is the fitness component corresponding to the life-cycle transition *k* (van Tienderen 2000; Caswell 2001). Consequently, an accurate estimation of the fitness of an organism requires the consideration of its entire life cycle (Arnold and Wade 1984a, 1984b; Campbell 1991; Charlesworth 1994; van Tienderen 2000; Caswell 2001). Unfortunately, the effect of pollinators is typically quantified on only one or a few intermediate, fecundity-related fitness components located in the transition between adults and seeds, such as the number of pollen grains deposited or removed or the number of seeds produced per pollinator visit (e.g., Dieringer 1992; Thompson and Pellmyr 1992; Vaughton 1992; Keys et al. 1995; Stone 1996; Olsen 1997; Gómez and Zamora 1999; Muchhala 2003). As a notable exception, Herrera et al. (2002) recently reported the effect of pollinators on *Helleborus foetidus* (Ranunculaceae) fitness in terms of recruited offspring.

To act as selective agents—besides affecting fitness—pollinators must produce a significant covariance between fitness and the value of a given trait (*fitness function* or *pollinator-mediated phenotypic selection;* table 7.1). If no such relationship exists, any plant phenotype will be selected with similar intensity and pollinators will not produce a differential fitness gain of certain phenotypes over others (Schemske and Bradshaw 1999; Armbruster et al. 2000; Gómez 2002).

Pollinators can exert phenotypic selection primarily by two mechanisms (table 7.1). First, pollinators may discriminate among conspecific plants, using several phenotypic traits as cues. A positive correlation between the frequency of interaction and the value of a given floral trait may confer a reproductive advantage to the preferred floral phenotypes; this mechanism acts primarily through the quantity component of the interaction. As a second mechanism, pollinators may pollinate some phenotypes more efficiently than others because they more accurately match the flowers (Wilson 1995). Flower types with a low probability of pollen transfer because of a poor fit to pollinator size or shape would be selected against; in this case, natural selection will optimize pollen transfer per pollinator visit, so this second mechanism acts on the quality component of the plant–pollinator interaction. The relative importance of these two mechanisms in explaining floral evolution is still unclear (Wilson 1995).

Each Pollinator Must Constitute a Distinctive Selective Agent

For a plant to increase its interaction with a specific pollinator while reducing its interaction with remaining flower visitors, another obligate condition is the existence of among-pollinator differences in the evolutionary effect on the plant (table 7.1). This can occur only when pollinators differ both in their particular effects on fitness (via per-visit effectiveness or abundance) and in the specific trait–fitness relationship that they produce (via preference or matching with floral traits).

Paraphrasing Schemske and Horvitz (1984), a precondition for mutualistic specialization is the existence of "variation among floral visitors in pollination ability." The establishment of a ranking of pollinators according to their effect on fitness allows adaptation to the pollinators that confer the best services (Schemske and Horvitz 1984, 1989; Stanton et al. 1991; Conner et al. 1995; Fishbein and Venable 1996; Waser et al. 1996; Galen 1999; for a different view see Aigner 2001, chap. 2 in this volume). Natural selection will favor those plant traits that attract the most efficient or abundant (most important) pollinators and will also favor the evolution of the phenotypes that cause the most abundant pollinators to also be the most effective. Consequently, deriving directly from Stebbins's "most effective pollinator principle," we expect that the abundance and effectiveness of the pollinators should be positively correlated for any plant species, something which rarely happens (Herrera 1996; Gómez and Zamora 1999; Mayfield et al. 2001).

A further precondition for the evolution of specialization is among-pollinator variations, not only in their average effects on fitness but also in the fitness functions that they generate (Wilson and Thompson 1996; Aigner 2001). A classical hypothesis in population ecology about the origin of specialization in resource use mirrors the old adage, "the jack-of-all-trades is a master of none" (MacArthur 1972). In modern terminology, specialization is facilitated by the existence of trade-offs: traits leading to increased fitness when using a given resource or living in a specific environment are detrimental when using other resources or environments (Futuyma 2001; Via and Hawthorne 2002). This argument is used widely to explain the ubiquitous host specificity in herbivorous insects (Futuyma 2001; Via and Hawthorne 2002). The argument applies equally in pollination biology: pollinator-mediated fitness trade-offs are necessary for specialization to occur. Aigner (2001, chap. 2 in this volume) has even suggested that among-pollinator difference in fitness function rather than in mean effectiveness is the root cause of specialization. Trade-offs between pollinators can occur when the pollinators differ in preference for floral traits or in mechanical fit with flowers (Waser 1998; Galen 1999; Schemske and Bradshaw 1999).

Back to the Real World: Main Reasons Accounting for Generalization in Pollination Systems

Having already described the theoretical pathways to specialization, in this section we examine several factors that can preclude the evolution of specialization in pollination systems. Several biologists have suggested that, although most plants are visited by multiple pollinators, these systems can be also specialized due to the selection supremacy exerted by the most important pollinators (Ollerton 1996; Johnson and Steiner 2000). The mere compilation of pollinator lists is insufficient to discard this critical proposal. Instead, a thorough understanding of the fitness functions imposed by different pollinators as well as other elements of the environment is essential to accurately explain the evolution toward specialization or generalization. Such an approach suggests several causes that would fuel the evolution of generalization.

Unpredictability of the Most Important Pollinators: Spatiotemporal Variability in the Pollinator Assemblage

Spatiotemporal variability in the abundance and composition of the floral visitor assemblage is doubtless the most-cited reason to explain the existence of generalized pollination systems (Herrera 1988, 1996; Waser et al. 1996; Armbruster et al. 2000). Temporal variation in the identity of the most important pollinators can cause a concomitant fluctuation in the magnitude and sign of phenotypic selection (Kalisz 1986; Schemske and Horvitz 1989; Widén 1991; Andersson 1992; Domínguez and Dirzo 1995; Maad 2000), which dilutes the overall selection exerted by these pollinators (Herrera 1988, 1996; Waser et al. 1996). Although more long-term studies are still needed to determine the extent and scale of temporal variation in pollinator assemblages, most studies that span more than two years have demonstrated temporal fluctuation (e.g., Herrera 1988; Petterson 1991; Gómez and Zamora 1992, 1999; Obeso 1992; Vaughton 1992; Traveset and Sáez 1997; Bingham 1998; Fenster and Dudash 2001; Thompson 2001). The pollinator assemblage can vary not only among years but also seasonally within years (Ashman and Stanton 1991). For example, *Lobularia maritima* (Cruciferae) is a perennial herb that flowers during eight months of the year. In some coastal areas of southwestern Spain, its flowers are visited mainly by flies during autumn and winter, whereas they are visited and pollinated in summer by the ant *Camponotus micans* (Formicidae; Gómez 2000a, 2000b).

Similarly, because natural selection is a populational phenomenon, spatial variation in the pollinator assemblage can also dilute the overall selection intensity (Schemske and Horvitz 1989; Johnston 1991; O'Connell and Johnston 1998; Gómez and Zamora 2000). Selection is diluted when two populations linked by gene flow are under different selection regimes. Spatial variation in pollinator identity is common in most plant species studied at intermediate spatial scales (Herrera 1988; Horvitz and Schemske 1990; Eckhart 1991, 1992; Guitián et al. 1996; Gómez and Zamora 1999; Thompson 2001). Differences in the identity of the most abundant pollinators can appear even at smaller spatial scales. The identity of insects visiting *Pinguicula vallisnerifolia* (Lentibulariaceae) flowers vary at an intrapopulational scale; the beetle *Eusphalerum scribae* (Staphylinidae) visits flowers in the shade, whereas the thrips *Taeniothrips meridionalis* is more active on flowers in sunlight (Zamora 1999).

Spatial variation in pollinator assemblage results in a dilution of the selection process only when there is gene flow among populations. By contrast, with genetic isolation, the occurrence of adaptations to local pollinators is possible. This scenario could produce a geographic mosaic of specialization, where the plant is a generalist at the species level but a specialist at the population or regional level (Thompson 1994).

Similarity among Pollinators in Selective Role

Specialization, as defined here, is impossible when different pollinator species play the same role as selective agents. Similarity among pollinators can occur when different pollinators have equivalent abundance and above all comparable effectiveness. In this case, pollinators of different species are functional equivalents from the plant perspective (Gómez and Zamora 1999; Zamora 2000). In fact, several pollination biologists have found that taxonomically different pollinators can forage at plants in similar ways and thereby have similar per-visit effectiveness. This occurs among pollinators that belong to the same family or order (e.g., Herrera 1987; Petterson 1991; Thøstesen and Olesen 1996) as well as those belonging to different higher-order taxa (e.g., Motten et al. 1981;

Dieringer 1992; Vaughton 1992; Gómez and Zamora 1999; Gómez 2000a). Functionally equivalent pollinators can visit not only open, actinomorphic flowers (radially symmetrical), but also complex, restrictive zygomorphic flowers (bilaterally symmetrical). For example, Thøstesen and Olesen (1996) showed that the specialist bumblebee *Bombus consobrinus* (Apidae) is as efficient as the generalist bumblebees *B. jonellus, B. pratorum,* or *B. wurfleinii* when removing pollen from *Aconitum septentrionale* (Ranunculaceae). Moreover, it is more probable that two pollinators have the same effectiveness per visit when the plant bears few ovules per flower and, thus, requires few pollen grains to fertilize all the ovules (Johnson et al. 1995). In those systems in which pollinators have similar per-visit effectiveness, the importance of a given floral visitor as pollinator is produced mainly by its abundance at flowers (Gómez and Zamora 1992, 1999). Consequently, there is still a possibility for evolutionary specialization on the most abundant pollinator—a neglected idea which is implicit in the principle of the most effective pollinator (Waser et al. 1996). The lack of a positive correlation between pollinator abundance and per-visit effectiveness is, therefore, circumstantial evidence for the absence of specialization (Vaughton 1992; Gómez and Zamora 1999; Mayfield et al. 2001; Potts et al. 2001).

Specialization is further constrained when different pollinators, even with different effectiveness, generate similar trait–fitness covariation because they display similar fit with floral parts or preference for plant traits. The modification of floral traits to attract effective pollinators can indirectly attract less-effective pollinators (Wilson and Thompson 1996; Armbruster et al. 2000). Despite its potential importance in understanding the evolution of pollination systems, variation among pollinator species in response to floral traits has received little attention (Conner and Rush 1996; Waser et al. 1996; Galen 1999; Schemske and Bradshaw 1999), although several examples are currently available. In a pioneering study, Galen et al. (1987) showed that bumblebees and flies generate contrasting relationships between *Polemonium viscosum* (Polemoniaceae) flower size and pollination efficiency; whereas bumblebees favor larger flowers, flies favor smaller flowers. Subsequently, Lee and Snow (1998) found that syrphid flies and bumblebees differ in their preferences for flower color in *Raphanus sativus* (Cruciferae) and *R. raphanistrum,* and Thompson (2001) reported that different types of pollinators display different preferences for *Jasminum fruticans* floral traits. Nevertheless, similarity in preference or match of floral morphology has also been demonstrated in a number of studies: for example, solitary bees and syrphid flies with respect to flower size of *R. raphanistrum* (Conner and Rush 1996); bees and butterflies with respect to flower size of *Wurmbea dioica* (Liliaceae; Vaughton and Ramsey 1998); several species of bumblebees with respect to stigma exsertion of *Erythronium grandiflorum* (Liliaceae) or to overall floral morphology of *Impatiens pallida* (Balsaminaceae; Wilson and Thomson 1996); 14 species of bees with respect to the morphology of *Collinsia heterophylla* (Scrophulariaceae; Armbruster et al. 2000); and muscid flies and ants with respect to flower size, display, and density of *L. maritima* (Gómez 2000b). However, empirical evidence is still scant; therefore, we strongly advocate a new research agenda to discover the prevalence of pollinator-imposed trade-offs in nature (see also Aigner, chap. 2 in this volume). Criteria and methods of this agenda should include the following: (1) working with the complete group of pollinator species and avoiding an automatic focus on the most obvious interactors, (2) exploring the preference patterns or the mechanical fits of the different co-occurring pollinators at flowers, and (3) seeking both the interspecific differences and the similarities in the mechanisms of plant–pollinator interaction (Waser et al. 1996; Zamora 2000).

A main consequence of the similarity among pollinators as selection agents is that any spatiotemporal variability in pollinator identity is dampened and selection is similar in most locations and during most seasons (Zamora 2000). Geographic scale studies are necessary to test whether there is a spatial variation in plant–pollinator interactions, as proposed by the theory of the geographic mosaic of interactions (Thompson 1994), or whether there are repeatable spatial or temporal patterns of selection on the floral phenotype.

The Real Effect of Pollinators on Plant Fitness and the Consequences of Adopting a Life-cycle Approach to the Study of Plant–Pollinator Interactions

A life-cycle and demographic approach to the study of the interactions between plants and pollinators has two main implications for the evolution of specialization: (1) modification of the estimation of pollinator importance and effectiveness by using more complete estimates of plant fitness and (2) consideration of extrinsic factors potentially constraining the plant evolutionary response to the selection exerted by pollinators.

The Importance of Using Most Complete Estimates of Pollinator Importance

Selection involving one fitness component can be irrelevant in evolutionary terms when significant trade-offs among fitness components lead to a low overall effect on total fitness (Ehrlén 2002). For example, because larger offspring typically have greater fitness, and a trade-off often occurs between the number and size of seeds produced, a pollinator may benefit seed production by fertilizing many ovules but reduce seedling survival because it causes the ripening of many low-quality seeds. Similarly, a large increase in seed production one year can be offset by a decrease in survival of the mother plants in successive seasons (Ehrlén 2002). Campbell (1991) acknowledged these concerns, stating that the ultimate goal when studying the actual evolutionary effect of pollinators on plants is "to estimate male and female fitness during all stages of reproduction, since there may be trade-offs between pollination success and other life-cycle components of fitness."

The effect of pollinators on total fitness can also be weakened when other factors acting on the same or another part of the plant life cycle account for a greater effect on fitness and concomitantly decrease the opportunity for selection and the variation in fitness from pollinators (Herrera 1996). For example, the variability in fitness explained by pollinators is barely 2% in *Viola cazorlensis* (Violaceae), less than 0.2% in *Moricandia moricandioides* (Cruciferae), and about 0.05% in *Hormathophylla spinosa* (Cruciferae) when the effect of some vertebrate and invertebrate herbivores are included in the phenotypic selection analyses (Herrera 1993; Gómez 1996; Gómez and Zamora 2000; J. M. Gómez, unpublished data). Furthermore, Herrera and co-workers have demonstrated that pollinators significantly affect *Paeonia broteroi* (Paeoniaceae) and *H. foetidus* fitness only when herbivores are experimentally excluded (Herrera 2000a; Herrera et al. 2002). More important, recent studies suggest that the fitness of some perennial plants depends more heavily on annual survivorship than on fecundity (Crone 2001), which reinforces the call for caution when inferring pollinator effect on fitness via seed production.

The question of whether differences in effectiveness among pollinators remain relevant when fitness is measured on stages beyond seed production is largely unexplored, and the few existing studies reach differing conclusions. Herrera (2000b) has shown that *Lavandula latifolia* (Labiatae) flowers pollinated by butterflies produced not only many seeds but also seedlings and juvenile plants of high vigor, which survived well after three years. In contrast, Gómez (2000a) reported that *Lobularia maritima* flowers pollinated by the ant *Camponotus micans* produce many seeds but with low germination rate and vigor, presumably because they are produced by geitonogamy (within-plant self-pollination). Most important, the difference in effectiveness between ants and flies when quantified at the seed-production stage completely disappeared when measured over the total life cycle of the plants (Gómez 2000a). These examples illustrate the value of considering as much of the life cycle as possible in assessing pollinator effectiveness.

Disruption of Pollinator-mediated Selection by Conflicting Pressures

Pollination biologists tend to focus attention on the plant–pollinator interaction; however, the plant life cycle is embedded in an ecological context in which many hazards may affect from one to several different stages of the process by operating through certain phenotypic traits of the plant (fig. 7.1). Selection cannot be derived from the study of any single component because a given phenotypic trait may be related to different fitness components in opposing ways. Figure 7.2 represents a hypothetical situation in which there is a positive covariance between the value of the trait z and a given fitness component $W_1(z)$ due to pollinator-mediated selection. However, this selection vanishes when

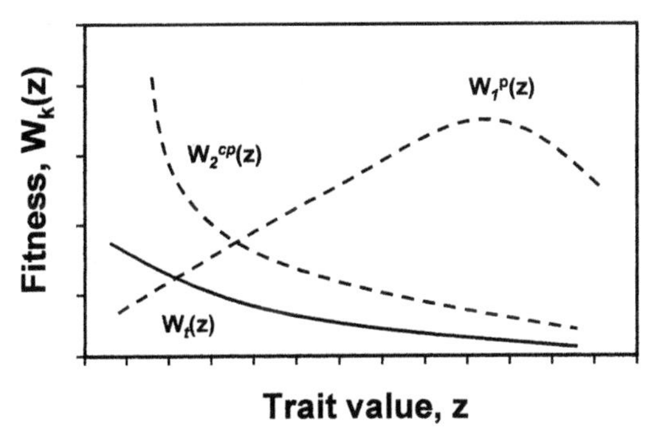

Figure 7.2 A hypothetical example that shows the reversal of pollinator-mediated selection by conflicting selection. The contrasting covariances occurring between the same trait *z* and two fitness components (1 and 2) produced by two different agents (pollinators *p* on component 1 and an unknown conflicting pressure *cp* on component 2) are shown. The covariance between total fitness, $W_t(z) = \prod W_k(z)$, and the trait *z* is negative, which negates any potential selective role played by pollinators.

considering total fitness $W_t(z)$, which is due to the conflicting selection by another agent (fig. 7.2).

Herbivores sensu lato (folivores, seed predators, florivores, nectar robbers, etc.) are among the most conspicuous organisms to interact with plants through several life-cycle stages, and nowadays many ecologists admit that pollinators and herbivores rarely operate independently of each other. For example, pollinators show widespread ability to discriminate between plants based on previous damage, preferring nondamaged flowers over damaged ones (Karban and Strauss 1993; Strauss et al. 1996, 1999; Strauss 1997; Krupnick et al. 1999; Mothershead and Marquis 2000; Adler et al. 2001; Hambäck 2001). Furthermore, herbivores and pollinators often affect the same plant traits, such as flowering phenology (Gómez 1993; Strauss et al. 1996; Brody 1997), flower number (Karban and Strauss 1993; Ehrlén 1997; Gómez and Zamora 2000; Herrera et al. 2002), floral morphology (Strauss et al. 1996; Lehtilä and Strauss 1999; Leege and Wolfe 2002), quantity and quality of floral nectar (Krupnick et al. 1999), pollen production per flower (Frazee and Marquis 1994; Strauss et al. 1996), or pollen performance (Delph et al. 1997). When pollinators and herbivores display similar preference patterns for particular plant traits, a trade-off between the fitness functions generated by each kind of organisms can appear. Consequently, any pollinator-mediated selection can be canceled by an overwhelming influence of herbivory. Empirical evidence for the occurrence of herbivore-pollinator selective conflicts is accumulating rapidly (Herrera 1993, 2000a; Quesada et al. 1995; Gómez 1996, 2003, 2005a; Brody 1997; Delph et al. 1997; and references therein), and several plant traits are currently thought to be the evolutionary result of conflicting selection exerted by these two kinds of organisms (Euler and Baldwin 1996; Ehrlén 1997; Strauss 1997; Galen and Cuba 2001; Lara and Ornelas 2001;

Ehrlén et al. 2002). Ashman (2002) has even recently proposed that herbivores can influence the evolution of sexual systems in plants. In all these cases, adaptations to avoid herbivory can constrain the evolution of plant–pollinator interactions, because the advantages associated with the latter function are countered by costs associated with the former (Fineblum and Rausher 1997; Strauss 1997; Gómez and Zamora 2000; Galen and Cuba 2001).

The capacity to exert conflicting selection is not confined to herbivores. Any other organism or factor mediating the trait-fitness covariance has the possibility to generate conflicting selection on plants (Galen 2000). For example, Shykoff et al. (1997) have shown that *Dianthus silvester* (Caryophyllaceae) individuals displaying larger flowers attract more pollinators but are also at higher risk of receiving infecting spores of the anther smut *Microbotryum violaceum* (Ustilaginales).

Another mechanism that causes trade-offs is the genetic correlation between traits selected by pollinators and other phenotypic traits. For example, the genes affecting floral pigments in many plant species, such as *Hypericum calycinum* (Guttiferae), *Dalechampia* spp. (Euphorbiaceae), *Acer* spp. (Aceraceae), or *Ipomoea purpurea* (Convolvulaceae), have pleiotropic effects and in some cases affect the same biochemical pathways to plant secondary compounds that function as defenses against herbivores (Simms and Bucher 1996; Armbruster 1997, 2002; Armbruster et al. 1997; Fineblum and Rausher 1997; Gronquist et al. 2001). In these cases, herbivores may indirectly affect the evolution of floral pigmentation when causing direct selection for specific defensive compounds. Indirect selection can be a mechanism causing the nonadaptive (and non-pollinator-mediated) evolution of floral traits (Gómez 2000b; Armbruster 2002).

A Case Study: Conflicting Selection on *Erysimum mediohispanicum*

Erysimum mediohispanicum (Cruciferae) is a short-lived monocarpic herb found in many montane regions of southeastern Spain, where it is distributed from 1100 to 2000 m above sea level. In the Sierra Nevada, plants usually grow for two to four years as vegetative rosettes then die after producing one to eight reproductive stalks that can display between a few and several hundred bright yellow, hermaphroditic, slightly protandrous flowers. Flowers are visited in the study area by several species of insects, particularly the pollen beetles *Meligethes maurus* (Nitidulidae) and several species of beetles, bumblebees, solitary bees, and syrphids (Gómez 2005a). Although this crucifer is self-compatible, it requires pollen vectors to produce a full seed set (Gómez 2005a).

Reproductive individuals are fed upon in the Sierra Nevada by many different species of herbivores. Several species of sapsuckers (particularly the bugs *Corimeris denticulatus, Eurydema fieberi,* and *E. ornata*) feed on the reproductive stalks, during both flowering and fruiting. In addition, the weevil *Ceutorhynchus chlorophanus* (Curculionidae) develops inside fruits, living on developing seeds, and another weevil (presumably *Lixus ochraceous,* Curculionidae) bores into stalks, consuming the inner tissues. However, the main herbivore of *E. mediohispanicum* in the study area is the Spanish ibex (*Capra pyrenaica* Bovidae), an ungulate that consumes flowers and, above all, green fruits by browsing on the reproductive stalks, which affects both reproductive output and population dynamics of the plants (Gómez 2005b).

The interaction between selection exerted by pollinators and ungulates was experimentally studied by excluding ungulates from part of a large plant population and comparing the selective scenario produced by pollinators in the absence and presence of these mammals (Gómez 2003, 2005a). Number of seeds dispersed per plant was the fitness component used to test conflicting selection between ungulates and pollinators.

Ungulates and pollinators displayed similar preferences for some plant traits. Ungulates preferred larger plants, with many flowers, and with more, taller, and wider stalks. During some years, pollinators also preferred plants with more and taller stalks, with many flowers, and with shorter corollas (fig. 7.3A; Gómez 2003, 2005a). In addition, ungulates imposed an ecological cost to the plants, quantified as a decrease in pollinator abundance at flowers in those parts of the plant population where the ungulates were present (fig. 7.3B; Gómez 2005a). A first consequence of ungulate presence was the elimination of any significant effect of pollinators on fitness. The pollinator visitation rate correlated significantly with plant fecundity inside, but not outside, the fences (Gómez 2003). Most important, the presence of ungulates altered the selective regime of the plants. When ungulates were absent, in a 2001 study we found significant selection on flower number, reproductive stalk height, basal diameter of the stalks, petal length, and inner diameter of the flowers, quantified both by selection gradients (Gómez 2003) and by structural equation modelling (fig. 7.3C). In contrast, when ungulates were present, selection on floral traits and stalk height completely disappeared and selection strength on flower number and other morphological traits decreased (fig. 7.3C; Gómez 2003). In brief, this study supports the idea that factors acting on the same fitness components as pollinators can prevent plant adaptation and specialization to pollinators by obscuring or counteracting the phenotypic selection exerted by them.

In 2002, we studied whether the selective scenarios vary depending on the fitness components considered. For this, we used not only number of seeds produced per plant (fecundity) but also the number of seeds successfully germinating and the number of seedlings surviving the first summer as fitness components. Although the data are still very preliminary, it seems that the only trait that had a significant selection gradient on fecundity in 2002 was the number of flowers, both inside and outside the exclosures ($p < .004$ in both cases). In contrast, when using the number of seeds germinating as a fitness component, there was a significant selection gradient on corolla tube length inside the exclosure,

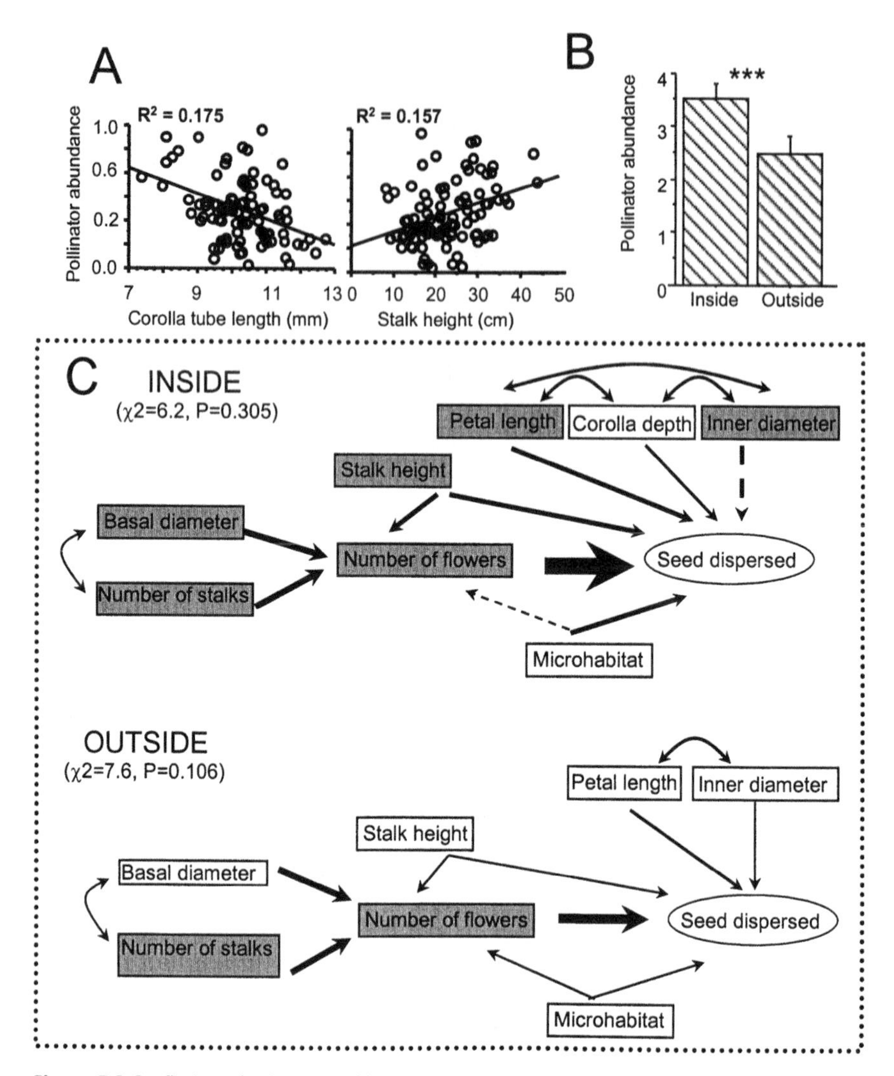

Figure 7.3 Conflicting selection exerted by mammalian herbivores and pollinators in *Erysimum mediohispanicum* (Cruciferae). (A) Preference of pollinators for increased stalk height and reduced corolla tube length (data from 2002, based on Gómez 2003). (B) Difference in pollinator abundance (insects min^{-1} $plant^{-1}$) depending on the presence of ungulates (outside the fences) or absence (inside); ***$p < .01$ after one-way analysis of variance (data from 2002, based on J. M. Gómez, unpublished data). (C) The most parsimonious structural equation models, showing phenotypic selection occurring on several plant traits inside and outside the fences. Gray boxes indicate significant total effects on fecundity (data from 2001, based on Gómez 2003).

indicating that plants with a longer corolla produced seeds with higher germination ability when protected from ungulates. The selection gradient on number of flowers almost disappeared outside the exclosures ($p \approx .02$ vs. $p < .0001$; J. M. Gómez, unpublished data), suggesting that the benefit of producing many flowers disappeared when fitness was quantified beyond seed production. Thus, a further conclusion derived from this study is that the type of traits selected by pollinators can change when different fitness components are considered.

Concluding Comments

In the preceding section, we have argued that the scenario suggested by Stebbins's most effective pollinator principle represents just one of multiple evolutionary solutions, which takes place only when some restrictive ecological conditions are met (table 7.1). We would like to finish this chapter by discussing two main implications emerging from the ideas that we have developed herein: the possibility for adaptive generalization and the necessity of integrative studies.

Adaptive and Nonadaptive Generalization

Specialization has traditionally been viewed as the most likely outcome of strong pollinator-mediated natural selection, generalization being, in all cases, the absence of strong or consistent selection. However, we believe that the evolution of generalization mediated by selection exerted by pollinators (*adaptive generalization*) is possible in some ecological scenarios. Although the adaptive significance of generalization has been recognized for many years (see, e.g., Grant and Grant 1965), we still lack an explicit theory of the mechanisms by which generalization evolves; we have suggested some mechanisms in this chapter.

Nonadaptive generalization occurs when pollinators do not constitute real selection agents. In this view, generalization is a consequence of the ecological forces constraining the development of adaptive specialization. This type of generalized system is by far the best studied at this point. By way of contrast, in those systems in which pollinators are strong agents of selection but have similar per-visit effectiveness and, most important, equal preference or mechanical fit, several pollinators can act in a concordant fashion to cause the evolution of floral adaptations; here the outcome is adaptive generalization of plants on pollinators instead of specialization on particular species (Armbruster et al. 2000; Gómez 2002).

Our point is that both adaptive and nonadaptive generalized pollination systems can occur in nature; indeed, many generalized systems presumably result from a mixture of adaptive and nonadaptive forces. A major challenge for the future is to determine the frequency of occurrence of the two types of generalization, and of their combination.

The Urgency of Integrative Studies

The ideas considered here suggest that we must go beyond studies that simply consider plants and pollinators in isolation from their complex ecological contexts. It is necessary not only to obtain unbiased estimates of the effectiveness and preferences of the whole floral visitor assemblage, but also to deeply explore the entire system in which focal organisms are imbedded, considering the myriad of other species interacting with the plant at different stages of the life cycle and recruitment process. It is absolutely crucial to determine the potential selective conflicts to be able to estimate the realized magnitude of pollinator-mediated selection and the evolution rate of plant–pollinator interactions. We have shown that, when selective conflicts are strong, pollinator-mediated selection based on a given fitness component may become irrelevant when one considers the total fitness of the plant. In these scenarios, the selection imposed by an effective pollinator does not always deliver the predicted evolutionary responses. We hypothesize that these selective conflicts are a major reason why generalization is more common than we would predict given the frequently observed ability of pollinators to generate strong phenotypic selection on plants.

Testing this hypothesis could be facilitated by research programs in which there are open and frequent collaborations among workers from different areas of plant population ecology, such as seed ecology or plant demography, and/or from different subdisciplines within the studies on plant–animal interaction that are not usually integrated in standard studies of pollination biology. It is our conviction that by integrating information from different fields we will be able to get a much more accurate picture of the evolution of plant–pollinator interactions and a much better understanding of their degree of specialization or generalization.

Acknowledgments

We thank Nick Waser and Jeff Ollerton for their invitation to participate in the "Specialization and Generalization in Pollination Systems" symposium held at the 2002 Ecological Society of America meeting in Tucson, Arizona (USA), and Nick Waser, Paul Aigner, and Rebecca Irwin for review of an early version of the manuscript. We are very grateful to the Sierra Nevada National Park for permitting the fieldwork. Our studies during the past 10 years have been supported by the Junta de Andalucia financial support to the RNM220 PAI Research Group, the University of Granada grant 2002-0001 and the Ministerio de Ciencia y Tecnología (MCYT) grant BOS2003-09045.

References

Adler, L. S., R. Karban, and S. Y. Strauss. 2001. Direct and indirect effects of alkaloids on plant fitness via herbivory and pollination. Ecology 82: 2032–2044.

Aigner, P. A. 2001. Optimality modeling and fitness trade-offs: When should plants become pollinator specialists? Oikos 95: 177–184.

Andersson, R. 1992. Phenotypic selection in a population of *Crepis tectorum* ssp. *pumila* (Asteraceae). Canadian Journal of Botany 70: 89–95.

Armbruster, W. S. 1997. Exaptations link evolution of plant–herbivore and plant–pollinator interactions: A phylogenetic inquiry. Ecology 78: 1661–1672.

Armbruster, W. S. 2002. Can indirect selection and genetic context contribute to trait diversification? A transition-probability study of blossom-colour evolution in two genera. Journal of Evolutionary Biology 15: 468–486.

Armbruster, W. S., V. S. Di Stilio, J. D. Tuxill, T. C. Flores, and J. L. Velásquez-Runk. 1999. Covariance and decoupling of floral and vegetative traits in nine Neotropical plants: A re-evaluation of Berg's correlation-pleiades concept. American Journal of Botany 86: 39–55.

Armbruster, W. S., C. B. Fenster, and M. R. Dudash. 2000. Pollination "principles" revisited: Specialization, pollination syndromes, and the evolution of flowers. Det Norske Videnskaps—Akademi. I. Matematisk Naturvidenskapelige Klasse, Skrifter, Ny Serie 39: 179–200.

Armbruster, W. S., J. J. Howard, T. P. Clausen, E. M. Debevec, J. C. Loquvam, M. Matsuki, B. Cerendolo, and F. Andel. 1997. Do biochemical exaptations link evolution of plant defense and pollination systems? Historical hypotheses and experimental test with *Delachampia* vines. American Naturalist 149: 461–484.

Arnold, S. J., and M. J. Wade. 1984a. On the measurement of natural and sexual selection: Theory. Evolution 38: 709–719.

Arnold, S. J., and M. J. Wade. 1984b. On the measurement of natural and sexual selection: Applications. Evolution 38: 720–734.

Ashman, T. L. 2002. The role of herbivores in the evolution of separate sexes from hermaphroditism. Ecology 83: 1175–1184.

Ashman, T. L., and M. Stanton. 1991. Seasonal variation in pollination dynamics of sexually dimorphic *Sidalcea oregana* ssp. *spicata* (Malvaceae). Ecology 72: 993–1003.

Bingham, R. A. 1998. Efficient pollination of alpine plants. Nature 391: 238.

Brody, A. K. 1997. Effects of pollinators, herbivores, and seed predators on flowering phenology. Ecology 78: 1624–1631.

Buchmann, S. L. 1987. The ecology of oil flowers and their bees. Annual Review of Ecology and Systematics 18: 343–369.

Campbell, D. R. 1991. Effects of floral traits on sequential components of fitness in *Ipomopsis aggregata.* American Naturalist 137: 713–737.

Caswell, H. 2001. Matrix population models: Construction, analysis, and interpretation, 2nd ed. Sinauer Associates, Sunderland, MA.

Charlesworth, D. W. 1994. Evolution in age-structured population, 2nd ed. Cambridge University Press, Cambridge.

Conner, J. K., R. Davis, and S. Rush. 1995. The effect of wild radish floral morphology on pollination efficiency by four taxa of pollinators. Oecologia 104: 234–245.

Conner, J. K., and S. Rush 1996. Effects of flower size and number on pollinator visitation to wild radish, *Raphanus raphanistrum.* Oecologia 105: 509–516.

Crone, E. E. 2001. Is survivorship a better fitness surrogate than fecundity? Evolution 55: 2611–2614.

Dafni, A., and P. Bernhardt. 1990. Pollination of terrestrial orchids of sourthern Australia and the Mediterranean region. Evolutionary Biology 24: 193–252.

Darwin, C. 1862. The various contrivances by which British and foreign orchids are fertilised by insects. Murray, London.

Darwin, C. 1876. The effects of cross and self fertilisation in the vegetable kingdom. Murray, London.

Delph, L. F., M. H. Johannsson, and A. G. Stephenson. 1997. How environmental factors affect pollen performance: Ecological and evolutionary perspectives. Ecology 78: 1632–1638.

Dieringer, G. 1992. Pollinator effectiveness and seed set in populations of *Agalinis strictifolia* (Scrophulariaceae). American Journal of Botany 79: 1018–1023.

Dilley, J. D., P. Wilson, and M. R. Mesler. 2000. The radiation of *Calochortus:* Generalist flowers moving through a mosaic of potential pollinators. Oikos 89: 209–222.

Domínguez, C. A., and R. Dirzo. 1995. Rainfall and flowering synchrony in a tropical shrub: Variable selection on the flowering time of *Erythroxylum havanense*. Evolutionary Ecology 9: 204–216.
Eckhart, V. M. 1991. The effects of floral display on pollinator visitation vary among populations of *Phacelia linearis* (Hydrophyllaceae). Evolutionary Ecology 5: 370–384.
Eckhart, V. M. 1992. Spatio-temporal variation in abundance and variation in foraging behavior of the pollinators of gynodioecious *Phacelia linearis* (Hydrophyllaceae). Oikos 64: 573–586.
Ehrlén, J. 1997. Risk of grazing and flower number in a perennial plant. Oikos 80: 428–434.
Ehrlén, J. 2002. Assessing the lifetime consequences of plant-animal interactions for the perennial herb *Lathyrus vernus* (Fabaceae). Perspectives in Plant Ecology, Evolution, and Systematics 5: 145–163.
Ehrlén, J., S. Käck, and J. Ågren. 2002. Pollen limitation, seed predation, and scape length in *Primula farinosa*. Oikos 97: 45–51.
Euler, M., and I. T. Baldwin. 1996. The chemistry of defense and apparency in corollas of *Nicotiana attenuata*. Oecologia 107: 102–122.
Faegri, K., and L. van der Pijl. 1979. The principles of pollination ecology, 3rd ed. Pergamon Press, Oxford.
Fenster, C. B., and M. R. Dudash. 2001. Spatiotemporal variation in the role of hummingbirds as pollinators of *Silene virginica*. Ecology 82: 844–851.
Fineblum, W. L., and M. D. Rausher. 1997. Do floral pigmentation genes also influence resistance to enemies? The W locus in *Ipomoea purpurea*. Ecology 78: 1646–1654.
Fishbein, M., and D. L. Venable. 1996. Diversity and temporal change in the effective pollinators of *Asclepias tuberosa*. Ecology 77: 1061–1073.
Fleming, T. H., and J. N. Holland. 1998. The evolution of obligate mutualisms: The senita cactus and senita moth. Oecologia 114: 368–375.
Frazee, J. E., and R. J. Marquis. 1994. Environmental contribution to floral trait variation in *Chamaecrista fasciculata* (Fabaceae: Ceasalpinoideae). American Journal of Botany 81: 206–215.
Futuyma, D. J. 2001. Ecological specialization and generalization. Pp. 177–192 *in* C. W. Fox, D. A. Roff, and D. J. Fairbairn (eds.), Evolutionary ecology. Oxford University Press, Oxford.
Galen, C. 1999. Why do flowers vary? Bioscience 49: 631–640.
Galen, C. 2000. High and dry: Drought stress, sex-allocation trade-offs, and selection on flower size in the alpine wildflower *Polemonium viscosum* (Polemoniaceae). American Naturalist 156: 72–83.
Galen, C., and J. Cuba. 2001. Down the tube: Pollinators, predators, and the evolution of flower shape in the alpine skypilot, *Polemonium viscosum*. Evolution 55: 1963–1971.
Galen, C., K. A. Zimmer, and M. E. Newport. 1987. Pollination in floral scent morphs of *Polemonium viscosum:* A mechanism for disruptive selection on flower size. Evolution 41: 599–606.
Gómez, J. M. 1993. Phenotypic selection on flowering synchrony in a high mountain plant, *Hormathophylla spinosa* (Cruciferae). Journal of Ecology 81: 605–613.
Gómez, J. M. 1996. Predispersal reproductive ecology of an arid land crucifer, *Moricandia moricandioides:* Effect of mammal herbivory on seed production. Journal of Arid Environments 33: 425–437.
Gómez, J. M. 2000a. Effectiveness of ants as pollinators of *Lobularia maritima:* Effects on main sequential fitness components of the host plant. Oecologia 122: 90–97.
Gómez, J. M. 2000b. Phenotypic selection and response to selection in *Lobularia maritima:* Importance of direct and correlational components of natural selection. Journal of Evolutionary Biology 13: 689–699.
Gómez, J. M. 2002. Generalización en las interacciones entre plantas y animales. Revista Chilena de Historia Natural 75: 105–116.
Gómez, J. M. 2003. Herbivory reduces the strength of pollinator-mediated selection in the Mediterranean herb *Erysimum mediohispanicum:* Consequences for plant specialization. American Naturalist 162: 242–256.
Gómez, J. M. 2005a. Non-additive effects of ungulates on the interaction between *Erysimum mediohispanicum* and its pollinators. Oecologia 143: 412–418.
Gómez, J. M. 2005b. Long-term effects of ungulates on performance, abundance, and spatial distribution of two montane herbs. Ecological Monographs 75: 231–258.

Gómez, J. M., and R. Zamora. 1992. Pollination by ants: Consequences of the quantitative effect on a mutualistic system. Oecologia 91: 410–418.

Gómez, J. M., and R. Zamora. 1999. Generalization vs. specialization in the pollination system of *Hormathophylla spinosa* (Cruciferae). Ecology 80: 796–805.

Gómez, J. M., and R. Zamora. 2000. Spatial variation in the selective scenarios of *Hormathophylla spinosa* (Cruciferae). American Naturalist 155: 657–668.

Gómez, J. M., R. Zamora, J. A. Hódar, and D. García. 1996. Experimental study of pollination by ants in Mediterranean high mountain and arid habitats. Oecologia 105: 236–242.

Grant, K., and V. Grant. 1965. Flower pollination in the phlox family. Columbia University Press, New York.

Gronquist, M., A. Bezzerides, A. Attygalle, J. Meinwald, M. Eisner, and T. Eisner. 2001. Attractive and defensive functions of the ultraviolet pigments of a flower (*Hypericum calycinum*). Proceedings of the National Academy of Sciences (USA) 98: 13745–13750.

Guitián, J., P. Guitián, and L. Navarro. 1996. Spatio-temporal variation in the interactions between *Cornus sanguinea* and its pollinators. Acta Oecologica 17: 285–295.

Hambäck, P. A. 2001. Direct and indirect effects of herbivory: Feeding by spittlebugs affects pollinator visitation rates and seedset of *Rudbeckia hirta*. Ecoscience 8: 45–50.

Herrera, C. M. 1987. Components of pollinator "quality": Comparative analysis of a diverse insect assemblage. Oikos 50: 79–90.

Herrera, C. M. 1988. Variation in mutualism: The spatio-temporal mosaic of a pollinator asemblage. Biological Journal of the Linnean Society 35: 95–125.

Herrera, C. M. 1989. Pollinator abundance, morphology, and flower visitation rate: Analysis of the "quantity" component in a plant–pollinator system. Oecologia 80: 241–248.

Herrera, C. M. 1993. Selection of floral morphology and environmental determinants of fecundity in a hawk-moth pollinated violet. Ecological Monographs 63: 251–275.

Herrera, C. M. 1996. Floral traits and plant adaptation to insect pollinators: A devil's advocate approach. Pp. 65–87 *in* D. G. Lloyd and S. C. H. Barrett (eds.), Floral biology. Chapman and Hall, New York.

Herrera, C. M. 2000a. Measuring the effects of pollinators and herbivores: Evidence for nonadditivity in a perennial herb. Ecology 81: 2170–2176.

Herrera, C. M. 2000b. Flower-to-seedling consequences of different pollination regimes in an insect-pollinated shrub. Ecology 81: 15–29.

Herrera, C. M., M. Medrano, P. J. Rey, A. M. Sánchez-Lafuente, M. B. García, J. Guitián, and A. J. Manzaneda. 2002. Interaction of pollinators and herbivores on plant fitness suggests a pathway for correlated evolution of mutualism- and antagonism-related traits. Proceedings of the National Academy of Sciences (USA) 99: 1683–1688.

Horvitz, C. C., and D. W. Schemske. 1990. Spatiotemporal variation in insect mutualists of a neotropical herb. Ecology 71: 1085–1097.

Johnson, S. D., and K. E. Steiner. 2000. Generalization versus specialization in plant pollination systems. Trends in Ecology and Evolution 15: 140–143.

Johnson, S. G., L. F. Delph, and C. L. Elderkin. 1995. The effect of petal-size manipulation on pollen removal, seed set, and insect-visitor behavior in *Campanula americana*. Oecologia 102: 174–179.

Johnston, M. O. 1991. Natural selection on floral traits in two species of *Lobelia* with different pollinators. Evolution 45: 1468–1479.

Kalisz, S. 1986. Variable selection on the timing of germination in *Collinsia verna* (Scrophulariceae). Evolution 40: 479–491.

Karban, R., and S. Y. Strauss. 1993. Effects of herbivores on growth and reproduction of their perennial host, *Erigeron glaucus*. Ecology 74: 39–46.

Keys, R. N., S. L. Buchmann, and S. E. Smith. 1995. Pollination effectiveness and pollination efficiency of insects foraging *Prosopis velutina* in southeastern Arizona. Journal of Applied Ecology 32: 519–527.

Krupnick, G. A., A. E. Weis, and D. R. Campbell. 1999. The consequences of floral herbivory for pollinator service to *Isomeris arborea*. Ecology 80: 125–134.

Lara, C., and J. F. Ornelas. 2001. Preferential nectar robbing of flowers with long corolla: Experimental studies of two hummingbirds species visiting three plant species. Oecologia 128: 263–273.
Lee, T. N., and A. A. Snow. 1998. Pollinator preferences and the persistence of crop genes in wild radish populations (*Raphanus raphanistrum,* Brassicaceae). American Journal of Botany 85: 333–339.
Leege, L. M., and L. M. Wolfe. 2002. Do floral herbivores respond to variation in flower characteristics in *Gelsemium sempervirens* (Loganiaceae), a distylous vine? American Journal of Botany 89: 1270–1274.
Lehtilä, K., and S. Y. Strauss. 1999. Effects of foliar herbivory on male and female reproductive traits of wild radish, *Raphanus raphanistrum.* Ecology 80: 116–124.
Levin, D. A. 2000. The origin, expansion, and demise of plant species. Oxford University Press, Oxford.
Lippok, B., A. A. Gardine, P. S. Williamson, and S. S. Renner. 2000. Pollination by flies, bees, and beetles of *Nuphar ozarkana* and *N. advena* (Nymphaeaceae). American Journal of Botany 87: 898–902.
Maad, J. 2000. Phenotypic selection in hawkmoth-pollinated *Platanthera bifolia:* Targets and fitness surfaces. Evolution 54: 112–123.
MacArthur, R. H. 1972. Geographical ecology: Patterns in the distribution of species. Harper and Row, New York.
Mayfield, M. M., N. M. Waser, and M. V. Price. 2001. Exploring the "most effective pollinator principle" with complex flowers: Bumblebees and *Ipomopsis aggregata.* Annals of Botany 88: 591–596.
Mitchell, R. J., R. G. Shaw, and N. M. Waser. 1998. Pollinator selection, quantitative genetics, and predicted evolutionary responses of floral traits in *Penstemon centranthifolius* (Scrophulariaceae). International Journal of Plant Sciences 159: 331–337.
Mothershead, K., and R. J. Marquis. 2000. Fitness impact of herbivory through indirect effects on plant-pollinator interactions in *Oenothera macrocarpa.* Ecology 81: 30–40.
Motten, A. F., D. R. Campbell, D. E. Alexander, and H. L. Miller. 1981. Pollination effectiveness of specialist and generalist visitors to a North Carolina population of *Claytonia virginica.* Ecology 62: 1278–1287.
Muchhala, N. 2003. Exploring the boundary between pollination syndromes: Bats and hummingbirds as pollinators of *Burmeistera cyclostigmata* and *B. tenuiflora* (Campanulaceae). Oecologia 134: 373–380.
Obeso, J. R. 1992. Pollination ecology and seed set in *Asphodelus albus* (Liliaceae) in Northern Spain. Flora 187: 219–226.
O'Connell, L. M. O., and M. O. Johnston. 1998. Male and female pollination success in a deceptive orchid, a selection study. Ecology 79: 1246–1260.
Olesen, J. M. 2000. Exactly how generalised are pollination interactions? Det Norske Videnskaps—Akademi. I. Matematisk Naturvidenskapelige Klasse, Skrifter, Ny Serie 39: 161–178.
Ollerton, J. 1996. Reconciling ecological processes with phylogenetic patterns: The apparent paradox of plant-pollinator systems. Journal of Ecology 84: 767–769.
Ollerton, J. 1999. La evolución de las relaciones polinizador-planta en los Artrópodos. Boletín de la Sociedad Entomológica Aragonesa 26: 741–758.
Ollerton, J., and S. Watts. 2000. Phenotype space and floral typology: Towards an objective assessment of pollination syndrome. Det Norske Videnskaps—Akademi. I. Matematisk Naturvidenskapelige Klasse, Skrifter, Ny Serie 39: 149–160.
Olsen, K. M. 1997. Pollinator effectiveness and pollinator importance in a population of *Heterotheca subaxillaris* (Asteraceae). Oecologia 109: 114–121.
Pellmyr, O. 1997. Pollinating seed eaters. Why is active pollination so rare? Ecology 78: 1655–1660.
Petterson, M. W. 1991. Pollination by a guild of fluctuating moth populations: Option for unspecialization in *Silene vulgaris.* Journal of Ecology 79: 591–604.
Potts, S. G., A. Dafni, and G. Ne'eman. 2001. Pollination of a core flowering shrub species in Mediterranean phrygana: Variation in pollinator diversity, abundance, and effectiveness in response to fire. Oikos 92: 71–80.
Quesada, M., K. Bollman, and A. G. Stephenson. 1995. Leaf damage decreases pollen production and hinders pollen performance in *Cucurbita texana.* Ecology 76: 437–443.

Schemske, D. W., and H. D. Bradshaw. 1999. Pollinator preference and the evolution of floral traits in monkeyflowers (*Mimulus*). Proceedings of the National Academy of Sciences (USA) 96: 11910–11915.

Schemske, D. W., and C. C. Horvitz. 1984. Variation among floral visitors in pollination ability: A precondition for mutualism specialization. Science 225: 519–521.

Schemske, D. W., and C. C. Horvitz. 1989. Temporal variation in selection on a floral character. Evolution 43: 461–465.

Shykoff, J. A., E. Bucheli, and O. Kaltz. 1997. Anther smut disease in *Dianthus sylvester* (Caryophyllaceae): Natural selection on floral traits. Evolution 51: 383–392.

Simms, E. L., and M. A. Bucher. 1996. Pleiotropic effects of flower-color intensity on herbivore performance on *Ipomoea purpurea*. Evolution 50: 957–963.

Stanton, M., H. J. Young, N. C. Ellstrand, and J. M. Clegg. 1991. Consequences of floral variation for male and female reproduction in experimental populations of wild radish, *Raphanus sativus* L. Evolution 45: 268–280.

Stebbins, G. L. 1970. Adaptive radiation of reproductive characteristics in angiosperms, I: Pollination mechanisms. Annual Review of Ecology and Systematics 1: 307–326.

Steiner, K. E., and V. B. Whitehead. 2002. Oil secretion and the pollination of *Colpias mollis* (Scrophulariaceae). Plant Systematics and Evolution 235: 53–66.

Stone, J. L. 1996. Components of pollination effectiveness in *Psychotria suerrensis*, a tropical distylous shrub. Oecologia 107: 504–512.

Strauss, S. Y. 1997. Floral characters link herbivores, pollinators, and plant fitness. Ecology 78: 1640–1645.

Strauss, S. Y., J. K. Conner, and S. L. Rush. 1996. Foliar herbivory affects floral characters and plant attractiveness to pollinators: Implications for male and female plant fitness. American Naturalist 147: 1098–1107.

Strauss, S. Y., D. H. Siemens, M. B. Decher, and T. Mitchell-Olds. 1999. Ecological costs of plant resistance to herbivores in the currency of pollination. Evolution 53: 1105–1113.

Thompson, J. D. 1994. The coevolutionary process. University of Chicago Press, Chicago.

Thompson, J. D. 2001. How do visitation patterns vary among pollinators in relation to floral display and floral design in a generalist pollination system? Oecologia 126: 386–394.

Thompson, J. D., and O. Pellmyr. 1992. Mutualism with pollinating seed parasites amid copollinators: Constraints on specialization. Ecology 73: 1780–1791.

Thøstesen, A. M., and J. M. Olesen. 1996. Pollen removal and deposition by specialist and generalist bumblebees in *Aconitum septentrionale*. Oikos 77: 77–84.

Traveset, A., and E. Sáez. 1997. Pollination of *Euphorbia dendroides* by lizards and insects: Spatiotemporal variation in patterns of flower visitation. Oecologia 111: 241–248.

Van Tienderen, P. H. 2000. Elasticities and the link between demographic and evolutionary dynamics. Ecology 81: 666–679.

Vaughton, G. 1992. Effectiveness of nectarivorous birds and honeybees as pollinators of *Banksia spinulosa* (Proteaceae). Australian Journal of Ecology 17: 43–50.

Vaughton, G., and M. Ramsey. 1998. Floral display, pollinator visitation, and reproductive success in the dioecious perennial herb *Wurmbea dioica* (Liliaceae). Oecologia 115: 93–101.

Via, S., and D. J. Hawthorne. 2002. The genetic architecture of ecological specialization: Correlated gene effects on host use and habitat choice in pea aphids. American Naturalist 159: S76–S88.

Waser, N. M. 1983. The adaptive nature of floral traits: Ideas and evidence. Pp. 241–285 *in* L. A. Real (ed.), Pollination biology. Academic Press, Orlando, FL.

Waser, N. M. 1998. Pollination, angiosperm speciation, and the nature of species boundaries. Oikos 81: 198–201.

Waser, N. M., L. Chittka, M. V. Price, N. M. Williams, and J. Ollerton. 1996. Generalization in pollination systems, and why it matters. Ecology 77: 1043–1060.

Widén, B. 1991. Phenotypic selection on flowering phenology in *Senecio integrifolius*, a perennial herb. Oikos 61: 205–215.

Wilson, P. 1995. Selection for pollination success and the mechanical fit of *Impatiens* flowers around bumblebee bodies. Biological Journal of the Linnean Society 55: 355–383.

Wilson, P., and J. D. Thomson. 1996. How do flowers diverge? Pp. 88–111 *in* D. G. Lloyd and S. C. H. Barrett (eds.), Floral biology. Chapman and Hall, New York.
Zamora, R. 1999. Conditional outcomes of interactions: The pollinator-prey conflict of an insectivorous plant. Ecology 80: 786–795.
Zamora, R. 2000. Functional equivalence in plant-animal interactions: Ecological and evolutionary consequences. Oikos 88: 442–447.

PART III

Community and Biogeographic Perspectives

Introductory Comments by Nickolas M. Waser and Jeff Ollerton

Ideas come and go in ecology, as in any field. Not so long ago, in the 1970s, community ecology held sway, with a goal of understanding the phenotypes and diversity of assemblages of trophically similar species at single sites and across sites. Community ecology was then eclipsed for a generation, following the severe critique by advocates of "null models," and only now is recrudescent in the new arenas of conservation and restoration ecology. It is interesting that, whereas Galápagos finches and other birds figure heavily in the perception of the null model debate in community ecology, phenological patterns in coexisting plant species, putatively driven by competition for pollination services among the plants, actually elicited an earlier shot across the bows. Stiles (1977) claimed regular spacing of phenologies (the implied expectation of the day being that competition would generate regularity), but Poole and Rathcke (1979) showed that spacing did not differ from random. (An interesting postscript is the demonstration by Cole [1981] that statistical regularity does exist if the data are subdivided seasonally.)

This example was not unique for the time; community-level studies of pollination enjoyed relative popularity in that earlier era. A common theme was phenological or other adaptation to minimize interspecific competition, revealed by patterns in community structure (e.g., Heithaus 1974; Feinsinger 1978; Pleasants 1980). But, in the footsteps of earlier workers such as Clements and Long (1923) and Robertson (1929), another trend was to simply catalog all the plants, the pollinators, and their interactions (e.g., Moldenke 1975, 1979a, 1979b; Moldenke and Lincoln 1979; O'Brien 1980). It was not feasible, however, either in the 1920s or the 1970s, to go much beyond cataloging. Thus, Moldenke and colleagues could attempt to characterize their impressive data in terms of species diversity, niche breadths, or "pollination syndromes" (Waser, chap. 1 in this volume) but could no more examine the properties of entire plant–pollinator assemblages than could Robertson a half century earlier. The consequence of these constraints of computational power (coupled with historical fads?) has been to

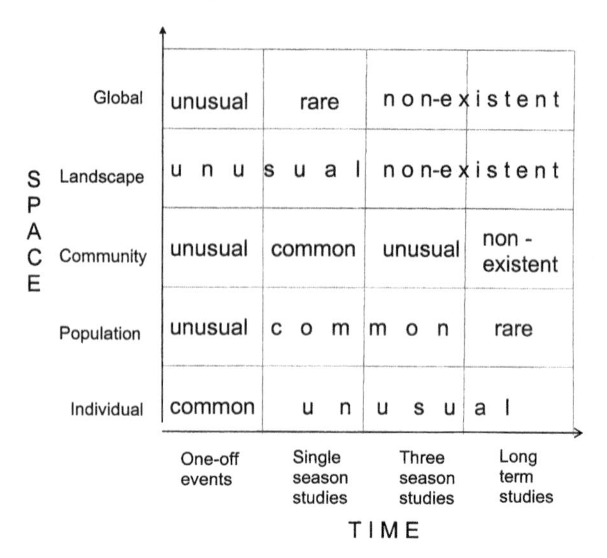

Figure P.1 The distribution of studies in pollination biology until very recently, on axes of space and time

leave large parts of the interesting territory of pollination biology unexplored (fig. P.1), a dilemma shared more generally throughout ecology. No wonder that Jordano (1987) had only fragments of pollination systems to analyze using a food-web approach!

The six chapters that follow illustrate how radically this situation is changing. Jordano et al. (chap. 8) start off part 3 by reviewing how entire plant–pollinator communities can be treated as networks of interactions. The multivariate statistical methods that these authors apply to pollination networks reveal properties far beyond those a traditional food-web analysis would have considered, and a number of these properties seem to be critical for the dynamics of pollination interactions. Indeed, Jordano et al. contend that a perspective that takes in pollination systems in their entirety is essential for answering many ecological questions, such as response of plants and pollinators to anthropogenic perturbations—in this regard, see Memmott and Waser (2002) and Memmott et al. (2004) for some of what is revealed from computer analysis of the data of Robertson (1929)—as well as evolutionary questions, such as the interpretation of floral adaptations (see also Aigner, chap. 2 in this volume). And these authors stress that pollination networks (and networks of other mutualistic interactions studied to date) appear to have common topological features that include patterns of specialization and generalization, which are critical to their dynamics.

One question now springs to mind: what evolutionary and ecological processes generate these features of pollination networks? Vázquez and Aizen (chap. 9) explore some themes very similar to those of chapter 8, but with the emphasis on comparing network patterns to those generated by two types of null models: one assuming random interactions among species and another assuming random encounters among individual plants and animals (hence, encounters influenced by relative abundances of different species). Vázquez and Aizen show that specialists and generalists are more common than predicted by the first null model—especially in larger pollination networks—which suggests that larger communities produce more opportunities for the evolution and persistence of extreme generalists and specialists. Larger networks also show more pronounced "asymmetric specialization"—the tendency of specialist pollinators to interact with generalist plants and vice versa (see also Jordano et al., chap. 8 in this volume; Petanidou and Potts, chap. 10 in this volume). Lurking behind these patterns is a strong correlation between relative abundance of a species and its apparent degree of generalization. Therefore, Vázquez and Aizen use rarifaction methods to explore whether this correlation is explained by sampling effort (few interactions are observed when one looks at a rare species, so it appears specialized) or by "real biology"; they come down on the side of the latter. This is a promising start toward a scenario for the evolutionary and ecological assembly of pollination networks.

Simulations of rarifaction aside, one's view of a plant–pollinator community *must* show some improvement as one samples over a longer time, and one's view of the resource use by component species must also improve as one examines these species on larger spatial scales. Petanidou and Potts (chap. 10) illustrate these points with the example of three Mediterranean systems. The best studied is in the Greek phrygana, and the substantial annual fluctuation in the pollinator fauna in this system is just one reason to treat studies of short duration or taxonomic scope with caution. Petanidou and Potts also introduce an index of "selectivity" in which breadth of resource use is weighted by the availability of mutualistic species. Whereas the three Mediterranean systems appear different in distribution of niche breadths (i.e., of specialization and generalization), and pollinators appear more specialized than plants, some of these differences seem to be due to sampling and lack of weighting by resource availability. Thus, pollinators and plants in the phrygana appear quite similar from the selectivity index. Furthermore, selectivity is greatest in the peak of the season, when competition among plants and among pollinators is expected to be greatest, which suggests greater niche partitioning during that period. Aha! A hint of true biology (and classical community ecology) in explaining specialization!

Pollinator (and plant) abundances are not only likely to fluctuate across years, but also across seasons. Medan et al. (chap. 11) discuss some of the consequences of this for characterizing pollination networks, using the Argentinian Talar forest and the Greek phrygana as examples. Both of these are "year-long" systems (i.e., lacking a short, discrete growth season); thus, it is possible to apply a varying temporal window in sampling them and to calculate web statistics either over short subsets of the year or over longer periods up to the entire year. Which approach is best? Medan et al. show how web connectance declines with increasing length of the sampling window because many interactions are impossible due to phenological separation of species. Therefore, as a connectance value for the entire year-long system, they advocate the average of values from shorter periods in which species actually overlap in phenology. They also explore the Talar and phrygana with both a "resource use" index and an index of evenness of resource use that takes account of *quantitative* information on interaction strength (see also Memmott 1999). The resource use and evenness indices are not correlated; therefore, Medan et al. examine how species in the Talar occupy a two-dimensional space where the two indices are the axes. Among the results, not all portions of the space are occupied, this varies by season of the year, and individual species shift position through the year (translation: specialization and generalization are context specific).

The final two chapters of this part expand to a global scope and take up a burning question: are the tropics more specialized than other latitudinal zones? Armbruster (chap. 12) brings to this quest a highly personal perspective based on the unparalleled geographic scope of his own studies, ranging from tropical to subarctic. Here he concentrates mainly on four plant-centered examples from different latitudes, augmenting this with additional natural history observations. The four plant groups and their pollination systems are cast into a phylogenetic framework, a powerful approach for exploring "evolutionary specialization" (as distinct from "ecological specialization"; see also the introduction to part 2 in this volume). To augment the clues revealed by mapping pollination systems onto phylogeny, Armbruster develops explicit natural selection scenarios for the evolution of specialization (which he defines broadly; e.g., to include precise pollen placement on insects) and the features that promote it, such as type of floral reward or basic flower *Bauplan*. The upshot is that the tropics do appear to have a greater *range* in degree of specialization than temperate zones, that specialization is quite reversible in evolutionary time, and that certain floral features do seem to provide opportunities for specialization.

Armbruster's approach is full of insights and challenges for further work, but he admits it may be idiosyncratic, a result of $N = 4$ systems. In chapter 13, Ollerton et al. take a different approach. The authors mine available data from around the world for trends in the diversity of pollination systems and in the range of their specialization or generalization. Tropical habitats do appear to harbor more pollination systems than those at any higher latitude, but the evidence is mixed that tropical pollination interactions are more specialized; the key here seems to be whether sampling effort is taken into account. Habitat complexity also has some effects in intuitively pleasing directions (i.e., complexity is positively correlated with diversity of pollination systems), although complexity is confounded with latitude in the data. Finally, Ollerton et al. choose two widespread plant taxa and show that their tendency to specialize versus generalize in pollination varies dramatically across regions—a caution against drawing con-

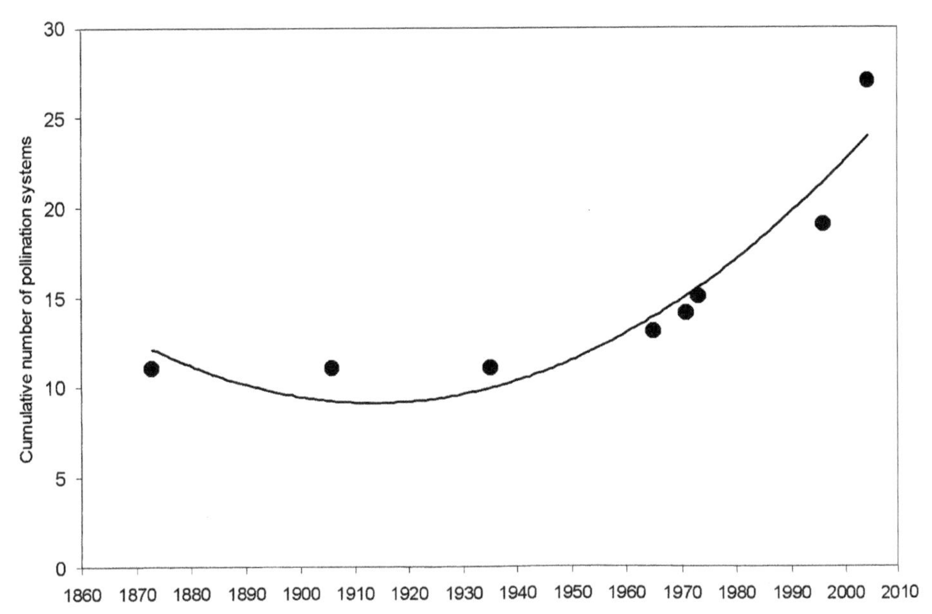

Figure P.2 The historical discovery of distinct pollination systems. Data points correspond to the cumulative number of pollination systems recognized by the following authors: Müller (1883), Knuth (1906), James and Clapham (1935), Percival (1965), Faegri and van der Pijl (1971), Proctor and Yeo (1973), and Proctor et al. (1996). The final data point is a post-1996 survey of the literature by J. Ollerton (unpublished data). "Distinct pollination systems" are defined as plant–pollinator interactions that are restricted to a taxonomically narrow set of pollinators and to flowers that (usually) show specific adaptations to that particular pollinator, for example, bird pollination or moth pollination.

clusions from a single region. This last is a main message of chapter 13 and indeed this entire part: there is no reason to expect ecological systems to adhere everywhere to a single pattern.

This point may seem trivial, but, if so, we must wonder why regional differences in outcome sometimes lead to heated disputes in ecology! In the case of specialization and generalization in pollination systems, the outcomes thus far do seem to differ by region and taxon; however, in our view this does not signal a "debate," which would imply that one view is "correct." Instead, a great deal of demanding and fascinating work remains to be done in characterizing what Ollerton and others term the "biogeography of species interactions." As the South African pollination biologists have been saying for some time, their part of the world harbors a fantastic diversity of pollination systems, many quite specialized. What other surprises are in store from poorly studied regions (e.g., southwestern Australia, with a flora as rich as South Africa's)? A rough historical survey of the discovery of unique pollination systems (fig. P.2) suggests that we are far from cataloging the full diversity of pollination interactions. The challenge for the future—and it is a big one—is not only to more completely characterize the pollination systems of the biosphere but also to mechanistically determine *why* some regions and taxa differ from others, for example, in patterns of specialization and generalization.

References

Clements, R. E., and F. L. Long. 1923. Experimental pollination: An outline of the ecology of flowers and insects. Carnegie Institution of Washington, Washington, DC.

Cole, B. J. 1981. Overlap, regularity, and flowering phenologies. American Naturalist 117: 993–997.

Faegri, K., and L. van der Pijl. 1971. The principles of pollination ecology, 2nd ed. Pergamon Press, Oxford, UK.

Feinsinger, P. 1978. Ecological interactions between plants and hummingbirds in a successional tropical community. Ecological Monographs 48: 269–287.

Heithaus, E. R. 1974. The role of plant–pollinator interactions in determining community structure. Annals of the Missouri Botanical Garden 61: 675–691.

James, W. O., and A. R. Clapham 1935. The biology of flowers. Oxford University Press, Oxford, UK.

Jordano, P. 1987. Patterns of mutualistic interactions in pollination and seed dispersal: Connectance, dependence asymmetries, and coevolution. American Naturalist 129: 657–677.

Knuth, P. 1906. Handbook of flower pollination, vol. 1. Clarendon Press, Oxford, UK.

Memmott, J. 1999. The structure of a plant–pollinator food web. Ecology Letters 2: 271–280.

Memmott, J., and N. M. Waser. 2002. Integration of alien plants into a native flower–pollinator visitation web. Proceedings of the Royal Society of London B 269: 2395–2399.

Memmott, J., N. M. Waser, and M. V. Price. 2004. Tolerance of pollination networks to species extinctions. Proceedings of the Royal Society of London B 271: 2605–2611.

Moldenke, A. R. 1975. Niche specialization and species diversity along a California transect. Oecologia 21: 219–242.

Moldenke, A. R. 1979a. Pollination ecology as an assay for ecosystemic organization: Convergent evolution in Chile and California. Phytologia 42: 415–454.

Moldenke, A. R. 1979b. Pollination ecology within the Sierra Nevada. Phytologia 42: 223–282.

Moldenke, A. R., and P. G. Lincoln. 1979. Pollination ecology in montane Colorado: A community analysis. Phytologia 42: 349–379.

Müller, H. 1883. The fertilization of flowers. Macmillan, London.

O'Brien, M. H. 1980. The pollination biology of a pavement plain: Pollinator visitation patterns. Oecologia 47: 213–218.

Percival, M. S. 1965. Floral biology. Pergamon Press, Oxford, UK.

Pleasants, J. M. 1980. Competition for bumblebee pollinators in Rocky Mountain plant communities. Ecology 61: 1446–1459.

Poole, R. W., and B. J. Rathcke. 1979. Regularity, randomness, and aggregation in flowering phenologies. Science 203: 470–471.

Proctor, M., and P. Yeo. 1973. The pollination of flowers. Collins, London.

Proctor, M., P. Yeo, and A. Lack. 1996. The natural history of pollination. HarperCollins, London.

Robertson, C. 1929. Flowers and insects: Lists of visitors to four hundred and fifty-three flowers. C. Robertson, Carlinville, IL.

Stiles, F. G. 1977. Coadapted competitors: The flowering seasons of hummingbird-pollinated plants in a tropical forest. Science 198: 1177–1178.

CHAPTER EIGHT

The Ecological Consequences of Complex Topology and Nested Structure in Pollination Webs

Pedro Jordano, Jordi Bascompte, and Jens M. Olesen

To me the most important thing in composition is disparity . . . Anything suggestive of symmetry is decidedly undesirable, except possibly where an approximate symmetry is used in a detail to enhance the inequality with the general scheme.
—Alexander Calder, "A Propos of Measuring a Mobile"

The extraordinary series of mobiles created by Alexander Calder provide a vivid illustration of how the dynamics of interconnected parts depends on the way they are connected or linked to each other. Calder's mobiles are complex structures of pieces of metal connected by wires or ropes that keep the massive sculptures in equilibrium while they move suspended in air (Calder and Davidson 1966). This equilibrium depends on both the number and the size of pieces and the way they are connected—not only pairwise, but collectively.

In nature, networks of species interactions are the architecture of biodiversity, because community dynamics rely deeply on the way species interact. Pollination by animals is the most common means of fertilization in higher plants, and the mutualism involved in the process illustrates the pervasiveness of complex networks of interaction. For example, tropical forests harbor woody floras in which more than 80% of species rely on animal pollinators for reproduction (Gentry 1982). Most pollination interactions are not specific and do not involve tight mutualisms between species pairs, yet pollination interactions are paradigmatic examples of coevolved interactions among animals and plants. Despite evidence for highly diversified interactions, the well-known precise adjustments between flowers and their pollinator visitors to ensure efficient pollination and adequate handling of the floral rewards led to the prevailing notion of highly specialized interactions. Darwin (1862) advanced a hypothesis of flower morphology evolution based on a highly specialized interaction between a long-spurred orchid and the specialized pollinator it ought to have, later found to be a long-tongued sphingid moth. Since then, textbooks have presented pollination interactions between animals and plants as paradigmatic of mutual speciali-

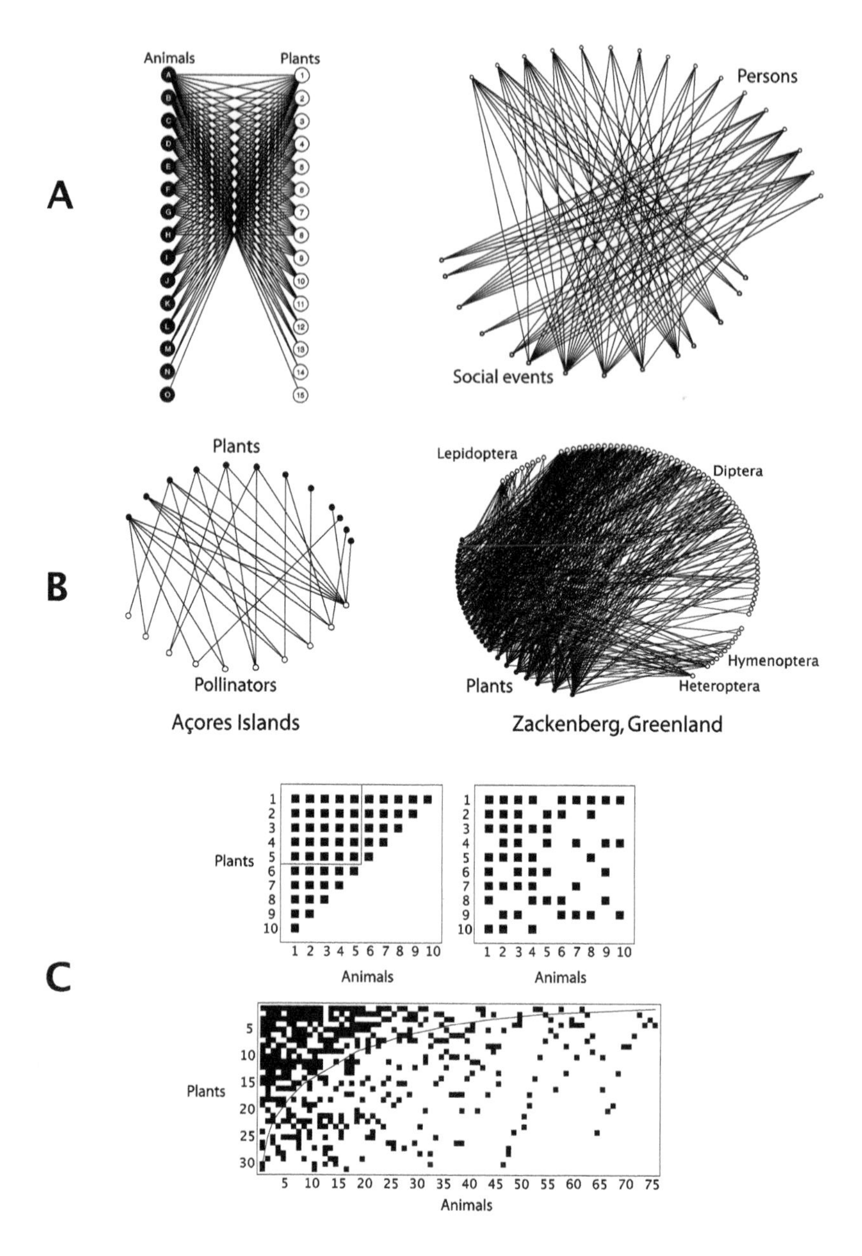

zation. However, when one considers communitywide patterns, for example, including all the flower species and all the pollinator taxa that interact in a particular location, a wide range of generalization in the mode of interaction emerges as a prevalent pattern. This illustrates the lasting debate about generalization versus specialization of pollination interactions. The debate stems from the difficulty of assessing the totality of biotic interactions within highly diversified communities, and it leads into the difficulties of quantifying generalization at

Figure 8.1 Bipartite graphs as representations of networks, illustrating plant–pollinator interaction networks. (A) Bipartite graph representation of the interactions among $A = 15$ pollinator taxa and $P = 15$ plant taxa, with $k = 120$ interactions. Species are nodes, or vertices, in such a graph, and the pairwise interactions among them are represented by lines connecting two nodes of different sets. Pollinator species A visits all 15 plants; plant species 15 is visited only by A. These bipartite networks are similar to, for example, sociological networks representing the relationships among people attending a series of social events (Davis et al. 1941). (B) Examples of the interaction patterns in relatively simple plant–pollinator webs in the Açores Islands (Olesen et al. 2002) and Zackenberg, Greenland (Olesen and Elberling, unpublished data). (C) Matrix representation of bipartite data. Rows represent plant species; columns represent pollinator species, and black boxes indicate actually documented pairwise interactions. The top left panel shows a perfectly nested matrix, where interactions of the more specialized species are a proper subset of the more generalized interactions; the top right panel shows a matrix of random interactions; and the bottom panel shows the actual dataset of Zackenberg, Greenland, in matrix form. The curved line shows the isocline of perfect nestedness; that is, all interactions (black boxes) would lie to the left of the isocline if the matrix were perfectly nested (see Bascompte et al. 2003).

the community level (Waser et al. 1996; Johnson and Steiner 2000; Olesen 2000). As stated by Thompson (1994), we need more than the analysis of pairwise interactions to understand the evolution of diversified mutualisms such as animal-mediated pollination.

Pairs of pollinator and plant species do not interact in an ecological vacuum, and the outcome of their interaction is best viewed within the network of community-level interactions. For instance, the possibility for a rare species to persist might depend on its ability to develop specialized interactions with a specialist pollinator or on the sharing of pollination services from generalists (Memmott and Waser 2002). The robustness of a network of interactions (i.e., the ability of the component species to persist given the extinction of a partner) may depend on the pattern of shared interactions, not uniquely on pairwise interaction with the extinct species. These issues, among others, require understanding of the web of plant–pollinator interactions.

The study of complex networks has flourished in recent years (Strogatz 2001; Albert and Barabási 2002), and general patterns are starting to emerge which point to interesting properties shared by many types of networks. Recent findings reveal consistent patterns in structure, irrespective of the type of network, for example, similarities between abiotic and biotic networks (Newman 2003). These networks share a fundamental structure or architecture of nodes (elements or parts) linked by connections (fig. 8.1). The frequency distribution of the number of links per node has generally been reported to decay as power-law (scale-free), broad-scale (i.e., truncated power-law), or faster-decaying functions (i.e., exponential; Amaral et al. 2000; Strogatz 2001). Power-law distributions of connectivity are charaterized by a high frequency of nodes with few connections and a few highly connected nodes. These generalized patterns have implications for the evolution, stability, and resilience to perturbations of these networks (Barabási and Albert 1999; Albert and Barabási 2002; Dorogovtsev and

Mendes 2002). For example, exponential functions describe randomly assembled networks, whereas power-law distributions result from predictable build-up processes (Barabási and Albert 1999). Thus, the comparative statistical analysis of complex networks sheds lights on their dynamics. Similar patterns have been documented in the ecological literature in recent years, yet few data are available that provide sufficient resolution (Williams and Martinez 2000; Dunne et al. 2002; Montoya and Solé 2002). We have only a limited sample of the complex and diversified patterns of interaction among species in natural ecosystems because most of the previous work on ecological networks focused on food webs and predator–prey interactions. Mutualistic, parasite–host, facilitation, and commensalism interactions are best represented by bipartite graphs of species interactions (Jordano 1987; Poulin 1996; Poulin and Guegan 2000; Jordano et al. 2003) and share both topological (connectivity) and structural patterns (Bascompte et al. 2003) with great potential for influencing species coevolution.

Considering plant–pollinator interactions at the community level is important for several reasons. First, the evolution of pollination adaptations in floral traits most likely results from community-level processes that involve the interaction of groups of species, and not exclusively from the sum of pairwise interactions between plant and animal species. Second, the evolutionary robustness of plant–pollinator interactions depends on properties best viewed at the community level, such as the resilience after extinctions of taxa or the resistance to invasions by exotic species. Third, the manner in which multispecies interactions are organized probably influences the possibilities of rare species for persistence (i.e., how they get reliable pollination services or floral rewards from other taxa). It is only by considering quantitative techniques for complex networks analysis—by characterizing the interactions among species (nodes or vertices) in animal and plant communities—that we can address the potential for variations in network topology to influence coevolutionary processes in high-diversity mutualistic webs.

In this chapter, we review recent advances in the analysis of complex interaction networks and apply them to the study of plant–pollinator interactions. In addition, we explore future avenues of research such as robustness to the loss of species. Taken together, these findings point to very general patterns of generalization–specialization gradients that rest on shared topological and structural properties of how interactions among complex species assemblages are built.

Definitions and Methods

Plant-pollinator records are typical two-mode data (Borgatti and Everett 1997), where the relations between two sets of entities (here, the sets of plant and pollinator species) are described (fig. 8.1A). Most, if not all, of the included interactions can be considered mutualistic interactions, where both animal and plant partners obtain a benefit; of course, a gradient of types of effects exists (Thompson 1982) in such a diverse array of interactions, ranging from completely beneficial to almost antagonistic (see also Renner, chap. 6 in this volume). A community-level analysis of mutualism-driven coevolution has to account for the full range of interactions and their outcomes.

Entities contain nodes or vertices, which in our case are interacting species; vertices are also called actors in the sociological literature (Newman 2003). Lines that connect two vertices are called edges or links. In the ecological literature, nodes are species and links are interactions among them. The degree of a node is the number of edges connected to it (i.e., the number of interactions per species). The links in these networks only run between nodes of the different sets. A plant–pollinator interaction network is thus defined by an adjacency matrix R which describes the reproductive and trophic interactions between communities of P plant species and A plant-visiting animal species within a well-defined habitat (Jordano et al. 2003):

$$R = [a_{ij}]_{A \times P},$$

where $a_{ij} = 0$ if there is no interaction observed between species i and j, or $a_{ij} = 1$ if an interaction has actually been recorded (figs. 8.1A, 8.1C, 8.2).

Thus, this matrix has k nonzero elements (a_{ij}) wherever plants are pollinated by flower-visiting animals that harvest pollen or nectar. The matrix R would have $A + P$ nodes or vertices (species) and k links among them (figs. 8.1A, 8.1C). These typically are sparse matrices (Duff et al. 1986; Boisvert et al. 1997); that is, matrices with a significant number of zero elements (fig. 8.2). The matrix of interactions captures the essence of interaction patterns at the species level within a given community. Whenever two species are recorded as interacting, the elements $a_{ij} = 1$ when only the qualitative interaction is recorded. If quantitative information is available (e.g., frequency of visitation), for elements with $a_{ij} \neq 0$, we have some estimate of reliance of the pollinator on the plant (e.g., fraction of the pollinator's visits to the plant species relative to the total number of visits) or reliance of the plant on the pollinator (e.g., fraction of visits by the pollinator relative to the total visits by all pollinators or fruit set level resulting from pollinator visitation; Jordano 1987; see also Laska and Wootton 1998; Vázquez and Aizen, chap. 9 in this volume). In this case, the matrix would be valued and, in the case of a bipartite graph representation, would have two values, one describing the dependence or strength of the plant on the pollinator and another one for the pollinator on the plant (see Jordano 1987; Dicks et al. 2002; Vázquez and Aizen, chap. 9 in this volume). Such networks can evolve over time, with links among plants and pollinators appearing and disappearing according to phenological variation or even changing in their strength values.

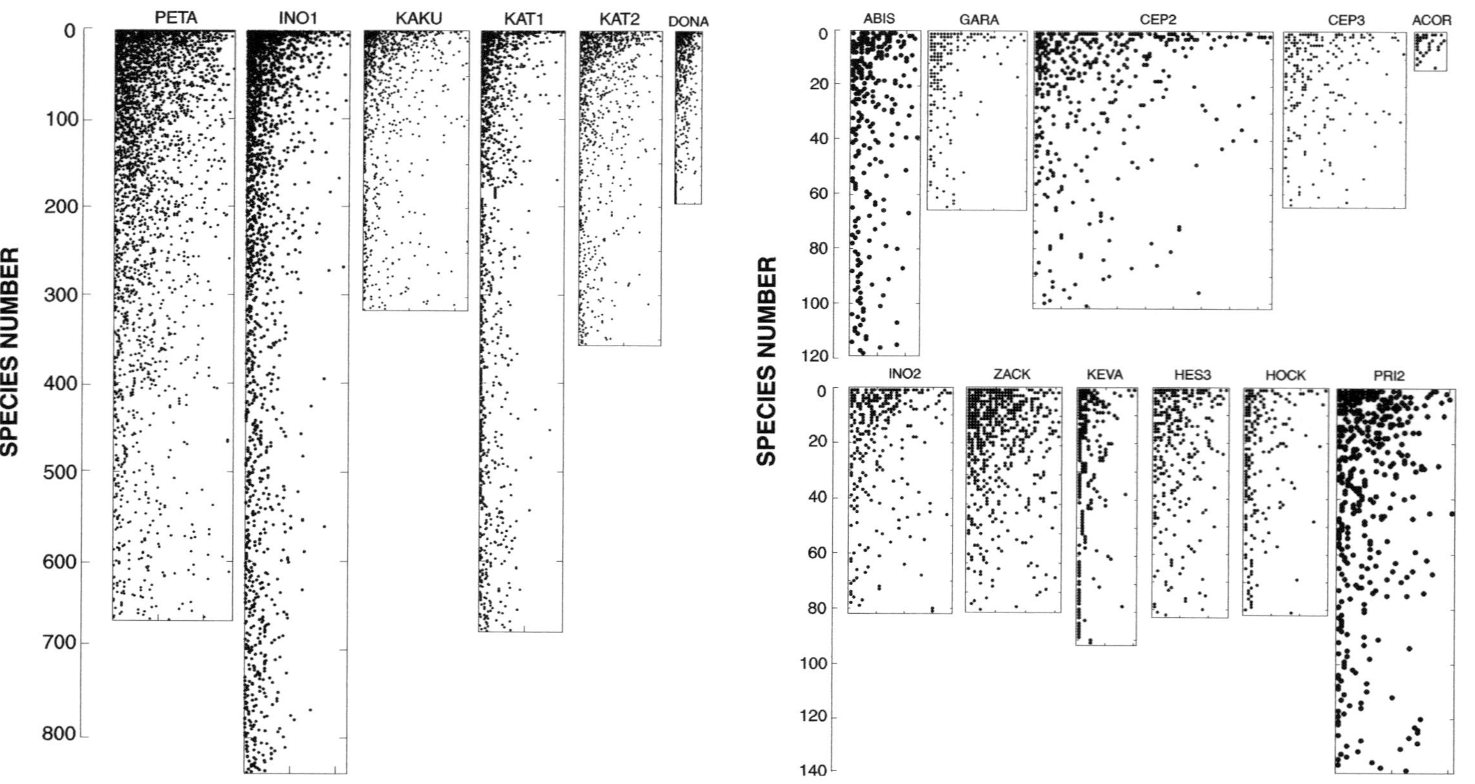

Figure 8.2 Examples of community matrices of plant–pollinator interactions. In each matrix, rows represent pollinator species and columns represent plant species. Dots indicate the presence of a particular pairwise interaction between pollinator species *i* and plant species *j*; that is, dots represent the nonzero elements of the interaction matrices (fig. 8.1C). Matrices are sorted by column and row in decreasing number of interactions per species. These are typical sparse matrices, with a significant number of elements, $a_{ij} = 0$. Abbreviations for each network are as shown in table 8.1. The left axis indicates the scale of the matrices in terms of number of species, by either row or column.

Here we examine total networks, generally compiled during the whole reproductive season and thus including a reasonably complete representation of the interactions. Compared to data available for food webs (Goldwasser and Roughgarden 1997; Bersier et al. 1999; Pimm 2002), these bipartite webs have very high resolution—down to the species level. However, potential biases introduced by variation in sampling effort have been discussed in detail by Jordano (1987), Olesen and Jordano (2002), and Vázquez and Aizen (2003; chap. 9 in this volume).

We use additional variables to characterize the interaction matrices and associated graphs. First, we analyze topological patterns defined by the way interactions are distributed among species—the so-called connectivity distribution (i.e., the probability density function of the number of interactions per species; Jordano et al. 2003). Second, we study structural patterns in the networks, mainly their nested structure and the presence of different compartments (i.e., whether the identity of interacting species is randomly established or defines a nonrandom, well-defined, subset; see Bascompte et al. 2003). Thus, our first approach aims to establish the number of interactions per species, and our second approach to determine the identity of each species' partners. Extending the latter, we also examine where the most-connected species are in the network and how they are connected with other generalists.

In general, multivariate methods have previously been used to represent sparse matrices such that the distances between rows and columns (vertices of the graphs) are meaningful in describing the pattern of presence/absence of interactions in the original matrix. The approach is to compute the geodesic distances between all pairs of nodes in the matrix and to subject the resulting distance matrix to ordination techniques. We used multidimensional scaling (MDS) to represent the pattern of relations among the species in the matrices, such that groupings that depend on the pattern of interactions can be visualized; however, we used the MINLEN modification routine to improve visualization (Borgatti and Everett 1997). We used both PAJEK (Batagelj and Mrvar 2003) and UCINET (Borgatti et al. 1999) packages to analyze the plant–pollinator network datasets; the main variables used were the following.

Density

Density involves the count of the number of links present. This is usually normalized by dividing by the maximum possible number of links, which for our bipartite graphs amounts to $A \times P$. This variable is frequently called connectivity or connectance of the network (see also Petanidou and Potts, chap. 10 in this volume; Medan et al., chap. 11 in this volume). Large sparse matrices illustrating plant–pollinator interactions usually have low density (i.e., only a small fraction of all possible interactions is actually recorded), even in intensively and adequately sampled studies.

Centrality and Connectivity Distribution

Centrality and connectivity distribution are used to measure different aspects of how a given network is centered on particular nodes—whether "central" nodes exist to which others are connected. Generalist species represent nodes of plant-pollinator networks with high centrality; they exhibit many interactions, both with other generalists (nodes which also have high k values) and with specialists (nodes with low k values), thus resulting in high centrality. Here we focus on two measures of centrality. First, the degree centrality of a node is the number of edges incident on (connected to) that node; thus, the degree of a pollinator is the number of plant species it pollinates. In the case of bipartite graphs, the maximum degree of a node is the number of nodes in the opposite set; therefore, degrees are normalized and we used the two-mode normalization proposed by Borgatti and Everett (1997). Second, the eigenvector centrality of a node is its associated eigenvector of the interaction matrix describing the network; it can be considered a weighted degree measure in which the centrality of a node is proportional to the sum of the centralities of the nodes it is connected to. Thus, a species with higher eigenvector centrality will be a generalist interacting with other generalists, located at a more central position of the network, if compared with more specialized species. We sort out the central species in a given network by examining the largest eigenvector centralities in a way similar to how we examine the largest eigenvalues of a multivariate dataset to sort out the main variables influencing covariation.

In a previous paper (Jordano et al. 2003), we examined the cumulative distributions $P(k)$ of the number of interactions per species, k_i, by fitting three different models: (1) exponential, $P(k) \sim \exp(-\gamma k)$; (2) power law, $P(k) \sim k^{-\gamma}$; and (3) truncated power law, $P(k) \sim k^{-\gamma}\exp(-k/k_x)$, where γ is the fitted constant (degree exponent) and k_x is the truncation value (see fig. 8.3B). The variable k_x is a critical number of interactions/species beyond which $P(k)$ decays faster than expected from a power-law function; k_x can be visualized (fig. 8.3B) as the k value in the abscissa beyond which $P(k)$ departs from the straight-line fit to the power law. In general, the cumulative distributions of connectivity, or degree distributions, reveal interesting patterns of the way networks are built. Random networks have characteristically exponential degree distributions; they are single-scale distributions because the distribution of links per node can be fully characterized by a single value, or scale: the mean number of links/node. Complex networks deviate markedly from this pattern and show link distributions that fit either power-law or truncated power-law models. These distributions are not fully described by a characteristic scale and are called scale-free and broad-scale distributions, respectively. They are more heterogeneous than random networks, because the cumulative distributions of k_i have longer tails. Thus, despite the fact that the bulk of species has few interactions in these networks, a few species have many more interactions than randomly expected. We found the best fit to different models

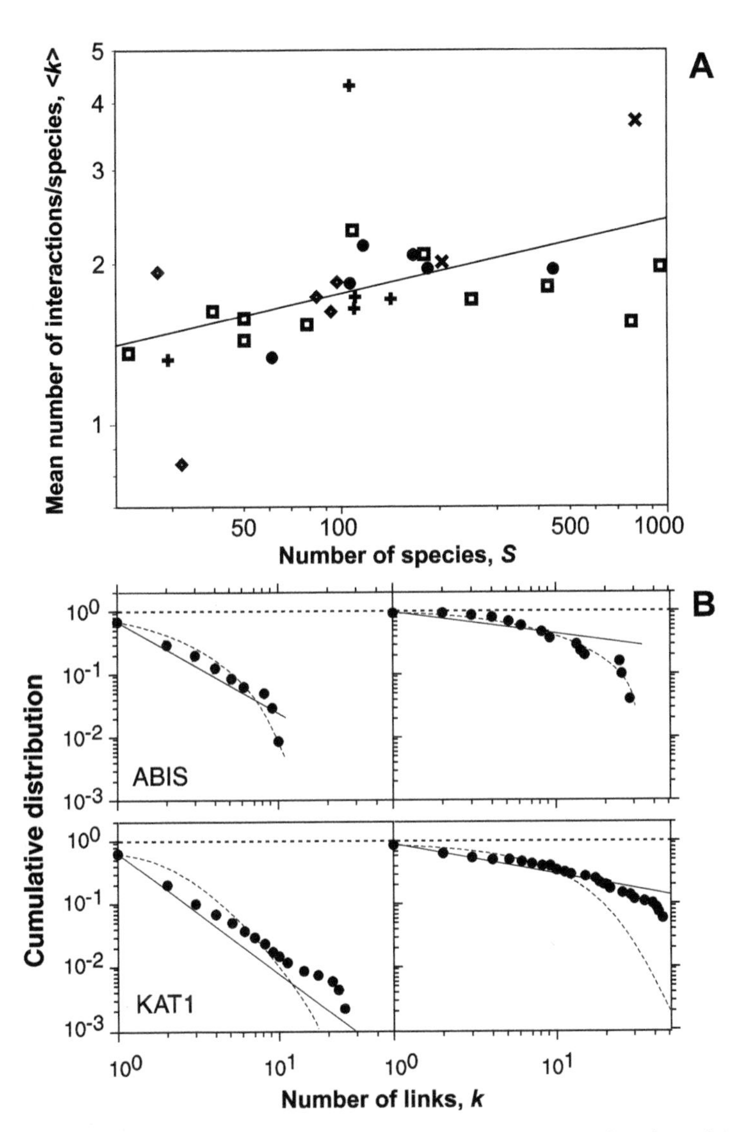

Figure 8.3 (A) Relationship between the mean number of interactions per species, <*k*>, and the total number of species (species richness, *S*) in plant–pollinator networks (table 8.1). Data are presented for tropical (●), arctic (+), alpine (◆), Mediterranean (x), and temperate (□) communities. The line is the least-squares fit to the log-transformed data. (B) Cumulative distributions of the number of interactions per species, or degree, k_i, for the ABIS and KAT1 networks (see table 8.1). The distributions of interactions are given separately for the pollinator and plant sets of species (left and right panels, respectively). Plots show the decay in the probability $P(k)$ of finding a species with *k* interactions as *k* increases. The observed data are plotted as dots, then the fits of the distribution to a power-law or truncated power-law model are represented by continuous or dashed lines, respectively. The best fit for the analyzed networks is given in table 8.1 (see also Jordano et al. 2003). Typically, the distributions of $P(k)$ depart from the straight-line fit to the power law beyond a certain value of k_x ($k_x < k$), so that there is a "truncation" at large values of *k*.

(Jordano et al. 2003) by examining the F values and associated adjusted R^2 values.

We also examine three additional structural properties of the plant–pollinator networks, namely, centralization, *k*-cores or cliques, and nestedness.

Centralization

The preceding variable of centrality characterizes the location of individual nodes or species, whereas the measure of centralization (Everett and Borgatti 1999) gives the extent to which a network has a highly central node or species around which peripheral species collect. A network with a high centralization value would resemble a star (e.g., a community with a single pollinator species interacting with all the plant species).

k-Cores

There are many ways to identify the internal heterogeneity of the network (i.e., the extent to which groups of nodes exist that share more links among themselves than with the remaining nodes). In the ecological and sociological literature, these have been called cliques (Pimm 2002) or cores (Everett and Borgatti 1999); *k*-cores are subsets of nodes with at least *k* interactions among them. Larger *k*-cores identify larger components of the network—groups of species that show a maximum number of interactions among them in comparison with other species. We used the *k*-cores routines in PAJEK and UCINET to identify subgroups of taxa in the plant–pollinator networks.

Nestedness

Imagine that we sort the interaction matrix from the most generalist pollinator species to the most specialist (i.e., matrix sorting by row); then we sort by column from the most generalist to the most specialized plant species (fig. 8.1C). The distribution of interactions among species yields nested patterns whenever species with fewer interactions appear to be "included" within those with more generalized interactions (fig. 8.1C); that is, the interactions of the more specialized species tend to be a proper subset of (i.e., nested within) the interactions already observed among the more generalized species.

To measure nestedness, one needs a quantitative measure and a benchmark to compare an observed value to check for significance. A quantitative measure was provided by Atmar and Patterson (1993). Their nestedness calculator provides a measure of disorder understood as a physical measure of "temperature." When temperature is zero, the system is totally ordered; in this case, species would be organized in the matrix to achieve maximum nestedness. In a situation of perfect nestedness, one could draw an isocline that separates the matrix into two parts. In the left-hand side of the matrix, all pairs of species would interact;

on the right-hand side of the matrix, no interactions at all would occur (see fig. 8.1C). Now imagine that temperature (or randomness) increases. Random noise would move some interaction away and we would depart from the perfect nested scenario to a random one in which all interactions are randomly distributed. The nested calculator measures the global distance to the situation of perfect nestedness: it works by calculating the distance of all the unexpected presences and absences to the isocline of perfect nestedness, and it averages this value. Bascompte et al. (2003) use a value of nestedness which is the inverse of the temperature T. Because temperature ranges from zero to one hundred, nestedness can be defined as $N = (100 - T)/100$; thus, nestedness ranges between zero and one (table 8.1). Nested patterns typically yield a core of species acting as a pivot cluster for other peripheral species (fig. 8.1C).

We used the FACTIONS and GENFAC2 routines in the UCINET package to identify the central and peripheral groups of species according to the distributions of interactions. Species with high eigenvector centrality are generally included in the core cluster of the network (Borgatti et al. 1999).

The Complexity of Plant–Pollinator Interaction Networks

Network Topologies

Plant–pollinator networks typically have sparse matrices, which best describe their topology (i.e., the way interactions occur among species; fig. 8.2). Thus, most interactions are simply not observed and only a fraction of the maximum possible number actually occurs. The connectance varies widely among networks and is strongly and negatively associated to species richness (see table 8.1; Olesen and Jordano 2002). Interactions rarify with increasing species richness and connectance decreases despite the fact that, when comparing networks, the number of interactions increases with the number of species (Jordano 1987; Olesen and Jordano 2002; Bascompte et al. 2003). Supergeneralists are hard to find; therefore, the probability of encountering a species with k interactions drops as k increases. The mean number of interactions per species increases with increasing species richness across networks ($\langle k \rangle = -0.08 + 0.139S$; for the log-transformed data, $F_{1,27} = 7.86$, $P = .009$), but the rate of increase is relatively low and even levels off beyond 150 species (fig. 8.3A). This may suggest a bound on the number of potential interactions a species can develop and may ultimately explain the decrease in connectance with S (see Pimm 2002 for discussion).

But supergeneralists (species with very large k) do exist; this differs from randomly built networks, where nodes with large k values simply do not exist (Albert and Barabási 2002; Vázquez and Aizen 2003). What biotic networks have that is special in contrast to other complex networks is that the frequency of these supernodes, with an extremely large number of connections, is lower than expected if the network has a scale-free distribution of k values. Thus, the proba-

Table 8.1 Summary statistics of plant–pollinator interaction networks

No.	Code	S	A	P	M	k	$\langle k \rangle$	$\langle k_a \rangle$	$k_{max}A$	$\langle k_p \rangle$	$k_{max}P$	$\gamma_{Pollinators}$	Fit[a]	γ_{Plants}	Fit[a]
1	ABIS	142	118	24	2832	242	1.70	2.05	10	10.08	28	−1.35	a	−0.89	b
2	ACOR	22	12	10	120	30	1.36	2.50	6	3.00	8	−1.02	b	−1.26	c
3	CEP1	61	25	36	900	81	1.33	3.24	34	2.25	25	−1.01	a	−0.97	b
4	CEP2	185	101	84	8484	361	1.95	3.57	15	4.30	7	−0.93	a	−1.24	b
5	CEP3	107	64	43	2752	196	1.83	3.06	16	4.56	14	−1.13	b	−0.92	b
6	DONA	205	179	26	4654	412	2.01	2.30	17	15.85	86	−1.33	a	−0.68	b
7	GALA	32	22	10	220	27	0.84	1.23	4	2.70	6	−1.76	b	−0.83	b
8	GARA	84	55	29	1595	145	1.73	2.64	16	5.00	24	−1.14	a	−0.76	a
9	HES1	50	40	10	400	79	1.58	1.98	7	7.90	17	−1.45	b	−0.78	b
10	HES2	50	42	8	336	72	1.44	1.71	5	9.00	17	−1.51	d	−0.83	c
11	HES3	108	82	26	2132	249	2.31	3.04	19	9.58	31	−1.10	b	−0.92	b
12	HOCK	110	81	29	2349	179	1.63	2.21	10	6.17	40	−1.46	b	−0.74	a
13	INO1	952	840	112	94,080	1876	1.97	2.23	37	16.75	119	−1.36	b	−0.62	b
14	INO2	117	81	36	2916	253	2.16	3.12	21	7.03	25	−1.08	a	−0.47	b
15	KAKU	428	314	113	35,482	774	1.81	2.46	26	6.85	68	−1.25	a	−0.85	b
16	KAT1	770	679	91	61,789	1193	1.55	1.76	25	13.11	188	−1.62	a	−0.62	a
17	KAT2	446	356	90	32,040	865	1.94	2.43	24	9.61	65	−1.26	a	−0.79	b
18	KATO	251	187	64	11,968	430	1.71	2.30	17	6.72	40	−1.36	a	−0.77	b
19	KEVA	111	91	20	1820	190	1.71	2.09	7	9.50	60	−1.39	b	−0.58	b
20	MAUR	27	13	14	182	52	1.93	4.00	12	3.71	8	−0.89	b	−1.23	b
21	MOSQ	29	18	11	198	38	1.31	2.11	7	3.45	9	−1.22	a	−0.90	b
22	PERC	97	36	61	2196	178	1.84	4.94	31	2.92	13	−1.00	b	−1.10	b
23	PETA	797	666	131	87,246	2933	3.68	4.40	104	22.39	124	−0.99	a	−0.89	b
24	PRI1	78	60	18	1080	120	1.54	2.00	9	6.67	12	−1.44	b	−0.89	b
25	PRI2	180	139	41	5699	374	2.08	2.69	16	9.12	43	−1.22	b	−0.73	b
26	PRI3	167	118	49	5782	346	2.07	2.93	26	7.06	43	−1.11	a	−1.00	c
27	RAMI	93	46	47	2162	151	1.62	3.28	17	3.21	10	−1.01	a	−1.24	c
28	SCHE	40	33	7	231	65	1.63	1.97	6	9.29	23	−1.33	b	−0.52	a
29	ZACK	107	76	31	2356	456	4.26	6.00	20	14.71	32	−0.89	c	−1.36	b

bility of finding a species with k interactions drops suddenly for a relatively large value of k (fig. 8.3B; see also fig. 2 of Jordano et al. 2003). In fact, the upper limit for k (k_{max}; table 8.1) is much lower for pollinator species than for plant species, although this might relate to the fact that plant–pollinator records are typically obtained with "phytocentric" surveys (i.e., surveys focused on plant species and documenting their interactions with pollinators). Although the plant–pollinator networks examined thus far are reasonably robust to sampling artifacts (Jordano 1987; Vázquez and Aizen, chap. 9 in this volume), future research should explore potential biases derived from sampling designs focused on particular sets of these bipartite networks. In addition, caution should be taken in the interpretation of results because of variable completeness of the data (Olesen and Jordano 2002; see Goldwasser and Roughgarden 1997; D. Vázquez, personal communication) and assumptions inherent to the analyses (Vázquez and Aizen, chap. 9 in this volume).

The distribution of number of interactions per species is markedly skewed in pollination networks (table 8.1). Most species have k_a or k_p values greater than 5; these networks share a general pattern of a dense core of species which interact

Table 8.1 *(continued)*

No.	Eigenvector	Centralization	*C*	*N*	*p*	Locality	Source[b]
1	4.01	20.35	0.0854	0.8602	*	Latnjajaure, Abisko, Sweden	Elberling and Olesen 1999
2	26.01	48.68	0.2500	0.6707	ns[c]	Flores, Açores Islands	J. M. Olesen, unpublished data
3	14.25	9.86	0.0900	0.9604	**	Cordón del Cepo, Chile	Arroyo et al. 1982
4	6.90	27.67	0.0425	0.9098	**	Cordón del Cepo, Chile	Arroyo et al. 1982
5	9.95	19.02	0.0712	0.9250	**	Cordón del Cepo, Chile	Arroyo et al. 1982
6	6.84	50.38	0.0885	0.9114	**	Doñana National Park, Spain	Herrera 1988
7	15.92	12.79	0.1227	0.7016	ns	Galapagos Islands	McMullen 1993
8	5.49	30.72	0.0909	0.9523	**	Garajonay, Gomera Island, Spain	Olesen MS/laurisilva
9	7.57	43.52	0.1975	0.6279	ns	Hestehaven, Denmark	Olesen MS/bog
10	7.60	29.08	0.2143	0.5938	ns	Hestehaven, Denmark	Olesen MS/forest
11	5.08	24.42	0.1168	0.8283	**	Hestehaven, Denmark	Olesen MS/fallow
12	4.62	40.96	0.0762	0.9454	**	Hazen Camp, United States	Hocking 1968
13	0.49	0.55	0.0199	—		Kibune, Kyoto, Japan	Inoue et al. 1990
14	9.29	20.46	0.0868	0.9041	*	Snowy Mountains, Australia	Inouye and Pyke 1988
15	2.04	23.11	0.0218	—		Kyoto City, Japan	Kakutani et al. 1990
16	0.62	38.12	0.0193	—		Ashu, Kyoto, Japan	Kato et al. 1990
17	2.02	19.97	0.0270	0.9746	**	Mt. Kushigata, Japan	Kato et al. 1993
18	2.79	21.88	0.0359	0.9551	**	Nakaikemi, Japan	Kato and Miura 1996
19	9.82	65.25	0.1044	0.9550	**	Hazen Camp, Canada	Kevan 1970
20	11.77	28.45	0.2857	0.8743	**	Mauritius Island	Eskildsen et al., unpublished data
21	6.29	18.08	0.1919	0.7808	ns	Melville Island, Canada	Mosquin and Martin 1967
22	10.38	36.01	0.0810	0.9254	**	Jamaica	Percival 1974
23	3.11	14.69	0.0336	—		Daphní, Athens, Greece	Petanidou 1991
24	5.85	23.55	0.1111	0.9397	**	Arthur's Pass, New Zealand	Primack 1983, AP
25	3.53	30.79	0.0656	0.9252	**	Cass, New Zealand	Primack 1983, Cass
26	7.56	35.30	0.0598	0.7363	**	Craigieburn, New Zealand	Primack 1983, Craigieb.
27	8.61	25.17	0.0698	0.8710	*	Canaima National Park, Venezuela	Ramirez 1989
28	8.98	50.04	0.2814	0.8668	**	Brownfield, Illinois, United States	Schemske et al. 1978
29	5.37	9.66	0.1935	0.7420	**	Zackenberg, Greenland	Elberling and Olesen MS

Source: See Olesen and Jordano 2002; Jordano et al. 2003; and Bascompte et al. 2003 for details.

Note: Column headings are as follows: (*S*) species richness; (*A*) number of pollinator species; (*P*) number of plant species; (*M*) matrix size (total number of potential interactions); (*k*) number of interactions recorded; (<*k*>) mean number of interactions per species (degree); ($<k_o>$) mean number of interactions per pollinator species; ($k_{max}A$) degree of most-connected pollinator species; ($<k_p>$) mean number of interactions per plant species; ($k_{max}P$) degree of most-connected plant species; ($\gamma_{Pollinators}$) gamma exponent for the fit of the cumulative frequency distribution of *P*(*k*) for pollinator species; (γ_{Plants}) gamma exponent for the fit of the cumulative frequency distribution of *P*(*k*) for plant species; (Fit) best fit of the cumulative frequency distribution of *P*(*k*) to a model; (Eigenvector) mean eigenvector centrality of the species (plants and pollinators pooled); (Centralization) network centrality value; (*C*) connectance, or density of the network; (*N*) nestedness.

[a]Power law (a); truncated power law (b); exponential (c); not available (d). Best fit determined by the highest *F* and adjusted-R^2 values.

[b]For reference list, see Bascompte et al. 2003 and Jordano et al. 2003.

[c]No significance.

$*p < .05$; $**p < .01$

with each other, surrounded by many species with few interactions, connected with those in the core (fig. 8.2). We found a number of networks that fit power-law distributions of k values (table 8.1; e.g., KAT1 in fig. 8.3B), but most were better described by a truncated power law (Jordano et al. 2003). In a truncated power-law distribution, the probability of a given value of k for a species drops with increasing k following a power-law function; then, beyond a certain value of k_x, the observed data depart from the power law and show a steep decay (ABIS network in fig. 8.3B; Amaral et al. 2000).

Therefore, from the perspective of connectivity distribution, plant–pollinator networks share many features irrespective of the ecological setting. These features are also shared with other plant–animal mutualisms (Jordano et al. 2003) and other complex networks (Newman 2003) and probably reveal very basic processes of the way species are arranged in mutualistic assemblages. Irrespective of the size of the network, plant–pollinator mutualisms center around a core of generalist species with a high density of interactions. The interactions of the core involve not only other generalists but also the more specialist species, and this pattern gives the characteristic aspect to the sparse matrices that describe these interactions (fig. 8.2). The pattern depends not only on the distribution of k values among individual species but also on "structural" patterns that define the distribution of interactions throughout the network, as described next.

Biological Patterns: Beyond the Topology of Interactions

The truncated distributions of $P(k)$ are not exclusive to pollination networks but occur whenever constraints are imposed in the way nodes establish links. In the presence of constraints, highly connected nodes would be less likely to occur than the frequency expected for a scale-free network (Mossa et al. 2002). The ubiquity of truncation in the distribution of the number of interactions per species in plant–pollinator and plant–disperser mutualisms led us to suggest (Jordano et al. 2003) that biological constraints are the main factor explaining truncation. Constraints occur because of the biological attributes of the species; if a plant and a pollinator differ in phenology (e.g., an early blooming herb and a late-summer migratory pollinator), their interaction cannot occur (see also Medan et al., chap. 11 in this volume). This translates into "structural" zeros in the interaction matrices, that is, pairwise interactions that will never be recorded despite intensive study. We can thus expect a sizeable fraction of the nonobserved interactions to be caused by these types of constraints. We defined these nonobservable interactions as "forbidden interactions" and hypothesized that they are the main cause of the patterns we observe in the distributions of $P(k)$ in plant–animal interaction networks in general.

What is the reason behind truncation of the cumulative degree distributions? We recently provided evidence for generalized truncation in plant–animal mutualistic networks (Jordano et al. 2003) and argued that, whenever a complex network evolves (i.e., by the addition of species), the new species are constrained in the way they set up interaction with partners. Species-specific traits set limits to the possibilities of successful interaction. This is readily evident from the sparse matrices that typically describe plant–pollinator networks (fig. 8.2): actual interactions are relatively "rare." Moreover, the number of interactions observed increases with species richness, but at a relatively low rate that results in a low fraction of the possible interactions realized at high species-richness values (fig. 8.3A). Therefore, forbidden interactions are a major component of the sparse interaction matrices (fig. 8.4).

The example of the Snow and Snow (1972) dataset (fig 8.4) exemplifies the ubiquity of forbidden interactions. These authors studied interactions between hummingbirds and plants in Arima Valley, Trinidad (10°40′ N), for almost two years. Connectance is relatively high ($C = 0.354$), which is typical of subnetworks that only include a subset of the pollinator fauna (Jordano 1987). However, there are only 185 interactions out of 522 possible, with 337 not recorded; figure 8.4 outlines the reasons for not observing these 337 pairwise interactions. In most cases (29%), habitat uncoupling between the plant species and the pollinator causes the interaction not to occur (matrix elements marked H in fig. 8.4). This chiefly occurs between subcanopy foragers like the hermit hummingbird species (*Phaethornis* spp. and *Glaucis hirsuta*) and canopy trees, and among *Anthracothorax nigricollis* and *Florisuga mellivora,* which avoid lower strata (Snow and Snow 1972). A relatively small fraction (13%) of forbidden interactions is due to uncoupling of corolla or flower characteristics (tube length, reward, or color) and pollinator (fig. 8.4). Thus, 6% of the interactions are not observed because of size restrictions (i.e., beak is too short relative to the corolla tube length), 4% can be attributed to the reward per flower being too small relative to the size of the bird; and 3% can be attributed to apparent color restrictions (e.g., *Phaethornis guy* and *G. hirsuta* only forage on red-flowered species).

Obviously, a sizeable fraction of the unrecorded interactions cannot be accounted for and might be related to unknown factors (*U*, 24%; fig. 8.4) which, among others, include chance effects and limited sampling effort. In fact, for the figure 8.4 dataset, interactions recorded only once were excluded in the original table; this might explain the relatively high frequency of *U* values. These unknowns are also found in other well-studied systems (Jordano et al. 2003; P. Jordano, unpublished data). They may simply result from an extremely low probability that the interaction actually occurs in nature despite an obvious cause. For instance, when two species are very rare, their probability of interaction is likely to be low. We believe that future explorations of the cause of forbidden interactions will shed light on important factors in the evolution of complex patterns of interaction in species-rich systems. A categorization of forbidden "types" may indicate repeated patterns that are independent of the ecological setting and

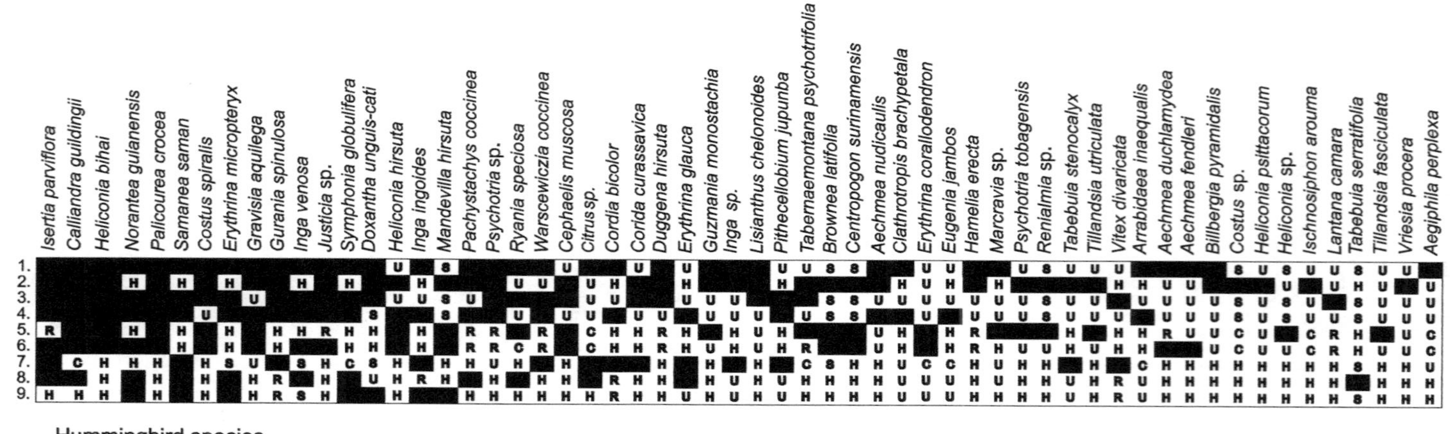
Plant species
Isertia parviflora
Calliandra guildingii
Heliconia bihai
Norantea guianensis
Palicourea crocea
Samanea saman
Costus spiralis
Erythrina micropteryx
Gravisia aquilega
Gurania spinulosa
Inga venosa
Justicia sp.
Symphonia globulifera
Doxantha unguis-cati
Heliconia hirsuta
Inga ingoides
Mandevilla hirsuta
Pachystachys coccinea
Psychotria sp.
Ryania speciosa
Warscewiczia coccinea
Cephaelis muscosa
Citrus sp.
Cordia bicolor
Corida curassavica
Duggena hirsuta
Erythrina glauca
Guzmania monostachia
Inga sp.
Lisianthus chelonoides
Pithecellobium jupunba
Tabernaemontana psychotrifolia
Brownea latifolia
Centropogon surinamensis
Aechmea nudicaulis
Clathrotropis brachypetala
Erythrina corallodendron
Eugenia jambos
Hamelia erecta
Marcravia sp.
Psychotria tobagensis
Renialmia sp.
Tabebuia stenocalyx
Tillandsia utriculata
Vitex divaricata
Arrabidaea inaequalis
Aechmea duchlamydea
Aechmea fendleri
Billbergia pyramidalis
Costus sp.
Heliconia psittacorum
Heliconia sp.
Ischnosiphon arouma
Lantana camara
Tabebuia serratifolia
Tillandsia fasciculata
Vriesia procera
Aegiphila perplexa
1.
2.
3.
4.
5.
6.
7.
8.
9.
Hummingbird species
1. Saurottia tobaci
2. Phaethornis longuemareus
3. Chlorestes notatus
4. Amazilia chionopectus
5. Phaethornis guy
6. Glaucis hirsuta
7. Chrysolampis mosquitus
8. Anthracothorax nigricollis
9. Florisuga mellivora

Figure 8.4 Patterns of forbidden interactions in a plant–hummingbird subnetwork (Snow and Snow 1972). Rows indicate hummingbird species ($A = 9$) and columns are their foodplants ($P = 58$). Forbidden interactions are those never observed in interaction matrices; that is, for each zero element of the interaction matrix, we note the potential cause for not having recorded that particular pairwise interaction. Black cells are the observed interactions in the matrix ($a_{ij} = 1$). For the nonobserved interactions ($a_{ij} = 0$), letters indicate the potential cause for not encountering that interaction: size restrictions (S); habitat restrictions (H), due to habitat uncoupling of birds and plants; reward limitation (R); flower color restrictions (C); and unknown reason (U; see text for detailed descriptions). There are general reasons for the actual interaction between a pair of plant and pollinator species being impossible to record in a given habitat, for instance, when the flowering period of the plant does not match the period of presence of the pollinator in the area, as in the case of migratory pollinators, or when the size of the pollinator mouthparts restricts access to the nectar and pollen.

may help to explain the invariant properties we document. It may also be possible to tease apart the relative importance of phylogenetic composition of the interaction partners and their ecological traits in causing forbidden interactions. In any case, forbidden interactions illustrate the types of constraints that are peculiar to these biotic interactions and that cause network patterns that severely deviate from other complex networks, especially the abiotic networks.

In the preceding section, we argued that the distribution of interactions among species indicates the presence of a central core of taxa showing the highest density of interactions. The centrality parameters, such as the eigenvector centrality (table 8.1), quantify to what extent a particular species has a central role in the network, that is, located as a central actor relative to others that link with it. The mean eigenvector centrality is negatively correlated across networks with species richness ($r = -0.581, P = .0009, N = 29$), meaning that increasingly diverse communities have a lower number of central species. Relatively simple communities have higher connectance, which means each species in one set is connected with a relatively large fraction of the species in the other set. This also means that simple plant–pollinator networks tend to be less centralized (fig. 8.5) and to be structured as a single core with no central actor. In most cases, however (fig. 8.5), a number of species can be identified as having the highest density of interactions. CEP3 has 16 species (10 plants, 6 pollinators) with eigenvector centrality greater than 20.0 (fig. 8.5; see table 8.1); among these, 9 species define a central core as identified by clustering algorithms in UCINET. This algorithm takes the bipartite graph and uses a combinatorial procedure to assign nodes to two clusters, one central and the other peripheral, such that it maximizes the fit to the expected situation where the density of links within each group is maximal and nonexistent between groups (Borgatti and Everett 1997, 1999). Thus, the algorithm finds the two groups of nodes that maximize the separation between a core and a periphery within the network. In larger networks, such as ZACK and PETA (table 8.1; fig. 8.5), a relatively small fraction of species forms the core. In ZACK, there are 20 species out of 107 with an eigenvector centrality

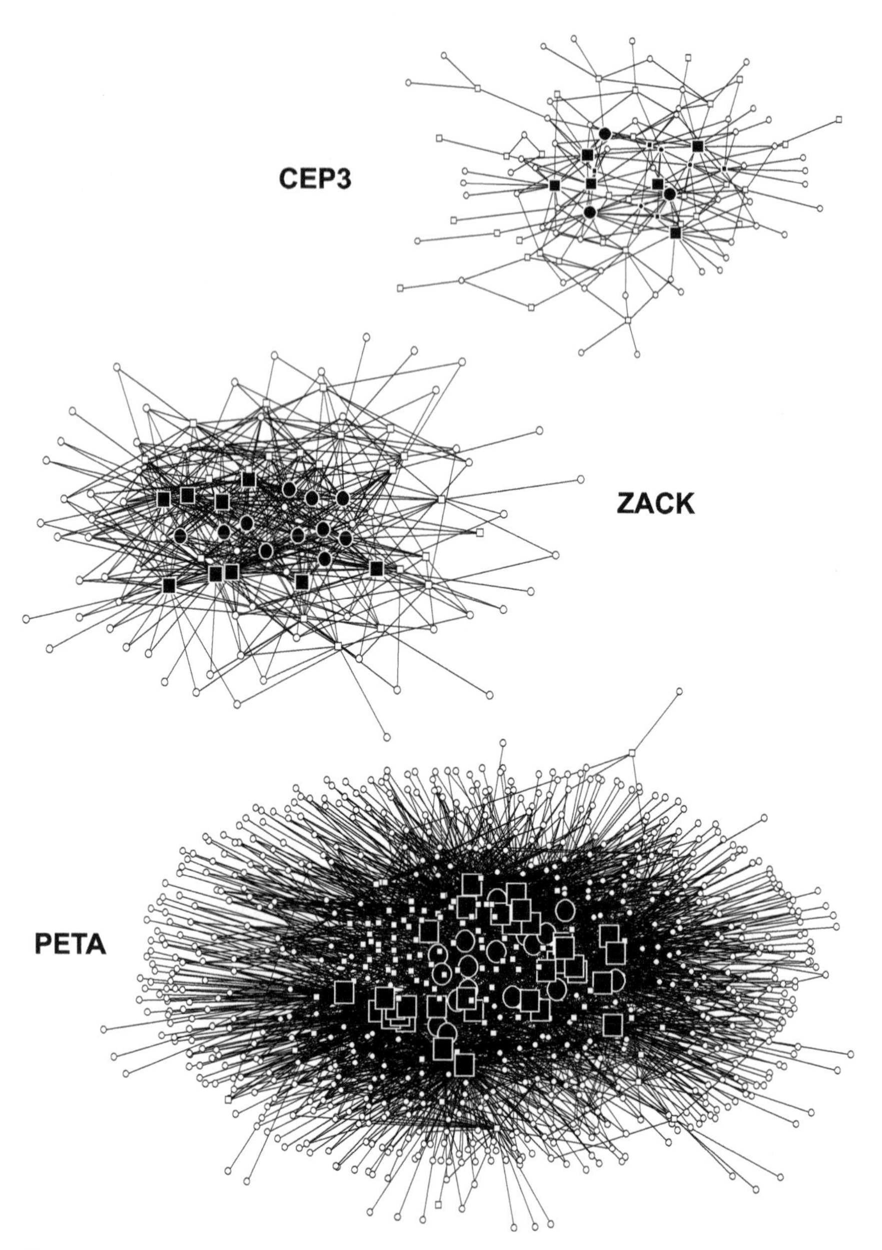

Figure 8.5 Examples of plant–pollinator networks with the core of species outlined, as identified by the eigenvector centrality value (two-mode normalized) of each node. Species with eigenvector centrality greater than 10.0 (PETA network) or greater than 20.0 (in CEP3 and ZACK networks) are outlined with larger symbols: circles indicate pollinator species; squares indicate plant species. The three networks have cores with 42 (PETA), 16 (CEP3), and 20 (ZACK) species. A pollinator's centrality is proportional to the sum of centralities of the plants it interacts with. A central pollinator species is more central by being a generalist—interacting with other generalist plants but also with specialized species. See table 8.1 for a description of the networks.

greater than 20.0 that form the core. In PETA, the core is composed of 42 species with eigenvector centralities greater than 10.0. In CEP3, flies (chiefly syrphids) dominate the core and also *Bombus* spp.; the PETA network core is composed of flies and bees in roughly the same proportions. In the plant set, in general, the most abundant species are included in the core. The ZACK and PETA cores also share a similar structure: the MDS ordination locates two distinct clusters within the core of plant species (groups of squares in fig. 8.5) at each of the two sides of the pollinator cluster in the center. Thus, ecological factors such as phenological variation presumably contribute to the location of a particular species within the complex network of interactions. A future research line would be to explore the ecological correlates of differences in these locations among species and whether there are predictable traits shared by the core species.

Network Structure: Nested Patterns

In the preceding sections, we have described patterns in connectivity distribution. This is a first step toward a description of the structure of plant–pollinator networks. To some extent, this has revolved around the level of generalization and specialization in these networks. As noted, the pollination webs are more heterogeneous than random webs; that is, there are species more connected than what would be expected to randonly occur. However, nothing has been said about the likelihood of interaction between two species (i.e., a generalist and specialist). For example, consider two focal species and their interactions. Are the interactions common in both subsets? Our next step in the description of the pattern of plant–pollinator assembly is not just to quantify the number of connections but also to look at their identity (Dicks et al. 2002). This is related to one of the classic questions in community ecology: whether networks of ecological interactions (e.g., food webs) are compartmentalized (Pimm and Lawton 1980).

One concept that captures network structure, and that recently has been introduced to the study of mutualisms, is nestedness. However, the concept of nestedness is not new in ecology. It was developed in the context of island biogeography to describe a specific, nonrandom pattern by which a set of species is distributed within a set of islands (Atmar and Patterson 1993). Bascompte et al. (2003) introduced this concept to the study of mutualistic interactions by imagining that plants are "islands" that a certain number of animal species "inhabit."

Nested matrices are organized as in Chinese boxes, with sets of species within larger sets of species. This nested structure has two important features. First, it generates highly asymmetric interactions. This can be seen by the fact that, as indicated in figures 8.1A–8.1C, specialist species tend to interact with the most generalist species (see also Minckley and Roulston, chap. 4 in this volume; Vázquez and Aizen, chap. 9 in this volume; Petanidou and Potts, chap. 10 in this volume). Second, nestedness implies that there is a core of taxa with a high density of interactions. In other words, generalist plants and generalist animals tend to interact among themselves. Thus, nestedness implies asymmetry at the level of specialists but symmetry at the level of generalists. The fact that generalist species interact among themselves creates a very cohesive structure—understood as a structure with redundancy, that is, multiple ways to connect the species within this "core." In figure 8.5, we plot the plant and animal species that constitute the core of specific mutualistic networks. As noted, a few species contain the bulk of interactions and cohesively build the rest of the network around themselves; they are "central" to the network and, thus, have high centrality values.

Our quantitative measures of nestedness for natural networks are summarized in table 8.1 (see Bascompte et al. 2003). Nestedness values range between 0 and 1, as measured with the nestedness calculator (Atmar and Patterson 1993). Once a measure is provided to characterize each community, we have to put this measure in context. How nested is a given community? Is it more nested than expected? Or is its value of nestedness similar to what we would expect for a randomly assembled matrix? This is a crucial question. If the value of nestedness is nothing more than what we would expect to occur by chance, then there is no biological pattern to explain. Answering this question depends on having an appropriate null model. Null models have been widely used in community ecology as a way to check whether an observed level of structure can be reproduced by simple rules (Gotelli 2000). Null models have been used in the context of plant-pollinator networks to explore whether levels of generalization and specialization are higher than expected by chance (see Vázquez and Aizen 2003, chap. 9 in this volume). Two different null models have been used to test the significance of the nested patterns. In null model 1 (the one provided by the nestedness calculator), all cells in the matrix have the same probability of being occupied—a probability estimated as the number of interactions divided by the total number of possible interactions (i.e., the connectivity). On average, each replicate will have the same number of connections but these will be randomly distributed. Note that this null model assumes that each species has the same probability of having an interaction; clearly this is not the case in plant-pollinator networks. In the preceding section, we have seen that the degree of distribution is highly skewed. How can we incorporate this fact into a null model? In null model 2, the probability of two particular species interacting is the average of the probability of interaction of both the plant and the animal; that is, the probability of a link is proportional to the degree of both plant and animal species. These two null models are very similar to those of Vázquez and Aizen (2003). Null model 1 is the same with the small difference that Vázquez and Aizen (2003) have the additional constraint that all species must have at least one interaction. Their null model 2 also has a specific probability of interaction for each pair of species—an average of the "presence" of both species. However, they use the frequency of visits by each species (as opposed to the degree) as a measure of "abundance."

Because there is a strong relationship between degree and frequency of visits (Vázquez and Aizen 2003, chap. 9 in this volume; D. Vázquez, personal communication; see also Jordano 1987), their null model 2 is essentially the same as that of Bascompte et al. (2003). However, from a conceptual point of view, Vázquez and Aizen (2003) define their null model at the individual level instead of at the species level; that is, individuals are the key elements involved in mutualisms. Generally the ideal situation would be to test significance of results with a suite of null models with increasing levels of complexity. This exercise would tell us what is important and what is irrelevant in producing an observed pattern.

The study by Bascompte et al. (2003) showed that the bulk of mutualistic networks are significantly nested; that is, they are much more structured than similar, randomly constructed networks based on either null model (fig. 8.2). Second, there were no significant differences between the level of nestedness for both plant–pollinator and plant–disperser networks, which, together with the result outlined in preceding sections about the pattern of connectivity distributions, suggests invariant properties in these two types of mutualisms. That is, there are conservative patterns of network assembly independent of the biological detail of the interaction, network size, latitude, and other differences—these patterns are very robust.

The implications of the nested pattern can be seen from the points of view of both community assembly and coevolution. From the point of view of community assembly, these patterns unambiguously show that mutualistic networks are neither randomly assembled nor compartmentalized. This is probably the best evidence for a pattern in networks of ecological interactions. The nonrandom pattern of these webs may be very relevant. From the pioneering work by May (1972) and Pimm and Lawton (1980), it was clear that the structure of food webs highly affects their stability. May (1972) used randomly assembled food webs in his influential study about the relationship between stability and complexity. However, at the end of the paper, he assumed that real food webs are probably not random and suggested that they may be organized in compartments. This structure assumes that species within the compartments are highly interactive, whereas there are almost no interactions among different compartments.

May (1972) explored compartmentalized food webs and concluded that they were more stable than random ones—a result challenged by Pimm and Lawton (1980), who found the opposite result when food webs are more realistically built. Interestingly enough, the concept of compartmentalization became entrenched, and subsequent papers have looked for compartmentalization in real food webs, but with poor results. For example, both Pimm and Lawton (1980) and Raffaelli and Hall (1992) failed to find compartmentalization, although some limited evidence exists (e.g., Dicks et al. 2002; Corbet, chap. 14 in this volume). Nestedness can be understood as the most significant and widely observed nonrandom pattern in networks of ecological interactions.

The two properties of nestedness (asymmetry and the core of interactions) may greatly affect the robustness of the mutualistic networks (see also Memmott et al. 2004). First, because specialist (and generally rare) species interact with generalist (and generally abundant) species, nestedness provides higher chances for the persistence of rare species. Second, due to the cohesive role of the core of species, with its redundancy of interactions, nestedness provides alternative routes for system responses after perturbations such as the elimination of a species or a link. Another element for robustness is the generalized broad-scale distribution of the number of interactions per species that we report, which seems to be a general pattern in plant–animal interaction networks (Jordano et al. 2003). Networks with broad-scale distributions of connectivities are generally thought to be more robust to loss of highly connected nodes than scale-free networks (Amaral et al. 2000). Figure 8.6 shows a simple simulation of the effects of species loss on the persistence of connectivity patterns in two plant–pollinator networks. We simulated loss of either plants or pollinators in decreasing order of their number of interactions (i.e., in decreasing order of their eigenvalue centrality). For each node (species) removed, we estimated the preserved connectance as a fraction of the original connectance. The results show that increasing the fraction of nodes removed can dramatically affect the connectivity of the network and its persistence. Removal of a relatively small fraction of the most connected nodes (more than 20%) can cause a collapse of the network (KAT1, fig. 8.6) or the loss of almost 50% of the interactions (CEP3, fig. 8.6). In both cases, the network is more robust to loss of pollinator species and more sensitive to loss of plant species. It is interesting that KAT1, a scale-free network, was less robust to the loss of plant species, confirming expectations of models for abiotic networks (Barabási and Albert 1999; Albert et al. 2000; Barabási et al. 2000; Jeong et al. 2000; Albert and Barabási 2002). CEP3, a broad-scale network, appeared more robust, especially to the loss of plant species. These preliminary results suggest ways to explore the robustness of plant–pollinator networks to species loss, to invasion by exotics, or to overall simplification due, for example, to agricultural practices or human intervention (Kearns et al. 1998; Memmott and Waser 2002; Memmott et al. 2004).

Concluding Remarks

Plant–pollinator networks are complex webs that share many properties with other types of networks, both abiotic and biotic. The most characteristic property is that interactions among species are not distributed at random, but, surprisingly, the nonrandom pattern we found is largely invariant across different ecological settings. This reveals very general patterns in the way interactions are assembled in these communities and suggests important clues to understand their evolution. Moreover, it demonstrates that these networks are more than the addition of pairwise interactions: it is the whole set of pairs of species in both

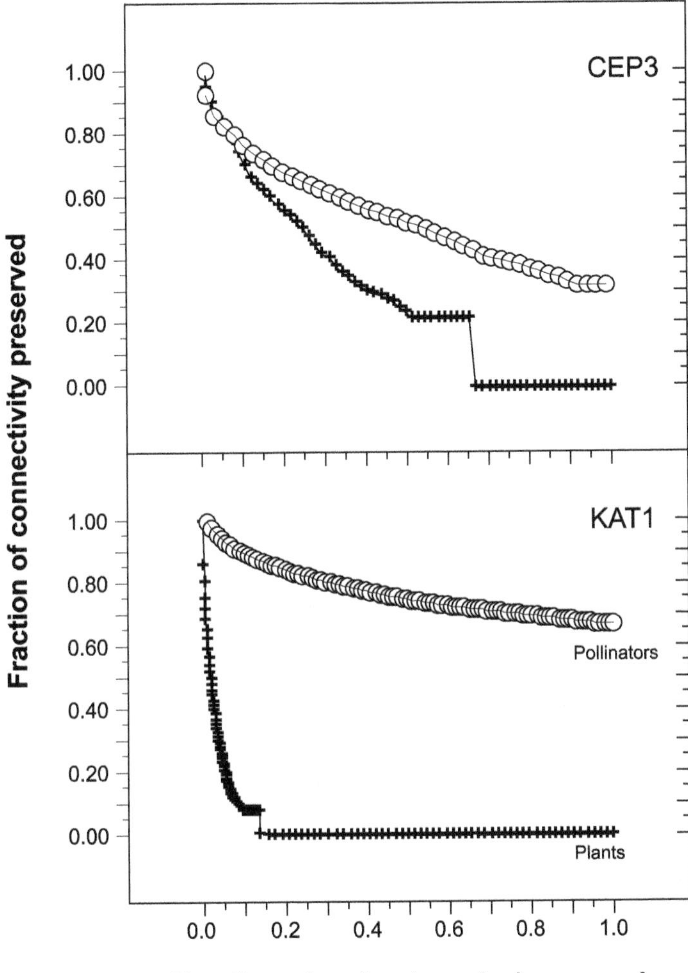

Figure 8.6 Decay in connectance as a function of removal of species in plant–pollinator networks. We simulate the loss of an increasing fraction of either plant (+) or pollinator (○) species (abscissa) by removing species, step by step, according to their decreasing k_i value, starting with the most generalist species. The ordinate represents how well the original connectivity of the network is preserved; it is estimated as the fraction of the actual connectance (degree) that the network would have after the loss of a given fraction of species (nodes). We use two examples of networks which illustrate the most general pattern: CEP3—Cordón del Cepo, in the Andes, Chile—with $S = 107$ species ($A = 64$, $P = 43$); and KAT1—Ashu, Kyoto, in temperate Japan—with $S = 770$ species ($A = 679$, $P = 91$).

the pollinator and plant sets that becomes organized in a complex way. This organization has both topological and structural aspects relevant to understanding its evolution.

First, a wide range of interactions per species occurs, but these interactions are predictably distributed according to truncated power-law or power-law models.

That is, the probability of finding a particular species interacting with k other species decays as k increases. Therefore, it is unlikely to find supergeneralists, but they do exist. The presence of these highly connected nodes is less frequent than what would be expected had plant–pollinator networks evolved similarly to other complex abiotic networks. Whenever such a network evolves by preferentially attaching new nodes to the already well-connected ones, a power-law (scale-free) distribution of connectivities emerges. But plant–pollinator networks differ from these because the probabilities for the most-generalist species lie below those expected from a scale-free network. We found few plant–pollinator networks that fit the power-law distribution of k values, and this was generally for the pollinator interactions, not the plant interactions. Therefore, biases due to sampling design (e.g., plant centered vs. pollinator centered) probably should be taken into account in future studies.

Second, a pervasive feature of complex plant–pollinator networks is that they are not randomly built but show a characteristic distribution of interactions throughout the matrix: interactions pivot around a core of species generated by the fact that interactions show a markedly nested pattern. From the point of view of coevolution, the nested assembly has very important implications. It clearly shows that mutualistic interactions are neither organized in specific pairwise interactions, as the ones expected for symbiotic mutualisms, nor organized as a "diffuse" assembly that precludes any analytic approximation. Traditionally, scientists have expected to find the pattern of pairwise specialization observed in symbiotic mutualisms when dealing with nonsymbiotic mutualisms. The lack of such evidence has led to the alternative view that plant–pollinator systems are "diffuse." Nestedness illustrates a highly structured assembly pattern that does not correspond to either of these two extreme views. The core of interactions may drive the coevolution of the rest of the species attached to it. It is a coevolutionary "vortex" sensu Thompson (1994). Bascompte et al. (2003) have reported a pattern in which specialists interact with generalists and generalists in turn interact among themselves. The finding of a nested pattern greatly advances the knowledge of plant–pollinator systems obtained simply by counting how many species are specialists and how many are generalists. Thus, viewed from a network perspective, the centrality of a given species relates not only to its own generalization level but also to how central are the other species with which it interacts. Combined with the results on the connectivity distribution (Jordano et al. 2003), this creates a scenario in which plant–pollinator communities are highly structured. The observed pattern delineates their "topology" and "architecture."

The nested structure of mutualisms contributes to other recent approaches, such as the geographic mosaic of coevolution (Thompson 1994), to bring tractability to the complexity of coevolutionary interactions. Whereas the emphasis in the geographic mosaic theory is in the geographic structure, with possible specific interactions at local scales but global interactions with a larger number of species at a global scale, our results indicate structure within local communities. Both views are in fact related, nestedness being eminently a geographic idea (Patterson and Atmar 1986). Further studies should elucidate how the geographic (i.e., among communities) and the local (within communities) nestedness patterns are related and contribute to the maintenance of biodiversity.

A future avenue for research should explore the phenotypic and phylogenetic correlates of variation in *k* among species, the phylogenetic diversity of core species, and whether there are repeated patterns among networks. For instance, are the species at the core of interactions, with the highest centrality, a random subset of the morphospace in the community? Thus, do the species at the core define a distinct morphological type, either in the range of corolla or pollinator morphologies? In addition, a network-based approach to plant–pollinator interactions could increase our predictive power for the effects of exotic species in the networks and in the evolutionary dynamics of these communities (Memmott and Waser 2002). These aspects are central to our understanding of the resilience of these mutualisms to species loss. Our simple simulations showed that networks might collapse with the loss of even a small fraction of species, especially if these are plants (for a somewhat contrasting view see Memmott et al. 2004). Therefore, understanding the geographic variation of interaction matrices will greatly contribute to clarifying potential effects of fragmentation on plant–pollinator communities. Techniques for rapid assessment of plant–pollinator interaction matrices will be rewarding for the design of conservation priorities for preserving the whole network of interactions. Differential robustness to the loss of plants or pollinators can ultimately be related to their positions within the network and their role as core species in these mutualisms. All these findings point to interactions as a major component of ecosystem biodiversity—indeed, they themselves are perhaps the architecture of biodiversity.

Acknowledgments

We are grateful to Carlos Melián, Miguel A. Fortuna, Yoko Dupont, Thomas Lewinshon, Paulo-Inácio Prado, Nick Waser, Jeff Ollerton, Wesley Silva, Ruben Alarcón, Diego Vázquez, Paulo Guimarães Jr., and two anonymous reviewers for discussions, help with analyses, and comments on drafts of this chapter. L. I. Eskildsen and H. Elberling shared unpublished data. Thomas Lewinshon suggested key ways to improve the analysis of bipartite graphs and was a continuous source of support and advice. Both John N. Thompson and Judith Bronstein helped us with insight and useful suggestions. The study was supported by the Spanish Ministerio de Ciencia y Tecnología and FEDER funds (BOS2000-1366-C01, BOS2000-1366-C02, REN2003-00273) and the Junta de Andalucía (PAI group RNM305) to PJ and JB and by the Danish Natural Science Research Council (94-0163-1) to JMO. Special thanks go to Myriam Márquez from PJ for her help during manuscript preparation. We thank Nick Waser and Jeff Ollerton for inviting us to participate in this book and in the symposium at the 2002 Annual Meeting of the Ecological Society of America.

References

Albert, R., and A. L. Barabási. 2002. Statistical mechanics of complex networks. Reviews of Modern Physics 74: 47–97.

Albert, R., H. Jeong, and A. L. Barabási. 2000. Error and attack tolerance of complex networks. Nature 406: 378–382.

Amaral, L. A. N., A. Scala, M. Barthélémy, and H. E. Stanley. 2000. Classes of small-world networks. Proceedings of the National Academy of Sciences (USA) 97: 11149–11152.

Atmar, W., and B. D. Patterson. 1993. The measure of order and disorder in the distribution of species in fragmented habitat. Oecologia 96: 373–382.

Barabási, A. L., and R. Albert. 1999. Emergence of scaling in random networks. Science 286: 509–512.

Barabási, A. L., R. Albert, and H. Jeong. 2000. Scale-free characteristics of random networks: The topology of the World Wide Web. Physica A 281: 69–77.

Bascompte, J., P. Jordano, C. J. Melián, and J. M. Olesen. 2003. The nested assembly of plant–animal mutualistic networks. Proceedings of the National Academy of Sciences (USA) 100: 9383–9387.

Batagelj, V., and A. Mrvar. 2003. PAJEK: Analysis and visualization of large networks. Pp. 77–103 *in* M. Jünger and P. Mutzel (eds.), Graph drawing software. Springer, Berlin.

Bersier, L. F., P. Dixon, and G. Sugihara. 1999. Scale-invariant or scale-dependent behavior of the link density property in food webs: A matter of sampling effort? American Naturalist 153: 676–682.

Boisvert, R. F., R. Pozo, K. Remington, R. Barrett, and J. J. Dongarra. 1997. The Matrix Market: A web resource for test matrix collections. Pp. 125–137 *in* R. F. Boisvert (ed.), Quality of numerical software, assessment and enhancement. Chapman and Hall, London.

Borgatti, S. P., and M. G. Everett. 1997. Network analysis of 2-mode data. Social Networks 19: 243–269.

Borgatti, S. P., and M. G. Everett. 1999. Models of core/periphery structures. Social Networks 21: 375–395.

Borgatti, S. P., M. G. Everett, and L. C. Freeman. 1999. UCINET 6.9 version 1.0. Natick: Analytic Technologies, Cambridge, MA.

Calder, A., and J. Davidson. 1966. Calder, an autobiography with pictures. Pantheon Books, New York.

Darwin, C. 1862. On the various contrivances by which British and foreign orchids are fertilised by insects, and on the good effects of intercrossing. Murray, London.

Davis, A., B. B. Gardner, and M. R. Gardner. 1941. Deep South: A social and anthropological study of caste and class. University of Chicago Press, Chicago.

Dicks, L. V., S. A. Corbet, and R. F. Pywell. 2002. Compartmentalization in plant–insect flower visitor webs. Journal of Animal Ecology 71: 32–43.

Dorogovtsev, S. N., and J. F. F. Mendes. 2002. Evolution of networks. Advances in Physics 51: 1079–1187.

Duff, I. S., A. M. Erisman, and J. K. Reid. 1986. Direct methods for sparse matrices. Oxford University Press, Oxford.

Dunne, J. A., R. J. Williams, and N. D. Martinez. 2002. Network structure and biodiversity loss in food webs: Robustness increases with connectance. Ecology Letters 5: 558–567.

Everett, M. G., and S. P. Borgatti. 1999. The centrality of groups and classes. Journal of Mathematical Sociology 23: 181–201.

Gentry, A. H. 1982. Patterns of neotropical plant species diversity. Pp. 1–84 *in* M. K. Hecht, B. Wallace, and G. T. Prance (eds.), Evolutionary biology, vol. 15. Plenum Press, New York.

Goldwasser, L., and J. Roughgarden. 1997. Sampling effects and the estimation of food-web properties. Ecology 78: 41–54.

Gotelli, N. J. 2000. Null model analysis of species co-occurrence patterns. Ecology 81: 2606–2621.

Jeong, H., B. Tombor, R. Albert, Z. N. Oltval, and A. L. Barabási. 2000. The large-scale organization of metabolic networks. Nature 407: 651–654.

Johnson, S. D., and K. E. Steiner. 2000. Generalization versus specialization in plant pollination systems. Trends in Ecology and Evolution 15: 140–143.

Jordano, P. 1987. Patterns of mutualistic interactions in pollination and seed dispersal: Connectance, dependence asymmetries, and coevolution. American Naturalist 129: 657–677.

Jordano, P., J. Bascompte, and J. M. Olesen. 2003. Invariant properties in coevolutionary networks of plant–animal interactions. Ecology Letters 6: 69–81.

Kearns, C. A., D. W. Inouye, and N. M. Waser. 1998. Endangered mutualisms: The conservation of plant–pollinator interactions. Annual Review of Ecology and Systematics 29: 83–112.

Laska, M. S., and J. T. Wootton. 1998. Theoretical concepts and empirical approaches to measuring interaction strength. Ecology 79: 461–476.

May, R. M. 1972. Will a large complex system be stable? Nature 238: 413–414.

Memmott, J., and N. M. Waser. 2002. Integration of alien plants into a native flower–pollinator visitation web. Proceedings of the Royal Society of London B 269: 2395–2399.

Memmott, J., N. M. Waser, and M. V. Price. 2004. Tolerance of pollination networks to species extinctions. Proceedings of the Royal Society of London B 271: 2605–2611.

Montoya, J. M., and R. V. Solé. 2002. Small world patterns in food webs. Journal of Theoretical Biology 214: 405–412.

Mossa, S., M. Barthélémy, H. E. Stanley, and L. A. N. Amaral. 2002. Truncation of power law behavior in "scale-free" network models due to information filtering. Physical Review Letters 88: 138701.

Newman, M. E. J. 2003. The structure and function of complex networks. SIAM Review 45: 167–256.

Olesen, J. M. 2000. Exactly how generalised are pollination interactions? Det Norske Videnskaps—Akademi. I. Matematisk Naturvidenskapelige Klasse, Skrifter, Ny Serie 39: 161–178.

Olesen, J. M., L. I. Eskildsen, and S. Venkatasamy. 2002. Invasion of pollination networks on oceanic islands: Importance of invader complexes and endemic super generalists. Diversity and Distributions 8: 181–192.

Olesen, J. M., and P. Jordano. 2002. Geographic patterns in plant–pollinator mutualistic networks. Ecology 83: 2416–2424.

Patterson, B. D., and W. Atmar. 1986. Nested subsets and the structure of insular mammalian faunas and archpielagos. Biological Journal of the Linnean Society 28: 65–82.

Pimm, S. L. 2002. Food webs. University of Chicago Press, Chicago.

Pimm, S. L., and J. H. Lawton. 1980. Are food webs divided into compartments? Journal of Animal Ecology 49: 879–898.

Poulin, R. 1996. Richness, nestedness, and randomness in parasite infracommunity structure. Oecologia 105: 545–551.

Poulin, R., and J. F. Guegan. 2000. Nestedness, anti-nestedness, and the relationship between prevalence and intensity in ectoparasite assemblages of marine fish: A spatial model of species coexistence. International Journal for Parasitology 30: 1147–1152.

Raffaelli, D., and S. J. Hall. 1992. Compartments and predation in an estuarine food web. Journal of Animal Ecology 61: 551–560.

Snow, B. K., and D. W. Snow. 1972. Feeding niches of hummingbirds in a Trinidad valley. Journal of Animal Ecology 41: 471–485.

Strogatz, S. H. 2001. Exploring complex networks. Nature 410: 268–276.

Thompson, J. N. 1982. Interaction and coevolution. John Wiley and Sons, New York.

Thompson, J. N. 1994. The coevolutionary process. University of Chicago Press, Chicago.

Vázquez, D. P., and M. A. Aizen. 2003. Null model analyses of specialization in plant–pollinator interactions. Ecology 84: 2493–2501.

Waser, N. M., L. Chittka, M. V. Price, N. M. Williams, and J. Ollerton. 1996. Generalization in pollination systems, and why it matters. Ecology 77: 1043–1060.

Williams, R. J., and N. D. Martinez. 2000. Simple rules yield complex food webs. Nature 404: 180–183.

CHAPTER NINE

Community-wide Patterns of Specialization in Plant–Pollinator Interactions Revealed by Null Models

Diego P. Vázquez and Marcelo A. Aizen

Understanding the causes and consequences of specialization in species interactions is central to ecology and evolutionary biology. This knowledge is important because specialization may have profound ecological and evolutionary consequences (Brown 1984; Thompson 1994; Waser et al. 1996; Vázquez and Simberloff 2002). A primary question is how common specialization (or generalization) is in nature. Plant–pollinator interactions have frequently been regarded as tightly coevolved and highly specialized, with a general evolutionary trend toward increased specialization (Waser, chap. 1 in this volume). But the extent to which such specialization actually occurs in most, or even some, of the species in a given community has recently been questioned (Ollerton 1996; Waser et al. 1996).

A limitation of recent discussions about the specialized/generalized nature of plant–pollinator interactions (e.g., Ollerton 1996; Waser et al. 1996; Johnson and Steiner 2000; Gómez 2002) is the lack of null hypotheses with which to contrast the observed patterns of specialization. Without such null hypotheses, it is difficult to ascertain if either extreme specialization or extreme generalization prevail in nature. For example, Waser et al. (1996) noted in two early community-wide studies (Clements and Long 1923; Robertson 1929) that plants were visited on average by 33.5 and 9.4 pollinator species, respectively. These figures represented a small fraction of all the pollinator species present in those two communities (ca. 220 and 1150 pollinator species, respectively, assuming 80% of visitors pollinate; Memmott and Waser 2002; Memmott et al. 2004), although there is a huge spread around this mean, with several species that interact with only one pollinator and others that interact with more than 50. However, just looking at these figures does not tell us much about whether extreme specialization or extreme generalization is unusually common.

Another limitation of recent discussions about specialization in plant–pollinator interactions is that our characterization of specialization is still rudimentary. One particular aspect that has received surprisingly little attention is the degree to which specialization is reciprocal between pairs of interacting species. Biologists have frequently assumed symmetric specialization: either specialists interact with specialists or generalists with generalists (see Thompson 1994; Renner 1998; Vázquez and Simberloff 2002). Under this scenario, reciprocally specialized interactions are more likely to lead to coupled ecological dynamics and coevolutionary change than reciprocally generalized interactions. However, this is not necessarily the case: only a fraction of specialists are likely to interact with specialists, while the rest are expected to interact with moderate-to-extreme generalists. The existence of asymmetric specialization between interaction partners may have important consequences: rather than reciprocal influences of interacting species affecting each other's ecological and evolutionary processes, it can lead to a situation in which specialists track generalists. However, some degree of asymmetry may be expected from the structural constraints inherent to the assembly of interaction networks, even if species interactions are solely the result of a random process. Thus, as just argued, the existence of a pattern does not mean that the pattern is unusual; we need to compare it with some null expectation.

A possible approach to provide a null hypothesis for specialization patterns in plant–pollinator interactions is the use of null models: the generation of randomized datasets in the absence of a hypothesized mechanism. Null models have played an important role in many areas of community ecology and biogeography (Gotelli and Graves 1996; Gotelli 2001). This approach may allow us to answer not only whether the occurrence of specialization or generalization is different from some specified null hypothesis, but also whether a particular mechanism may be a likely explanation for such an occurrence. Until recently, no null model analysis of specialization patterns in plant–pollinator interactions had been undertaken. In the following, we describe recent work we have conducted using this approach, present some new results, and identify future avenues of research. Although much remains to be done, our null model approach provides a rigorous way of studying community-wide patterns of specialization in plant–pollinator interactions.

Defining Specialization

When studying specialization, it would be ideal to use a definition that is ecologically and evolutionarily relevant. For a plant, pollinators represent one of many components of their niche—one that has direct consequences in their reproductive performance and, ultimately, in their demography. For a pollinator, flowers usually represent food sources or, in some circumstances, shelter. Therefore, a meaningful definition of specialization would require, for example, measuring the reproductive consequences of each of the interactions between a given plant species and its flower visitors or the nutritional consequences of each plant species visited by a given pollinator. This type of study is feasible, and there are many examples in the literature, at least for plants (e.g., Schemske and

Horvitz 1984; Herrera 1989; Fishbein and Venable 1996; Olsen 1997; Gómez and Zamora 1999); however, conducting such studies for large assemblages of plants and pollinators may be prohibitive, and none of the community-wide datasets currently available in the literature include such information. Most available studies include information only about the occurrence of pairwise interspecific interactions (i.e., which plant species interact with which flower visitor species), although some studies also include information about the frequency of interaction between pairs of species (i.e., how many times a given plant species was visited by a given pollinator species during the course of the study; see, e.g., Armbruster, chap. 12 in this volume; Kwak and Bekker, chap. 16 in this volume). Thus, our definition of specialization will necessarily ignore most biological details of the interactions, focusing only on their occurrence.

Another important point regarding specialization is the distinction between what we call "fundamental" and "realized" specialization, in direct reference to the niche concept proposed by G. E. Hutchinson (1957). Fundamental specialization refers to the potential interactions that would lead to positive fitness for a given species under any possible ecological circumstances; thus, fundamental specialization ultimately depends on the genetic background of a species. Conversely, realized specialization refers to the actual specialization attained under a particular ecological context. For example, most fig species are fundamentally specialized on one or a few species of pollinating wasps, and they fail to reproduce if they are moved to different environments unless their specialized pollinators are introduced as well. Conversely, many other plant species that are pollinated by a single pollinator in a particular habitat are often pollinated by a different species when they are introduced in other habitats (Richardson et al. 2000). In the context of our study, we will be restricted to the realized niche, because in most cases we know little about what happens with the interactions outside the arbitrary boundaries of the study.

Given the preceding limitations, the definition of specialization we use in our analyses is a very simple one: the inverse of the number of interaction partners of a given species (i.e., the fewer interaction partners, the more specialized a species). Of course, this definition ignores much of the biology and ecology of interactions; it assumes all interactions are equal, regardless of their relative frequency and quality. However, we believe the use of such a grossly simplified measure of specialization is still useful, for at least two reasons: (1) many previous and current studies have used a similar measure of specialization, and, thus, we are able to revisit some conclusions with our null model perspective; and (2) using such simple measures is, at least at present, the only way of studying community patterns of plant–pollinator specialization. Our definition has the additional advantage of allowing us to treat specialization and generalization as two extremes in a continuum, thus getting away from the dichotomy that has pervaded the literature.

Null Models for Patterns of Specialization in Plant–Pollinator Interactions

We have recently developed null models to study patterns of specialization in plant–pollinator interaction networks (Vázquez and Aizen 2003, 2004). The rationale behind the approach is that, by randomly generating datasets that share some characteristics with the observed data (such as number of species of plants and pollinators and total number of links among interacting species), it is possible to obtain patterns that can be compared with the observed data (Gotelli and Graves 1996; Gotelli 2001). If the randomized data differ substantially from the real data, and assuming that the null model is "correct," then it is possible to invoke some sort of mechanism responsible for the observed pattern. Conversely, if the randomized data do not differ from the observed data (and, again, assuming that the null model is correct), it is possible to reject the hypothesis that there is a mechanism other than randomness responsible for the observed pattern. It is also possible to compare the explanatory power of different models of increasing complexity that explicitly incorporate particular mechanisms.

We start with an "interaction matrix" compiled from various datasets (see table 9.1), in which columns represent plant species and rows flower visitor species found in a particular community; a cell with a 1 indicates that a given plant and a pollinator species interact, whereas a cell with a 0 indicates no interaction between a pair of species. In addition, some datasets contain information on the frequency of interaction between pairs of species (either total number of pollinator visits recorded per unit time or number of collected individuals per insect species for each plant species), which we use to construct an "interaction frequency matrix." We use simple measures of specialization, which can be readily calculated from the limited information available in the datasets: s, the total number of species with which a given species interacts, obtained from the binary interaction matrix as the sum of columns or rows for plants and pollinators, respectively; and p, the average value of s of the interaction partners of a given species. We also calculate f, the frequency of interaction of plant and pollinator species, obtained as the row and column sums in the interaction frequency matrix, respectively.

The null models we have used are based on randomization procedures. Our simplest null model is one that assumes random interactions among species. Under this model, every species is assigned an interaction by randomly selecting another species from the pool of possible interaction partners. Once every species has been assigned one (and only one) interaction, the remaining interactions are assigned by randomly selecting plant and pollinator species pairs. Thus, in this model the expected value of s (the number of interaction partners) is equal to that probability multiplied by the number of potential interaction partners (n) and is the same for all species, independent of their frequency of interaction (fig. 9.1). Thus, this model assumes neutrality at the species level

Table 9.1 Datasets that describe plant–pollinator interaction webs used to conduct analyses reported in the text

Dataset	Habitat type	Location	Interaction data	*m*	*n*	*C*
Barrett and Helenurm 1987	Boreal forest	Central New Brunswick, Canada	Individuals caught	12	102	0.14
Clements and Long 1923	Montane forest and grassland	Pikes Peak, CO, United States	Binary	96	276	0.03
Elberling and Olesen 1999	Alpine subarctic community	Latnjajaure, Sweden	Visits	23	118	0.09
Hocking 1968	Arctic community	Ellesmere Island, NWT, Canada	Binary	29	86	0.07
Inouye and Pyke 1988	Montane forest	Kosciusko National Park, NSW, Australia	Individuals caught	42	91	0.07
McMullen 1993	Multiple communities	Galápagos Islands	Binary	106	54	0.04
Medan et al. 2002	Xeric scrub	Laguna Diamante, Mendoza, Argentina	Binary	21	45	0.09
Medan et al. 2002	Woody riverine vegetation and xeric scrub	Río Blanco, Mendoza, Argentina	Binary	23	72	0.08
Memmott 1999	Meadow	England	Visits	25	79	0.15
Mosquin and Martin 1967	Arctic community	Melvile Island, NWT, Canada	Individuals caught	11	18	0.19
Motten 1982	Deciduous forest	Durham and Orange Counties, NC, United States	Visits	13	44	0.25
Olesen et al. 2002	Coastal forest	Île aux Aigrettes, Mauritius	Visits	14	13	0.29
Olesen et al. 2002	Rocky cliff and open herb community	Flores Island, Azores	Visits	10	12	0.25
Ramírez and Brito 1992	Palm swamp community	Central Plains, Venezuela	Binary	33	53	0.06
Schemske et al. 1978	Maple/oak woodland	Brownfield Woods, IL, United States	Visits	7	32	0.26
Small 1976	Peat bog	Mer Bleue, Ottawa, Canada	Individuals caught	13	34	0.32
C. Smith-Ramírez et al. (unpublished data)	Temperate rainforest	Chiloé Island, Chile	Binary	24	111	0.11
Vázquez and Simberloff 2002, 2003	Evergreen montane forest	Nahuel Huapi National Park, Argentina	Visits	14	93	0.13

Note: All datasets, with the exception of the unpublished data by C. Smith-Ramírez et al., are available through the Interaction Web Database (online URL http://www.nceas.ucsb.edu/interactionweb). Interaction data is the information on plant–pollinator interactions included in each publication: number of individuals caught, number of visits observed, or presence/absence of interaction (binary); *m:* number of plant species; *n:* number of flower visitor species; *C:* connectance of the interaction matrix.

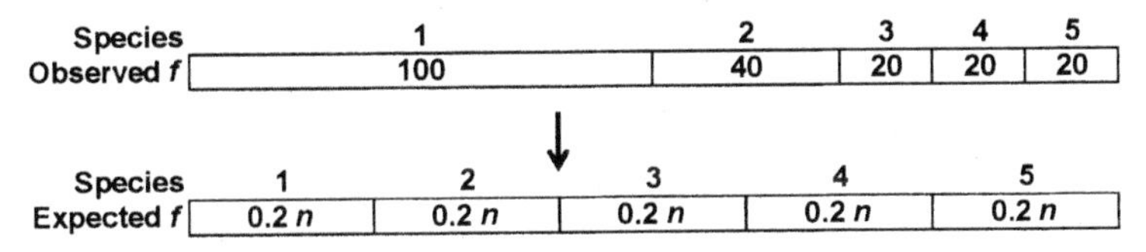

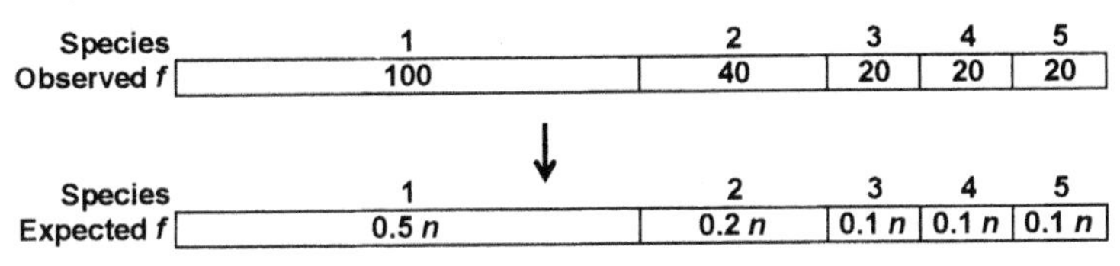

Figure 9.1 Schematic representation of null models used by Vázquez and Aizen (2003, 2004) to study patterns of specialization in plant–pollinator interaction networks. In null model 1, the probability of being assigned an interaction in each random draw is equal for all species. The expected value of s (the number of interaction partners) is equal to that probability multiplied by the number of potential interaction partners, n. Thus, if there are 5 plant and 20 pollinator species in the community, the expected value of s for each plant species will be 4. In null model 2, the probability for each species of being assigned an interaction in each random draw is proportional to their observed frequency of interaction, f. Therefore, the expected value of s is not the same for all species. Thus, for a community with 5 plant and 20 pollinator species, and with the values of f shown in the figure, the expected s for each plant species will be 10, 4, 2, 2, and 2, respectively.

(i.e., all species are equal, regardless of their identity) and allows us to test whether observed patterns of specialization could arise from random interaction among species, independent of their characteristics.

The second null model we have used assumes random interactions among individuals (rather than species) in the spirit of recent neutral models of community structure (Bell 2000; Hubbell 2001). Here every species is assigned an interaction by randomly selecting a species from the pool of possible interaction partners. Once every species has been assigned one (and only one) interaction, the remaining interactions are assigned proportionally to a species' observed interaction frequency, so that frequently interacting species have a higher probability of being assigned an interaction than rarely interacting species (fig. 9.1). Thus, this model allows us to test a mechanism responsible for the patterns observed in plant–pollinator interactions: that it is individuals (and not species) that interact randomly, which results in species with more individuals (or more frequently interacting individuals) interacting with more other species than rare species.

Models similar to the ones we developed have recently been used by Bascompte et al. (2003) to study nested patterns in plant–animal mutualistic networks (see also Jordano et al., chap. 8 in this volume). These authors also define two null models. One model makes the probability of a species to be assigned an interaction homogenous for all species, and is thus equivalent to our model 1. Their second null model assumes that such probability varies among species and is proportional to the number of links per species. As the number of links per species changes while the interaction matrix is filled during the randomization process, the species probabilities are changed accordingly. Because frequency of interaction is positively correlated with the number of interaction partners, Bascompte et al.'s null model 2 should be roughly equivalent to our null model 2. Another aspect in which Bascompte et al.'s models differ from ours is that not all species are required to have at least one interaction. Such a difference can affect the variability of randomized patterns but should not affect the mean. Thus, our null models are generally roughly equivalent to those used by Bascompte et al., and conclusions reached with any of the two approaches should be qualitatively similar.

Patterns of Specialization in Plant–Pollinator Interactions

As discussed earlier, plant–pollinator interaction networks include species with a broad range of degree of specialization, from very specialized to very generalized. We have recently shown (Vázquez and Aizen 2003) that the number of extreme specialists and extreme generalists is significantly higher than expected under the simplest of our null models, which assumes random interaction among species (fig. 9.2). These results suggest that one or more processes, missing in this simple null model, generate the observed pattern.

The finding that most communities have a high number of extreme specialists and extreme generalists does not mean that such number is equally high in all communities. For example, the percentage of plant species interacting with only one partner species among the five plant–pollinator interaction webs analyzed in Vázquez and Aizen (2003) ranged from 0% in Motten's (1982) dataset to 18% in Clements and Long's (1923) dataset; whereas only 13% of plant species interacted with 20 or more pollinators in Clements and Long's dataset, 23% did so in Motten's study. Is it possible to explain such variation in the prevalence of extreme specialization and generalization in communities? One possibility is that such prevalence is determined by community characteristics such as community size (i.e., number of species of plants and pollinators) or number of links (i.e., pairs of interacting species).

We used *d*, a statistic that measures the difference between either observed or each of the randomized distributions and the expected degree of specialization in the null model (see Vázquez and Aizen 2003), as a measure of the prevalence of extreme generalization and specialization. The larger the value of *d*, the higher the prevalence of species in these extremes of the continuum. Thus, if specialization and generalization increase with, say, community size, the value of *d*

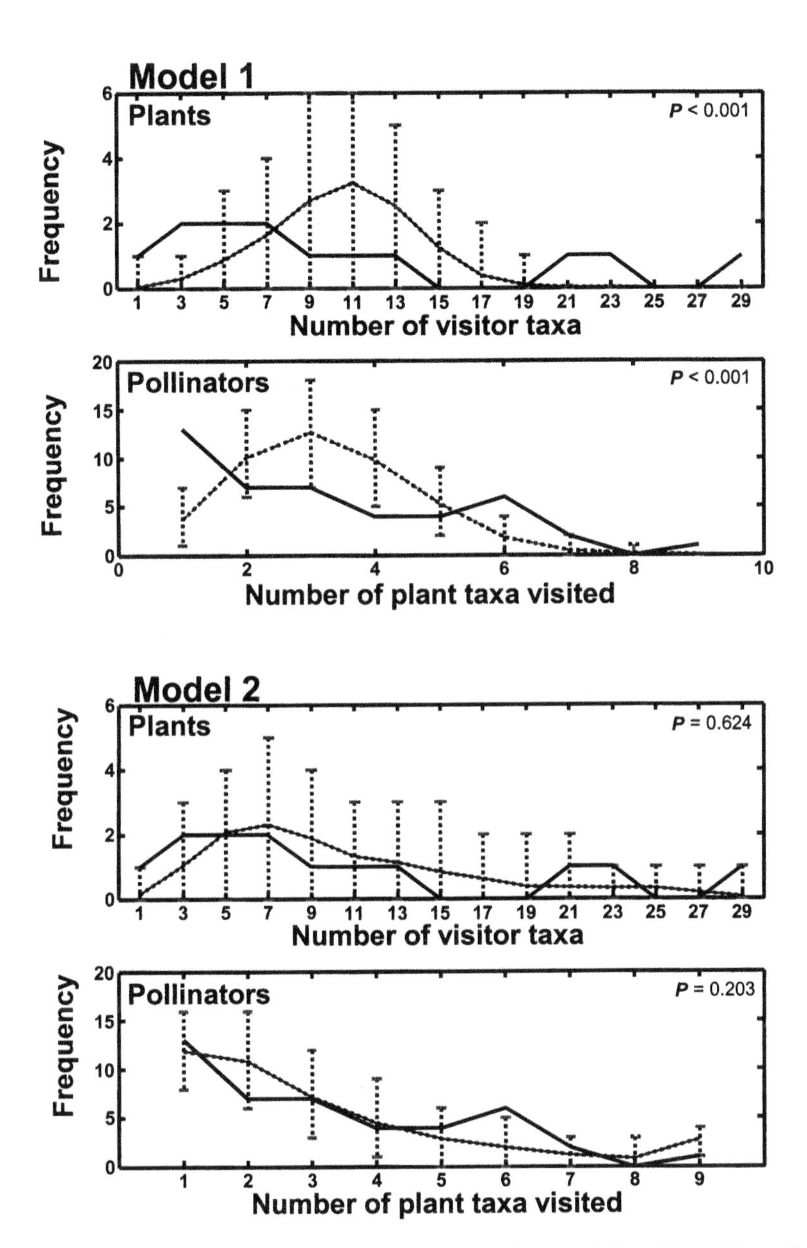

Figure 9.2 Comparison between observed (solid lines) and expected (dotted lines with error bars) distributions of *s*, the number of species of interaction partners (an estimate of degree of specialization). Mean ±95% confidence intervals are given for expected distribution. Numbers at the upper-right corner of plots are the *P* values of the difference test comparing expected and observed distributions (curves are considered significantly different with $P < .05$). Data are from Motten (1982), and results of analyses are from Vázquez and Aizen (2003).

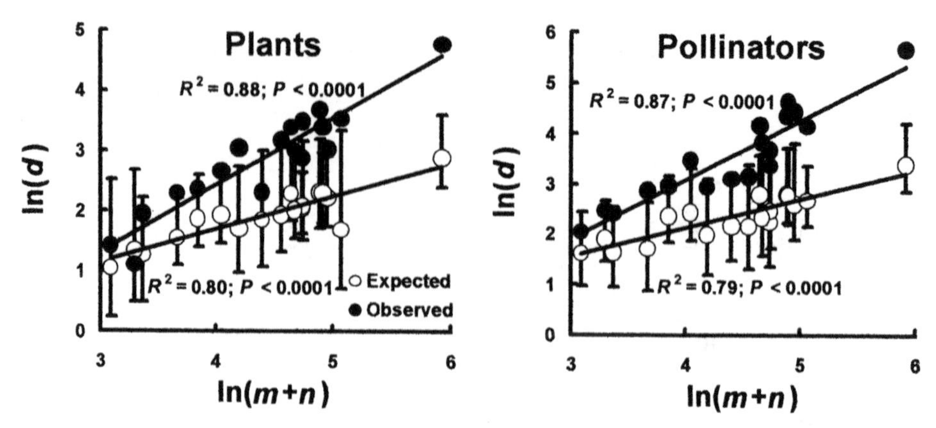

Figure 9.3 Difference statistic (d) versus species richness of plants and pollinators ($m + n$). The difference statistic measures the difference between the expected and observed number of species in each category of s. Thus, d can be taken as a measure of the occurrence of extreme specialization and extreme generalization (see text). Each dot represents a community (see table 9.1 for list of datasets). Solid circles represent the d statistic calculated for observed data; open circles represent expected values of the d statistic calculated for 1000 randomized communities; randomizations for each community were conducted separately, and, thus, each dot represents an independent data point. Lines represent the fitted regression line to each set of data; the coefficient of multiple determination and the probability value associated with each regression are given next to each line. The difference in slopes was tested using Wilks's lambda statistic in the REG procedure in SAS (SAS Institute 1999): $F_{1,16} = 26.56$; $P < .0001$.

should also increase. Figure 9.3 presents the results of this test for null model 1, for the 18 datasets listed in table 9.1. (We used model 1 to be able to use all 18 datasets; however, similar results are obtained if model 2 is used instead.) Because number of species and number of links are highly correlated ($r = 0.96$; $P < .0001$), we present results for number of species only. The magnitude of the d statistic increases with community size for both the randomized and the observed datasets. However, the slope for the observed datasets is significantly steeper than that for the randomized datasets (see fig. 9.3): the increase in the prevalence of extreme specialization and generalization in these communities is greater than that expected under our simple null model and implies that more diverse communities support an increasingly higher number of specialists and generalists than should be expected by chance. This result opens the intriguing possibility that high species richness favors the evolution of extreme specialization and generalization in pollination systems.

An interesting feature of most datasets we have examined thus far is a positive correlation between observed frequency of interaction of species, f, and number of interaction partners they have, s (Vázquez and Aizen 2003, 2004; see also fig. 9.6 below). In other words, plant species that are visited by many individual pollinators, and pollinators that visit many flowers, tend to interact with more species than do rarely interacting species. We hypothesized that this correlation could be partly responsible for the pattern of specialization observed in the datasets. To examine this hypothesis, we conducted two different tests (Vázquez and Aizen 2003). In the first one, we removed the correlation between *f* and *s* and used the residuals as a corrected estimate of specialization. In the second test, we incorporated the correlation into a null model by assigning interactions proportionally to the frequency of interaction (i.e., null model 2 described earlier). In both cases, the pattern generated by the null model was closer, and not significantly different from, the observed pattern (fig. 9.2). The important message here is that simple random interaction among individuals (which does not assume any difference among species besides differences in frequency of interaction) does an extremely good job of explaining observed patterns of specialization in plant–pollinator networks.

Patterns of Asymmetric Specialization in Plant–Pollinator Interactions

As we argued earlier, the degree of specialization between pairs of interacting species should not necessarily be symmetric: specialists could, and often do, interact with generalists, and the opposite (generalists interacting with specialists) is also possible and often observed (Petanidou and Ellis 1996; Renner 1998; Vázquez and Simberloff 2002; Minckley and Roulston, chap. 4 in this volume; Petanidou and Potts, chap. 10 in this volume). A comparison of the distribution of *s* and *p* observed in communities with that generated by our simplest null model shows that the prevalence of asymmetric specialization in real communities is substantially higher than expected (Vázquez and Aizen 2004; fig. 9.4). This asymmetry is a consequence of specialists specializing on generalists or of species being extreme generalists themselves (which necessarily means that they interact with many specialists in the community).

As before, we asked whether the contrasting patterns between observed and randomized data could depend partly on community characteristics such as number of links or number of species. As a measure of the agreement between predicted and observed asymmetry, we used the percentage of species in the original dataset that fell inside confidence intervals generated by the null model. This percentage decreases with increasing community size and number of links (Vázquez and Aizen 2004). In other words, prevalence of asymmetric specialization in plant–pollinator interaction webs increases with web size. This result matches our previous result that prevalence of extreme specialization and generalization increases with community size. Thus, larger communities seem to have proportionally more extreme specialists, extreme generalists, and asymmetrically specialized species than smaller communities.

As we did in the preceding section, we now ask whether the observed correlation between *f* and *s* may partly account for the distribution of asymmetric specialization observed in the datasets. Incorporating such a correlation into the null model (i.e., null model 2) results in a pattern that is, in most cases, strikingly similar to that observed in the data and generally provides a much better fit to the data

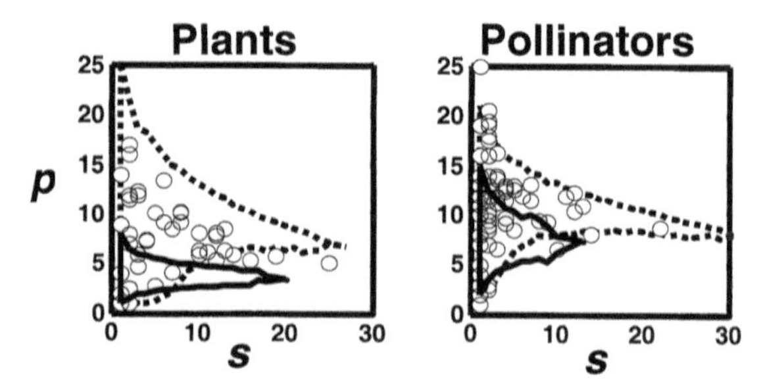

Figure 9.4 Asymmetric specialization in plant–pollinator interactions; degree of specialization, *s*, and average specialization of interaction partners, *p*, are shown. Circles represent observed *s*–*p* values, solid lines represent null space for null model 1, and dotted lines represent null space for null model 2. Data are from Inouye and Pyke (1988), and results of analyses are from Vázquez and Aizen (2004).

than does model 1 (Vázquez and Aizen 2004; fig. 9.4). Thus, the asymmetry observed in most datasets can be partly accounted for by the relationship between species' frequency of interaction and their estimated degree of specialization.

It is important to mention that two recent studies have independently reached conclusions similar to those reached in our analysis of asymmetric specialization. Jordano et al. (2003; see also Jordano et al., chap. 8 in this volume) studied the distribution of the number of interactions per species using a large database with datasets on plant–pollinator and plant–seed disperser interactions (which substantially overlaps with the one used by us, listed in table 9.1). They found that the cumulative frequency of the number of links per species (called "degree" in the network literature) decays as a power law as the number of links increases, although in most cases the power law is "truncated" at large values of links per species and rapidly approaches zero. These types of degree distributions in networks are expected when a few nodes (i.e., species) have many links and many nodes have a few links, which necessarily implies that highly connected nodes will be connected to many nodes with a low number of links (Albert and Barabási 2002; Jordano et al. 2003). In a subsequent paper, Bascompte et al. (2003; see also Jordano et al., chap. 8 in this volume) have elegantly shown that in most cases these plant–animal mutualistic networks have a nested structure, whereby specialized species always interact with a subset of interaction partners of those that interact with the most generalized species. A nested pattern of interactions necessarily indicates high asymmetry in the degree of specialization of interaction partners.

A Conceptual Framework for Understanding Community Patterns of Plant–Pollinator Specialization

Our null model analyses of plant–pollinator interaction networks from around the world show that both extreme specialization and extreme generalization are common; they also show that asymmetric specialization between pairs of inter-

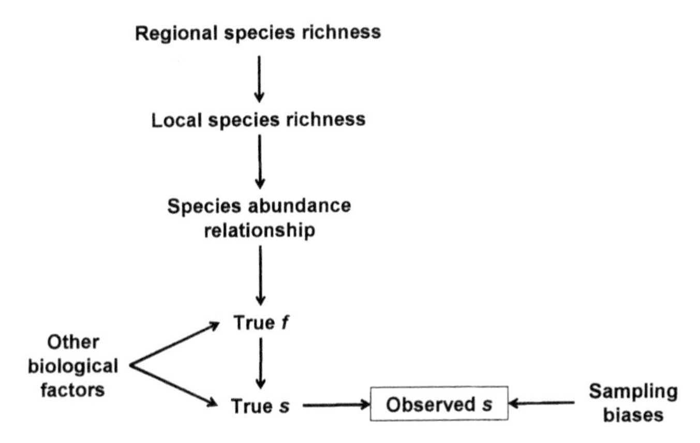

Figure 9.5 A conceptual framework for understanding observed community-wide patterns of specialization in plant–pollinator interactions (see text for explanation).

acting species is not uncommon. Furthermore, our results suggest that these patterns are largely explained by the characteristics of the community (particularly the number of interacting species) and by a correlation between the number of interaction partners of species, s, and their frequency of interaction, f. Therefore, a critical question to understand patterns of specialization in plant–pollinator interaction networks is what factors are behind the f–s correlation.

Frequency of interaction may be thought of as a function of species abundance. This assumption is particularly appropriate for flower visitors. Abundant pollinators are likely to be observed visiting flowers more often than rare pollinators. However, other factors may also affect f, such as pollinator floral preferences, mobility, and dependence on flowers as food sources. In the same way, f will be a function of plant species' abundance and of other factors, including flower attractiveness to pollinators or researchers' choice. (That is, sampling of plant species to estimate pollinator visitation frequencies may be conducted irrespective of plant abundance to obtain a similar number of samples for all species.) We can define α as a parameter that summarizes these different factors that affect f independently of abundance. Thus, the frequency of interaction of species i may be described as $f_i = \alpha_i A_i$, where A_i is the abundance of species i. And, since the logarithms of f and s are linearly related (see fig. 9.6 below), we have $\log s = b \log \alpha A$, where b is a constant that determines the strength of the relationship between s and f. The interesting fact about this simple model is that now we can relate patterns of specialization to well-known patterns of species abundance in communities.

We can now introduce a conceptual model that will help visualize the aforementioned findings and hypotheses and help identify what factors should be incorporated in future models aimed at advancing our understanding of patterns of specialization in plant–pollinator interactions (fig. 9.5). We assume that local species richness of plants and pollinators are partly determined by regional species richness (see Ricklefs 1987; Lewinsohn 1991; Cornell 1999). In turn, local species richness will influence the shape of the species abundance relationship (which will usually approximate a lognormal or log-series distribution; Preston 1948; May 1975; Hubbell 2001) so that species-poor communities have steeper rank-abundance curves than their species-rich counterparts. As argued earlier, local species abundance will partly determine the frequency of interaction among plants and pollinators, *f*. Finally, the true degree of specialization, *s*, will be determined by *f* and by other biological factors (see earlier discussion); however, observed *s* will also be affected by sampling biases. In the following we discuss each of these factors and their potential contribution to observed patterns of *s*.

Proximate Determinants of Observed Degree of Specialization

Given the regional and local community contexts in which plant–pollinator interactions occur, it is desirable to understand the role played by each of the three proximate determinants of observed patterns of specialization identified earlier. These factors are not mutually exclusive, and all are likely contributors to the observed patterns.

Observed *s* could be determined by the random interaction among individual plants and pollinators, whereby species that interact more frequently do it with more species than rarely interacting species; this is the basic assumption of our null model 2. The fact that, for most datasets, null model 2 provides a better fit to the data than model 1 supports this explanation. We are not the first to invoke this kind of mechanism to explain patterns of interaction among species. For example, Southwood (1961) and Janzen (1968) proposed a similar idea to explain the observation that abundant plants tended to host a greater number of species of herbivorous insects than rare plants (i.e., the frequency-of-encounter hypothesis; see also Strong et al. 1984). Neutrality at the individual level rather than at the species level is also central to recent models of community structure (Bell 2000; Hubbell 2001). Random interaction among individuals would produce a positive correlation between *f* and *s* like the one observed in most of the datasets we have thus far examined.

Other biological factors could also influence the number of interaction partners related to a given species. As we discussed earlier, pollinator characteristics such as floral preferences, mobility, and dependence on flowers as food sources and plant characteristics such as flower attractiveness to pollinators could be important. These traits could be the result of adaptive processes that favor coadaptation between pairs of interacting species. However, high reciprocal specialization could mean—for most species and under most circumstances—low fitness. For example, Waser et al. (1996), using simple models of plant and pollinator fitness, showed that when the abundance, per-capita visitation rates, or quality of interaction partners vary strongly over time, it should be more beneficial to be a generalist than to be a specialist. Specialization on a single species should be beneficial only when the abundance and efficiency of the interaction partner are constant in time, which is more likely to occur when the partner is an abundant generalist. On the other hand, high symmetric specialization could mean high risk of extinction (Renner 1998; Vázquez and Simberloff 2002; see also Memmott et al. 2004), which could also affect the distribution of degree of specialization in communities.

Finally, observed *s* could be the result of sampling artifacts. First, observed patterns of specialization could be influenced by data aggregation. The information usually available in most datasets does not typically include phenological or spatial patterns in the distribution of abundance (see also Medan et al., chap. 11 in this volume). For example, consider a pollinator that interacts sequentially with several specialist plants throughout the flowering season; although this pollinator species may interact with many plant species, the number of species with which it interacts at any given time may be much smaller. Likewise, spatial aggregation of data can occur when a widespread species interacts with different species at different locations; if data for such a species are pooled into a single dataset, species may appear more generalized than they actually are at each locality (Fox and Morrow 1981; Thompson 1994).

Second, the observed patterns, particularly the *f–s* correlation, could simply be a result of a sampling bias, so that frequently interacting species simply appear more generalized than rare species (Petanidou and Ellis 1996; Vázquez and Aizen 2003, 2004). Ollerton and Cranmer (2002; see also Ollerton et al., chap. 13 in this volume) have pointed out a similar sampling bias: studies that have been conducted for longer periods tend to suggest a higher degree of generalization than those with shorter study periods. Similar sampling artifacts have been invoked to explain other ecological and biogeographic patterns. For example, one explanation of the ubiquitous positive relationship between local abundance and geographical range among species is simply that rare species are less likely to be found at any given site; thus, their geographical range appears smaller than that of locally abundant species (Hanski et al. 1993; Gaston et al. 1997). Similarly, the observation that abundant and widespread plants tend to harbor more species of phytophagous insects (Strong et al. 1984; Leather 1991) has also been attributed to sampling artifacts (Kuris et al. 1980; Stevens 1986; Lewinsohn 1991).

Uncovering the factors determining the observed *s* is thus key to understanding patterns of specialization in plant–pollinator interactions. Although teasing apart biological mechanisms from sampling artifacts is difficult, we believe there are several possible ways of tackling this problem. In the next section, we illustrate a possible approach that can be used to disentangle these multiple causes of observed degree of specialization.

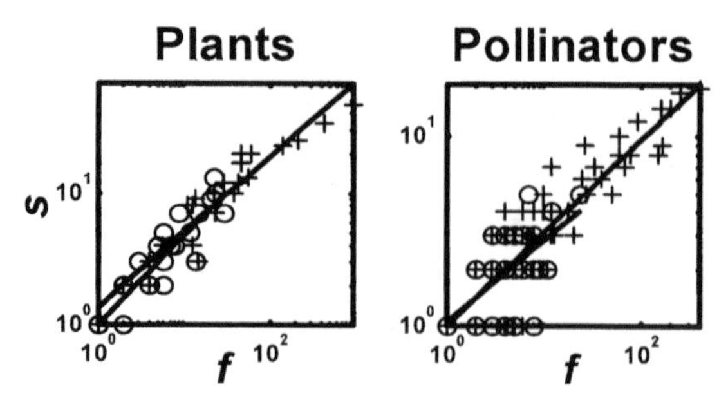

Figure 9.6 Relationship between species' frequency of interaction, *f*, and the number of interaction partners with which they interact, *s*, in Memmott's (1999) dataset. Data are shown for two proportional sampling intensities, $\psi = 1$ (+) and $\psi = 0.1$ (○). Lines represent the fitted regression lines of ln *s* versus ln *f*.

The *f–s* Relationship: Biological Reality or Sampling Artifact?

Although *f* and *s* are linearly related in a log-log scale (see fig. 9.6), the relationship should reach an asymptote in an arithmetic scale at high values of *f* (because of a finite number of species in the community). Thus, if all species were sampled enough, a flattening of the relationship should be observed, provided that the correlation is the result of a sampling bias; this argument would hold if increased sample size were accompanied by a decrease in the proportion of rarely interacting species. Conversely, decreasing sampling effort should increase the slope of the relationship. However, if the *f–s* correlation was not a result of a sampling bias, modifying sampling intensity should not affect the slope of the relationship.

We conducted a simulation to answer the question of whether sampling intensity can affect the *f–s* correlation. Because it is difficult to know what would have happened if we had increased sampling, we asked the opposite: what would have happened if we had sampled with a lower intensity? We used an algorithm that sampled interactions from the interaction frequency matrix until a proportion ψ of the data had been sampled. This simulation can be thought of as a reduction in the time spent in the field by an investigator conducting observations of pollinator visits to flowers. So, for example, we can ask: what would have been the pattern observed in the data (and, particularly, the *f–s* relationship) had we spent only 50% of the time conducting observations? (See Ollerton and Cranmer 2002 for a similar example of using a sensitivity analysis to explore the effects of sampling effort on the results of plant-pollinator surveys.)

The results of this exercise are very clear. First, reducing the sampling effort to 10% of the original effort does not change the *f–s* relationship: the fitted regression lines for the original and subsampled datasets are virtually undistinguishable from each other (fig. 9.6). The interesting point about this result is that we can argue that sampling effort will be unlikely to affect the *f–s* relationship; even if we increased sampling effort, this relationship is unlikely to change. Thus, this relationship is unlikely to simply be a result of low sampling effort.

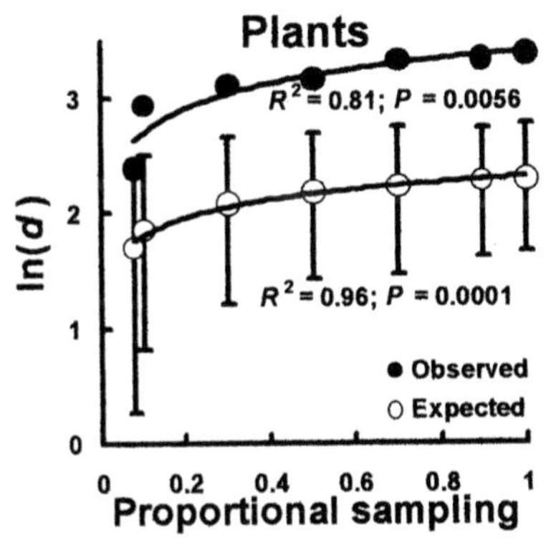

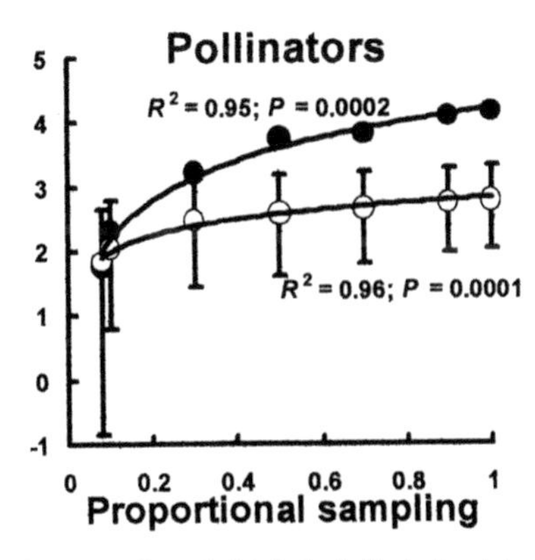

Figure 9.7 Difference statistic d versus proportional sampling intensity ψ. Solid circles indicate the d statistic calculated for observed data; open circles represent the expected value of the d statistic calculated for 1000 randomized communities. Randomizations for each community were conducted separately; thus, each dot represents an independent data point. Lines represent the regression line fitted to the equation $\ln d = a + b \ln \psi$. The coefficient of multiple determination and probability value associated to each regression are given next to each line.

Second, decreasing sampling effort results in a decrease of the difference between the observed and predicted values of the d statistic (fig. 9.7). Thus, less-intense sampling could have led to the conclusion that the observed and expected distributions of s under null model 1 are not significantly different. However, for this to happen, we must decrease sampling effort to a small fraction of the original (i.e., to 8% or less of the original effort for plants and 25% or less for pollinators). This result suggests that observed patterns in the distribution of specialization should be robust to moderate variations in sampling effort and that increasing sampling effort would only exacerbate patterns of extreme specialization and generalization.

An important caveat about the preceding results is that our simulation makes several simplifying assumptions. First, it assumes that pollinator visits are independent of each other, which is arguably unrealistic: if we observe an individual of pollinator species A visiting a flower of plant species B, there is a high chance that the next visit we observe is also of pollinator species A to plant species B, because of temporal and spatial autocorrelation in the distribution of plants and pollinators. Second, the preceding analysis also assumes that the expected decrease in sampling effort is proportionally the same for all plant and pollinator species. However, this need not be the case; for example, an investigator could choose to increase the proportional sampling on rarely interacting plant species while reducing overall sampling effort, to observe at least one visit in every plant species in the community. Future studies should explore the robustness of the preceding results to these simplifying assumptions.

Concluding Remarks

Null models reveal the existence of nonrandom structure in plant–pollinator interaction networks. Our simplest model—one that assumes random interactions among species—shows that extreme specialization, extreme generalization, and a high degree of asymmetry characterize this type of network. These results suggest that, despite the fact that species are distributed along a continuum of specialization/generalization, the extremes of the specialization gradient could represent two alternative strategies favored by natural selection. In addition, the fact that specialists tend to interact with generalist partners beyond random expectations could relate to a type of coevolutionary dynamics that generate and maintain asymmetric interactions. In turn, asymmetry could constitute an important feature of natural webs, contributing to their robustness, resilience, and persistence over time (Melián and Bascompte 2002; Memmott et al. 2004). The result that species-rich webs harbor a larger proportion of extreme specialists and generalists and proportionally more asymmetric interactions than species-poor webs is intriguing and may help explain apparently contradictory claims of high specialization and generalization in tropical latitudes (Bawa 1990, 1994; Renner and Feil 1993; Renner 1998; Olesen and Jordano 2002; Ollerton and Cranmer 2002; Vázquez and Stevens 2004; see also Armbruster, chap. 12 in this volume).

However, that most of the variation of these structural features can be accounted for by simple community properties, such as species richness, and differences among species, particularly frequency of interaction, suggests that these ecological factors could represent more proximate determinants of patterns of specialization/generalization or asymmetry without the need to invoke coadaptation among interacting species resulting from coevolution. Future efforts to develop null models of increasing complexity may help to elucidate the role of the aforementioned different factors in shaping the structure of plant–pollinator interaction webs.

Acknowledgments

We thank Nick Waser and Jeff Ollerton for inviting us to participate in this book, Diego Medan and Jane Memmott for providing files with their published data, and C. Smith-Ramírez for kindly allowing us to use her unpublished data. Pedro Jordano, Jeff Ollerton, and an anonymous reviewer made useful comments on the manuscript. This work was written while DPV was a Postdoctoral Associate at the National Center for Ecological Analysis and Synthesis, a Center funded by the National Science Foundation (grant DEB-0072909), the University of California, and the Santa Barbara campus. MAA was supported by the National Research Council of Argentina (CONICET) and the FONCyT program (PICT 01-07320).

References

Albert, R., and A. L. Barabási. 2002. Statistical mechanics of complex networks. Reviews of Modern Physics 74: 47–97.

Barrett, S. C. H., and K. Helenurm. 1987. The reproductive biology of boreal forest herbs. I. Breeding systems and pollination. Canadian Journal of Botany 65: 2036–2046.

Bascompte, J., P. Jordano, C. J. Melián, and J. M. Olesen. 2003. The nested assembly of plant–animal mutualistic networks. Proceedings of the National Academy of Sciences (USA) 100: 9383–9387.

Bawa, K. S. 1990. Plant-pollinator interactions in tropical rain forests. Annual Review of Ecology and Systematics 21: 399–422.

Bawa, K. S. 1994. Pollinators of tropical dioecious angiosperms: A reassessment? No, not yet. American Journal of Botany 81: 456–460.

Bell, G. 2000. The distribution of abundance in neutral communities. American Naturalist 155: 606–617.

Brown, J. H. 1984. On the relationship between abundance and distribution of species. American Naturalist 124: 255–279.

Clements, R. E., and F. L. Long. 1923. Experimental pollination. An outline of the ecology of flowers and insects. Carnegie Institution of Washington, Washington, DC.

Cornell, H. V. 1999. Unsaturation and regional influences on species richness in ecological communities: A review of the evidence. Écoscience 6: 303–315.

Elberling, H., and J. M. Olesen. 1999. The structure of a high latitude plant–flower visitor system: The dominance of flies. Ecography 22: 314–323.

Fishbein, M., and D. L. Venable. 1996. Diversity and temporal change in the effective pollinators of *Asclepias tuberosa*. Ecology 77: 1061–1073.

Fox, L. R., and P. A. Morrow. 1981. Specialization: Species property or local phenomenon? Science 211: 887–893.

Gaston, K. J., T. M. Blackburn, and J. H. Lawton. 1997. Interspecific abundance-range size relationships: An appraisal of mechanisms. Journal of Animal Ecology 66: 579–601.

Gómez, J. M. 2002. Generalizaci6n en las interacciones entre plantas y polinizadores. Revista Chilena de Historia Natural 75: 105–116.

Gómez, J. M., and R. Zamora. 1999. Generalization vs. specialization in the pollination system of *Hormathophylla spinosa* (Cruciferae). Ecology 80: 796–806.

Gotelli, N. 2001. Research frontiers in null model analysis. Global Ecology and Biogeography 10: 337–343.

Gotelli, N. J., and G. R. Graves. 1996. Null models in ecology. Smithsonian Institution Press, Washington, DC.

Hanski, I., J. Kouki, and A. Halkka. 1993. Three explanations of the positive relationship between distribution and abundance of species. Pp. 108–116 *in* R. E. Ricklefs and D. Schluter (eds.), Species diversity in ecological communities: Historical and geographical perspectives. University of Chicago Press, Chicago.

Herrera, C. M. 1989. Pollinator abundance, morphology, and flower visitation rate: Analysis of the "quantity" component in a plant–pollinator system. Oecologia 80: 241–248.

Hocking, B. 1968. Insect–flower associations in the high Arctic, with special reference to nectar. Oikos 19: 359–388.

Hubbell, S. P. 2001. The unified neutral theory of biodiversity and biogeography. Princeton University Press, Princeton, NJ.

Hutchinson, G. E. 1957. Concluding remarks. Cold Spring Harbor Symposia on Quantitative Biology 22: 415–427.

Inouye, D. W., and G. H. Pyke. 1988. Pollination biology in the Snowy Mountains of Australia: Comparisons with montane Colorado, USA. Australian Journal of Ecology 13: 191–210.

Janzen, D. H. 1968. Host plants as islands in evolutionary and contemporary time. American Naturalist 102: 592–595.

Johnson, S. D., and K. E. Steiner. 2000. Generalization versus specialization in plant pollination systems. Trends in Ecology and Evolution 15: 140–143.

Jordano, P., J. Bascompte, and J. M. Olesen. 2003. Invariant properties in coevolutionary networks of plant-animal interactions. Ecology Letters 6: 69–81.

Kuris, A. M., A. R. Blaustein, and J. J. Alio. 1980. Hosts as islands. American Naturalist 116: 570–586.

Leather, S. R. 1991. Feeding specialization and host distribution of British and Finnish *Prunus* feeding macrolepidoptera. Oikos 60: 40–48.

Lewinsohn, T. M. 1991. Insects in flower heads of Asteraceae in southeast Brazil: A case study on tropical species richness. Pp. 525–559 *in* P. W. Price, T. M. Lewinsohn, G. W. Fernandes, and W. W. Benson (eds.), Plant-animal interactions: Evolutionary ecology in tropical and temperate regions. Wiley Interscience, New York.

May, R. M. 1975. Patterns of species abundance and diversity. Pp. 81–120 *in* M. L. Cody and J. M. Diamond (eds.), Ecology and evolution of communities. Belknap Press of Harvard University Press, Cambridge, MA.

McMullen, C. K. 1993. Flower-visiting insects of the Galápagos Islands. The Pan-Pacific Entomologist 69: 95–106.

Medan, D., N. H. Montaldo, M. Devoto, A. Mantese, V. Vasellati, and N. H. Bartoloni. 2002. Plant-pollinator relationships at two altitudes in the Andes of Mendoza, Argentina. Arctic, Antarctic, and Alpine Research 34: 233–241.

Melián, C. J., and J. Bascompte. 2002. Complex networks: Two ways to be robust? Ecology Letters 5: 705–708.

Memmott, J. 1999. The structure of a plant-pollinator food web. Ecology Letters 2: 276–280.

Memmott, J., and N. M. Waser. 2002. Integration of alien plants into a native flower-pollinator visitation web. Proceedings of the Royal Society of London B 269: 2395–2399.

Memmott, J., N. M. Waser, and M. V. Price. 2004. Tolerance of pollination networks to species extinctions. Proceedings of the Royal Society of London B 271: 2605–2611.

Mosquin, T., and J. E. H. Martin. 1967. Observations on the pollination biology of plants on Melville Island, N.W.T., Canada. Canadian Field Naturalist 81: 201–205.

Motten, A. F. 1982. Pollination ecology of the spring wildflower community in the deciduous forests of piedmont North Carolina. PhD dissertation, Duke University, Durham, NC.

Olesen, J. M., L. I. Eskildsen, and S. Venkatasamy. 2002. Invasion of pollination networks on oceanic islands: Importance of invader complexes and endemic super generalists. Diversity and Distributions 8: 181–192.

Olesen, J. M., and P. Jordano. 2002. Geographic patterns in plant-pollinator mutualistic networks. Ecology 83: 2416–2424.

Ollerton, J. 1996. Reconciling ecological processes with phylogenetic patterns: The apparent paradox of plant-pollinator systems. Journal of Ecology 84: 767–769.

Ollerton, J., and L. Cranmer. 2002. Latitudinal trends in plant-pollinator interactions: Are tropical plants more specialised? Oikos 98: 340–350.

Olsen, K. M. 1997. Pollination effectiveness and pollinator importance in a population of *Heterotheca subaxillaris* (Asteraceae). Oecologia 109: 114–121.

Petanidou, T., and W. N. Ellis. 1996. Interdependence of native bee faunas and floras in changing Mediterranean communities. Pp. 201–226 *in* A. Matheson, M. Buchmann, C. O'Toole, P. Westrich, and I. H. Williams (eds.), The conservation of bees. Linnean Society Symposium Series, no. 18, Academic Press, London.

Preston, F. W. 1948. The commonness, and rarity, of species. Ecology 29: 254–283.

Ramirez, N., and Y. Brito. 1992. Pollination biology in a palm swamp community in the Venezuelan central plains. Botanical Journal of the Linnean Society 110: 277–302.

Renner, S. S. 1998. Effects of habitat fragmentation on plant pollinator interactions in the tropics. Pp. 339–360 *in* D. M. Newbery, H. H. T. Prins, and N. D. Brown (eds.), Dynamics of tropical communities. Blackwell Science, London.

Renner, S. S., and J. P. Feil. 1993. Pollinators of tropical dioecious angiosperms. American Journal of Botany 80: 1100–1107.

Richardson, D. M., N. Allsopp, C. M. D'Antonio, S. J. Milton, and M. Rejmánek. 2000. Plant invasions: The role of mutualisms. Biological Reviews 75: 65–93.

Ricklefs, R. E. 1987. Community diversity: Relative roles of local and regional processes. Science 235: 167–171.
Robertson, C. 1929. Flowers and insects: Lists of visitors to four hundred and fifty-three flowers. C. Robertson, Carlinville, IL.
SAS Institute. 1999. The SAS system for Windows, version 8e. SAS Institute, Cary, NC.
Schemske, D. W., and C. C. Horvitz. 1984. Variation among floral visitors in pollination ability: A precondition for mutualism specialization. Science 225: 519–521.
Schemske, D. W., M. F. Willson, M. N. Melampy, L. J. Miller, L. Verner, K. M. Schemske, and L. B. Best. 1978. Flowering ecology of some spring woodland herbs. Ecology 59: 351–366.
Small, E. 1976. Insect pollinators of the Mer Bleue peat bog of Ottawa. Canadian Field Naturalist 90: 22–28.
Southwood, T. R. E. 1961. The number of species of insect associated with various trees. Journal of Animal Ecology 30: 1–8.
Stevens, G. C. 1986. Dissection of the species–area relationship among wood-boring insects and their host plants. American Naturalist 128: 35–46.
Strong, D. R., J. H. Lawton, and T. R. E. Southwood. 1984. Insects on plants. Blackwell, Oxford.
Thompson, J. N. 1994. The coevolutionary process. University of Chicago Press, Chicago.
Vázquez, D. P., and M. A. Aizen. 2003. Null model analyses of specialization in plant–pollinator interactions. Ecology 84: 2493–2501.
Vázquez, D. P., and M. A. Aizen. 2004. Asymmetric specialization: A pervasive feature of plant–pollinator interactions. Ecology 85: 1251–1257.
Vázquez, D. P., and D. Simberloff. 2002. Ecological specialization and susceptibility to disturbance: Conjectures and refutations. American Naturalist 159: 606–623.
Vázquez, D. P., and D. Simberloff. 2003. Changes in interaction biodiversity induced by an introduced ungulate. Ecology Letters 6: 1077–1083.
Vázquez, D. P., and R. D. Stevens. 2004. The latitudinal gradient in niche breadth: Concepts and evidence. American Naturalist 164: E1–E19.
Waser, N. M., L. Chittka, M. V. Price, N. M. Williams, and J. Ollerton. 1996. Generalization in pollination systems, and why it matters. Ecology 77: 1043–1060.

CHAPTER TEN

Mutual Use of Resources in Mediterranean Plant–Pollinator Communities: How Specialized Are Pollination Webs?

Theodora Petanidou and Simon G. Potts

Until the 1980s, much of the literature relating to pollination ecology overestimated the frequency of specialization in plant–pollinator interactions and advocated the existence of "pollination syndromes" (Faegri and van der Pijl 1979). However, the first rigorous community analysis of pollination systems (Jordano 1987) suggested that plant–pollinator networks are very much more generalized than was previously thought and are characterized by asymmetric patterns of resource use. Today, most researchers agree that extreme specialization is relatively rare and that generalization may be an advantageous evolutionary outcome under particular circumstances (Waser et al. 1996; Olesen and Jordano 2002). Olesen and Jordano (2002) found a trend of increased plant specialization on islands, which they suggest is probably associated with the impoverishment of potential animal pollinators in these systems. This finding may appear confusing to island biologists, because island plants are generally thought to be generalists (Barrett 1996). The reason for this "inconsistency" arises from the difference in absolute and relative specialization: island plants are "specialized" because few potential pollinator species are found on islands relative to mainland areas. On the other hand, they are "generalized" (or nonselective, according to the terminology of this chapter) because they will use whatever animals are available as pollinators. Indeed, contrary to the prevailing view that tropical ecological interactions tend toward higher specificity compared to their temperate counterparts, Ollerton and Cranmer (2002) found no evidence for a more pronounced specialization in tropical plant–pollinator networks. On the other hand, Olesen and Jordano (2002) found a higher specificity of the plants in the tropics relative to higher latitudes; however, this was not mirrored by their insect counterparts (see also Ollerton et al., chap. 13 in this volume).

The most widely used parameter in the aforementioned type of studies is *connectance* (C), which is defined as the percentage of all possible interactions within a network (M) that are actually established (I). Connectance has been repeatedly used in the literature as an indicator of the degree of interconnection within a mutualistic system of available plants and their flower visitors, that is, the degree of saturation of the existing plant–pollinator food web (Jordano 1987; Elberling and Olesen 1999; Olesen and Jordano 2002; Olesen et al. 2002). Similarly, Olesen (2000) used the concept of *linkage:* the number of taxa within a community with which a given species (plant or insect) interacts. Linkage is in fact the same index used by Ollerton and Cranmer (2002): number of species of flower visitors per plant species. In the earlier literature, a similar parameter has been widely used, namely, *phily* and *tropy* for plants and insects, respectively (Moldenke and Lincoln 1979; Petanidou 1991a, 1993; Petanidou and Ellis 1996; see also Cane and Sipes, chap. 5 in this volume).

Although connectance has been shown to be a good basis for the comparative analysis of different plant–pollinator systems (Jordano 1987; Olesen and Jordano 2002), the calculated value largely depends on the sampling effort and the size (taxonomic breadth) of the food web defined by the investigators. However, both of these factors are frequently biased by time and resource restrictions (Ollerton and Cranmer 2002); as a result, connectance, when used alone, may lead to inappropriate estimations of structure and mutual use of resources within the plant–pollinator food webs. For instance, according to Olesen and Jordano (2002), species-rich systems show low connectance (i.e., poor niche saturation in plant–pollinator partnerships). Such systems, however, may not necessarily show high levels of specialization; instead they may be composed of many oligotropous (oligolectic) to mesotropous species, but not specialists. To overcome this apparent problem it is therefore necessary to intensively sample the mutual use of pollination resources across the full extent of the web and through several seasons.

In addition, temporal variation in the degree of connectance within a system will severely bias the results of any analysis. Flower-visitor fauna may vary tremendously in composition between years of study (C. M. Herrera 1988; Petanidou and Ellis 1993, 1996; Williams et al. 2001); therefore, one year should only be considered a preliminary insight into the study of specialization at the community level. Exhaustive field studies are required in order to have high confidence in the connectance measures calculated, and the intensity of observations will need to reflect the level of unpredictability of the focal system.

In this study, we analyze in detail the plant–pollinator food-web structure of three lowland shrub communities of the Mediterranean Basin. Two extensive datasets are from studies undertaken by the authors: a system in Greece (T. Petanidou) and one in Israel (S. Potts). These are compared with a third system in Spain using published data (J. Herrera 1988). All three communities are characteristic of the region, share broadly similar taxonomic compositions, and are subject to the same ecological drivers (fire, grazing, and changing land-use patterns; Petanidou et al. 1995b; Blondel and Aronson 1999). They are geographically spread along the Mediterranean Basin, from the far east (Israel) to central eastern (Greece) through to the west (Spain), and represent homologous ecosystem types (batha, phrygana, and garriga, respectively).

Our study explores the organization and function of the pollination food webs in these systems, with particular reference to the following:

1. the mutual use of pollination resources, in terms of the overall degree of specialization/generalization for each one of the pollination partners (monolateral) and the incidence of one-to-one (mutual) specialized relationships;
2. the relative importance of specialization indices (e.g., connectance), assessed by comparing communities with different invested sampling efforts and different sample sizes and taxa included;
3. the utility and discriminant ability of "traditional" specialization indexes (e.g., tropy and phily). We introduce a new index of "selectivity" as a better descriptor of specialization, which takes into account the *temporal* variation within a system; and
4. the effect of seasonal and annual variations on the measurement of specialization/generalization in pollination partnerships, with special reference to the interyear variation in plant-host selection.

These four approaches demonstrate that specialization is a phenomenon of observable mutual dependence between a plant species and a (restricted number of) pollinator species, in which both space and time dimensions matter. In this study and other similar ones that compare plant–pollinator communities, we can distinguish between "true" specialization, which is not merely a result of choice by necessity or coincidence, and a predetermined intrinsic preference brought about by a mutual coevolution.

Methods

The Datasets

The data used in the study come from three separate pollination studies: in Greece (Daphní, Athens), Israel (Mt. Carmel, Haifa), and Spain (Doñana). All three systems are characterized by similar constraints imposed mainly by the harsh climatic regime (very dry and hot summers; wet and mild-to-relatively-cool winters). The main vegetation type of the sites is Mediterranean low scrub dominated by entomophilous plants. Physiognomy is determined by woody plants, but annuals constitute the predominant life form in terms of species diversity (Petanidou et al. 1995b; Potts et al. 2003a). Each area can be considered an ecosystem that is blocked in its succession toward a climax vegetation (i.e., a Mediterranean forest); the factors responsible for arresting its successional development are shallow soil, fire, grazing, low precipitation rate, and strong winds (especially in areas close to the sea).

Although the foci of all three studies differed, they all involved intensive collection of plant–pollinator associations using a rigorous and repeatable methodology. In all cases, pollinators were considered to be any animal that repeatedly visited floral units for any reward and contacted any part of a flower's reproductive organs; such behavior may result in pollination of the visited flower (Faegri and van der Pijl 1979). Each study was based on the observation/collection of pollinator species and aimed to sample the insect fauna as extensively as possible with the resources available. However, it was unrealistic to collect all the species present because (1) Mediterranean systems are very dynamic and species turnover is particularly rapid through time so that any survey would be unable to keep pace through the peak season (Petanidou and Ellis 1996; Williams et al. 2001), and (2) the Mediterranean bee fauna is dominated by very rare species (e.g., ca. 50% of species in the Israel survey were represented by singletons) and, hence, it would be impossible to capture/observe every species. Among all three studies, the most complete is that of the Greek phrygana—one of the best-studied communities—because it considered the continuous turnover of plant–pollinator interactions throughout four successive years ("the most complete study we know," Buchmann and Nabhan 1997, 78; "a remarkable exception," Olesen and Jordano 2002). The other two studies entail more restricted datasets; however, at present they provide the best data available for the Mediterranean.

Greece

The Daphní site is located approximately 10 km west of the center of Athens, and comprises a 30 ha section of the Diomedes Botanical Garden of Athens University Nature Reserve. It is situated on the slopes of Mt. Aegaleo (altitude between 135 and 215 m above sea level [a.s.l.], inclination between 20 and 30°, with a northern and eastern aspect). The vegetation was characteristic of regenerating phrygana which had been partly burned in 1975. Detailed descriptions of the site and of the methods of insect collection and plant flowering are given by Petanidou and Ellis (1993, 1996) and Petanidou et al. (1995b).

Fieldwork was carried out continuously for four years (April 1983 through May 1987) and consisted of systematic monitoring of the visiting fauna of all entomophilous plants of the community throughout their period of flowering. Over 18,000 insect specimens were collected, including bees (except that honeybees were not collected), butterflies, anthophilous flies, Heteroptera, and Symphyta, but excluding spiders and ants. The insects (665 species) are listed by Petanidou (1991b). Voucher specimens are available in the entomological collection of the Goulandris Natural History Museum in Athens.

All 133 angiosperm species that occurred in the study site and exhibited a biotic pollination syndrome were included in the study. Each was followed throughout its flowering period for at least 2 out of every 20 days. Observations were made at approximately every third half hour for each of the co-flowering plant species and they were extended until no new visiting species were recorded. The number of pollinator-directed observation hours totaled over 5000. The observations started at sunrise and continued until after sunset, when no pollinators were active, except for the cases of nocturnal plant anthesis, where additional nocturnal observations were made (e.g., on *Capparis spinosa* L.). The authors recognize that the pollinating fauna contains some nocturnal species that were not surveyed; however, for the Greek and Israeli phrygana, based on long-term studies, these species represent a very small proportion of the total floral visitors. Voucher specimens were deposited in the Herbarium of the Institute of Systematic Botany and Phytogeography of the Aristotle University in Thessaloniki.

Israel

Mount Carmel National Reserve is situated 5 km northeast of Haifa and comprises ca. 150 km^2 of highly disturbed Aleppo pine forest that forms a complex mosaic of postfire regenerating scrub habitats. Areas surveyed were all between 130 and 340 m a.s.l. with a southern or western aspect and average slope of less than 20°. Full descriptions of the sites can be found in the literature (Potts et al. 2003a, 2003b).

Previous studies had shown that the pollinating fauna of core flowering species of the area were predominantly bees (Hymenoptera: Apoidea), which accounted for the majority of all floral visits (Potts et al. 2001); therefore, the surveys of flower visitors were restricted to this taxon. Between February and August in 1999 and 2000, intensive surveys of bees were made between dawn and dusk, totaling more than 7000 person-hours. Additional observations were also made at other times. We recorded 340 species of bees visiting 170 species of flowering plants. Voucher specimens were deposited at the Oxford University Museum of Natural History, and details of the studies undertaken are available in the literature (Potts et al. 2001, 2003a, 2003b). Using the abundance data for the 116 species collected in 1999 gave a "jacknife" estimate of total species richness of 174 species for all the habitats through the entire season. Thus, we estimate that we captured representatives of approximately two-thirds of the species present; no data were specifically collected on phenology of the floral community.

Spain

The third dataset comes from J. Herrera (1988) and focuses on Doñana National Park in southwestern Spain. The site is in a sandy coastal area dominated by scrub; the vegetation is described fully by Rivas-Martínez et al. (1980). The author focused only on woody perennial species and recorded all potential pollinators. Observations were made on a weekly basis between January 1982 and February 1984. The lack of observations associated with annual flowers is a potential weakness in the dataset; however, this study still represents one of the best currently available. Therefore, we have exercised some degree of caution when interpreting the findings relating to the analysis of the Spanish data. Detailed data on the methodology employed are given by J. Herrera (1988).

Data Analysis: Interactions, Phily, Tropy, Selectivity, and Connectance

We define an *interaction* as any observed relationship between an insect species and a plant, where the insect has visited the flowers of a plant species for any reward and made contact with the flowers' reproductive structures.

The number of plant species hosting an insect species and the number of insect species visiting a plant species are traditionally referred to as the insect's tropy and the plant's phily, respectively. Both of these indices may be indicative of the level of specialization of the pollination partners at the community level; however, neither of them considers the actual flora/entomofauna available during the period of activity of an insect species or the flowering period of a plant species. Relating the number resources used to the absolute number available at that particular time in the system will give relative phily and tropy values that more accurately reflect the degree of specialization at that time. This value, henceforth referred to as the *selectivity*, is defined as

$$S = (E - T)/E,$$

where E is the number of plant species that a pollinator species may encounter during its period of activity (i.e., all interactions possible) and T is the species actually encountered (i.e., the species' actual tropy). This index ranges from 0 to 1. The selectivity of a plant species is defined analogously. In practice, selectivity is an indicator of how "choosy" the species are with respect to the actual availability of their partners during their period of activity/flowering. Because of the lack of time data in most community studies, selectivity has only been explored for the Greek dataset.

Connectance is defined as $C = (I \times 100)/M$, where I is the total number of interactions observed and M is the food web size, that is, the maximum number of interactions possible. ($M = A \times P$, where A and P are the total number of interacting animal and plant species in the community.)

Phenological Data

Phenological analyses were carried out for the Greek phrygana system because this was the only study in which detailed floral phenologies were available (Petanidou 1991a). The period of activity of an insect species was considered to be the time between the earliest and the latest observations in a calendar year. Species that were observed for less than 14 days were arbitrarily assumed to be active for 14 days around the midpoint of the available date (cf. Petanidou and Ellis 1996, for bees only). The period of plant flowering was taken from Peta-

Table 10.1 Parameters related to the structure of the plant–pollinator food web in the three communities considered in this study

Parameters studied	Communities in different geographical areas				
	Greece		Israel	Spain	
	All insects	Only bees	Bees	All insects	Only bees
No. of plant species visited (*P*)	132	129	179	26	24
No. of visiting insect species (*A*)	665	262	340	180	55
No. of interactions (*I*)	3006	1390	921	415	155
Plant phily					
Range	1–123	1–48	1–43	1–86	1–25
Mean value ± SE	22.77 ± 1.67	10.78 ± 0.80	5.15 ± 0.53	15.96 ± 3.23	6.46 ± 1.01
Insect tropy					
Range	1–103	1–103	1–20	1–17	1–14
Mean value ± SE	4.52 ± 0.26	5.31 ± 0.52	2.71 ± 0.16	2.31 ± 0.18	2.82 ± 0.36
Connectance: $(I\times100)/(A\times P)$	3.42	4.11	1.51	8.87	11.74

nidou et al. (1995b). Based on the periods of activity of the individual species, we calculated a daily diversity for insects and plants as the number of insect species present (for the plants that were in flower) on any calendar day. Similarly, we calculated the mean selectivity, tropy and phily for every calendar day of the year.

Results

Diversity of Entomophilous Plants and Their Insect Visitors

In Greece, of the 133 plant species monitored in the community with observed entomophilous pollination syndromes, only one was not visited by insects (*Romulea linaresii* subsp. *graeca*) and is therefore not included in the calculations. Analysis of interaction values for plant–insect partnerships are shown in table 10.1, which gives comparative scores from the three communities considered in this study.

In the Israeli community, a total of 340 bee species visited 179 plant species (table 10.1). The total recorded number of plant–pollinator partnerships was 921. In Spain, 26 plant species, all woody perennials, were visited by 180 insect species (table 10.1). These numbers do not include the nonidentified insects recorded as "other families" by the author (J. Herrera 1988). Figures 10.1 and 10.2 show that the tropy and phily values for both Israel and Spain were much lower than those of Greece, which may reflect the fact that surveys were restricted either only to bees and mainly during the peak flowering period (Israel) or only to woody perennials (Spain).

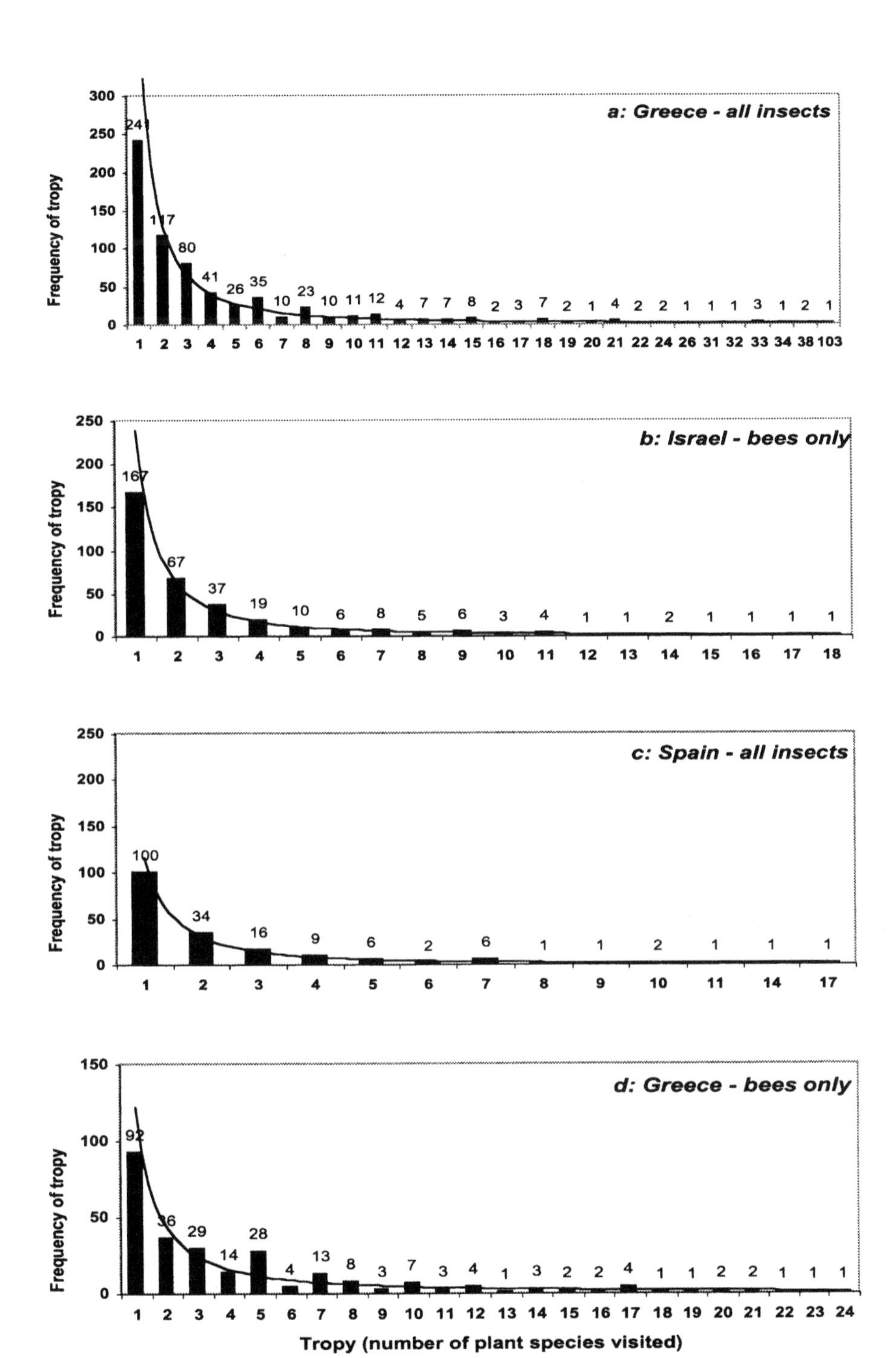

Figure 10.1 Distribution of the tropy values (number of plant species visited by an insect) in the three Mediterranean communities considered in this study. Numbers above the bars show frequencies.

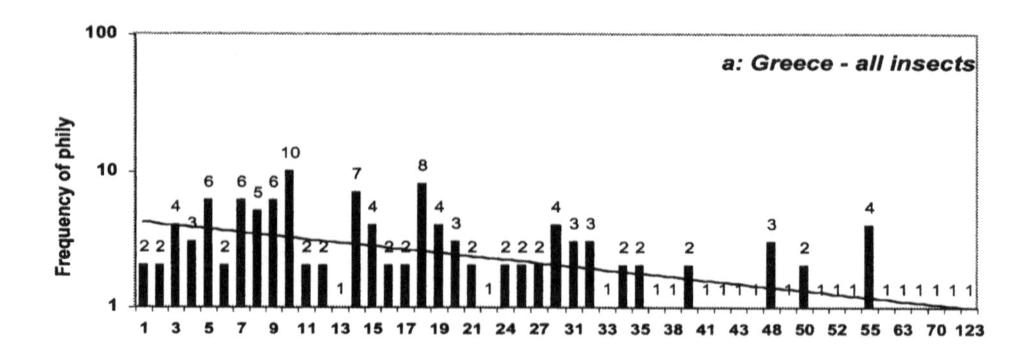

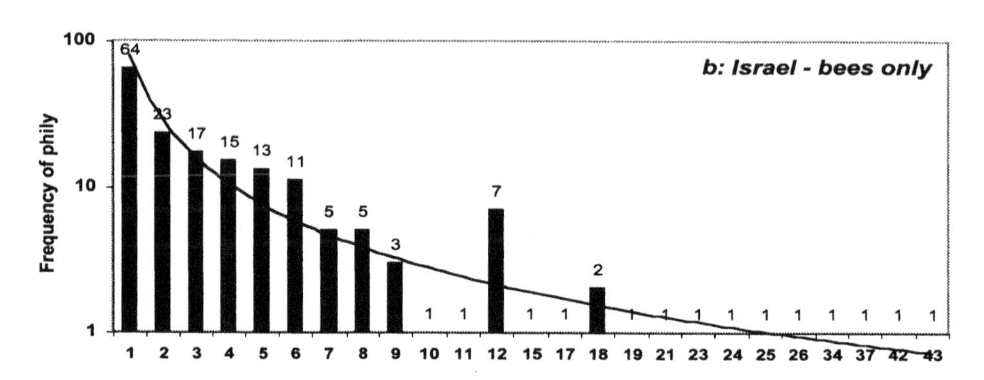

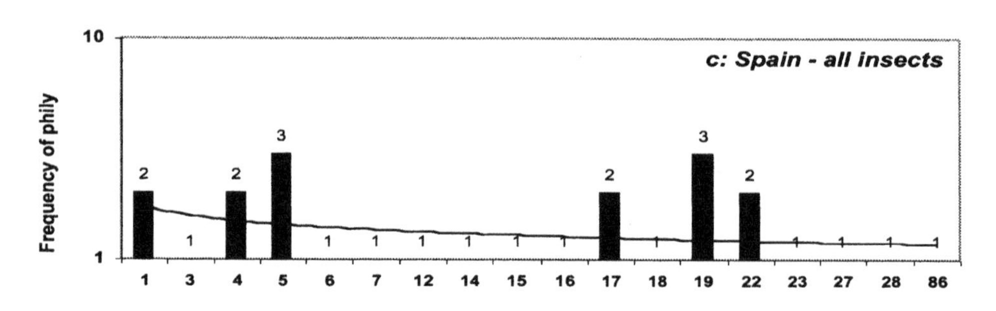

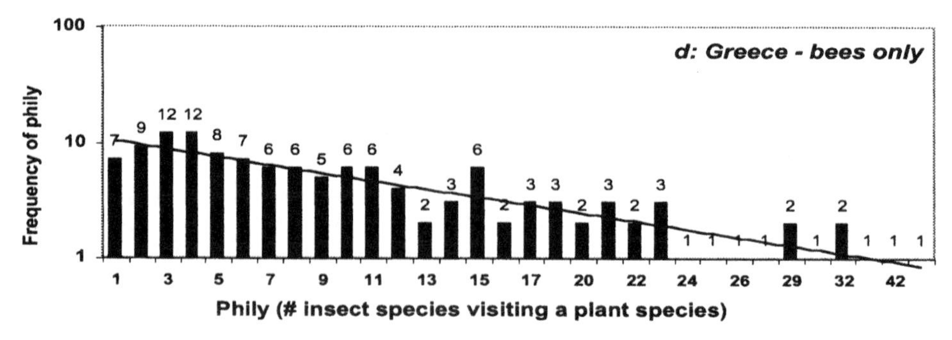

Figure 10.2 Distribution of the phily values (number of insect species visiting a plant species) in the three Mediterranean communities considered in this study. Numbers above the bars show frequencies.

Table 10.2 Flower-visiting insects of the three Mediterranean communities studied

	Number of species observed		
	Greece	**Israel**	**Spain**
All insects	665	340	180
Bees	262 (39.4)[a]	340	55 (30.6)[a]
Andrenidae	49 (18.7)[b]	86 (25.3)[b]	8 (14.5)[b]
Anthophoridae	63 (24.0)[b]	114 (33.5)[b]	14 (25.5)[b]
Apidae	3 (1.1)[b]	1 (0.3)[b]	2 (3.6)[b]
Colletidae	11 (4.2)[b]	22 (6.5)[b]	5 (9.1)[b]
Halictidae	51 (19.5)[b]	44 (12.9)[b]	17 (30.9)[b]
Megachilidae	84 (32.5)[b]	72 (21.2)[b]	7 (12.7)[b]
Melittidae	1 (0.4)[b]	1 (0.3)[b]	2 (3.6)[b]
Formicidae	—	—	7
Wasps + Symphyta	80 (12.0)[a]	—	21 (11.7)[a]
Beetles	72 (10.8)[a]	—	30 (16.7)[a]
Flies	188 (28.3)[a]	—	28 (15.6)[a]
Bombyliidae	45 (23.9)[b]	—	9 (32.1)[b]
Syrphidae	49 (26.1)[b]	—	16 (57.1)[b]
"Muscoid" flies	88 (46.8)[b]	—	3 (10.7)[b]
Butterflies	31 (4.7)[a]	—	39 (21.7)[a]
Heteroptera	31	—	—
Neuroptera	1	—	—

[a]Number in parentheses is the percentage over the total number of insect species of the site.
[b]Number in parentheses is the percentage over the total species number of bees or flies.

Table 10.2 partitions the flower visitor community by taxon. Bees constituted a large proportion of the whole fauna, both in the Greek and Spanish communities (39.4 and 30.6%, respectively). The bee faunas have a similar family structure (table 10.2), although the Megachilidae are more dominant in Greece. Both communities contained similar proportions of "wasps" (non-Apoidea aculeates + Symphyta), beetles, and flies. The composition of the fly communities differed; syrphids predominated in Spain and "muscoid flies" in Greece. However, they differed markedly with regard to the number of butterflies (4.7% in Greece and 21.7% in Spain).

Connectance

A summary of the relationships between plants and insects in the three Mediterranean communities is given in table 10.1. There is a remarkable difference between the values of connectance found: the Greek and Israeli systems have among the lowest values ever recorded (Olesen and Jordano 2002). In contrast, the value for Spain is much higher, which may be an artifact due to the restricted floral group chosen. Indeed, when considering the Greek and Spanish scrub, connectance values are higher for the "bee-only" fauna than for the complete insect fauna. When joining our dataset to those data used by Olesen and Jordano

(2002), we found that connectance decreased exponentially with total number of species ($A + P$) as $C = 277.8(A + P)^{-0.740}$ ($R^2 = 0.841$) and also decreased with the maximum number of possible interactions ($M = A \times P$) as $C = 193.1M^{-0.403}$ ($R^2 = 0.884$). This agrees with the finding by Olesen and Jordano (2002) that connectance decreases with community size.

Insect Tropy

The distribution of the insect tropy within the three communities is summarized in figure 10.1. It is interesting that the tendency of insects in choosing floral resources is similar in all three communities, with a very high proportion of monotropous insects recorded: 36.2% in Greece, 49.1% in Israel, and 55.6% in Spain. Greece appears to have the least-specialized fauna when considering only bees: the proportion of monotropous bees is 35.1% in Greece, 49.1% in Israel, and 47.3% in Spain. The higher average value of insect tropy in Greece relative to the other two communities (table 10.1) can be attributed to the broader taxonomic spectrum considered in this study, and perhaps the higher sampling effort; therefore, it probably represents the best estimate of a real community value. Bees have a slightly higher tropy than the total insect fauna in both the Greek and Spanish sites (table 10.1). The difference between bee tropy and non-Apoidea insect tropy was significant in both sites (analysis of variance [ANOVA]: for Greece, $F_{2,663} = 104.8$, $P < .0001$, mean values 5.3 [$N = 262$] and 4.0 [$N = 403$] for bees and non-Apoidea, respectively; for Spain, $F_{2,178} = 826.3$, $P < .0001$, mean values 2.7 [$N = 55$] and 2.1 [$N = 125$] for bees and non-Apoidea, respectively). The differences remain even when the supergeneralist *Apis mellifera* is removed from the calculations ($F_{2,662} = 153.0$, $P < .0001$; $F_{2,177} = 942.4$, $P < .0001$ for Greece and Spain, respectively).

Plant Phily

The distribution of the phily values of the three communities are given in figure 10.2. Unlike the tropy values, the results differ across the three communities: the proportion of monophilous plant species are 1.5, 35.8, and 7.7% for Greece, Israel, and Spain, respectively. This may be attributed, in part, to the dissimilarity of the data (all insects and plants in Greece; only bees in Israel; selected plants in Spain). When considering only the bees of Greece (fig. 10.2), the distribution of phily values is more similar to that of Israel; yet the proportion of monophilous plant species in Greece (5.4%) is still much lower than in Israel (35.8%) and Spain (16.7%). The Spanish data do not reveal a clear distribution pattern for phily values (fig. 10.2), which may reflect the absence of the annual plant component. Overall, the use of pollinator resources by plants is a more complex phenomenon compared to the use of resources by their insect partners. The average value of plant phily was much lower in Israel, which may relate to the reduced period of insect collection (table 10.1). When only bees were considered in

Greece and Spain, the respective phily values were much lower. Indeed, the phily value of Spanish bees did not differ significantly from that in Israel (ANOVA: $F_{1,201} = 0.76, P = .38$).

Selectivity

Greece was the only site for which appropriate data were available for calculating selectivity, and values for the flower-visiting insects and entomophilous plants are summarized in figure 10.3. Unlike the distribution of tropy/phily values, these graphs indicate similar trends for plants and insects in selecting partners within the community; in other words, there is a strong tendency for both plants and insects to specialize in the community. Large proportions of both plant (69.7%) and insect species (79.8%) have a selectivity greater than 0.90. Again, the higher value for insects indicates a greater tendency for specialization in insects than plants, although the degree of specialization of some specific plants is higher than that of insects (see following discussion).

Insect selectivity is significantly correlated to tropy (Spearman correlation coefficient: $R_s = -0.765, N = 665, P < .0001$) and, similarly, plant selectivity to phily ($R_s = -0.819, N = 132, P < .0001$). The selectivity data show that not only the insects but also the plants are remarkably choosy as pollination partners, though the choice is less strong in plants than insects (figs. 10.1–10.3). It is interesting that some plant species showed a higher degree of selectiveness than the insects, with values greater than 0.99 (fig. 10.3). These species are *Sarcopoterium spinosum* (visited by 1 insect species), *Psoralea bituminosa* (5), *Linum strictum* subsp. *strictum* (1), *Hymenocarpos circinnatus* (2), *Centaurea raphanina* subsp. *mixta* (4), *Alcea pallida* (3), and *Fumaria amarysia* (3). There is no obvious floral trait shared by these species that could explain why these plants are the choosiest in the system.

Reciprocal Use of Pollination Resources

For each partnership, we used the paired phily (of the plant) and tropy (of the insect) to generate a community phily–tropy dataset. These phily–tropy values are negatively correlated in all three communities studied (Greece, all insects: $R_s = -0.183, N = 3006, P < .0001$; Greece, bees only: $R_s = -0.168, N = 1390, P < .0001$; Israel, bees only: $R_s = -0.132, N = 921, P < .0001$; Spain, all insects: $R_s = -0.262, N = 406, P < .0001$).

Figures 10.4 and 10.5 illustrate the mutual use of pollination resources by both plants and visiting insects. Tropy categories are all significantly negatively correlated with the mean phily values of the plant species visited, as shown in figure 10.4 (Greece, all insects: $R_s = -0.823, N = 30, P < .0001$; Greece, bees only: $R_s = -0.732, N = 25, P = .0001$; Israel, bees only: $R_s = -0.853, N = 18, P < .0001$; Spain, all insects: $R_s = -0.907, N = 13, P < .0001$).

Phily categories are also all negatively correlated with the average tropy values

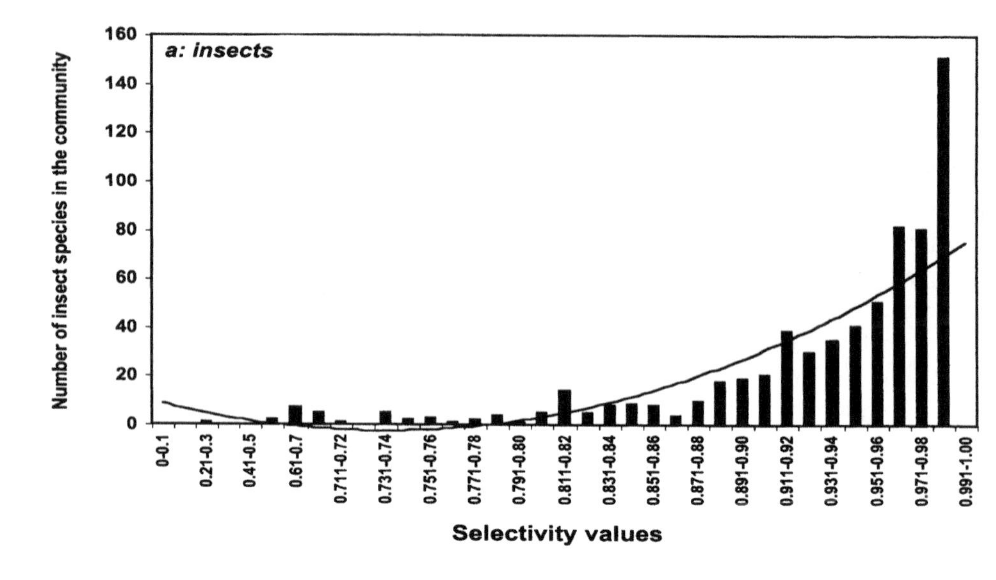

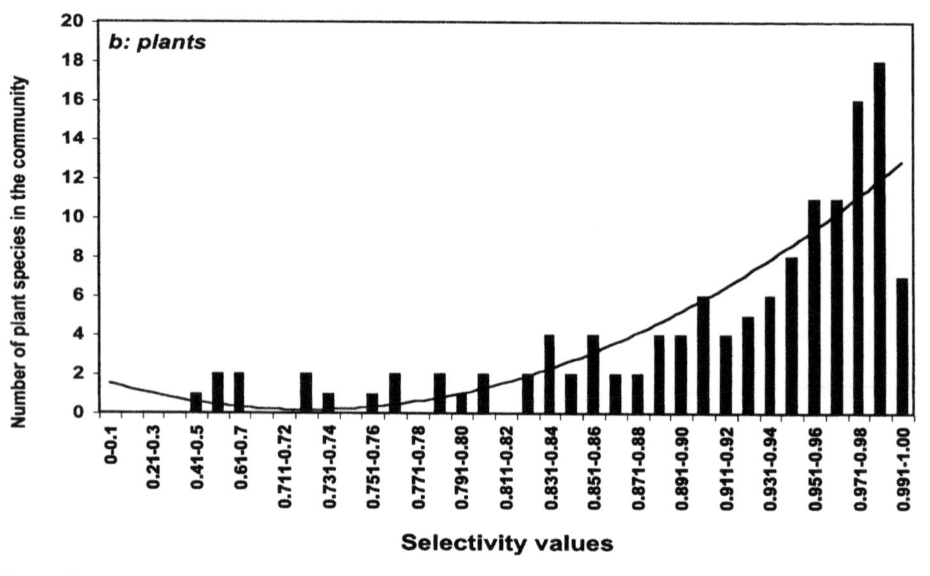

Figure 10.3 Distribution of selectivity values for insects and plants in the phryganic community of Greece. Selectivity $(S) = (E - T)/E$, where E represents available resources (number of partner species) during the activity period of a given species, and T represents actual resources used (in species). Each bar represents values of selectivity within a specific interval (i.e., 0–0.1, 0.11–0.2, 0.21–0.3, etc.). To avoid visual confusion only every other interval is labeled. Note that the intervals become smaller for higher values of selectivity.

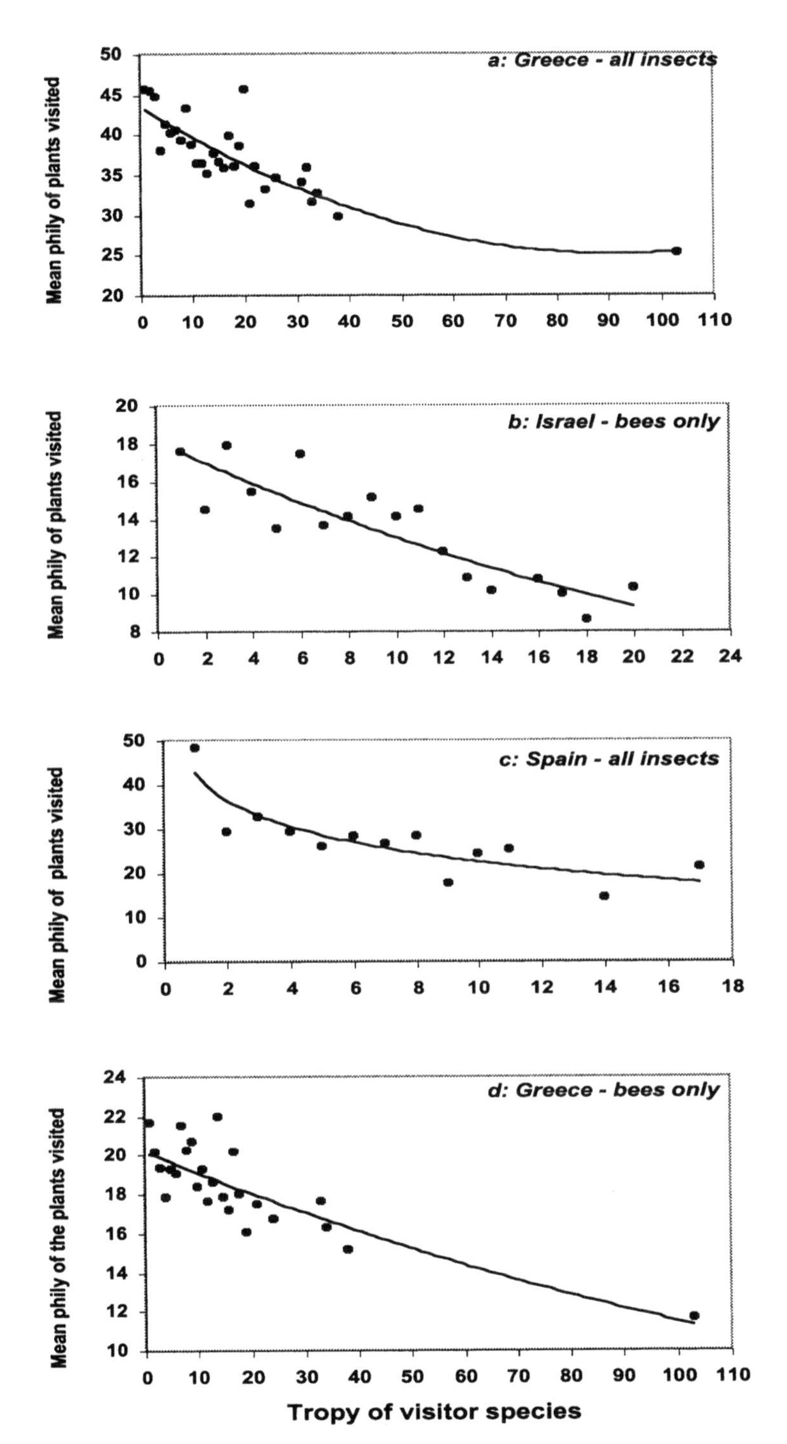

Figure 10.4 Reciprocal use of plant resources by visiting insects (mean phily of plants visited by insects of different tropy values).

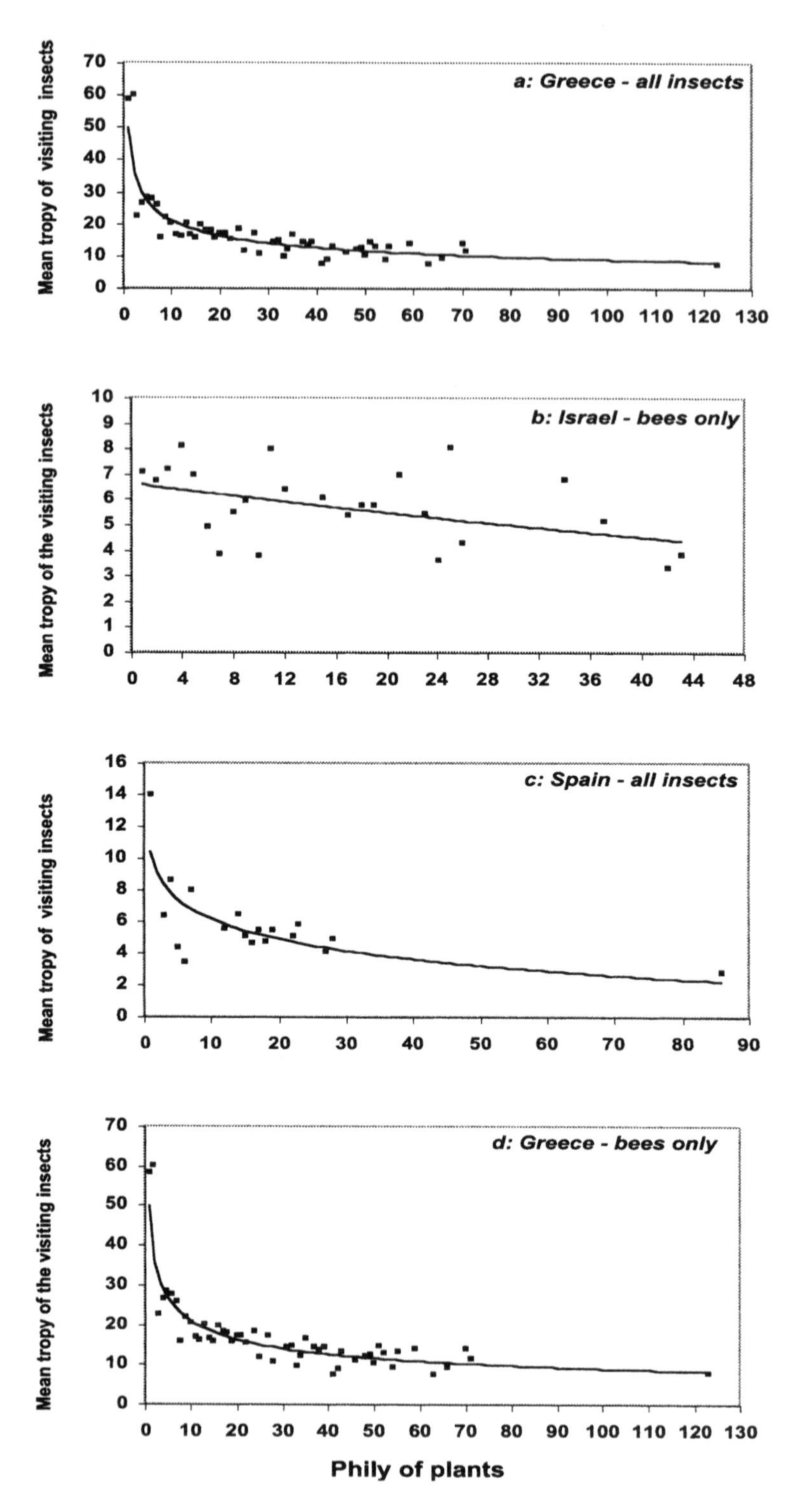

Figure 10.5 Reciprocal use of insect resources by plants (mean tropy of insects visiting plants of different phily values).

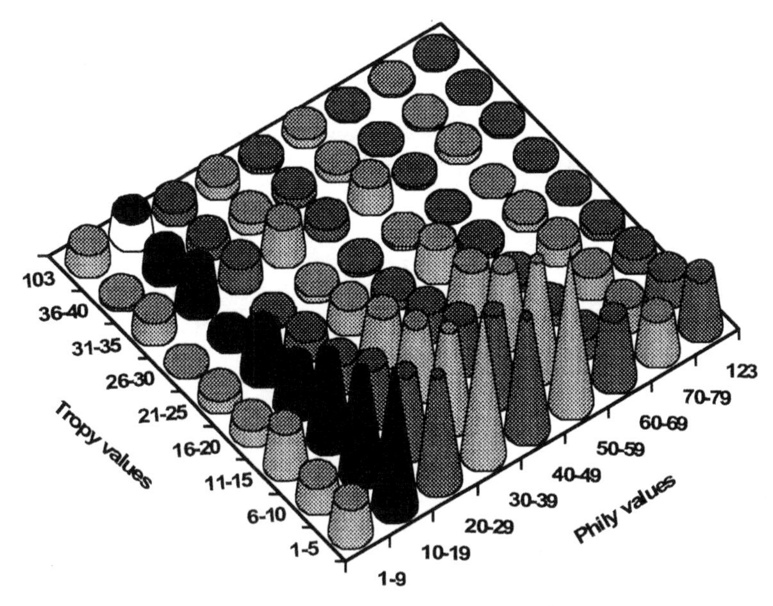

Figure 10.6 Distribution of interactions between plants and insect species in the phryganic community of Greece as a function of specialization level: phily (*x* axis), tropy (*y* axis), and frequency (*z* axis).

of insects in all cases, as shown in figure 10.5 (Greece, all insects: $R_s = -0.850$, $N = 51$, $P < .0001$; Greece, bees only: $R_s = -0.888$, $N = 32$, $P < .0001$; Israel, bees only: $R_s = -0.436$, $N = 25$, $P = .0293$; Spain, all insects: $R_s = -0.527$, $N = 18$, $P = .0245$).

The overall levels of generalization/specialization in plant–pollinator relationships are shown in figure 10.6 for the Greek phrygana. The graph is based on the total number of interactions in the community and clearly shows that insects with low tropy values predominantly tend to visit oligophilous to polyphilous plants, whereas those with large tropy values are linked specifically with oligophilous plants. There are no one-to-one plant–insect relationships in the Greek community; similarly, one-to-one plant–pollinator relationships were absent from the communities of Israel and Spain.

Temporal Variation in Host–Visitor Specialization

The Greek data, collected continuously for several consecutive years, allow us to estimate how insect specialization varies from year to year. Out of the 665 insect species, 40.5% appeared only once out of the 4.3 years of the survey. Details of the temporal dimension of insect activity are given in table 10.3. The mean number of years an insect species was active relates significantly to the level of specialization, namely, the number of plant species visited ($R_s = 0.948$, $N = 14$, $P < .0001$).

Among all monotropous species of Greece, only 29 (12.0%) were active for

Table 10.3 Relation between level of specialization and years of activity within the Greek phrygana community

Level of tropy	Total no. of insects in the system	Percentage of insects appearing >1 year	Years of appearance	
			Mean ± SE	Range
Monotropous	241	12.0	1.13±0.024	1–3
2-tropous	117	65.0	1.86±0.070	1–3
3-tropous	80	87.5	2.31±0.079	1–4
4-tropous	41	87.8	2.61±0.148	1–4
5-tropous	26	96.2	3.66±0.172	1–4
6-tropous	35	100.0	2.96±0.142	2–5
7-tropous	10	100.0	3.90±0.314	2–5
8-tropous	23	100.0	4.22±0.188	2–5
9-tropous	10	100.0	3.80±0.249	3–5
10-tropous	11	100.0	3.73±0.237	3–5
11-tropous	12	100.0	4.08±0.149	3–5
12–15-tropous	26	100.0	4.50±0.114	3–5
16–20-tropous	15	100.0	4.33±0.126	4–5
21–38-tropous	17	100.0	4.76±0.106	4–5
103-tropous	1	100.0		5
Total	665	59.5		1–5

more than one season of study. Nonmonotropous insect species showed an irregular pattern of host selection, either by visiting all their host plants in only one year or by visiting them over many years albeit without constancy to all of them (e.g., one year visiting one to few plants, the next year the rest; alternatively, insects were associated with mainly one plant over the years but occasionally visited other plants in other years); we consider these species constant (with respect to one season) rather than specialized. We conclude that, of all insects of the Greek community, only the 29 monotropous species appearing for more than one year are really specialized; to these we can add the di- and tritropous species that appeared to have a constant visitation profile.

Discussion

Structure of Pollination Food Webs in Mediterranean Communities: Flower Visitors

All three communities share an insect pollinator fauna of similar overall taxonomic composition, which is in agreement with earlier findings suggesting that the pollinating faunas of Mediterranean-type ecosystems (including those of Chile and California) are analogous (Moldenke 1976; Petanidou and Ellis 1993). Apparent differences observed in this study can be attributed to biases of the methodologies employed, namely, selection of the study area, choice of focal group of plants or insects, invested taxonomic effort, as well as the intensity and duration of the survey. All of these may have affected total diversity and abundance of insects and their recording as pollinators. For instance, the dominance of Lepidoptera in the Spanish dataset is easily explained by the nighttime censuses undertaken by J. Herrera (1988).

The composition of the bee faunas varied to some degree between studies, which could be explained by numerous biogeographic factors. Greece had the highest proportion of megachilids, which are generally considered late-season flyers; and this was the study with the most intensive sampling during this period. Similarly, the early season families, Andrenidae and Anthophoridae, were dominant in the Israeli fauna, which focused more on that time of the year. Anthophorids are also associated with freshly burned and mature postfire habitats, which were common in the Israel study (Potts et al. 2003a).

Structure of Pollination Food Webs in Mediterranean Communities: Mutual Use of Resources

All three systems exhibit a high proportion of insect monotropy (insects visiting only one plant species: 36.2, 49.1, and 55.6% for Greece, Israel, and Spain, respectively) and oligotropy (one to eight plant species visited: 86.2, 93.8, and 96.7%, respectively). Overall, all three communities show similar patterns of insect specialization/generalization frequencies (fig. 10.1); however, these findings are not mirrored in the specialization/generalization displayed by the plants (fig. 10.2). The Israeli plant community shows the same negative exponential trend as the insects, the Greek plants show a negative and linear pattern, but there is no clear pattern in the Spanish community. These differences may again be attributed to the sampling protocols; however, they all share a general negative association. The frequency of oligophilous plants in Spain is higher when only bees are considered, showing a pattern similar to that of the Israeli plants. The monophilous plant species in the Greek community (considered a complete reference community) are relatively rare (1.5%), which is also true for the oligophilous plants (each visited by one to eight insect species; 22.7%) when compared with their insects partners. Evidently, plants do not share the same degree of specialization as their flower-visitor counterparts (figs. 10.1 and 10.2).

Our results show that there is a clear pattern in mutual pollination resource use by both plants and animals that is characterized by extreme asymmetry. Such asymmetry has been also found in other plant–pollinator systems (Bascompte et al. 2003; Dupont et al. 2003; Vázquez and Aizen, chap. 9 in this volume) and indicates that mono- to oligospecialized plants and animals are associated with more generalized partners (figs. 10.4 and 10.5). This applies not only to all three communities considered in the study but also to communities that are broken down by plant/insect groups. In fact, in all three communities, polyphilous plants tend to host most of the monotropous insects, whereas supergeneralist insects such as honeybees are associated with the most poorly visited plants of the whole community. For instance, in the Greek community, *Apis mellifera* visits

103 plant species, among them the monophilous *Sarcopoterium spinosum* and diphilous (*Anchusa variegata* and *Hymenocarpos circinnatus*). Conversely, in the same community, the "cornucopia" plant species *Chrysanthemum coronarium* (71-tropous) and *Thymus capitatus* (123-tropous) host 9 and 16 species of monotropous insect species, respectively. There was no case of bilateral (one-to-one) species specialization in any of the systems analyzed in this study. This finding, together with the strong asymmetry in resource utilization in all sites, suggests that pollination partnerships in the Mediterranean communities tend to avoid cospecialization. However, there may be some individual species that do tend to specialize, more so for insects than for plants.

The tendency of flower visitors to specialize is also known in other Mediterranean communities, for example, two Californian chaparrals (one mature and another one in postfire stage) and also a Chilean matorral (Moldenke 1979a). However, this distribution is not characteristic of Mediterranean-type communities: the preponderance of flower-visiting species is found in association with one to three plant species (Heithaus 1974; Moldenke 1975, 1979a; Moldenke and Lincoln 1979). In a comparative study of several bee faunas, Moldenke (1976) argued that bees tend to generalize when their diversity decreases. This also applies to the datasets of Greece and Israel analyzed in this study. Furthermore, our analysis shows that the rate of bee specialization may increase in dry climates, which also agrees with Moldenke (1979b). In conclusion, we argue that niche partitioning may be the driver for increased specialization, not merely resource heterogeneity as suggested by Michener (1974). However, the high degree of bee specialization is certainly supported by the high diversity of floral resources available in the area, as Michener (1974) proposed. This prediction, however, is not corroborated by the Spanish dataset, although the relatively small sample size may account for this.

Is Connectance a Reliable Index for the Study of Specialization?

Our results agree with those of Olesen and Jordano (2002) in that connectance decreases with community size. This is apparent even when large datasets are partitioned into functional subsets (e.g., plant life habit or insect taxon) as shown by Medan et al. (chap. 11 in this volume). Consequently, connectance may be largely biased by the methodology employed: selection of the species spectrum and sample size, the ecosystem breadth and heterogeneity, and the time invested. However, connectance is not a measure without meaning. For instance, in our analysis of the Israeli bee dataset, connectance indicates the degree of relative specialization within the system. Why, then, does connectance score so low in this dataset? In fact, the results of table 10.1 predict a much smaller value for the entire Israeli insect community, so we can confidently assume that this system is extremely rich in species, with many empty niches resulting in a low (loose) connectance. This is emphasized by the high monotropy and monophily rates of this system (49.1 and 35.8% of the total insect and plant species, respectively), the highest that have been found in our study. In general terms, the extreme diversity of Israel can be attributed to its biogeographical position as a "junction" between Asia, Europe, and Africa, and Mt. Carmel provides a landscape where Mediterranean and semidesert habitats adjoin (which is also seen in other Mediterranean-type ecosystems in South Africa, Chile, California, and Australia). In addition, the area is prone to a wide variety of moderate disturbance factors including fire, grazing, and other anthropogenic perturbations. Together these have produced a "hotspot" for bee and floral diversity (Potts et al. 2003a).

So is connectance really an indicator of specialization or highly specialized communities? Our results show that, in general terms, connectance may be a good preliminary indicator of the *average* tendency for specialization at the ecosystem level but may say little about interaction patterns of the species involved. This implies that connectance is not a reliable measure of specialization of either partners or partnerships. Furthermore, connectance values may vary greatly depending on the actual size of the dataset or the system being studied (table 10.1; see also Medan et al., chap. 11 in this volume), a phenomenon already discussed by Fonseca and John (1996). We conclude that postconnectance analyses are necessary to provide a more meaningful assessment of whether communities with low connectance contain partnerships and species that are really specialized.

Seasonal and Annual Variations in Insect–Host Selection: Are They Important in Specialization?

Plant–pollinator systems are stochastic and insect abundance is a major element of unpredictability (table 10.3); however, such high variability in interyear phenology was not found for their plant counterparts (Petanidou et al. 1995b). The variable character of insect visitor phenology has been also discussed by C. M. Herrera (1988) and Petanidou and Ellis (1993, 1996), who attributed this phenomenon to the highly variable climate of the Mediterranean area. Climate unpredictability may have similar effects on the phenology of the parasitic bees, resulting in a much lower percentage of bee parasites in the Mediterranean than in other seasonal systems (Petanidou et al. 1995a).

The unpredictable nature of insect abundance has an enormous effect on the degree of certainty in estimating specialization in pollination partnerships. Therefore, to reduce the level of uncertainty and incompleteness, such studies in the Mediterranean (and other systems) should be carried out over an extended period of time, which for the case of Greece was found to be at least four consecutive years (Petanidou and Ellis 1993, 1996). In studies of specialization, this is particularly important because the probability of missing an insect species is much higher in specialized partnerships (table 10.3); such insects are short lived and have small populations, resulting in an underestimation of the number of specialists in the community. In addition, because some insects may change partner choices between years, they may be counted as oligotropous or specialized if undersampled. In fact, these factors may compromise the validity of conclusions if not taken into account.

Seasonal constancy is a key element of specialization, at least in microecological terms. But, according to our results, there are few pollinator cases that can be reliably characterized as specialized: they are all (29) monotropous, of very high selectivity, and have been consequently "choosy" for the same plant species over at least two years of study. A further, and safer, approach in clarifying the sense of specialization is a macroecological one. For instance, *Nemognatha chrysomelina* (Meloidae), a beetle species specializing on *Echinops microcephalus* in Greece, and *Nemognatha* sp., specializing on the congener *Echinops ritro* in northern Europe (Kugler 1979, cited by Willemstein 1987), constitute a very good example of real specialization on this continent. We suggest that only cases beyond the ecosystemic level (i.e., cases of evolutionary importance in the framework of wider areas or regions) merit being characterized as truly specialized. In a similar vein, Armbruster et al. (2000) defined the terms *evolutionary* versus *ecological specialization* (process vs. state, respectively). However, our results here and elsewhere (Petanidou and Ellis 1996) show that the degree of ecological specialization is strongly affected by both spatial (macro- vs. microecology) and temporal (issue of sampling effort or of the community time succession) factors.

How Specialized Are Mediterranean Systems?

We conclude that, because of the very strong asymmetry in use of pollination resources, Mediterranean communities are characterized overall by very limited specialization within their partnerships. However, there are some species within the community that show a tendency to specialize, and this is more pronounced for the insect visitors than the plants. This trend is much even more evident when one considers the selectivity values calculated for the Greek scrub system (fig. 10.3). Even though Greece appears to be the least specialized of all three communities considered in this study, it encompasses species that are highly selective during their period of flowering (namely, plants) and activity (namely, insects). In addition, there is a small group of both plants and insects that tend to generalize within the system (figs. 10.1 and 10.2), and, among plants and insects, insects tend to be relatively more selective on the whole, but it is the plants that have the highest absolute selectivity values.

It is evident that in Greece both pollination partners tend to be highly selective. On the other hand, a small fraction of both animal and plant partners are very generalized, which might compensate for the ecosystem cost of specialization. Such pollination partners, which may host many insect visitors or visit a large number of plants, have been described several times in the literature and are referred to as cornucopia species (Mosquin 1971; Moldenke and Lincoln 1979; Petanidou 1991a). In the Greek community, such species are the plants *Thymus capitatus* (123 insect species visiting), *Chrysanthemum coronarium* (71 insect species), and the insect *Apis mellifera* (visiting 103 plant species). Equivalent cornucopia plants (supergeneralists according to Olesen et al. 2002) are present in the Israeli (Potts et al. 2001) and Spanish communities (J. Herrera 1988). Such supergeneralists are very important for the function of a system characterized by high specialization levels, such as the Mediterranean systems. Petanidou and Ellis (1996) referred to the "bridge function" of such plants operating as "sequential mutualists" in accordance with the idea of Waser and Real (1979).

Plants have high selectivity values associated with them, and this is particularly marked during the peak of the flowering period and to a lesser extent during the secondary flowering period. This implies that selectivity is highest during peak flowering, when insect resources are copiously available (cf. Petanidou and Ellis 1996). The activity of highly selective species contributes to a reduction in the level of competition for either floral rewards or for pollinator resources during the main flowering period. Conversely, low levels of selectivity are indicative of reduced competition systems: in Mediterranean scrublands this occurs during the secondary flowering period (Dafni and Dukas 1986). Whether or not selectivity is an intrinsic property of plants and animals that is associated with particular ecosystems remains an open question.

Why are there trends for particular species to specialize within Mediterranean communities? There certainly does not appear to be evidence for a genetic predisposition, which Michener (1974) proposed to explain the high specialization of tropical bees. It is more likely that the high diversity of the resources available in the Mediterranean for both plants and insects drives the observed specialization (Blondel and Aronson 1999). The fact that insects tend to be more specialized than plants may be related to the extraordinarily large number of insect species in these communities compared to plants; according to a Michener's (1979) axiom, the Mediterranean is one of the world's centers of bee speciation.

However, contrary to the simple specialization trends shown by some species, the strong asymmetry in resource use at the community level is still somewhat puzzling. Indeed, why don't trends for species specialization ultimately converge into a highly specialized system? We argue that high specialization keeps competition reduced and contributes to the sustainability of the pollination resources in the Mediterranean communities. Being a specialist is a risky strategy, especially in the Mediterranean because of the vicissitudes and unpredictability of the climate as well as the changing character of the communities due to anthropogenic and natural catastrophes, such as fires. But both plants and insects are adapted to such perturbations, after which they successively and repeatedly reestablish themselves (Petanidou and Ellis 1996; Potts et al. 2003a). Specialized relationships, on the other hand, are much more difficult to reestablish, depending on a variety of different stochastic events. This is why we believe that, although Mediterranean species are well adapted and can afford the implications and complications of specialization, a specialized plant–pollinator partnership itself is too complex for a high-risk mutual specialization. It appears that in Mediterranean systems such as those in Greece and Israel, the insects have a greater choice in selecting a plant partner and are perpetual opportunists able to specialize on a given floral species. Plants, however, are more dependent on seasonal factors and are less likely to specialize.

Acknowledgments

TP wishes to thank Willem Ellis for the first discussions on the issue of selectivity, and Gabriel Hiotellis for his valuable help in calculating this parameter. She also thanks the staff of the Diomedes Botanical Garden of Athens University and the numerous insect taxonomists that identified the material collected during her PhD study. SGP thanks the following for their contributions to the various projects undertaken on Mt. Carmel, Israel: Amots Dafni, Gidi Ne'eman, Pat Willmer, Betsy Vulliamy, Chris O'Toole, and Stuart Roberts. Research on Mt. Carmel was funded by the Natural Environment Research Council, United Kingdom (GR3/11743). This chapter has greatly benefited from comments made by Diego Medan, Jeff Ollerton, and two anonymous reviewers.

References

Armbruster, W. S., C. B. Fenster, and M. R. Dudash. 2000. Pollination "principles" revisited: Specialization, pollination syndromes and evolution of flowers. Det Norske Videnskaps—Akademi. I. Matematisk Naturvidenskapelige Klasse, Skrifter, Ny Serie 39: 179–200.

Barrett, S. C. H. 1996. The reproductive biology and genetics of island plants. Philosophical Transactions of the Royal Society of London B, 351: 725–733.

Bascompte, J., P. Jordano, C. J. Melián, and J. M. Olesen. 2003. The nested assembly of plant–animal mutualistic networks. Proceedings of the National Academy of Sciences (USA) 100: 9383–9387.

Blondel, J., and J. Aronson. 1999. Biology and wildlife of the Mediterranean region. Oxford University Press, Oxford.

Buchmann, S. L., and G. P. Nabhan. 1997. The forgotten pollinators. Island Press, Washington DC.

Dafni, A., and R. Dukas. 1986. Insect and wind pollination in *Urginea maritima* (Liliaceae). Plant Systematics and Evolution 154: 1–10.

Dupont, Y. L., D. M. Hansen, and J. M. Olesen. 2003. Structure of a plant–flower visitor network in the high altitude sub-alpine desert of Tenerife, Canary Islands. Ecography 26: 301–310.

Elberling, H., and J. M. Olesen. 1999. The structure of a high latitude plant–flower visitor system: The dominance of flies. Ecography 22: 314–323.

Faegri, K., and L. van der Pijl. 1979. The principles of pollination ecology, 3rd ed. Pergamon, Oxford.

Fonseca, C. R., and J. L. John. 1996. Connectance: A role for community allometry. Oikos 77: 353–358.

Heithaus, E. R. 1974. The role of plant–pollinator interactions in determining community structure. Annals of the Missouri Botanical Garden 61: 675–691.

Herrera, C. M. 1988. Variation in mutualisms: The spatio-temporal mosaic of a pollinator assemblage. Biological Journal of the Linnean Society 35: 95–125.

Herrera, J. 1988. Pollination relationships in southern Spanish Mediterranean shrublands. Journal of Ecology 76: 274–287.

Jordano, P. 1987. Patterns of mutualistic interactions in pollination and seed dispersal: Connectance, dependence asymmetries, and coevolution. American Naturalist 129: 657–677.

Kugler, H. 1979. Zur Bestäubung von *Echinops ritro* L. Bericht der Deutschen Botanischen Gesellschaft 92: 637–643.

Michener, C. D. 1974. The social behavior of bees: A comparative study. Harvard University Press, Cambridge.

Michener, C. D. 1979. Biogeography of the bees. Annals of the Missouri Botanical Garden 66: 277–347.

Moldenke, A. R. 1975. Niche specialization and species diversity along an altitudinal transect in California. Oecologia 21: 219–242.

Moldenke, A. R. 1976. Evolutionary history and diversity of the bee faunas of Chile and Pacific North America. Wassmann Journal of Biology 34: 147–178.

Moldenke, A. R. 1979a. Pollination ecology as an assay for ecosystem organization: Convergent evolution in Chile and California. Phytologia 42: 415–454.

Moldenke, A. R. 1979b. Host-plant coevolution and the diversity of bees in relation to the flora of North America. Phytologia 43: 357–419.

Moldenke, A. R., and P. G. Lincoln. 1979. Pollination ecology in montane Colorado: A community analysis. Phytologia 42: 349–379.

Mosquin, T. 1971. Competition for pollinators as a stimulus for the evolution of flowering time. Oikos 22: 398–402.

Olesen, J. M. 2000. Exactly how generalised are pollination interactions? Det Norske Videnskaps—Akademi. I. Matematisk Naturvidenskapelige Klasse, Skrifter, Ny Serie 39: 161–178.

Olesen, J. M., L. I. Eskildsen, and S. Venkatasamy. 2002. Invasion of pollination networks on oceanic islands: Importance of invader complexes and endemic super generalists. Diversity and Distributions 8: 181–192.

Olesen, J. M., and P. Jordano. 2002. Geographic patterns in plant-pollinator mutualistic networks. Ecology 83: 2416–2424.

Ollerton, J., and L. Cranmer. 2002. Latitudinal trends in plant-pollinator interactions: Are tropical plants more specialised? Oikos 98: 340–350.

Petanidou, T. 1991a. Pollination ecology in a phryganic ecosystem. PhD dissertation, Aristotle University, Thessaloniki, Greece.

Petanidou, T. 1991b. Pollinating fauna of a phryganic ecosystem. Verslagen en Technische Gegevens Amsterdam 59: 1–12.

Petanidou, T. 1993. Bee pollination in phrygana: Facts and actions. Pp. 37–47 *in* E. Bruneau (ed.), Bees for pollination: Proceedings of a workshop of the community programme of research and technological development in the field of competitiveness of agriculture and management of agricultural resources (1989–1983). Commission of the European Communities, Luxembourg.

Petanidou, T., and W. N. Ellis. 1993. Pollinating fauna of a phryganic ecosystem: Composition and diversity. Biodiversity Letters 1: 9–22.

Petanidou, T., and W. N. Ellis. 1996. Interdependence of native bee faunas and floras in changing Mediterranean communities. Pp. 201–226 *in* A. Matheson, S. L. Buchmann, C. O'Toole, P. Westrich, and I. H. Williams (eds.), The conservation of bees. Linnean Society Symposium Series 18. Academic Press, London.

Petanidou, T., W. N. Ellis, and A. C. Ellis-Adam. 1995a. Ecogeographical patterns in the incidence of brood parasitism in bees. Biological Journal of the Linnean Society 55: 261–272.

Petanidou, T., W. N. Ellis, N. S. Margaris, and D. Vokou. 1995b. Constraints on flowering phenology in a phryganic (East Mediterranean) ecosystem. American Journal of Botany 82: 607–620.

Potts, S. G., A. Dafni, and G. Ne'eman. 2001. Pollination of a core flowering shrub species in Mediterranean phrygana: Variation in pollinator abundance, diversity, and effectiveness in response to fire. Oikos 92: 71–80.

Potts, S. G., B. Vulliamy, A. Dafni, G. Ne'eman, C. A. O'Toole, S. Roberts, and P. G. Willmer. 2003a. Response of plant–pollinator communities following fire: Changes in diversity, abundance, and reward structure? Oikos 101: 103–112.

Potts, S. G., B. Vulliamy, A. Dafni, G. Ne'eman, and P. G. Willmer. 2003b. Linking bees and flowers: How do floral communities structure pollinator communities? Ecology 84: 2628–2642.

Rivas-Martínez, S. M., M. Costa, S. Castroviejo, and E. Valdés. 1980. Vegetación de Doñana (Huelva, España). Lazaroa 2: 5–19.

Waser, N. M., L. Chittka, M. V. Price, N. M. Williams, and J. Ollerton. 1996. Generalization in pollination systems, and why it matters. Ecology 77: 1043–1060.

Waser, N. M., and L. A. Real. 1979. Effective mutualism between sequentially flowering plant species. Nature 281: 670–672.

Willemstein, S. C. 1987. An evolutionary basis for pollination ecology. Leiden Botanical Series, vol. 10. E. J. Brill, Leiden, The Netherlands

Williams, N. M., R. L. Minckley, and F. A. Silveira. 2001. Variation in native bee faunas and its implications for detecting community change. Conservation Ecology 5: 57–89.

CHAPTER ELEVEN

Measuring Generalization and Connectance in Temperate, Year-long Active Systems

Diego Medan, Alicia M. Basilio, Mariano Devoto, Norberto J. Bartoloni, Juan P. Torretta, and Theodora Petanidou

Plant–animal mutualistic interactions have a pervasive influence on community dynamics and diversity. Since Darwin (1859), biologists have been increasingly aware of the importance and complexity of pollination links in communities (Kearns and Inouye 1997; Kearns et al. 1998; Memmott 1999; Traveset 1999). This complexity is the result of plant–pollinator (p–p) networks which embed not only the trophic relationships among mutualistic partners (Jordano 1987) but also the complexities of the evolutionary effects on each other that drive coevolutionary processes (Thompson 1994, 1998). Thus, understanding the structure and function of p–p systems requires issues to be addressed in both ecological and evolutionary contexts.

Several approaches to the study of p–p systems have promoted a vigorous discussion on the issue of generalization/specialization of interactions between partners (e.g., Jordano 1987; Waser et al. 1996 and references therein; Johnson and Steiner 2000; Bascompte et al. 2003; Jordano et al. 2003; Vázquez and Aizen 2003). The following discussion is centered on the adaptive significance of floral traits in relation to pollination. Encouraged by available examples of tight coevolution and specialization, early evolutionary and ecological theory speculated that p–p systems should, as a rule, be highly specialized. Specialization was considered critical for plant speciation and evolutionary radiation. This gave rise to the concept of pollination syndromes, which are suites of floral traits proposed to reflect adaptations to one or another pollinator type roughly at the level of orders (e.g., beetles vs. flies) or above (e.g., bats vs. birds). Recently this view has been challenged and there is a growing skepticism about the specialized nature of pollination syndromes (Ollerton 1996, 1998; Waser et al. 1996; Kearns et al. 1998; Aigner 2001, chap. 2 in this volume; Gómez 2002; but see Johnson and Steiner 2000). This "paradigm shift" has an obvious impact on theoretical sciences such as evolution and ecology, but it also has important implications for applied disciplines such as conservation biology and agriculture (Bond 1994;

Traveset 1999; Waser et al. 1996 and references therein; Corbet, chap. 14 in this volume; Steffan-Dewenter et al., chap. 17 in this volume).

Several recently published studies of total pollination systems, most of them reviewed by Jordano et al. (2003), Ollerton and Cranmer (2002), and Waser et al. (1996), have rekindled traditional debates in community ecology and biogeography, and the generalization/specialization issue was not an exception (Gómez 2002; Vázquez and Aizen 2003). The degree of generalization in p–p systems has been addressed at both species and community levels. At the species level, the most commonly applied measure of generalization is the number of interaction partners, *S,* whereas at the community level the connectance (*C,* the ratio between the actual number and the potential number of links in a network) has been repeatedly used to describe the mean level of generalization of a whole system. These two measures, *S* and *C,* have been used for many years without being seriously criticized (but see Waser et al. 1996; Olesen and Jordano 2002; Petanidou and Potts, chap. 10 in this volume). One good reason for this is probably the relative simplicity of obtaining the field data required to calculate these measures.

However, some concerns have been raised regarding the use of *C* and *S* as measures of generalization. For instance, recent studies of pollination webs highlight the influence of system size (i.e., the number of plants and animals in a given community) on connectance values (Olesen and Jordano 2002), which limits the utility of comparing the degree of generalization across systems of different size. Besides the obvious physical influence of size on connectance of the web (i.e., the greater the number of actors, the smaller the possibility of interaction between them), there is an additional issue. Most published studies, when calculating *C* of a p–p network, assume all partners to coexist and to be simultaneously active in pollen presentation or pollen transport; however, in several temperate and some tropical systems, this assumption does not hold, because many p–p interactions occur that do not temporally overlap. With the traditional approach, therefore, a "missing" interaction in such a system may be due to the failure of a particular potential visitor species to visit a given plant species or to a simple lack of coincidence in time (Jordano 1987; Martinez and Dunne 1998; Jordano et al. 2003).

Provided that all interactions were simultaneous, and the benefits were similar for all partners, *S* would be a good measure of generalization; however, one or more animal mutualists may not be available through the plant's entire flowering span, and even similar reward quantities may have unequal fitness effects on different visitors. Similarly, pollinator service may differ in quality and intensity among visitors (Waser et al. 1996); and even if that were not the case, pollen transfer needs would probably not be equal for all plants (Gómez 2002). Thus, any attempt to improve *S,* for example, by weighting each interaction by a fitness or efficiency measure, would require substantial effort. Applying such an index to anything but a very small community would demand an inordinate amount of work and resources.

In this chapter, we use data from two year-long active systems to assess the extent to which phenological noncoincidence can impair the value of connectance as a generalization measure. Furthermore, we present *G*, an innovative, two-dimensional index of generalization, and apply it to data of one of the two study systems. Both approaches are aimed at improving presently applied measures of generalization and connectance of p–p systems.

Two Examples of Year-long Active Systems Analyzed by Means of Periodical Sampling

We are aware of only two cases to date in which year-long active systems were studied by means of systematic, periodical sampling of interactions (i.e., in which the scale of analysis was consistently smaller than that of the entire system). Before going into the analysis of data from these studies, we briefly present the scope, methods, and main conclusions of each.

Some of us have analyzed pollination interactions in the Talar, a xeric temperate forest located at sea level some 125 km southeast of Buenos Aires, Argentina (35° S, 57°30′ W). The objective was to depict changes in the properties of a complete p–p network using periodical sampling over multiple years. The existence of seasonality in the local climate, year-long pollination activity, and the moderate level of richness of the interacting communities offered a reasonable compromise between variability of system parameters and amount of field work required. Field sites were visited with an average frequency of 1.2 times a month over a period of 41 months. Approximately 3500 observations of plant–flower visitor interactions, reflecting some 560 hours of field work, were compiled into 12 one-month periods and were subsequently used to construct monthly plant–flower visitor interaction matrices. Interaction data, additional noninteraction observations, and data from insect trapping were used to construct phenograms depicting flowering periods of plants and activity periods of insects. Admittedly, the pooling of data across years for each given month (January, February, etc.) obscures interannual variation in mutualist phenologies and thus the timing of their interactions. However, we believe that the results do reflect the "hard core" of mutualists and interactions that are characteristic of each monthly period. The existence of a "periphery" of mutualists and interactions that do not occur every year (or do so but not at the same time every year) is not expected to alter the conclusions that follow. The main conclusions of this study can be summarized as follows (Basilio et al., unpublished data). The plant community is dominated by trees and shrubs distributed in a fragmented pattern; grasses and mostly nonnative annuals occupy open areas. The animal community is codominated by Diptera and Hymenoptera. Among the 41 plant and 104 animal species, 318 unique interactions were recorded. The composition of mutualist assemblages and all system parameters, notably network size and number of interactions, varied across months by more than one order of magnitude. Maximum system activity occurred during October and April, when the number of mutualist species and interactions peaked, system asymmetry (number of plant species per number of animal species) was higher than average, visitor abundance was highest, and some of the characteristic, woody species of the Talar were in bloom.

Our second example is Petanidou's (1991) study on p–p relationships in the phryganic ecosystem on Mt. Aegaleo, near Athens, Greece. This is a larger system ($S = 797$), which was sampled weekly or biweekly over 50 months (for further details see Petanidou 1991; Petanidou and Ellis 1993; Petanidou et al. 1995; Petanidou and Potts, chap. 10 in this volume). The plant community is dominated by shrubs, where therophytes constitute the most species-rich life form, and the animal community is dominated by Hymenoptera, particularly apoid bees. As a result of over 5000 hours of pollinator-directed observations, 3006 unique interactions were recorded among the 132 plant and 665 insect species. For the purpose of this chapter, we compiled the interaction data into 12 monthly periods. Between-year phenological shifts do occur in this system (Petanidou et al. 1995), but, as reasoned earlier for the Talar, these are unlikely to seriously affect the current analysis.

Network Connectance in Cumulative and Consecutive Webs

Connectance is a scale-free measure of community coherence that has been used to make comparisons among biological and other types of networks (Jordano 1987; Martinez and Dunne 1998; Strogatz 2001; Olesen and Jordano 2002; Jordano et al. 2003). The analysis of a number of p–p and plant–seed disperser systems (Olesen and Jordano 2002; Jordano et al., chap. 8 in this volume) has shown that there is a negative-exponential relationship between connectance and system size.

A potential problem in computing connectance values arises from the fact that in year-long active systems there are noncoincident phenologies and, therefore, forbidden links (sensu Jordano et al. 2003, chap. 8 in this volume), that is, impossible interactions between partners that never overlap in time. Consequently, the number of mutualists actually interacting at any given time will generally be far lower than anticipated from the total number of species in the system, except perhaps at the midpoint of the system's time span. This problem has been recognized before (Jordano 1987; Olesen and Jordano 2002; Olesen et al. 2002; Bascompte et al. 2003; Jordano et al. 2003) but its impact has apparently not been analyzed in detail for p–p networks.

To assess the effect of nonoverlap of phenologies on C we needed high-quality data of fine-grained temporal resolution. By gradually aggregating such data to perform consecutive calculations of C, the problematic effect should

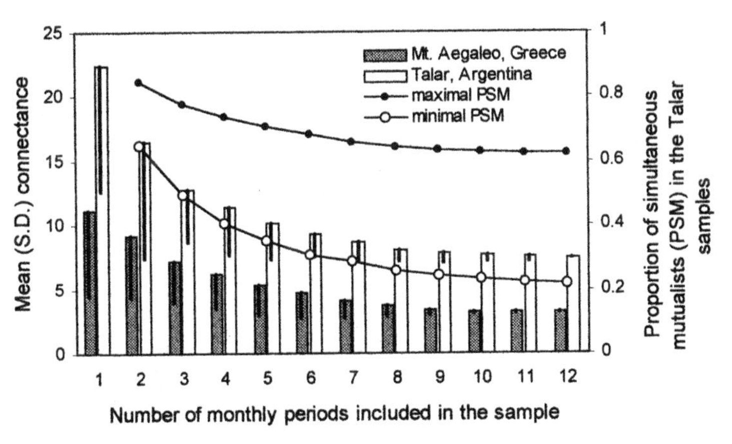

Figure 11.1 Variation of connectance in two year-long active p–p systems as a function of the number of monthly periods merged in the analyzed sample for calculation. Values shown are means (histograms) and standard deviations (bars included in histograms) of 12 samples, except for the 12-month sample size. For sample sizes 2–12 in the Talar system, curves show the maximal/minimal percentage of mutualists included in the sample that were simultaneously active in pollination interactions.

show itself. Unfortunately, published studies rarely provide detailed phenological information (but see Petanidou 1991). To overcome this problem, we performed an analysis on the Talar data, encompassing time periods of increasing length, from 1 to 12 months. We started with the C values for the 12 one-month periods, which vary from 41.0% when the system is near its smallest size, in June, to 10.6 and 15.0% at the spring and fall peaks of pollination activity, respectively. First, we computed the arithmetic mean and the standard deviation of C for these one-month periods (fig. 11.1). Then, we merged the mutualist assemblages of all possible pairs of consecutive months (January + February, February + March, etc.), removing repeated mutualists and interactions in the process, computed connectance values for the resulting 12 two-month periods, and obtained a mean and standard deviation. We repeated the process for three- to eleven-month periods. Finally, we computed overall connectance for the entire system, that is, the value that would result from assuming (as in most published studies) that all network mutualists were simultaneously active. The same analysis was applied to the data from Petanidou (1991; fig. 11.1).

We expected a decrease of connectance with increasing sample size, because the denominator in the calculation of C (the number of possible interactions) grows geometrically whereas the numerator (the number of realized interactions) grows at a slower rate (Olesen and Jordano 2002). However, beyond a certain point, C values could be affected by the phenology effect. As a way to assess this effect, in each sample that included two or more months, we first established the point at which the most plant and animal species were simultaneously active, then counted these mutualists, and finally divided this figure by the total number of mutualists in the sample. We called this value the maximum proportion of simultaneous mutualists (PSM_{max}). Note that a value of PSM_{max} equal to 1 indicates that all mutualists in the sample have coincident phenologies, whereas a value of 0.7 indicates that the largest set of coincident mutualists found in the sample includes only 70% of the mutualists. We also searched for the smallest set of coincident phenologies in all samples and calculated the minimum proportion of simultaneous mutualists (PSM_{min}). A value of PSM_{min} equal to 0.4 indicates that at least 40% of the mutualists will be found simultaneously active throughout the time span included in the sample. As with connectance values, we averaged values of PSM_{max} and PSM_{min} for each sample period.

A plot of connectance versus number of monthly periods included in the sample (fig. 11.1) shows that mean connectance values (open histograms) decrease rapidly in the Talar as sample size increases. Some implications of this decay are apparent. Taking samples of only four consecutive months in this system will produce, on average, half the connectance value produced by the average one-month sample. Compared with one-month samples, samples of three or more months will have statistically different connectances (two-tailed Student t test, $P < .01$). The system's overall connectance (12-month sample, $C = 7.4\%$) is 2.97 times lower than the average connectance of one-month periods ($C = 22.4\%$).

We analyzed the data from the phryganic system in the same fashion (fig. 11.1). The decay of connectances with increasing sample size is not as strong as that in the Talar. Therefore, in this system, a larger sample (five months) is needed to reach half the connectance value of an average monthly sample, and statistical differences between one-month and larger samples ($P < .05$) will only occur when four-month or longer samples are taken. Overall connectance in this system ($C = 3.42\%$) is 3.26 times lower than the average connectance of the one-month periods ($C = 11.1\%$).

In the Talar, the mean flowering period of plants is 5.1 ± 2.6 months (average ± standard error), and the mean flight period of insect species is 4.0 ± 3.2 months (A. Basilio et al., unpublished data). Corresponding values for the Greek system were 1.8 ± 0.07 and 1.6 ± 0.7 months. Thus, full overlap of phenologies could only be expected in the shortest sample sizes, and the values of PSM confirm this expectation. In two-month samples in the Talar, the largest set of coincident phenologies already includes, on average, fewer than 85% of the mutualists, although there are no sets that include fewer than 60% of these (fig. 11.1). PSM values decay exponentially, with PSM_{max} virtually stabilizing a bit above 0.6 at sample periods of 10 to 11 months; for these sample periods, PSM_{min} takes values near 0.2, indicating that such large samples may include really small, phenologically isolated groups of mutualists. PSM values for the phryganic system (not shown) are somewhat smaller and decay in a similar fashion with increasing sample size, starting with averages of 0.8 and 0.5 (PSM_{max} and

Table 11.1 Monthly variation of connectance values (%) of three pollinator-centered subsets of the p–p network in the Talar forest, Argentina

Pollinator group	W1	W2	W3	Sp1	Sp2	Sp3	Su1	Su2	Su3	F1	F2	F3
Apoid bees	30	50	100	20	20	30	40	20	20	30	60	60
Muscoid flies	60	50	40	30	30	100	0	100	0	30	50	60
Syrphid flies	30	50	60	20	30	30	100	100	0	60	100	100
Whole system	24	31	29	11	14	13	18	17	20	15	36	41

Note: Whole-system values are added for comparison. W = Austral winter, Sp = spring, Su = summer, F = fall; thus, within these seasons, W1 = June, Sp1 = September, Su1 = December, and F1 = March.

PSM_{min}, respectively) for 2-month samples and reaching values of 0.5 and 0.05 for 11-month samples.

A first conclusion of this exercise is that, in extended-activity systems, there does not seem to be a single connectance value that can unequivocally characterize the p–p network. Overall connectance may be seriously misleading if it is calculated for systems whose entire duration clearly exceeds the period of pollination activity of the average plant or animal mutualist. Note that, even when not compromised by the problem of different phenologies, overall connectance will simply be the lowest connectance value one can calculate for a system, but not necessarily the most biologically meaningful one. Except for steady systems whose composition does not change through time, connectance values higher than overall connectance will always be obtained for shorter sampling periods.

Perhaps more important is that the use of shorter sampling periods can reveal biologically relevant connectance fluctuations that a whole-system figure could obscure. Because connectance unequivocally expresses the actual degree of connectedness among pollination partners only in short-period networks, we propose that, whenever possible, connectance for entire systems should be calculated as the average connectance of such short-duration networks.

Last but not least, except in very simple and briefly active p–p networks, groups of mutualists whose degree of connectedness is above the network average are usually present, and this web substructuring is obscured when analyzing the whole-system connectance. Jordano (1987) compared the connectances of several such partial webs (for instance, of hummingbirds and their plants), but connectances for the corresponding complete systems were not available. Conversely, Dicks et al. (2002) showed the existence of compartments in complete webs in a formal way, but partial connectances were not computed. Here we offer examples of ecologically meaningful groups of mutualists whose connectances are compared to those of the total system. In the Talar (table 11.1), some pollinator-centered subsets (certain insect taxa with their visited plants) tended to have higher connectance values than that of the entire system. Group

Table 11.2 Connectance of the Greek phryganic community broken down into different plant-centered groups according to the plant species life form and flowering period

Plant group	*I*	*P*	*A*	*C* (%)
Frutescent perennials	633	22	340	8.46
Frutescent perennials (peak season)	551	19	310	9.35
Geophytes	416	26	196	8.16
Spring geophytes	274	16	143	11.98
Autumn geophytes	142	10	82	17.32
Herbaceous perennials	754	33	356	6.42
Herbaceous perennials (peak season)	624	29	318	6.77
Therophytes	1203	51	401	5.88
Whole system	3006	132	665	3.42

Note: I = number of interactions, *P* = number of plant species, *A* = number of insect species, and *C* = connectance.

Table 11.3 Connectance of the Greek phryganic community broken down into different pollinator-centered groups

Pollinator group	*I*	*P*	*A*	*C* (%)
Lepidoptera	231	68	31	10.96
Coleoptera	366	72	73	6.96
Diptera	762	100	188	4.05
Hymenoptera	1563	129	342	3.54
Symphyta	19	15	9	14.07
Wasps	161	45	63	5.68
Bees	1390	129	262	4.11
Whole system	3006	132	665	3.42

Note: I = number of interactions, *P* = number of plant species, *A* = number of insect species, and *C* = connectance.

connectance was highest for Syrphidae (year's average = 56%) and somewhat lower for muscoid flies (45%) and apoid Hymenoptera (40%). In the Greek phrygana (table 11.2), plant-centered subsets (plant species of similar life form and their insect visitors) also showed higher connectances than the whole system, as well as variation in connectance between periods of the year. Similarly, pollinator-centered subsets (taxonomically homogeneous pollinator groups and their plants) also showed higher connectances when considered alone compared with the values for the overall system (table 11.3). In both analyses, connectances of subsets clearly decreased as their system size (S = number of plant species + number of animal species) increased.

An Improved Generalization Measure for Among-web Comparisons

When analyzing the dynamics of p–p webs through their annual cycle, we felt the need for a measure of generalization that would allow for sound comparisons between species. The number of recorded mutualists of a plant or an animal, *S*, is a straightforward measure of generalization that immediately conveys a quantitative impression: other things being equal, plant species A that interacts with 11 insect species is obviously more generalized than another species, B, which is visited by just 3 species. However, comparisons between species belonging to different ecological contexts are less useful. For instance, the species to be compared may belong to communities of unequal size, which is what occurs when webs are compared from different periods of the year in the Talar. If species A belongs to a web with 50 available insect partners, and species B belongs to smaller web with only 7 available partners, A is interacting with only 22% of its potential mutualists (11/50), whereas B is interacting with 42.8%. In this view, species B is much more a generalist than species A.

Therefore, as a first attempt to characterize a pollination mutualist, we propose to define a generalist as any plant or animal that meets its needs by using a high proportion of the partners available in the community. Conversely, specialists are species that satisfy their needs using a low proportion of the potential mutualists. We therefore compute a resource usage (RU) index as follows:

$$\mathrm{RU} = \frac{\text{No. of effectively used mutualists}}{\text{No. of available mutualists}}.$$

Values of the RU index approach 0 for specialists, and may reach 1 for supergeneralists.

Mutualists with equal RU indices may still differ in their degree of generalization. An insect species C may visit five plant partners, all of them with equal frequency (say, 3:3:3:3:3), whereas another species, D, may also have five mutualists but with a far more unequal interaction schedule (say, 10:2:1:1:1). Although each species has the same number of mutualists, and the overall interaction "effort" is the same for both, species D is specialized on one of its plant partners, whereas C relies evenly on all of them (i.e., species C is a generalist).

Thus, as a second qualifying aspect, we propose to consider any plant or animal pollination mutualist that meets its needs by evenly resorting to its partners to be a generalist and to consider species using preferentially one or a few of their partners to be specialists. We quantified this aspect of generalization using the evenness measure of the Shannon–Wiener function (Krebs 1989):

$$E = \frac{-\sum_{i} p_i \cdot \ln p_i}{\ln S},$$

where p_i is the proportion of all interactions corresponding to the *i*th mutualist of a given plant or animal, and *S* is the total number of its mutualists. This index takes the value 1 when evenness is maximal and approaches 0 when the inter-

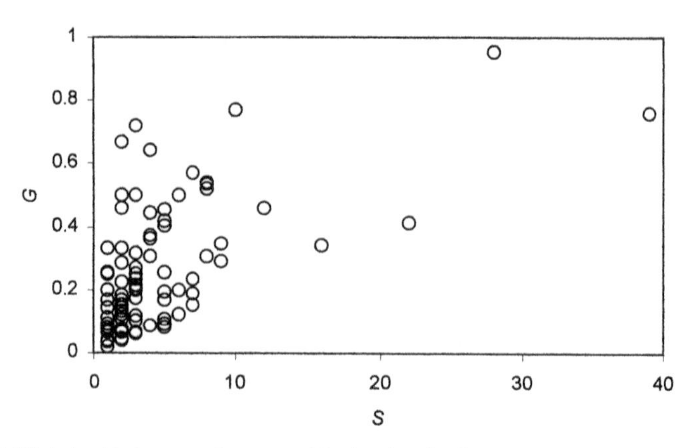

Figure 11.2 Relationship between *G* scores and *S* values for all pollination mutualists of the Talar system. Dots reflect mutualist scores at 1 of 12 monthly periods; thus, if a given mutualist's *G* and/or *S* values change across months, that mutualist will be represented by two or more dots.

action frequencies are very unequal among mutualists. When a plant or animal has only one mutualist, the index cannot be computed (because it requires division by zero), and in these cases we arbitrarily assumed maximum evenness and so assigned a score of 1.

To summarize, in our view a mutualist is a greater generalist the more its links with the available counterparts in the community are extended and the more equitably it uses the partners it interacts with. Because both RU and *E* have ranges fixed between 0 and 1, they allow for comparisons among different systems irrespective of their sizes (i.e., they are scale independent). The two indices can be multiplied to get a single generalization score *G*, which also ranges from values of 0 to 1 and is also scale independent compared to *S*, which is not. *G* may be used for comparions among mutualists and to characterize changes in the generalization of particular mutualists across their activity periods in the community.

We computed RU and *E* for all plants and animals involved in pollination interactions in the Talar. The two indices are very weakly correlated; that is, the information contained in them is not redundant ($R^2 = 0.02$ for a linear fit to an RU–*E* plot). We, thus, proceeded to apply RU and *E* in two ways. First, we multiplied them to calculate *G* values and plotted *G* against the corresponding *S* scores to search for possible redundancy of *G* (fig. 11.2). *G* and *S* are positively but very weakly correlated (R^2 of a linear fit = 0.027), indicating that *G* conveys information not contained in *S*.

Second, we plotted both components of *G* against one another (fig. 11.3). The area included in an *E*–RU graph may be viewed as the "generalization surface" that can be occupied by the pollination partners in a community. Data points located near the lower-left corner of any panel correspond to full specialists, and

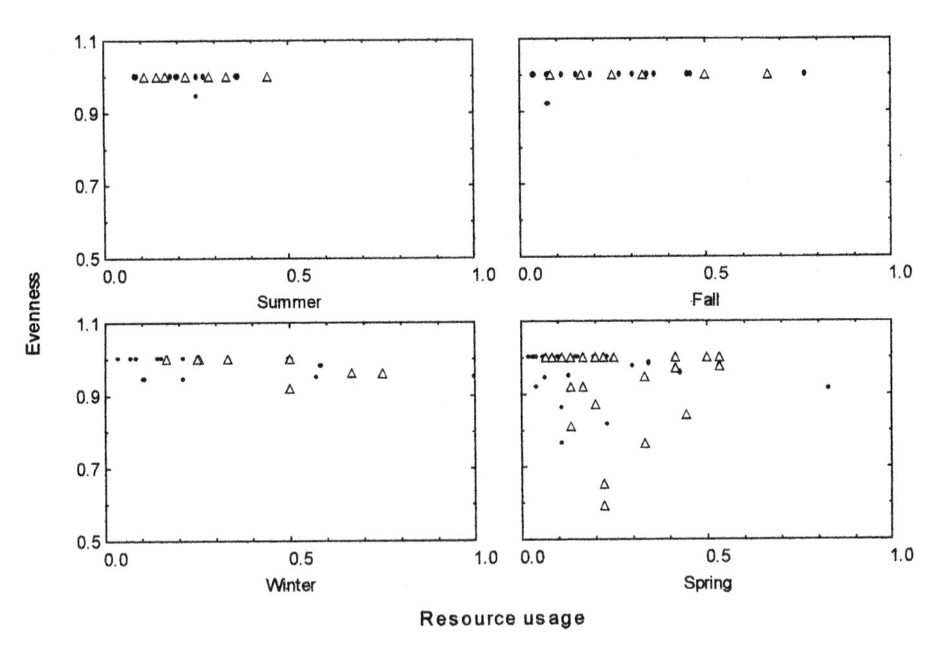

Figure 11.3 Values for two components of the generalization score, *G*, of all mutualists of the Talar system, grouped by seasons. Dots represent plants, and triangles represent animals. Symbols reflect mutualist scores at one of 12 monthly periods; thus, if a given mutualist's *E* and/or RU values change across months, that mutualist will be represented by two or more symbols at the same or different seasons.

those on the upper-right corner pertain to supergeneralists. In each graph, we included values for all plants and animals active during the three month-long periods corresponding to a year's seasons. Because many plants and animals were active over two or more months, if their generalization scores varied across the season they appear repeatedly on the same graph. The number of data points seems lower than the number of mutualists, because certain generalization scores are shared by several mutualists and data points overlap.

Except for the graph for spring, all graphs show appreciable variation only along the RU axis. In spring, there is also substantial variation along the *E* axis, particularly at low values of RU and for animals. In most seasons, ranges of RU are wider for plants. Moreover, ranges of both RU and *E* tend to increase with system size. The regression of the coefficient of variation of RU on *S* is significant for animals ($F_{1,10} = 6.36$; $P < .03$) but not so for plants ($P > .54$). Regressions of the coefficient of variation of *E* on *S* are highly significant ($F_{1,10} = 19.1$; $P < .0014$ for animals; $F_{1,10} = 15.9$; $P < .0025$ for plants). The overall picture suggests that, at low system sizes, mutualist strategies differ only quantitatively (i.e., in the amount of available resources effectively used), whereas strategies also start to diverge qualitatively as system size increases (different partners are "treated" in different ways).

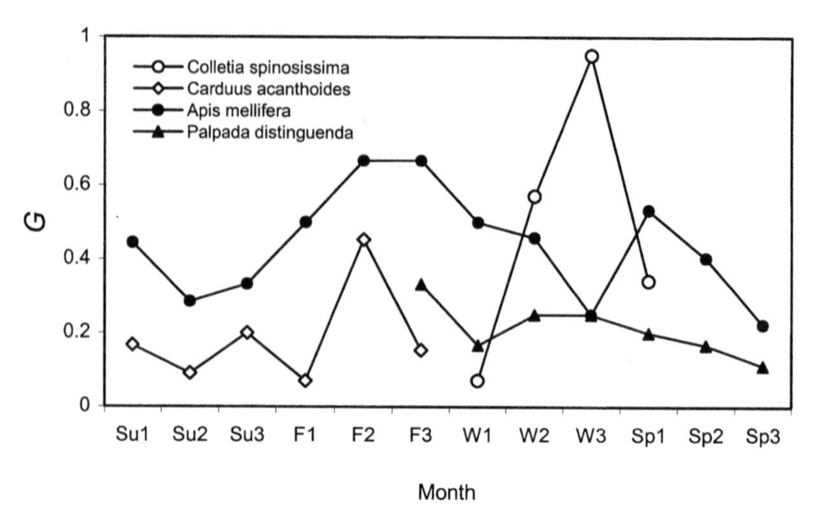

Figure 11.4 Values of the generalization score, *G*, for four mutualists of the Talar system throughout the year. Open symbols represent plant species; solid symbols represent animal species. Su = Austral summer, F = fall, W = winter, Sp = spring; thus, within these seasons, Su1 = December, F1 = March, W1 = June, and Sp1 = September.

Using *G* to Show Changes in a Single Species' Degree of Generalization

Turning our attention to individual species, it is interesting to analyze the dynamics of mutualists whose generalization scores changed throughout their periods of activity in the system. We chose two plant and two insect species whose interactions lasted four or more consecutive months in the Talar community (fig. 11.4). Our selection, which is admittedly arbitrary, is intended solely to illustrate contrasting patterns. All selected species had high *E* values; thus, their *G* scores mainly reflect variations of RU.

Among plant species, the shrub *Colletia spinosissima* experienced a sharp increase followed by a decrease in generalization score during winter and early spring, reflecting variation in the number of its insect partners, which in turn influenced RU values. *Colletia spinosissima* was relatively specialized at the start and the end of its period of activity ($G < 0.6$, $S \leq 16$) but rose to the level of a cornucopian species (i.e., one visited by most insects) at its flowering peak ($G = 0.95$, $S = 28$). On the other hand, the alien weedy thistle *Carduus acanthoides* behaved as a moderate specialist throughout its flowering period ($G \leq 0.45$), a result probably associated with this species' restrictive flower morphology and not unexpected for a plant recently integrated into the Talar native plant community (see Memmott and Waser 2002). The introduced honeybee *Apis mellifera* showed moderate generalist to moderate specialist scores (*G* varied from 0.66 in fall to 0.22 at the end of spring) throughout its extremely extended activity period. Because the honeybee is often characterized as a supergeneralist (e.g.,

Memmott and Waser 2002), it is interesting to learn that, at least in the Talar, such behavior does not actually occur at any time of the year. The long list of plant hosts (20, or 48.8% of all plant species, indeed suggesting supergeneralism) simply resulted from the addition of several, mostly noncontemporary, mutualists that honeybees visited through the year (maximum number of simultaneous plant hosts was 8). Generalization scores were generally lower for the native syrphid fly *Palpada distinguenda* which, in spite of being frequently seen in winter and spring, never had more than three hosts at a time (*G* values between 0.33 and 0.11).

Conclusions

This chapter has focused on some issues inherent to the analysis of p–p systems that are active during extended periods of time. We have shown that the most commonly used measure of generalization at the community scale, *C*, should be interpreted with caution whenever the system is active for an extended part of the year. Analysis of the two available examples of sequentially studied, year-long active networks unequivocally showed that *C*, as it is usually computed, underestimates the average connectance of one-month networks by a factor of about 3. We also showed that ecologically significant subsets of both systems display higher-than-average connectances, suggesting that alternative ways of measuring *C* may prove useful beyond providing a sounder interpretation of network structure.

We also presented a new way to measure the degree of generalization that integrates two independent aspects of the relationship between a particular mutualist and the remainder of the network: how many of its potential partners it actually interacts with and how uniformly it does so. Using data from one of the year-long active systems, we showed that the proposed index *G* is useful both for depicting community-wide changes in generalization along the annual cycle of the p–p network and for comparing specific patterns of generalization.

We conclude that information on network properties of many systems will not be fully captured by currently used generalization and connectance measures. The approaches advocated in this chapter clearly demand additional work: suitable phenological knowledge of mutualists must be in hand to decide if, and how, a whole system is to be partitioned into shorter periods for analysis; frequencies of all interactions are needed to compute our proposed *G* score. However, the examples from two year-long active systems showed that the additional effort can be amply repaid in a better understanding of interaction patterns, with far-reaching inplications for understanding ecology and evolution of p–p systems (Bascompte et al. 2003; Jordano et al., chap. 8 in this volume). We hope that future researchers will also conclude that these new insights are worth the effort.

Acknowledgments

We are grateful to Nick Waser for inviting us to collaborate in this book and for encouraging us to develop the thoughts presented here on generalization in plant-pollinator systems. The manuscript benefited from critical reading by Ruben Alarcón and Ingolf Steffan-Dewenter. This work was supported by grants from the University of Buenos Aires to D. Medan and A. M. Basilio. D. Medan is a member of CONICET, Argentina.

References

Aigner, P. A. 2001. Optimality modeling and fitness trade-offs: When should plants become pollinator specialists. Oikos 95: 177–184.

Bascompte, J., P. Jordano, C. J. Melián, and J. M. Olesen. 2003. The nested assembly of plant-animal mutualistic networks. Proceeding of the National Academy of Sciences (USA) 100: 9383–9387.

Bond, W. J. 1994. Do mutualisms matter? Assessing the impact of pollinator and disperser disruption on plant extinction. Philosophical Transactions of the Royal Society of London B 344: 3–90.

Darwin, C. 1859. On the origin of species by means of natural selection. Murray, London.

Dicks, L. V., S. A. Corbet, and R. F. Pywell. 2002. Compartmentalization in plant-insect flower visitor webs. Journal of Animal Ecology 71: 32–43.

Gómez, J. M. 2002. Generalizations in the interactions between plants and pollinators. Revista Chilena de Historia Natural 75: 105–116.

Johnson, S. D., and K. E. Steiner. 2000. Generalization versus specialization in plant pollination systems. Trends in Ecology and Evolution 15: 140–143.

Jordano, P. 1987. Patterns of mutualistic interactions in pollination and seed dispersal: Connectance, dependence asymmetries, and coevolution. American Naturalist 129: 657–677.

Jordano, P., J. Bascompte, and J. M. Olesen. 2003. Invariant properties in coevolutionary networks of plant-animal interactions. Ecology Letters 6: 69–81.

Kearns, C. A., and D. W. Inouye. 1997. Pollinators, flowering plants, and conservation biology. Bioscience 47: 297-307.

Kearns, C. A., D. W. Inouye, and N. M. Waser 1998. Endangered mutualisms: The conservation of plant-pollinator interactions. Annual Review of Ecology and Systematics 28: 83–112.

Krebs, C. J. 1989. Ecological methodology. Harper and Row, New York.

Martinez, N. D., and J. A. Dunne. 1998. Time, space and beyond: Scale issues in food-web research. Pp. 207–226 *in* D. L. Peterson and V. T. Parker (eds.), Ecological scale. Theory and applications. Columbia University Press, New York.

Memmott, J. 1999. The structure of a plant-pollinator food web. Ecology Letters 2: 276–280.

Memmott, J., and N. M. Waser. 2002. Integration of alien plants into a native flower-pollination visitation web. Proceedings of the Royal Society of London B 269: 2395–2399.

Olesen, J. M., L. I. Eskildsen, and S. Venkatasamy. 2002. Invasion of pollination networks on oceanic islands: Importance of invader complexes and endemic super generalists. Diversity and Distributions 8: 181–192.

Olesen, J. M., and P. Jordano. 2002. Geographic patterns in plant-pollinator mutualistic networks. Ecology 83: 2416–2424.

Ollerton, J. 1996. Reconciling ecological processes with phylogenetic patterns: The apparent paradox of plant-pollinator systems. Journal of Ecology 84: 767–769.

Ollerton, J. 1998. Sunbird surprise for syndromes. Nature 394: 726–727

Ollerton, J., and L. Cranmer. 2002. Latitudinal trends in plant-pollinator interactions: Are tropical plants more specialised? Oikos 98: 340–350.

Petanidou, T. 1991. Pollination ecology in a phryganic ecosystem. PhD dissertation, Aristotelian University, Thessaloniki.

Petanidou, T., and W. N. Ellis. 1993. Pollinating fauna of a phryganic ecosystem: Composition and diversity. Biodiversity Letters 1: 9–22.

Petanidou, T., W. M. Ellis, N. S. Margaris, and D. Vokou. 1995. Constraints on flower phenology in a phryganic (east Mediterranean shrub) community. American Journal of Botany 82: 607–620.

Strogatz, S. H. 2001. Exploring complex networks. Nature 410: 268–276.

Thompson, J. N. 1994. The coevolutionary process. University of Chicago Press, Chicago.

Thompson, J. N. 1998. The population biology of coevolution. Researches on Population Ecology 40: 159–166.

Traveset, A. V. 1999. Ecology of plant reproduction: Mating systems and pollination. Pp. 545–588 *in* F. I. Pugnaire and F. Valladares (eds.), Handbook of functional plant ecology. Marcel Dekker, New York.

Vázquez, D. P., and M. A. Aizen. 2003. Null model analyses of specialization in plant–pollinator interactions. Ecology 84: 2493–2501.

Waser, N. M., L. Chittka, M. V. Price, N. M. Williams, and J. Ollerton. 1996. Generalization in pollination systems, and why it matters. Ecology 77: 1043–1060.

CHAPTER TWELVE

Evolutionary and Ecological Aspects of Specialized Pollination: Views from the Arctic to the Tropics

W. Scott Armbruster

Interest in the evolution of ecological specialization has a long history, which can be traced back to Darwin's time, at least. Indeed, Darwin himself relied heavily on specialization in plant–pollinator interactions and floral morphology to help visualize, and later provide evidence for, how natural selection might operate (Darwin 1859, 1877a, 1877b). The specialized relationships between plants and their pollinators have thus been the subject of longer discussion and study than perhaps any other interspecific relationship; yet we are far from a scientific consensus on the importance of specialization in pollination, how specialization might vary geographically, or how it evolves. In this chapter I review hypotheses and observations concerning the evolution of specialized and generalized plant–pollinator relationships, with emphasis on the adaptation of plants to their pollinators.

One of the major macroevolutionary trends recognized in early evolutionary literature is the evolution of specialized interspecific relationships from more generalized ones (Cope 1896; see reviews by Futuyma and Moreno 1988; Thompson 1994). The irreversibility of specialization has often been assumed (Futuyma and Moreno 1988), but recent evidence indicates this is not always the case (e.g., Futuyma 1991; Lanyon 1992; Armbruster and Baldwin 1998). As Thompson (1994) points out, however, there are too few examples for us to draw general conclusions about whether specialization is largely irreversible. It has been generally concluded that generalized pollination relationships can evolve toward specialization but that extremely specialized pollination rarely reverts to a generalized state (e.g., Grant 1963; Grant and Grant 1965; Faegri and van der Pijl 1971; Endress 1994; but see discussion by Stebbins 1974; Armbruster 1993; Luckow and Hopkins 1995; Bruneau 1997; Armbruster and Baldwin 1998; Johnson et al. 1998; Goldblatt et al. 2000).

Although pollination has often been viewed as a series of textbook examples of extreme specialization (e.g., Campbell 1987), contemporary views are more balanced: most pollination relationships are seen to fall out along a continuum between generalization and specialization (see Waser et al. 1996; Johnson and Steiner 2000). Rather than viewing specialization and generalization as alternative models, we instead need to assess whether specialization is common or rare, how specialization and generalization evolve, and whether the degree of specialization in pollination relationships might vary from region to region.

I will focus on two general issues in this chapter. First, I consider factors that may favor the evolution of specialized or generalized pollination, and whether specialization is generally irreversible. Second, I consider whether there may be biogeographic patterns in the degree of specialization in pollination ecology. I address these questions by comparing the adaptations for pollination in four different plant systems, each from a different region and latitude. Four study systems form too small a sample to be representative or provide statistical tests for biogeographic trends; however, they represent a starting point for comparisons of evolutionary trends in specialization of pollination ecology. I should emphasize that plant trends toward specialized pollination do not necessarily imply specialization on the part of their pollinators; in fact, the reverse might often be the case (e.g., Petanidou and Ellis 1996; Vázquez and Aizen, chap. 9 in this volume; Petanidou and Potts, chap. 10 in this volume; Medan et al., chap. 11 in this volume).

Evolutionary and Ecological Specialization in Pollination

For purposes of this contribution, I will define *generalization* as the tendency for plants to use a large proportion of the available flower-visiting fauna as pollinators, as defined by species or functional groups. *Specialization* occurs when plants use a relatively small proportion of the available flower-visiting fauna as pollinators, as defined by species or functional groups (Armbruster et al. 2000; Johnson and Steiner 2000; see also Petanidou and Potts, chap. 10 in this volume). I find it useful to distinguish between *ecological specialization*—the state of being specialized—and *evolutionary specialization*—the process of evolving toward greater specialization in response to selection generated by pollinators. Because the former is a contemporary ecological state, it is more easily measured. The latter is a process that has occurred in the past (or in some cases may be ongoing) and, hence, is much more difficult to assess. This distinction may help us understand the apparent contradiction between ecological observations that generalization is quite common (e.g., Waser et al. 1996; but see Johnson and Steiner 2000) and the general perception that much of floral evolution has been in the direction of specialization on particular pollinators (Ollerton 1996; Johnson and Steiner 2000; see also Aigner 2001, chap. 2 in this volume).

The distinction between ecological and evolutionary specialization is also important in light of Aigner's (2001) recent work showing that plant evolutionary response to selection generated by one pollinator may sometimes occur without an increase in the amount of ecological specialization, at least if that evolutionary response does not compromise the plant's ability to use other pollinators (i.e., there are no trade-offs, or only weak ones). This leads to a refinement of our definition of evolutionary specialization: response to selection generated by a small proportion of potential selective agents. Thus, according to Aigner's (2001) model, evolutionary specialization may sometimes lead to a decrease in ecological specialization, for example, when an unimportant, minority pollinator becomes more important through the plant response to the selection it generates (increasing the evenness and statistical diversity of pollinators, although not their species richness). This counterintuitive result deserves more theoretical and empirical scrutiny. Specifically, we need to know if such trade-offs in effectiveness of pollinator use by flowers are ubiquitous, common, or rare (see Aigner, chap. 2 in this volume).

How and Why Do Specialization and Generalization Evolve?

Waser et al. (1996) developed a simple model that supports the intuitive expectation that generalization in pollination ecology should evolve whenever pollinators are rare or variable in abundance from year to year or when a specific pollinator is lost through extinction or plant migration (see also Armbruster 1990; Armbruster and Baldwin 1998). The conditions that favor the evolution of specialization are more difficult to model. The model presented by Waser et al. (1996) also showed that plants should specialize on their most important pollinators when there is little variance in pollinator abundance and behavior and when an increase in the fitness value of one pollinator is balanced by a corresponding decrease in value of an alternative pollinator. In contrast, Aigner (2001, chap. 2 in this volume) showed that plants may specialize on less-important pollinators (less common or less effective than others) when trade-offs in fitness value are weak or nonexistent. As alluded to earlier, this may result in the occurrence of evolutionary specialization without an increase in ecological specialization. In light of the results of these models, I suggest that two potentially common conditions in nature may lead to specialization: (1) specialization may result from selection generated by competition for pollination or (2) specialization may occur as a result of a "runaway-selection" trap.

Pollination has two components: quantity of visitation and quality of pollination (i.e., getting pollen to and from the right individuals; Waser 1983). When pollinators visit more than one species during foraging bouts or even longer periods of time, there is a good chance that pollen from one species will get lost to floral parts of the second species (including loss to alien stigmas). Pollen grains are copies of genes that will not show up in the next generation; hence, strong selection against visitation by "promiscuous" pollinators may be generated in such situations. It is interesting that traits favoring attraction and pollination only by less-promiscuous pollinators can be favored by selection even if some reduction of seed set occurs (because half the genes are transmitted at each generation through male function).

Runaway selection is a concept usually applied to the operation of sexual selection (e.g., Fisher 1930; see review by Futuyma 1997). A parallel runaway process may occur, however, when natural selection operates in a self-reinforcing fashion to favor traits that improve certain aspects of reproductive performance at a cost to other aspects; specialization in pollination ecology may evolve in this way. Indeed, Darwin (1877a) described just such a self-reinforcing process for the evolution of long floral tubes and long insect proboscides. A flower with a long floral tube will select for long tongues in the animal visitors (or, in some cases, long legs; see Steiner and Whitehead 1990, 1991a) because individuals with long tongues are best able to reach the reward. However, as the average pollinator-tongue length increases, many individual visitors may cease to contact the anthers and stigmas because the visitors no longer need to push tightly into the flower to reach the reward. This generates selection for longer tubes because longer tubes force visitors to push in more tightly and to contact the floral sexual parts (Darwin 1877a; Nilsson 1988). Evolutionary response on the part of the plant then starts the cycle over again. This runaway coevolutionary process is essentially an arms race between organisms in a mutually exploitative relationship. This process generates an increasingly specialized relationship between plant and pollinator (because shorter-tongued animal species are excluded from the nectar), but not because specialization is in itself adaptive relative to more generalized pollination. The process can be visualized graphically as a population being trapped on a very small, local adaptive peak—one that may actually get lower as the coevolutionary feedback cycle proceeds. In the absence of random jumps (Wright 1931) to nearby, higher adaptive peaks (e.g., using a broader range of pollinators), the population remains trapped in extreme specialization (e.g., Lindberg and Olesen 2001).

One view of how specialization may usually evolve treats specialization as proceeding along one to several of four or more axes of pollination niche space, roughly corresponding to the five W's of reporting: who, what, why, when, and where. *Who pollinates* is determined by which animals visit flowers (as influenced by *what* rewards are offered [quantity and quality], how the reward is presented [e.g., openly vs. in a tube], and *what* advertisements are employed) and which of these visitors contact anthers and stigmas (as influenced by the mechanical fit of flowers and visitors). *When in the season* pollination is achieved is determined by the flowering phenology and constraints on flowering time such as fruit and seed biology, seasonality, and so forth (e.g., Stiles 1975; see discussions of Waser 1983; Kochmer and Handel 1986). *When in the diel period* pollination occurs is determined by the ability of the flower to time its activities (flower opening/closing, anther dehiscence, nectar-secretion schedules, etc.): for example, nocturnal versus diurnal pollination, early morning versus afternoon opening of flowers, and so forth (Armbruster 1985; Armbruster and McCormick 1990; Stone et al. 1998). *Where pollen is placed on the pollinator* is influenced by the geometry of the flower and the flower's ability to orient the pollinator repeatably and to achieve precise and accurate pollen placement on, and receipt from, the animal. Common specialization along this axis includes nototribic (top) versus sternotribic (bottom) pollen placement (Faegri and van der Pijl 1971; see also Keller and Armbruster 1989), and placement in other more-specific locations on the body (e.g., Macior 1983). Orchids are especially well known for differentiating along this axis (e.g., Dressler 1968; Nilsson et al. 1987; Johnson 1997). Where pollen is placed on pollinators does not affect attraction of pollinators or competition for pollinator service, but it may seriously affect the quality of pollination.

In the following I address the issues introduced earlier in the context of taxonomic and latitudinal variation in specialization. Specifically, I examine the natural history of pollination across a small, haphazard sample of study systems: (1) *Saxifraga* (Saxifragaceae) of the high arctic and subarctic, (2) *Collinsia* and *Tonella* (Plantaginaceae) of temperate North America, (3) the Stylidiaceae of temperate and subtropical New Zealand and Australia, and (4) *Dalechampia* (Euphorbiaceae) of the Neo- and Paleotropics.

Materials and Methods

I observed pollination of *Saxifraga* species in Interior Alaska, during the summers of 1981–1989, and on Spitsbergen, Svalbard (Norway), in July 1999. I observed pollination of *Collinsia* species in California during most springs from 1976 to 2004. I observed pollination of *Stylidium* species in Western Australia, October–November 1992, and of *Dalechampia* species in Central and South America, Tanzania, Gabon, South Africa, and Madagascar intermittently from 1975 to 2001.

For *Collinsia, Dalechampia,* and the Stylidiaceae, the sequences of evolutionary changes in pollination relationships and specialization were inferred by optimizing (using parsimony and/or maximum likelihood) pollination traits onto phylogenies inferred from molecular data (internal transcribed spacer [ITS] of the nuclear ribosomal DNA, as corroborated by chloroplast and morphological data), or, in the case of the Stylidiaceae, from published molecular-phylogenetic studies.

Results

Saxifraga in the Arctic/Subarctic

All members of this genus have open, essentially actinomorphic (radially symmetrical) flowers, and exposed nectar rewards. As a result, they do not orient or position floral visitors; hence, there is little or no precision in the placement of pollen on pollinators. Such flowers are usually regarded as generalized (see Berg 1960; Armbruster et al. 1999), although confirmation of this requires detailed

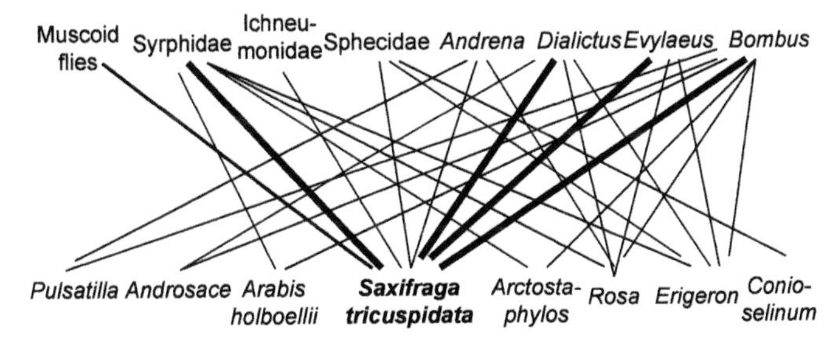

Figure 12.1 Partial visitation web for interior Alaskan steppe during the blooming period of *Saxifraga tricuspidata,* showing a high degree of generalization from both plant and visitor perspectives. The only sympatric, synchronic flower visitors *not* seen on *S. tricuspidata* were Coleoptera, Lepidoptera, Bombyliidae, *Megachile, Osmia,* and *Nomada.* Line width indicates frequency of visits.

data on the actual floral visitors versus those that are available (see Thompson 1994).

Observations on *S. cernua, S. cespitosa, S. hieracifolia, S. nivalis, S. oppositifolia,* and *S. rivularis* at Spitsbergen indicated low visitation rates by only a few species of flies (W. S. Armbruster and D. Rae, unpublished data). Pollination by only a few species of insects would be considered specialized on a species-number criterion. However, these visitors represented nearly all the flower-visiting animals observed at our study sites in Svalbard. Thus, by a faunal-proportion criterion, Spitsbergen *Saxifraga* had generalized pollination. Furthermore, all species of *Saxifraga* in Spitsbergen appeared to have the same pollinators.

In subarctic Alaska, I observed pollination of *S. tricuspidata* and *S. reflexa* by a wide variety of nectar- and/or pollen-feeding insects. The most numerous and effective were syrphid flies and halictid bees and, in the case of *S. reflexa,* bombyliid flies (figs. 12.1 and 12.2; see McGuire and Armbruster 1991). About half the insects that were observed pollinating the flora of these communities visited both *Saxifraga* species.

Saxifraga floral morphology is fairly uniform throughout the genus; there is relatively little variation in flower size, nectar quantity, or stamen number. Thus, one can infer similarly limited variation in pollination ecology, as confirmed by the preceding observations (but see Thompson 1994). Thus, mapping pollination ecology onto published phylogenies of *Saxifraga* (e.g., Soltis et al. 1996; Mort and Soltis 1999) would probably reveal very little evolution of pollination ecology; most lineages appear to have remained generalized in their pollination. This apparently limited evolutionary change in pollination ecology could be the result of pollination being so generalized, with evolution and speciation occurring in response to selection on vegetative traits, or because the morphological and developmental constraints preclude specialization and diversification (see Armbruster et al. 1994; Waser 1998).

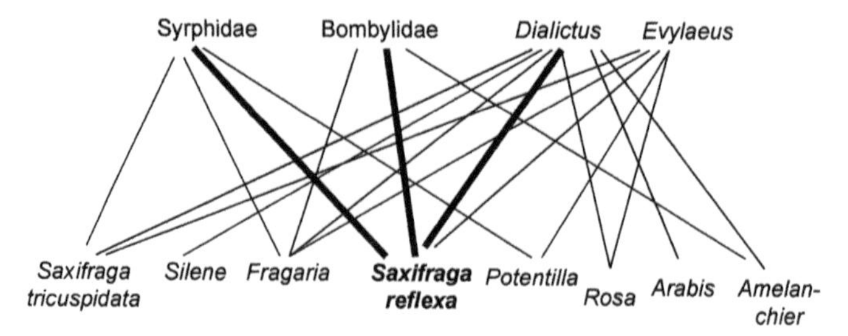

Figure 12.2 Partial visitation web for interior Alaskan steppe during the blooming period of *Saxifraga reflexa,* showing a medium degree of generalization from both plant and visitor perspectives. The only sympatric, synchronic flower visitors *not* seen on *S. reflexa* were Coleoptera, Lepidoptera, Sphecidae, *Hylaeus, Andrena, Megachile, Hoplitis, Osmia, Nomada,* and *Bombus.* Line width indicates frequency of visits.

Collinsia and *Tonella*

Collinsia and *Tonella* are closely related sister genera; they are characterized by nearly open to tubular flowers with bilateral symmetry, four stamens, and a thin style with a terminal stigma (see Kalisz et al. 1999). Nectar is secreted from the base of the corolla tube by a highly reduced staminode. The corolla of *Collinsia* is tubular, with a pea flower–like configuration of lobes: an erect banner, a folded keel enclosing the stamens and pistil, and a pair of wings forming a landing platform. *Collinsia* flowers are visited and pollinated primarily by long-tongued, nectar-feeding bees (including *Osmia* [Megachilidae], *Anthophora,* and *Bombus* [Apidae]), which also often collect pollen. The flowers of *Tonella* are essentially open and are probably visited by flies as well as a variety of small bees.

It appears that *Collinsia* is more specialized in pollination than *Tonella;* the former is pollinated only by long-tongued bees (Apidae, s.l., Megachilidae). However, this may involve quite a number of species, although they are all similar in their functional interactions with the flowers and the selective pressures they generate (see Armbruster et al. 2000). Although Lepidoptera and bombyliid flies can sometimes reach *Collinsia* nectar, they do not depress the keel and, hence, never pollinate. From a community perspective, only a small proportion of the diverse flower-visiting fauna of the western United States is attracted to, and pollinates, *Collinsia* flowers.

Phylogenetic studies have shown *Tonella* and *Collinsia* to be sister taxa; together they form a distinct monophyletic group in the Chelonieae, s.l. (Wolfe et al. 1997; Armbruster et al. 2002). I used the length of the floral tube (less than 0.5 mm in *Tonella,* 2–10 mm in *Collinsia*) as a proxy for specialization in pollination ecology under the assumption that more open flowers with short or nonexistent tubes would be pollinated by a wider variety of insects than flowers with long tubes that restrict nectar access to only those insects, in this case bees, with

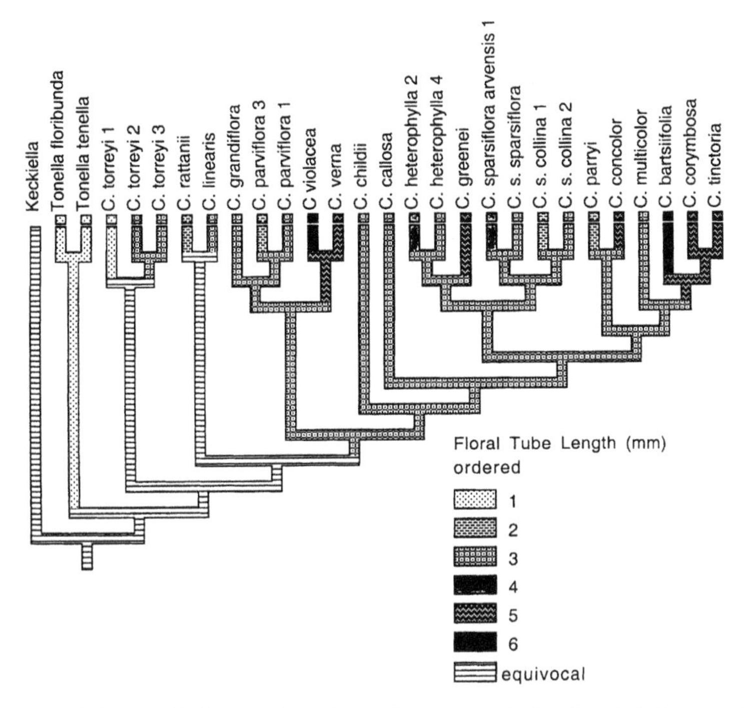

Figure 12.3 Phylogeny of Collinsieae (Plantaginaceae), showing evolution of specialization as measured by the length of the floral tube.

sufficiently long proboscides. Using linear parsimony I mapped nectar-tube length onto the estimated phylogeny of the species in these two genera (fig. 12.3). The same basic results were obtained using minimum squared change and maximum likelihood optimizations. Although the exact number of transitions between short- and long-tubed corollas is not certain, it does appear that longer tubes and greater specialization have evolved independently from short-tubed ancestors several times. It also appears that specialized pollination has mostly evolved from more generalized pollination rather than vice versa (fig. 12.3).

Stylidiaceae

Whereas specialization in *Collinsia* pollination has largely evolved along the "who" axis, specialization in the Stylidiaceae has evolved primarily along the axis of "where" pollen is placed. There is considerable variation in the precision and complexity of pollen delivery and pickup within the family. Whereas *Forstera, Phyllachne,* and *Oreostylidium* have radially symmetrical flowers, *Levenhookia* and *Stylidium* have bilateral symmetry, which allows more consistent orientation of pollinators and placement of pollen on them. *Levenhookia* have active pollen placement on pollinators with a motile petal, and *Stylidium* have even more precise pollen placement with a motile gyneocial/staminal column (ini-

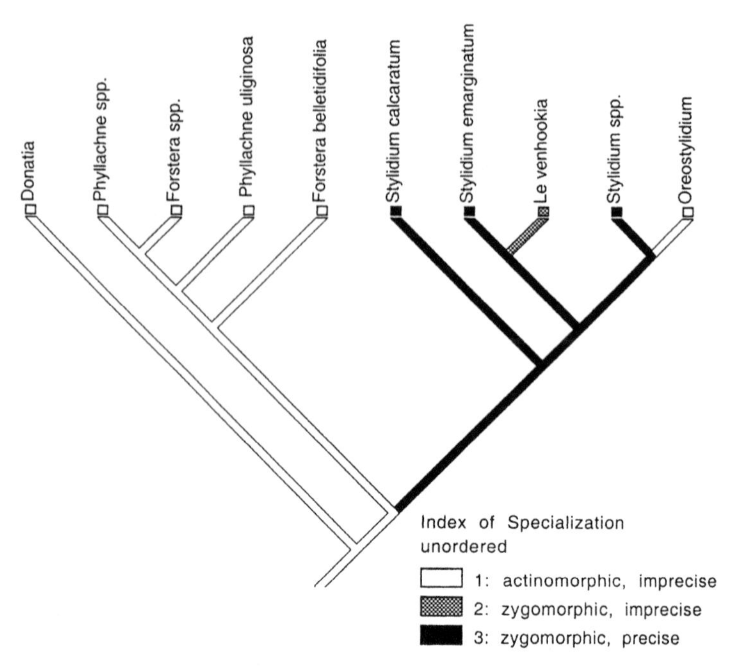

Figure 12.4 Phylogeny of the Stylidiaceae (maximum parsimony estimation based on sequence of the ITS region of the nuclear ribosomal DNA; based on Laurent et al. 1999; Wagstaff and Wege 2002), showing evolution of specialization as measured by the degree of precision in pollen placement on pollinators.

tially bearing pollen, and later the receptive stigma; see Erickson 1958; Armbruster et al. 1994; Darnowski 2002). The degree of precision in pollen placement on pollinators in *Stylidium* is almost unrivaled (except by some orchids) and is partly the result of the complete fusion of staminate and pistillate tissues into a single column (Armbruster et al. 2004). Because of this, the anthers and stigma of a flower contact pollinators in almost exactly the same location. Indeed, this precision and accuracy allows specialization of *Stylidium* populations on placing pollen in different locations on the bodies of shared pollinators and the associated reduction of the negative consequences of occurring sympatrically with similar, related species (Armbruster et al. 1994, 2004).

Phylogenetic analyses based on molecular or combined molecular/morphological data published by Laurent et al. (1999) and Wagstaff and Wege (2002) allowed me to consider this variation in specialization from a historical perspective. *Forstera-Phyllachne* turns out to be sister to the other three taxa mentioned earlier, whereas *Oreostylidium* and, possibly, *Levenhookia* are nested within *Stylidium* (fig. 12.4; Wagstaff and Wege 2002). An alternative reconstruction (not shown) indicates that *Levenhookia* is sister to *Stylidium* (Laurent et al. 1999; Wagstaff and Wege 2002). In either case, it appears that specialization has increased, with the origin of bilateral symmetry (zygomorphy) and motile pollen placement in the origin of *Stylidium* from a *Forstera*-like or *Levenhookia*-like ancestor. In at least one case, and possibly two, there have been reversals back to less-specialized pollination. In the unambiguous case of reversal in *Oreostylidium,* this is associated with dispersal to New Zealand.

Dalechampia

For a medium-sized genus, *Dalechampia* displays a surprising diversity of pollinator-reward systems. Some *Dalechampia* species and members of the sister genus *Tragia* are pollinated by pollen-collecting bees and/or other pollen-feeding insects. A few species of *Dalechampia* are "buzz" pollinated, wherein certain bees that can vibrate (buzz) their thoracic muscles at a high frequency can shake pollen out of nearly closed male flowers. Some *Dalechampia* species are pollinated by fragrance-collecting male euglossine bees (which apparently use the fragrances to impress female bees and increase their probability of mating). Most species of *Dalechampia,* however, are pollinated by resin-collecting female bees, which use resin as a building material in nest construction (Armbruster 1984, 1993).

Phylogenetic analyses suggest that pollination by resin-collecting bees originated via a shift from pollination by pollen-collecting insects that coincided with the evolutionary split between *Tragia* and *Dalechampia* (Armbruster 1993). This is a clear example of evolutionary specialization because pollination by pollen-feeding insects usually involves a variety of bees, flies, and often beetles. In contrast, relatively few species of bees collect resin.

Within the group of resin-producing species, there has been repeated evolution toward increasing or decreasing the amount of resin produced and, hence, the size of bee attracted (Armbruster 1993). Large resin glands offer large amounts of resin and attract both small and large bees; small resin glands attract only small bees (Armbruster 1984). Thus, we might expect evolution of smaller resin glands to reflect evolutionary specialization and that of larger glands to reflect generalization. However, concomitant with evolutionary changes in resin-gland size, changes have occurred in the distance between the gland and the stigmas and between the gland and the anthers. These distances determine the size of the visiting bees that deposit and pick up pollen, respectively (Armbruster 1988). These traits coevolve along an allometric axis of specialization (fig. 12.5), which presumably reflects the shape of the underlying adaptive surface (Armbruster 1990, 1991): low fitness in the upper left-hand corner is the result of low pollination rates (the only bees attracted are too small to contact the sexual parts) and low fitness in the lower right-hand corner is the result of interspecific pollination (large and small bees attracted, but one or the other is already being used by a sympatric species of *Dalechampia*). Thus, selection for specialization, if it is occurring, is probably driven by the ecology of species coexistence. There is some evidence that this axis of evolution is maintained by genetic correlations, at least over the short term (Hansen et al. 2003). This raises the possi-

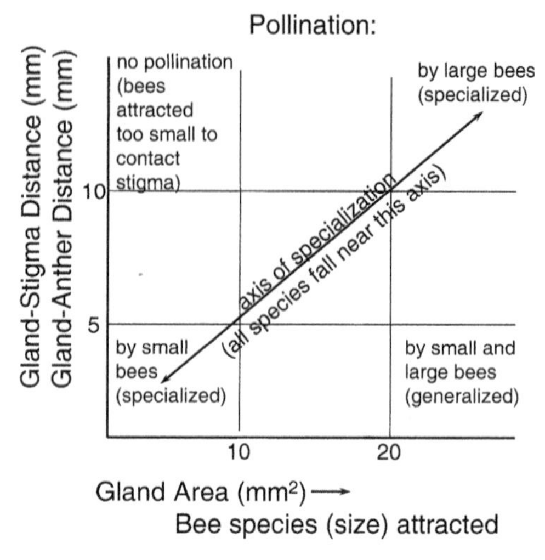

Figure 12.5 Axis of specialization in pollination for species of *Dalechampia* that produce resin as a pollinator reward. Most populations/species fall out close to the axis: the top left cell (low fitness) is completely empty and the bottom right cell (low specialization) contains only a few populations.

bility that specialization may sometimes reflect genetic constraints interacting with ecological trade-offs (cf. Aigner 2001).

Pollination by male euglossine bees has independently evolved from resin-reward ancestors at least three times (Armbruster 1993). Although these are major shifts in reward type and pollinator relationships, they represent little or no change in degree of specialization.

Pollination by pollen-collecting bees or other insects has independently evolved three times (Armbruster 1993; Armbruster and Baldwin 1998); each case probably represents a reversal to less specialization (from fewer species and functional groups of pollinators to more species and functional groups). In the only Paleotropical example, this shift is associated with dispersal from Africa to Madagascar (Armbruster and Baldwin 1998). Interestingly, there is a secondary reversal to more-specialized pollination in Madagascar, with the evolution of buzz pollination (fewer insect species involved) from generalized, pollen-reward blossoms (Armbruster and Baldwin 2003). Thus, in *Dalechampia,* there are several examples of evolutionary specialization (and reversals to generalization) along the "who pollinates" axis, which are determined by what rewards the pollinators collect.

Another line of specialization can be seen along the "when in the day" axis. This analysis is based on the assumption that blossoms that are open to pollination for shorter periods of the day are more specialized than those open all day or all day and all night. There appear to be four independent shifts to blossoms being open for shorter periods of time (fig. 12.6), but two of these involve nocturnal closure, which is more likely related to the exclusion of nocturnal flori-

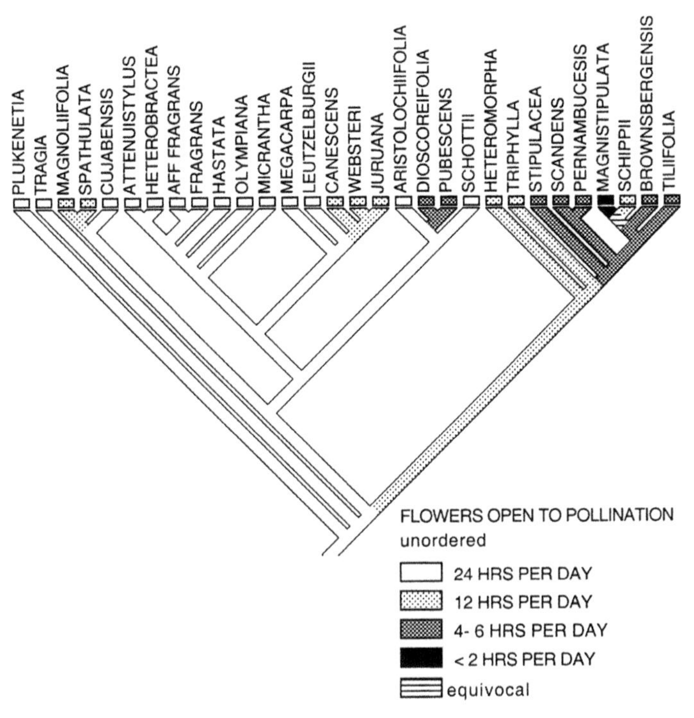

Figure 12.6 Phylogeny of *Dalechampia,* showing evolution of specialization as measured by the amount of time blossoms are open each day. More restricted opening allows access by fewer species of bees and is thus assumed to be more specialized.

vores rather than nocturnal pollinators (Armbruster 1997). This still leaves two independent origins of restricted opening, with one lineage showing a progression from being open all day, to less than six hours, to less than two hours (fig. 12.6). There is one apparent reversal from more specialized to less (*D. schippii*).

Discussion

The four study systems vary in their overall degree of specialization in pollination ecology: *Saxifraga* species are largely generalized, *Collinsia* species are somewhat more specialized, and *Stylidium* and *Dalechampia* species, for the most part, are considerably more specialized. Interestingly, specialization in *Stylidium* and *Dalechampia* has proceeded along very different axes. *Dalechampia* species have specialized along the axis of what reward is offered and which visitor species pollinate the flowers ("who"). Specialization has also occurred in terms of what time of day the flowers are available for pollination ("when"). In contrast, *Stylidium* species have specialized most spectacularly along the axis of where pollen is placed on pollinators ("where").

Comparison of these sets of natural history data, and other sets drawn from published and unpublished observations, allows us to address several general questions about the evolution of specialization in pollination ecology. What factors promote or inhibit specialization? Is specialization reversible? What is the relationship between specialization in pollination and speciation rates or species richness? Are there any biogeographic patterns in specialization of pollination relationships?

Factors Promoting Specialization

It has been repeatedly suggested that competition for pollination (the negative effects of sharing pollinator service and of sending pollen to heterospecific flowers and receiving heterospecific pollen) is a potent ecological and evolutionary force (see review by Waser 1983). Thus, selection generated by co-occurring congeners that share pollinators may lead to specialization in pollination ecology (Armbruster et al. 1994). This specialization can occur along one or more of the four axes described earlier. If accuracy and precision in pollen placement are high, populations can specialize by placing pollen in different locations on the same individual pollinators. Specialization along this axis is precluded in species with low precision, and specialization must take the form of utilizing different pollinator species or being pollinated in different seasons or times of day (see Armbruster et al. 2004).

If competition for pollination selects for specialization, we would expect plant species that regularly co-occur with similar congeners to be therefore more specialized. In all four study systems described earlier, species commonly co-occur with one to several sympatric congeners. Some degree of specialization has evolved in *Collinsia, Stylidium,* and *Dalechampia,* with apparently reduced overlap of "pollination niches." Yet this is notably absent in *Saxifraga,* which regularly occur sympatrically with one to four congeners. Lack of specialization in the face of interspecific pollination (see McGuire and Armbruster 1991) may reflect broad genetic constraints seen across most of the family (see following discussion). However, other interpretations are also possible: for example, selection for generalization may be strong in this group because of chronic pollinator limitation (Waser et al. 1996). Indeed the genus is largely Arctic/alpine in distribution; these are regions where pollinator shortages are probably common. Also, it should be noted that some sympatric species of *Saxifraga* have different flowering seasons, perhaps representing character displacement along the "when" axis (McGuire and Armbruster 1991).

In the case of *Dalechampia* and *Collinsia,* specialization and coexistence of congeners is often associated with blooming in different seasons (*Collinsia*) or blossoms opening at different times of day (*Dalechampia*), combined with utilizing pollinator species of different body size (e.g., Armbruster 1985, 1988). The latter is manifested as selection for blossoms of different sizes but similar shapes (scaling along a size axis). In contrast, *Stylidium*—with its extreme precision in pollen placement and pickup—has specialized, perhaps in response to sympatry with congeners, by placing pollen in different locations on the same insects. This is manifested through the evolution of new flower shapes without major changes in flower or pollinator sizes (Armbruster et al. 1994, 2004). Genetic/developmental constraints may limit evolution away from the axis of allometric scaling within a *Stylidium* species, but off-axis evolution has occurred repeatedly in the divergence of *Stylidium* species (Armbruster et al. 1994, 2004).

Differences in the nature and degree of specialization may also be influenced by the extent of developmental control and morphological precision, which is reflected in the degree of floral "integration" (Armbruster et al. 2004). *Saxifraga* has the least-integrated flowers: flowers have radial symmetry, they do not orient the pollinators along an axis, and all floral parts (except the ovaries) are unfused and developmentally/morphologically independent; there is no adnation (fusion among floral whorls) or connation (fusion among parts within a whorl). Because this combination of traits characterizes the whole family, it seems reasonable to think of it as a constraint on specialization in this group. *Collinsia, Stylidium,* and *Dalechampia* are, in contrast, bilaterally symmetrical and orient their pollinators along a single axis such that pollinators contact the floral parts with consistent parts of their bodies (Armbruster 1988; Armbruster et al. 1994, 2002, 2004). *Collinsia* and *Stylidium* have varying degrees of fusion: in *Collinsia,* petals are connate (fused to each other) and stamens are adnate (fused) to the corolla; in *Stylidium,* sepals and petals are adnate and connate at the base, and the stamens and pistil are fused into a motile column. The column acts with extreme precision in pollen placement and pickup (Armbruster et al. 1994, 2004). Thus, evolutionary specialization in pollination ecology is easier to imagine in these two groups. Specialization in *Dalechampia* is not related to developmental integration and fusion but rather to the chemical evolution of the pollinator rewards. Resin itself is a restrictive reward (relatively few genera of bees collect resin), and recurrent shifts to fragrance rewards (Armbruster et al. 1992; Armbruster 1993) may also contribute to specialization and species coexistence (Armbruster 1986).

Specialization in Pollination Ecology Appears to Be Evolutionarily Labile and Reversible

One recurrent theme that emerges from the evolutionary reconstructions presented here (figs. 12.3–12.6) is that specialization is evolutionarily labile and prone to both parallelism and reversal. If we accept floral-tube length as a measure of specialization in *Collinsia,* we observe about 10 independent increases in specialization but no unambiguous examples of reversal to more generalized pollination (fig. 12.3). If we accept precision of pollen placement and pickup as a measure of specialization, in the Stylidiaceae we observe one or two origins of increasing specialization and one or two reversals to less specialized pollination (fig. 12.4). In *Dalechampia,* we see evolutionary lability in the types of pollinator rewards and several independent reversals from specialized (resin reward) to less specialized (pollen reward; Armbruster 1993; Armbruster and Baldwin 1998). There is also at least one secondary increase in specialization (origin of buzz pollination in Madagascar; Armbruster and Baldwin 2003). Measuring *Dalechampia* specialization by how long the blossoms are open for pollination each day, we see five to six independent increases in specialization and one apparent reversal (fig. 12.6). Thus, specialization appears to be evolutionarily labile and reversible, at least after specialization has initially evolved.

Specialization, Species Richness, and Speciation

Is specialization in pollination ecology a key innovation that leads to clade success? The seminal work of Scott Hodges (1997) on *Aquilegia* and other groups certainly suggests that this is the case. He found that clades with long floral tubes (presumably more specialized in pollination ecology) were generally more species rich than clades without floral tubes. There is also a general association between animal pollination and clade species richness (Dodd et al. 1999). These and other authors (e.g., Johnson and Steiner 2000) have suggested that such associations are the result of higher speciation rates in clades with more specialized pollination. At the risk of repetition (Armbruster 1993), however, I should point out that species richness of clades is affected by extinction rates as well as by speciation rates; hence, specialized pollination might affect clade success by reducing extinction rates. For example, pollination-niche width and species packing might affect the probability of lineage survival (see Waser 1983; Armbruster et al. 1994).

If we address this question with the *Dalechampia* study system, however, we see no apparent association between species richness of the clade and degree of specialization in pollination ecology. *Dalechampia* (120 species), which has mostly specialized pollination, is no more species rich than its sister genus *Tragia* (150 species), which apparently has unspecialized pollination (open pollen reward). Comparing resin-reward clades (specialized) with sister pollen-reward clades (less specialized) within *Dalechampia,* we again fail to find the expected differences (see, e.g., trees presented by Armbruster 1993; Armbruster and Baldwin 1998): resin-reward clade in Africa (specialized: 2–8 species) versus pollen-reward sister clade in Madagascar (generalized: 10 species); *D. shankii* (pollen reward) versus *D. osana* (resin reward); *D. liesneri* (pollen reward) versus one or two sister species (specialized fragrance reward).

Data from *Dalechampia* support the idea that pollination systems sometimes influence extinction rates as much as they influence speciation rates. Male euglossine bee pollination (fragrance reward) almost certainly increases speciation rates due to the specificity of bee response to small changes in fragrance composition (Dressler 1968). In contrast, resin rewards are unlikely to promote speciation, because the traits that influences which bees are attracted vary quantitatively, and the bees respond probabilistically (Armbruster 1988, 1993). If speciation rate were the main driver of clade species richness, we would expect fragrance-reward clades to be much more species rich than sister resin-reward clades; however, this is not observed (see Armbruster 1993). There are three comparisons possible, and, in all cases, the species richness of the paired clades (with contrasting reward types) is similar (two vs. one; five vs. five; one vs. one). This suggests that any differences in speciation rates are masked by differential extinction rates. Indeed, the high attractiveness of resin rewards (Armbruster 1984) suggests that resin-reward clades may have lower rates of extinction (Armbruster 1993).

Comparison of species richness in the two sister clades in Collinsieae supports the hypothesis that specialization is related to speciation and/or extinction rates: *Tonella,* with only two species, is relatively less specialized in floral morphology and, presumably, pollination ecology. *Collinsia,* with some 20 species, is significantly more specialized in floral morphology, with pollination accomplished primarily by long-tongued bees.

Patterns of species richness in the Stylidiaceae also support the idea that specialization is related to speciation and/or extinction rates. Whereas *Stylidium,* with more specialized pollination ecology, contains some 220 species, its sister clade *Forstera/Phyllachne* (according to the ITS tree), with more generalized pollination, contains only 9 species. The alternative *rbc*L tree shows *Levenhookia* as sister to *Stylidium; Levenhookia* also has less specialized pollination than *Stylidium* and contains only 10 species. Some insights into the process of species accumulation in clades with specialized pollination may be gained from examining the evolution of pollination ecology in *Stylidium:* the precision and diversity of pollen placement and pickup mechanisms (see Erickson 1958; Armbruster et al. 1994; Darnowski 2002) seems particularly important in its diversification. Phylogenetic studies suggest that the position of pollen placement on pollinators has a dynamic evolutionary history, with repeated switches between dorsal, lateral, and ventral placement on pollinators (Wege 1999), as well as much diversification in longitudinal placement position (e.g., head, thorax, abdomen) within and among species (Armbruster et al. 1994). According to traditional models (e.g., Grant 1963), this evolutionary dynamic might be interpreted as the result of speciation being promoted by precise placement and specialized pollination. However, the degree of genetic isolation needed to promote speciation is probably greater than would be achieved even by precise pollen placement. In the absence of strong selection, homogenizing gene flow would probably overwhelm divergence (Armbruster et al. 1994; Waser 1998); however, a speciation model recently developed by Doebeli and Dieckmann (2003) suggests that the dynamics of gene flow and speciation may be more complex than assumed here (and this deserves scrutiny in the future). It seems possible that, instead of driving speciation, specialization in pollen placement decreases extinction rates by allowing tighter species to pack into pollination-niche space. In southern Western Australia, four to five species of *Stylidium* commonly co-occur on a small spatial scale and co-flower while keeping pollen flow reasonably (but not perfectly) segregated. This pattern appears to be the result of local evolution (character displacement) promoting species coexistence and, hence, potentially reducing extinction rates (Armbruster et al. 1994). Thus, the association between clade species richness and specialization in pollination might be due to low extinction rates rather than high speciation rates.

Another way that the association between species richness of clades and specialized pollination may come about is that species richness evolves for other reasons, sometimes leading to communities that contain many congeneric species. Co-occurrence of numerous congeners then selects for complementarity of pollination traits (character displacement). Some groups respond to this selection with increased specialization, whereas others are constrained by floral morphology and development and fail to respond. To summarize, although it seems likely that there is indeed an association between specialized pollination and species diversity, we are still a long way from understanding what processes generate this association.

Biogeographic Trends

Latitude

Broad biogeographic patterns of specialization in pollination have only been examined to a limited extent, primarily in the context of attempting to resolve disagreements about the commonness of specialization (e.g., Johnson and Steiner 2000, 2003; see also Olesen and Jordano 2002; Ollerton and Cranmer 2002; Ollerton et al., chap. 13 in this volume). It would be unrealistic to infer biogeographic trends with any confidence from the sample of plant genera presented here. Although together they include nearly 3000 species, they are drawn from only five plant families. Nevertheless, it is interesting that the species occurring at high latitudes/altitudes (*Saxifraga*) or in cool temperate regions (*Oreostylidium, Donatia, Forstera, Phyllachne*) have the most generalized pollination systems, whereas the warm-temperate species (*Collinsia, Stylidium*) have moderately to quite specialized pollination relationships. *Dalechampia* is perhaps the most specialized overall (with several interesting exceptions), and it is restricted to the tropics. Thus, the data presented here are consistent with (but do not rigorously test) a trend of decreasing specialization with increasing latitude.

To this small and idiosyncratic dataset, I will add additional impressions drawn from fieldwork in Svalbard, Norway, Alaska, California, Africa, and Central and South America. These haphazard observations probably should not be used for statistical analyses of biogeographic trends, such as those conducted by Olesen and Jordano (2002) and Ollerton and Cranmer (2002). However, I present them here because these observations have the advantage of being based on

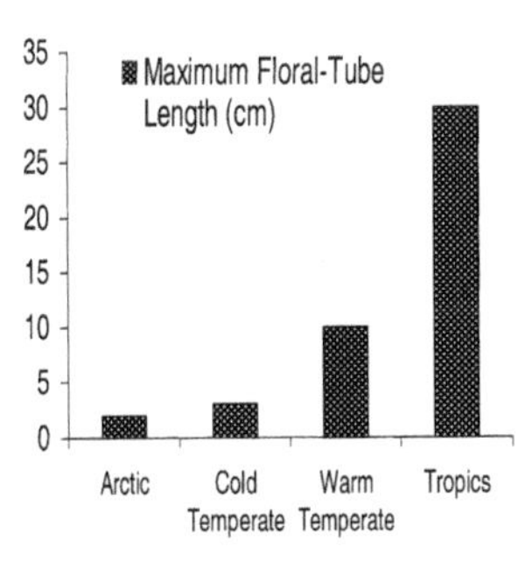

Figure 12.7 Latitudinal variation in maximum floral-tube length, assessed from haphazard samples of plants in Svalbard, interior Alaska, California, and French Guiana.

similar field studies by one investigator rather than heterogeneous studies by many different investigators. Thus emboldened, I will argue that there are indeed latitudinal trends in specialization but that the clearest trend is in variances rather than means.

First, consider some latitudinal trends in the opportunities for specialization. One thing that is clear from casual observation (or the literature) is that plants in Svalbard and northern Alaska do not have the opportunity to specialize on bird or bat pollinators because of the complete absence of these groups—even bumblebees are absent from Svalbard. Therefore, as the number of functional groups of pollinators declines with latitude, the opportunity for evolutionary specialization also declines. A similar trend is apparent in the latitudinal distribution of rewards. Nectar and pollen are ubiquitous rewards, but brood sites and oil are restricted to temperate and tropical regions. Fragrance and resin rewards are also restricted to tropical regions.

One major way in which flowers become more specialized in their relationships with pollinators is to restrict access to rewards by offering them from the base of long floral tubes or spurs. Only a small subset of floral visitors can access nectar, pollen, or oil hidden deeply in floral tubes. In the case of extremely long spurs or tubes, the access may be restricted to only one or a few species of moths with extremely long proboscides (Nilsson 1988). My scattered observations in the field and perusal of local floras (Munz 1959; Hultén 1968; Croat 1978; Lid 1985; Rønning 1996; Mori et al. 1997, 2002) suggest a distinct trend in maximum lengths of floral tubes. Although I would not hazard a guess about the trend in means, it is clear that the upper limit (and hence variances) on floral-tube length is higher in the tropics, at least on this transect, ranging from 2 cm nectar tubes in *Pedicularis* (Scrophulariaceae) on Spitsbergen and in Alaska to 30 cm nectar tubes in *Tocoyena longiflora* (Rubiaceae) in French Guiana (fig. 12.7; note that Darwin's *Angraecum* orchids with 40 cm nectar spurs are also, of course, from the tropics).

I believe this conclusion about the biogeography of floral tubes can be applied more broadly to latitudinal trends in specialization. We should be thinking more about *variances* in specialization than about *means*. I predict that with more extensive research we will find that (1) extremes of specialization are pretty uncommon nearly everywhere (except in the polar regions, where they are probably absent altogether) and (2) greater specialization exists at lower latitudes than higher ones, but this creates a stronger latitudinal trend in variances than in means.

Generalized Pollination on Islands?

A second biogeographic trend may be discerned from this small set of natural history stories. Pollination on islands is more generalized than that on the mainland (see Olesen et al. 2002), and plants with specialized pollination may often shift to generalized pollination after colonizing oceanic islands (see Baker 1965). Extreme simplicity and generalization of pollination on Spitsbergen (Svalbard) may be as much because it is an oceanic island as because it is in the high Arctic (even Ellesmere has bumblebee and lepidopteran pollinators; Kevan 1972). One of the three reversions to more generalized pollination by pollen-feeding insects in *Dalechampia* (and the most successful phylogenetically) was associated with colonization of Madagascar from mainland Africa (see Armbruster and Baldwin 1998). Finally, colonization of New Zealand by *Stylidium* from Australia was associated with a major shift in floral morphology and pollination ecology, involving reversion to an unspecialized form, perhaps through paedomorphosis (Laurent et al. 1999; Wagstaff and Wege 2002).

Conclusions

The impression I have gained from several decades of fieldwork, and in assembling these data, is that specialized features in pollination are actually moderately common (but hardly the rule) outside environments where they are precluded by harsh conditions and large year-to-year variation in growing-season climate and pollinator availability. These specialized features are the result of adaptation to certain pollinators, in most cases, through a process we can call evolutionary specialization. The results are not usually extreme ecological specialization and pollination by only a few species of animals, although this certainly is occasionally seen in warm-temperate and tropical regions. As Waser et al. (1996) have demonstrated, specialization often has large costs in terms of departure from the "global fitness optimum." The course of evolution is, nevertheless, influenced by the local rather than global topography of the governing fitness surface. Populations cannot generally jump to higher adaptive peaks (e.g., using more pollinator species) even though they sit (indeed are trapped) on relatively low local peaks. Specialized pollination relationships may often reflect this situation: an ecological relationship that does not result in the theoretically maximal fitness may, as a by-product, allow tighter species packing, coexistence of more related species, and potentially lower extinction and/or higher speciation rates.

In this contribution, I have undoubtedly raised more questions than I have answered. Nevertheless, I hope this collection of scattered observations provides enough evidence to stimulate further investigation into the microevolutionary dynamics, macroevolutionary patterns, and ecological consequences of specialization and generalization in pollination.

Acknowledgments

I thank numerous students, postdocs, and assistants for their help in the field, greenhouse, and laboratory; Bruce Baldwin for his collaboration in reconstructing the phylogenies of *Dalechampia* and *Collinsia;* Charlie Fenster, Paul Wilson, and James Thomson for numerous insightful discussions of specialization; and Nick Waser, Jeff Ollerton, and several anonymous reviewers for helpful comments on drafts of the manuscript. This research was supported by the U.S. National Science Foundation and the Norwegian Research Council.

References

Aigner, P. A. 2001. Optimality modeling and fitness trade-offs: When should plants become pollinator specialists? Oikos 95: 177–184.

Armbruster, W. S. 1984. The role of resin in angiosperm pollination: Ecological and chemical considerations. American Journal of Botany 71: 1149–1160.

Armbruster, W. S. 1985. Patterns of character divergence and the evolution of reproductive ecotypes of *Dalechampia scandens* (Euphorbiaceae). Evolution 39: 733–752.

Armbruster, W. S. 1986. Reproductive interactions between sympatric *Dalechampia* species: Are natural assemblages "random" or organized? Ecology 67: 522–533.

Armbruster, W. S. 1988. Multilevel comparative analysis of morphology, function, and evolution of *Dalechampia* blossoms. Ecology 69: 1746–1761.

Armbruster, W. S. 1990. Estimating and testing the shapes of adaptive surfaces: The morphology and pollination of *Dalechampia* blossoms. American Naturalist 135: 14–31.

Armbruster, W. S. 1991. Multilevel analyses of morphometric data from natural plant populations: Insights into ontogenetic, genetic, and selective correlations in *Dalechampia scandens*. Evolution 45: 1229–1244.

Armbruster, W. S. 1993. Evolution of plant pollination systems: Hypotheses and tests with the neotropical vine *Dalechampia*. Evolution 47: 1480–1505.

Armbruster, W. S. 1997. Exaptations link the evolution of plant-herbivore and plant-pollinator interactions: A phylogenetic inquiry. Ecology 78: 1661–1674.

Armbruster, W. S., and B. G. Baldwin. 1998. Switch from specialized to generalized pollination. Nature 394: 632.

Armbruster, W. S., and B. G. Baldwin. 2003. Pollination and evolution of euphorb vines in Madagascar. Pp. 391–393 *in* S. Goodman and J. Benstead (eds.), The natural history of Madagascar. University of Chicago Press, Chicago.

Armbruster, W. S., V. S. Di Stilio, J. D. Tuxill, T. C. Flores, and J. L. Velasquez Runk. 1999. Covariance and decoupling of floral and vegetative traits in nine neotropical plants: A reevaluation of Berg's correlation-pleiades concept. American Journal of Botany 86: 39–55.

Armbruster, W. S., M. E. Edwards, and E. M. Debevec. 1994. Character displacement generates assemblage structure of Western Australian triggerplants (*Stylidium*). Ecology 75: 315–329.

Armbruster, W. S., C. B. Fenster, and M. R. Dudash. 2000. Pollination "principles" revisited: Specialization, pollination syndromes, and the evolution of flowers. Det Norske Videnskaps—Akademi. I. Matematisk Naturvidenskapelige Klasse, Skrifter, Ny Serie 39: 179–200.

Armbruster, W. S., A. L. Herzig, and T. P. Clausen. 1992. Pollination of two sympatric species of *Dalechampia* (Euphorbiaceae) in Suriname by male euglossine bees. American Journal of Botany 79: 1374–1381.

Armbruster, W. S., and K. D. McCormick. 1990. Diel foraging patterns of male euglossine bees: Ecological causes and evolutionary response by plants. Biotropica 22: 160–171.

Armbruster, W. S., C. P. H. Mulder, B. G. Baldwin, S. Kalisz, B. Wessa, and H. Nute. 2002. Comparative analysis of late floral development and mating-system evolution in tribe Collinsieae (Scrophulariaceae, s.l.). American Journal of Botany 89: 37–49.

Armbruster, W. S., C. Pelabon, T. F. Hansen, and C. P. H. Mulder. 2004. Floral integration, modularity, and accuracy: Distinguishing complex adaptations from genetic constraints. Pp. 23–49 *in* M. Pigliucci and K. A. Preston (eds.), The evolutionary biology of complex phenotypes. Oxford University Press, Oxford.

Baker, H. G. 1965. Characteristics and modes of origin of weeds. Pp. 147–168 *in* H. G. Baker and G. L. Stebbins (eds.), The genetics of colonizing species. Academic Press, New York.

Berg, R. L. 1960. The ecological significance of correlation pleiades. Evolution 14: 171–180.

Bruneau, A. 1997. Evolution and homology of bird pollination syndromes in *Erythrina* (Leguminosae). American Journal of Botany 84: 54–71.

Campbell, N. A. 1987. Biology. Benjamin/Cummins, Menlo Park, CA.

Cope, E. D. 1896. The primary factors of organic evolution. Open Court Publishing, Chicago.

Croat, T. B. 1978. Flora of Barro Colorado Island. Stanford University Press, Stanford, CA.

Darnowski, D. W. 2002. Triggerplants. Rosenberg Publishing, Kenthurst, NSW, Australia.

Darwin, C. 1859. The origin of species. Republ. 1958, New American Library, New York.

Darwin, C. 1877a. The various contrivances by which orchids are fertilised by insects. Republ. 1984, University of Chicago Press, Chicago.

Darwin, C. 1877b. The different forms of flowers on plants of the same species. Republ. 1984, University of Chicago Press, Chicago.

Dodd, M. E., J. Silvertown, and M. W. Chase. 1999. Phylogenetic analysis of trait evolution and species diversity variation among angiosperm families. Evolution 53: 732–744.

Doebeli, M., and U. Dieckmann. 2003. Speciation along environmental gradients. Nature 421: 259–264.

Dressler, R. L. 1968. Pollination by euglossine bees. Evolution 22: 202–210.

Endress, P. K. 1994. Diversity and evolutionary biology of tropical flowers. Cambridge University Press, Cambridge.

Erickson, R. 1958. Triggerplants. Republ. 1981, University of Western Australia Press, Nedlands, Western Australia.

Faegri, K., and L. van der Pijl. 1971. The principles of pollination ecology, 2nd ed. Pergamon Press, Oxford.

Fisher, R. A. 1930. The genetical theory of natural selection. Clarendon Press, Oxford.

Futuyma, D. J. 1991. Evolution of host specificity in herbivorous insects: Genetic, ecological, and phylogenetic aspects. Pp. 431–454 *in* P. W. Price, T. M. Lewinsohn, G. W. Fernandes, and W. W. Benson (eds.), Plant-animal interaction: Evolutionary ecology in tropical and temperate regions. Wiley, New York.

Futuyma, D. J. 1997. Evolutionary biology, 3rd ed. Sinauer Associates, Sunderland, MA.

Futuyma, D. J., and G. Moreno. 1988. The evolution of ecological specialization. Annual Review of Ecology and Systematics 19: 207–233.

Goldblatt P., P. Bernhardt, and J. C. Manning. 2000. Adaptive radiation of pollination mechanisms in *Ixia* (Iridaceae: Crocoideae). Annals of the Missouri Botanical Garden 87: 564–577.

Grant, V. 1963. The origin of adaptations. Columbia University Press, New York.

Grant, V., and K. A. Grant. 1965. Flower pollination in the Phlox family. Columbia University Press, New York.

Hansen, T. F., W. S. Armbruster, M. L. Carlson, and C. Pelabon. 2003. Evolvability and genetic con-

straint in *Dalechampia scandens:* Genetic correlations and conditional evolvability. Journal of Experimental Zoology (MDE) 296B: 23–39.
Hodges, S. A. 1997. Floral nectar spurs and diversification. International Journal of Plant Sciences 158 (Supplement): 81–88.
Hultén, E. 1968. Flora of Alaska and neighboring territories. Stanford University Press, Stanford, CA.
Johnson, S. D. 1997. Insect pollination and floral mechanisms in South African species of *Satyrium* (Orchidaceae). Plant Systematics and Evolution 204: 195–206.
Johnson, S. D., H. P. Linder, and K. E. Steiner. 1998. Phylogeny and radiation of pollination systems in *Disa* (Orchidaceae). American Journal of Botany 85: 402–411.
Johnson, S. D., and K. E. Steiner. 2000. Generalization vs. specialization in plant pollination systems. Trends in Ecology and Evolution 15: 140–143.
Johnson, S. D., and K. E. Steiner. 2003. Specialized pollination systems in southern Africa. South African Journal of Science 99: 345–348.
Kalisz, S., D. Vogler, B. Fails, M. Finer, E. Shepard, T. Herman, and R. Gonzales. 1999. The mechanism of delayed selfing in *Collinsia verna* (Scrophulariaceae). American Journal of Botany 86: 1239–1247.
Keller, C. S., and W. S. Armbruster. 1989. Pollination of *Hyptis capitata* in Panama by eumenid wasps. Biotropica 21: 190–192.
Kevan, P. 1972. Insect pollination of high arctic flowers. Journal of Ecology 60: 831–867.
Kochmer, J. P., and S. N. Handel. 1986. Constraints and competition in the evolution of flowering phenology. Ecological Monographs 56: 303–325.
Lanyon, S. 1992. Interspecific brood parasitism in blackbirds (Icterinae): A phylogenetic perspective. Science 255: 77–79.
Laurent, N., B. Bremer, and K. Bremer. 1999. Phylogeny and generic relationships of the Stylidiaceae (Asterales), with a possible extreme case of paedomorphosis. Systematic Botany 23: 289–304.
Lid, J. 1985. Norsk, svensk, finsk flora, 2. oppgavet. Det Norske Samlaget, Oslo.
Lindberg, A. B., and J. M. Olesen. 2001. The fragility of extreme specialization: *Passiflora mixta* and its pollinating hummingbird *Ensifera ensifera.* Journal of Tropical Ecology 17: 323–329.
Luckow, M., and H. C. F. Hopkins. 1995. A cladistic analysis of *Parkia* (Leguminosae, Mimosoideae). American Journal of Botany 82: 1300–1320.
Macior, L. W. 1983. The pollination dynamics of sympatric species of *Pedicularis* (Scrophulariaceae). American Journal of Botany 70: 844–853.
McGuire, A. D., and W. S. Armbruster. 1991. An experimental test for reproductive interactions between two sequentially blooming *Saxifraga* species. American Journal of Botany 78: 214–219.
Mori, S., G. Creamers, C. Gracie, J.-J. de Granville, S. V. Heald, M. Hoff, and J. D. Michell (eds.). 1997, 2002. Vascular plants of central French Guiana, Parts 1 and 2. New York Botanical Garden, New York.
Mort, M. E., and D. E. Soltis. 1999. Phylogenetic relationships and the evolution of ovary position in *Saxifraga* section *Micranthes.* Systematic Botany 24: 139–147.
Munz, P. A. 1959. A California flora. University of California Press, Berkeley, CA.
Nilsson, L. A. 1988. The evolution of flowers with deep corolla tubes. Nature 334: 147–149.
Nilsson, L. A., L. Jonsson, L. Ralison, E. Randrianjohany. 1987. Angrecoid orchids and hawkmoths in central Madagascar: Specialized pollination systems and generalist foragers. Biotropica 19: 310–318.
Olesen, J. M., L. I. Eskildsen, and S. Venkatasamy. 2002. Invasion of pollination networks on oceanic islands: Importance of invader complexes and endemic super generalists. Diversity and Distribution 8: 181–192.
Olesen, J. M., and P. Jordano. 2002. Geographic patterns in plant-pollinator mutualistic networks. Ecology 83: 2416–2424.
Ollerton, J. 1996. Reconciling ecological processes with phylogenetic patterns: The apparent paradox of plant-pollinator systems. Journal of Ecology 84: 767–769
Ollerton, J., and L. Cranmer. 2002. Latitudinal trends in plant-pollinator interactions: Are tropical plants more specialised? Oikos 98: 340–350.

Petanidou, T., and W. N. Ellis. 1996. Interdependence of native bee faunas and floras in changing Mediterranean communities. Pp. 201–226 *in* A. Matheson, M. Buchmann, C. O'Toole, P. Westrich, and I. H. Williams (eds.), The conservation of bees. Linnean Society Symposium Series 18, Academic Press, London.

Rønning, O. 1996. The flora of Svalbard. Norwegian Polar Institute, Tromsø, Norway.

Soltis, D. E., R. K. Kuzoff, E. Conti, R. Gornall, and K. Ferguson. 1996. *mat*K and *rbc*L gene sequence data indicate that *Saxifraga* (Saxifragaceae) is polyphyletic. American Journal of Botany 83: 371–382.

Stebbins, G. L. 1974. Evolution above the species level. Belknap Press of Harvard University Press, Cambridge.

Steiner, K. E., and V. B. Whitehead. 1990. Pollinator adaptation to oil-secreting flowers: *Rediviva* and *Diascia*. Evolution 44: 1701–1707.

Steiner, K. E., and V. B. Whitehead. 1991a. Oil flowers and oil bees: Further evidence for pollinator adaptation. Evolution 45: 1493–1501.

Steiner, K. E., and V. B. Whitehead. 1991b. Resin collection and pollination of *Dalechampia capensis* by *Pachyanthidium cordatum* (Hymenoptera: Megachilidae) in South Africa. Journal of the Entomological Society of South Africa 54: 67–72.

Stiles, F. G. 1975. Coadapted competitors: The flowering seasons of hummingbird pollinated plants in a tropical forest. Science 198: 1177–1178.

Stone G. N., P. Willmer, and J. A. Rowe. 1998. Partitioning of pollinators during flowering in an African *Acacia* community. Ecology 79: 2808–2827.

Thompson, J. N. 1994. The coevolutionary process. University of Chicago Press, Chicago.

Wagstaff, S. J., and J. Wege. 2002. Patterns of diversification in New Zealand Stylidiaceae. American Journal of Botany 89: 865–874.

Waser, N. M. 1983. Competition for pollination and floral character differences among sympatric plant species: A review of evidence. Pp. 277–293 *in* C. E. Jones and R. J. Little (eds.), Handbook of experimental pollination biology. Van Nostrand Reinhold, New York.

Waser, N. M. 1998. Pollination, angiosperm speciation, and the nature of species boundaries. Oikos 82: 198–201.

Waser, N. M., L. Chittka, M. V. Price, N. M. Williams, and J. Ollerton. 1996. Generalization in pollination systems, and why it matters. Ecology 77: 1043–1060.

Wege, J. A. 1999. Morphological and anatomical variation within *Stylidium* (Stylidiaceae): A systematic perspective. PhD thesis, University of Western Australia, Nedlands, Australia.

Wolfe, A. D., W. J. Elisens, L. E. Watson, and C. W. dePamphilis. 1997. Using restriction-site variation of PCR-amplified cpDNA genes for phylogenetic analysis of tribe Cheloneae (Scrophulariaceae). American Journal of Botany 84: 555–564.

Wright, S. 1931. Evolution in Mendelian populations. Genetics 16: 97–159.

CHAPTER THIRTEEN

Geographical Variation in Diversity and Specificity of Pollination Systems

Jeff Ollerton, Steven D. Johnson, and Andrew B. Hingston

The biogeography of species interactions is a fledgling field in ecology and evolution. Given impetus mainly by Thompson's (1994) treatise on a geographical approach to coevolution, the field has progressed from studies of single species and their interactions to comparative studies of interactions at the community level (e.g., Olesen and Jordano 2002; Ollerton and Cranmer 2002; Bascompte et al. 2003). In this chapter, we pose several questions about how diversity and specificity of pollination systems vary geographically at the community and plant family level; we then search for answers from analyses of published and unpublished datasets. Although there have been many detailed studies of geographical variation in the pollination systems of individual plant species (e.g., Miller 1981; Armbruster 1985; Pellmyr 1986; Robertson and Wyatt 1990; Johnson 1997; Johnson and Steiner 1997; Galen 1999; Sánchez-Lafuente 2002), synthesis of these is beyond the scope of this chapter, because our focus is on broader-scale patterns.

Terrestrial plant communities vary enormously in their structural complexity and taxonomic diversity. At one end of this scale of variation are communities such as Arctic tundra, which contain few species of plants and are structurally very simple. At the other extreme are tropical rainforests with their high levels of alpha and beta diversity and their multilayered configuration of herbs, shrubs, lianas, trees, and epiphytes. In between are tropical and subtropical grasslands, which contain high levels of plant species richness yet are relatively simple in their structure, whereas the reverse can be true for temperate woodland.

Botanists and ecologists have often compared this variation in an attempt to discover general rules regarding the structure, diversity, and functioning of terrestrial communities. We know, for example, that Neotropical and Asian rainforests contain much higher levels of plant species richness than their African counterparts, but all are similar in their structural complexity (Whitmore 1998). In addition, there is a general trend toward increased diversity of species, genera, and families of plants as one moves from Arctic to temperate to tropical latitudes (e.g., Gaston and Williams 1996; Hillebrand 2004) and patterns of primary productivity also show similar predictable trends. This much is well documented, although the functional causes of these patterns remain elusive and have been the subject of much speculation (Rohde 1992; Hillebrand 2004).

One reason why global analyses of variation among communities have tended to focus on structure, species diversity, and patterns of primary productivity is that these data are relatively easy to obtain. On the other hand, global analyses of species interactions have seldom been undertaken because collecting the data requires time-consuming natural history observations. Unlike research on global patterns of primary productivity, such data cannot be collected remotely, although video cameras are being used more often to document rare interactions (e.g., Kay and Schemske 2003). And yet biologists have been documenting species interactions for the better part of two hundred years. This accumulated database is vast, though patchy in its coverage and usually scattered in its distribution. Careful synthesis of these data can result in useful compilations of information with which to test broad-scale hypotheses regarding the biodiversity of species interactions (Waser et al. 1996; Ollerton and Liede 1997; Olesen and Jordano 2002; Ollerton and Cranmer 2002; Bascompte et al. 2003; Johnson and Steiner 2003; Ollerton, chap. 18 in this volume).

Pollinators are a major niche axis for flowering plants, and, whereas niche partitioning and species packing have usually been approached from the perspective of plant physiology and growth form, differential use of pollinators can provide a set of circumstances by which species richness of plants can increase within a community (see Ollerton et al. 2003 and references therein). There are broadly three mechanisms which are relevant here: (1) increase the size of niche axes, that is, utilize pollen vectors that are not normally used; (2) decrease niche breadth, for example, by specializing on a narrower range of pollinator species; and, finally, (3) increase niche overlap such that pollinators are shared between taxa, perhaps facilitated by character displacement (Armbruster 1986, 1995, chap. 12 in this volume; Armbruster et al. 1994; Stone et al. 1998; Ollerton et al. 2003). In relation to the increase in plant species richness in the tropics, we may expect to find evidence that one or more of these mechanisms is important in structuring plant communities at different latitudes; we return to this question in the conclusions.

The past few years have witnessed some of the first attempts at broad-scale comparisons of terrestrial plant communities in terms of their levels of specificity in plant–pollinator interactions (Olesen and Jordano 2002; Ollerton and Cranmer 2002; Bascompte et al. 2003). A key element of this new research agenda is to understand how species interactions are influenced by the greater plant species diversity and primary productivity in tropical communities compared to that of other communities. For example, we may want to ask how this greater productivity moves between trophic levels, and whether species interactions are quantitatively different. Such questions are easy to frame but difficult to answer and require in-depth, detailed, long-term fieldwork (compared to, say, collection of biological specimens for taxonomic studies).

Biotic pollination is a ubiquitous class of species interaction that is important in almost all terrestrial communities, with the exception of those dominated by coniferous trees, wind-pollinated angiosperm trees, or grasses. Approximately 90% of all flowering plants are animal pollinated (Linder 1998) and this class of interaction was almost certainly an important factor that fueled the mid-Cretaceous explosive diversification of the angiosperms (Crepet 1983; Grimaldi 1999). This is despite the fact that plant–pollinator interactions are probably not a major direct link in the movement of primary productivity within most terrestrial communities. For example, in a semimature tropical forest in Panama, it was calculated that flower visitors consumed only 3.2% of above-ground net primary productivity (Roubik 1993). But, in addition to this, one must consider that the flower-visiting animals themselves form part of a food chain, as do the fruit and seeds resulting from successful pollination. To our knowledge, these links have never been fully quantified for any ecosystem, and the role of species interactions in ecosystem function remains a subject for which there are few data.

Thus it is important for us to understand the large-scale biogeography of such a vital biotic interaction. We have used existing and previously unpublished data on pollination systems to address the following questions:

1. Does the average number of pollination systems found in communities vary according to latitudinal zone, altitude, or habitat complexity?
2. Do communities of a given latitudinal zone, altitude, or habitat type show convergence in terms of the types of pollination systems that they support?
3. Does ecological specialization in plant pollination systems correlate significantly with latitude?
4. At a finer taxonomic level, do plant families vary between regions in their degree of ecological specialization in pollination systems?

Armbruster et al. (2000) discussed the multiple meanings of the word "specialization" in the context of pollination biology and the confusion that it has caused. In the following analyses and discussion, we have distinguished between *ecological specialization* and *functional specialization* (cf. Johnson and Steiner 2000; see Fenster et al. 2004). Thus a plant pollinated only by beetles is *functionally* specialized but may or may not be *ecologically* specialized at the level of pollinator species (it may be pollinated by one or many species of beetles). It also may or may not show specific adaptations to beetle pollination (scent, color, etc.), indicating *phenotypic specialization* (J. Ollerton et al., unpublished data).

Methods and Datasets

We have assembled a number of datasets with which to investigate the preceding questions. These datasets are drawn from the published literature and from our unpublished work; some of them have been previously published (Ollerton and Cranmer 2002; Johnson and Steiner 2003) but are presented here in summarized form to add to the completeness of this analysis.

Dataset 1 comprises 32 published community-level surveys of plant–pollinator interactions. We used this dataset to investigate geographical variation in the functional specialization of plant species. The plants in each survey were categorized into 16 exclusive, broad functional pollination systems (medium and large bee, small bee, euglossine bee, wasp, fig wasp, fly, beetle, butterfly, moth, hawkmoth, thrips, bird, bat, generalist insect, mixed insect/vertebrate, and wind) on the basis of their spectrum of pollinators, not on floral morphology (i.e., not using a pollination syndrome approach; e.g., Faegri and van der Pijl 1971). Only native plants and animals were recorded, and we only considered plant species that were observed for at least five days and were visited by at least five individuals or species of potential pollinators (or were wind pollinated). Species of plant were categorized as specialized on a particular group if more than 85% of visitors were from that group. Species of plants with no group comprising more than 85% of visitors were categorized as insect-pollinated generalists if all visitors were insects, or vertebrate/insect generalists if they were visited by insects and vertebrates. Bees were grouped as large (longer than 8 mm) or small species (shorter than 8 mm). If we were unsure of the size of a particular species of bee, we categorized the Andrenidae, Anthophoridae, Apidae, Megachilidae, and Mellitidae as large bees and the Halictidae and Colletidae as small bees.

For each survey, the proportion of plants in each broad class of pollination system was calculated, as well as the following statistics: number of plant species surveyed, number of broad pollination systems identified (see the caveats that follow about data quality), and the ratio of plant species to pollination systems. In addition, we categorized each study according to three criteria: latitudinal zone—*Arctic* (including subarctic; 90–65° latitude), *temperate* (64–40°), *subtropical* (39–30°), and *tropical* (29–0°); altitudinal zone—*low* (up to 300 m above sea level), *mid* (300–1900 m), or *high* (2000 m and above); and complexity of habitat structure—*simple* (grassland, tundra, etc.), *medium* (scrub and mixed grassland), and *complex* (forest and woodland). We recognize that other factors are likely to influence the range of pollination systems present in a community (e.g., successional status, anthropogenic disturbance), but these are harder to quantify and so are not included.

Categorizing plant species into broad pollination systems is fraught with difficulty and loses much of the subtlety of the distinction between, for example, different types of medium- and large-bee pollination (distinguished by variation in floral morphology and/or bee behavior) or hovering- versus perching-bird pollination. However, we feel that this approach ought to capture some of the variation between different communities, albeit at a coarse level.

The different surveys varied in their data quality, as might be expected. Three particular sources of variation were apparent: whether or not nocturnal surveys had been undertaken, whether only insects had been surveyed (ignoring birds and other vertebrates), and whether or not wind-pollinated plants had been included in the survey (many studies focused specifically on biotic interactions). We have taken account of this and have repeated our analyses using the following subsets of dataset 1: (a) only "complete" studies (surveys that included wind-pollinated plants, had some nocturnal sampling, and looked at vertebrates as well as insects); (b) the studies in subset a, minus wind pollination as a system and including those studies which did not survey wind-pollinated species; (c) all of the studies in subset b, minus nocturnal pollination systems and including those studies which did not undertake a nocturnal sample; and, finally, (d) all of the studies in subset c, minus vertebrate pollination systems and including those studies that ignored vertebrates in their survey. Although the sample sizes and exact number of pollination systems varied across these analyses, all were in broad agreement with regard to general outcomes. Therefore, we have included only the results from data subset c because it had the largest sample of surveys which included most pollination systems (though omitting wind pollination and the major nocturnal pollination systems, bat, moth and hawkmoth) and, therefore, was more statistically robust than data subsets a, b, or d. Communities with fewer than 10 species of plants were excluded from the final analysis. Results from the other analyses are available from the first author upon request.

Sampling effort has been shown to have a significant effect on observed levels of plant ecological specialization (see discussion that follows; Ollerton and Cranmer 2002). We suspected that the same might be true of counting numbers of discrete pollination systems (i.e., functional specialization). However, not all of the studies that we used in our analyses included a measure of sampling effort, nor could sampling effort be estimated from their methods. In addition, by removing pollination systems such as wind or nocturnal pollination from our analyses, we are effectively altering the results of those studies in relation to sampling effort. There does not seem to be an easy way to take account of this. However, we note that there is no correlation between sampling effort and number of observed pollination systems (Pearson's correlation: for all surveys, $r = 0.32$, $N = 19$, $P = .18$; for "complete" surveys [i.e., data subset a], $r = 0.20$, $N = 6$, $P = .71$; for data subset c, $r = 0.22$, $N = 17$, $P = .39$). It appears that, unlike the findings of

Ollerton and Cranmer (2002), sampling effort is not a significant factor affecting levels of variation between these surveys, at least at the level of recording discrete pollination systems.

Dataset 2 is also a set of community-level studies, including many used in dataset 1 (for survey details, see Ollerton and Cranmer 2002). We used this set of data to examine geographic variation in plant ecological specialization. For each community survey, the mean number of flower visitors per plant species was calculated and then analyzed in relation to the latitude at which the survey was performed and with regard to sampling effort.

Dataset 3 comprises species-level surveys of plant–pollinator interactions in the asclepiads (Apocynaceae subfamily Asclepiadoideae). This dataset was also used to investigate variation in plant ecological specialization with latitude, but this time at a more taxonomically constrained level compared to that of dataset 2. Once again, these surveys were used in the analyses of Ollerton and Cranmer (2002) and specific details may be found in their work.

Dataset 4 was also used to study variation in ecological specialization at a lower taxonomic level, contrasting an area of extremely high alpha and beta plant diversity (South Africa) with one of lower diversity (Europe and North America). The data are from published and unpublished studies of pollination systems in Orchidaceae ($N = 114$ species) and Apocynaceae subfamily Asclepiadoideae ($N = 44$). We calculated the number of species and orders of potential pollinators (pollen-carrying insects) recorded for each plant species, as well as the sampling effort in each study. The advantage of using these two particular groups of plants is that both disperse their pollen as discrete masses (pollinia) rather than as free pollen, making identification of those animals that actually serve as pollinators much more straightforward. Comparisons were made between the levels of ecological specialization for species native to North America and Europe with those species native to southern Africa (see Johnson and Steiner 2003). Sources of the data are given by Johnson and Steiner (2003) and Ollerton and Cranmer (2002).

Analysis

Data were analyzed using SPSS 8.0 for Windows (1997, SPSS, Inc., Chicago, IL) with the exception of variation among communities in dataset 1, which was explored using the ordination method semistrong hybrid multidimensional scaling (SSH MDS), within the computer program PATN (Belbin 1993). Similarities between communities in proportional composition of the various broad pollination systems (represented by their constituent plant species) were calculated using the Bray–Curtis index and were displayed as the distance between communities on the ordination plot. A Monte Carlo technique was used to determine the statistical significance of the correlations between each pollination system and the ordination of the communities, and the pollination systems were fitted as vectors on the ordination plot using the PCC module in PATN.

Results and Discussion

Using the four datasets just described, we have attempted to provide answers to the questions we outlined in the introduction. We take each question in turn.

Does the Average Number of Pollination Systems Found in Communities Vary According to Latitudinal Zone, Altitude, or Habitat Complexity?

There is a statistically significant difference in the mean number of pollination systems recorded for communities in different latitudinal zones (fig. 13.1A; table 13.1). However, this difference is accounted for only by an increased number of pollination systems in tropical communities; Arctic, temperate, and subtropical communities do not, on average, vary in the range of pollination systems that they support.

This step increase from the subtropics to the tropics is at odds with the linear increase in numbers of plant species and species richness of *most* pollinator groups with decreasing latitude (although see reason 4, below). We can suggest a number of non-mutually-exclusive reasons why this step increase should exist and below we evaluate the evidence for and against each reason.

1. There is in fact a step increase in plant diversity in tropical communities compared to those at higher latitudes. Thus, one should not expect a linear relationship between number of pollination systems and latitude: the greater-than-expected diversity of the tropics would lead to greater-than-predicted numbers of pollination systems. At the plant family level, there does appear to be a quite smooth, linear relationship between latitude and taxonomic richness, judging from the data presented by Williams et al. (1994). However, it is possible that at a lower taxonomic level (e.g., genus or species) the relationship is more complex and variable between different groups of plants, yielding a step function.

2. There are groups of pollinating animals in the tropics that do not exist in other zones, which leads to more opportunities for the evolution of functionally specialized systems. In our community comparison analysis (see discussion to follow), the only category of pollination system that involves animal taxa with a uniquely tropical distribution is that of euglossine bee pollination; the Euglossini is restricted to the Neotropics (Michener 1979). This in itself could not explain the step increase in tropical pollination systems, because it accounts for only one pollination system; the increase between the subtropics and the tropics involves an average of three or four systems.

3. The tropics are more climatically stable over the long term than other regions, which allows more types of pollination systems to persist. Ecological generalization in pollination systems is usually attributed to plants not evolving

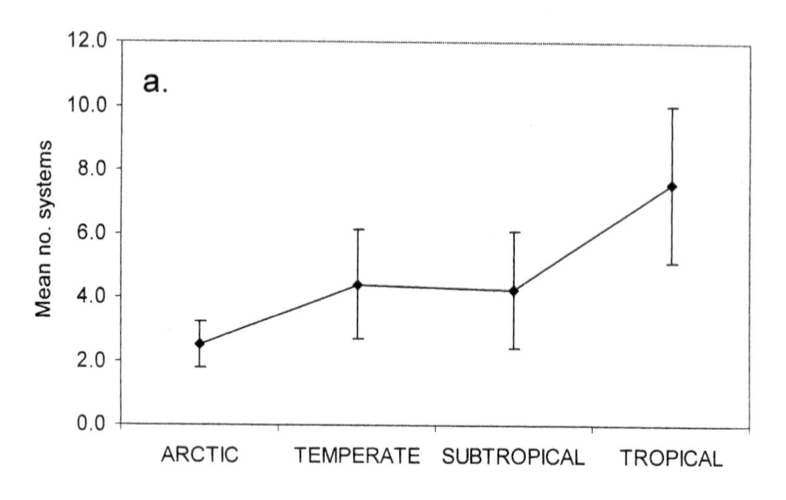

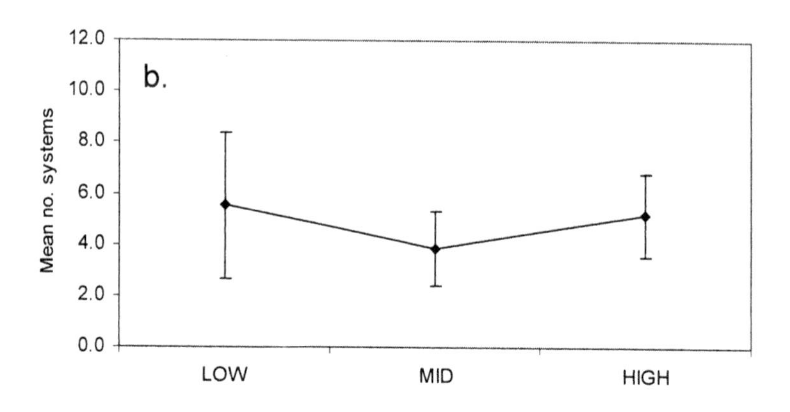

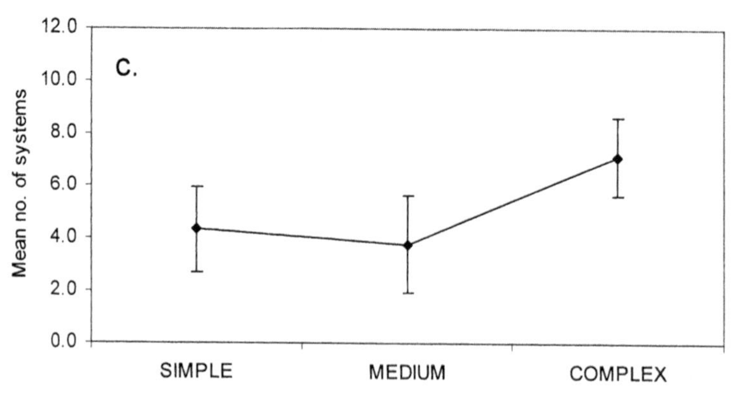

Figure 13.1 Mean number (± SD) of pollination systems per community in (A) the different latitudinal zones, (B) the different altitudinal zones, and (C) the habitats of different structural complexity; all results are from dataset 1.

Table 13.1 Results of univariate GLM analyses of variance that compare the effects of three factors on the number of pollination systems recorded per community in dataset 1

Factor	F	Overall significance	Post hoc results (Tamhane's T2 test, equal variances not assumed)
Latitudinal zone	$F_{3, 29} = 12.8$	$P < .0001$	Tropical > subtropical = temperate = arctic ($P < .005$)
Altitude zone	$F_{2, 29} = 0.78$	$P = .47$	Low = mid = high ($P > .05$)
Habitat structure	$F_{2, 29} = 9.1$	$P = .001$	Complex > medium = simple ($P < .025$)

specialization because of spatiotemporal variation in pollinator availability (Howe 1984; Herrera 1988, 1996; Pettersson 1991; Thompson 1994; Ollerton 1996; Waser et al. 1996; Gómez and Zamora 1999; Hingston and McQuillan 2000; Fenster and Dudash 2001; Fleming et al. 2001; Thompson and Cunningham 2002). If this is the major factor that influences the number of pollination systems per community, then it is probable that pollinators are more predictable in time and space in tropical latitudes than in other zones. Long-term temporal variation in the abundance of pollinators (associated with periods of climatic instability such as interglacial–glacial transitions) may therefore act to filter out more of the specialized pollination systems in the Arctic to subtropical zones compared to those in the tropics, resulting in more types of pollination systems in the tropics.

4. Bees are the dominant pollinators in most communities (Proctor et al. 1996) and their maximum diversity exists outside of the tropics: the highest levels of bee species richness are in areas with dry, mediterranean climates (Michener 1979; fig. 13.2A). This pattern is also striking when the relative numbers of bee and plant species in a region are considered (fig. 13.2B). The lower bee diversity in the (particularly wet) tropics may therefore allow a wider range of pollination systems to evolve if, rather than competing for the services of a limited number of bees, floral mutations evolve which attract alternative pollinators. However, as appealing as this hypothesis is, an analysis of our data suggests that it may not be correct—there is no statistically significant difference in the mean percentage of bee-pollinated plants in communities across the four latitudinal zones (one-way analysis of variance of $\log_N + 1$ transformed data: $F_{3, 28} = 2.2$, $P = .108$).

We conclude that the disproportionately higher diversity of pollination systems in the tropics is a phenomenon which is easy to recognize, but difficult to adequately explain—as is generally true for tropical biological richness!

Communities at different altitudes do not vary significantly in mean number of pollination systems (fig. 13.1B; table 13.1). However, habitat complexity does have an effect on the mean number of pollination systems: complex habitats are significantly richer in pollination systems than simple or medium habitats (fig. 13.1C; table 13.1). A multifactorial analysis to examine interactions between these factors was impossible due to the small sample size, and consequent low statistical power, for some combinations of factors.

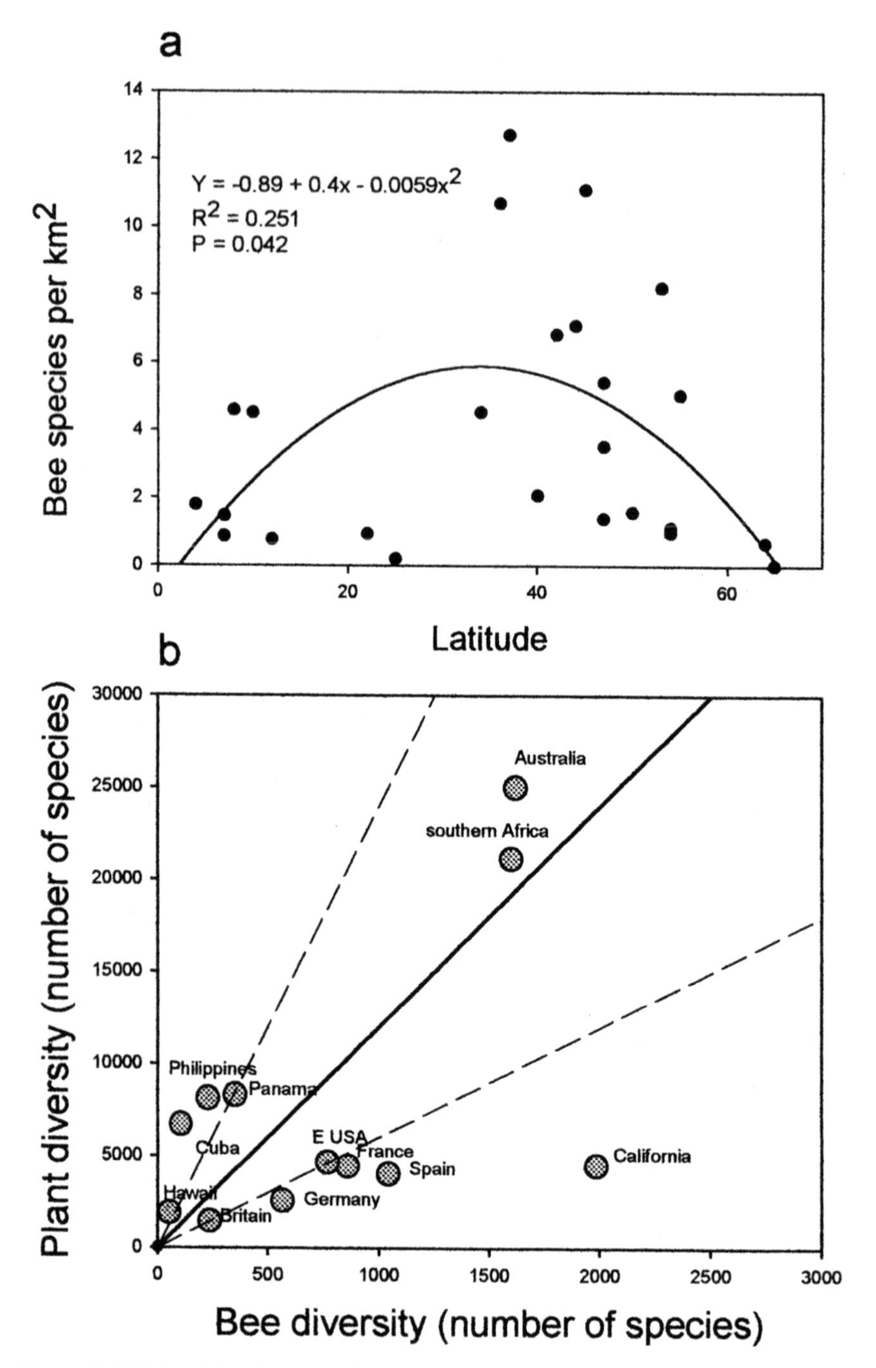

Figure 13.2 (A) The relationship between bee species richness (corrected for landmass area) and latitude in bees (Hymenoptera: Apoidea), based on data presented by Michener (1979). (B) The relationship between bee diversity and plant diversity in a number of regions across the globe, adapted from Johnson (2004). The solid line represents a plant-to-bee ratio of 10:1; the upper dashed line represents a ratio of 20:1, and the lower dashed line represents a ratio of 5:1.

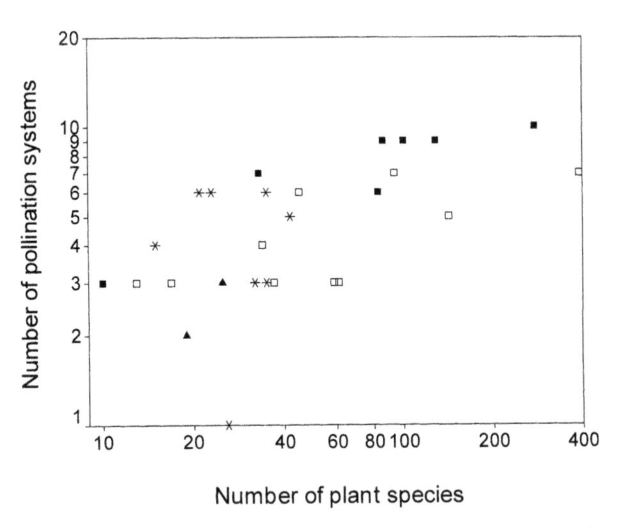

Figure 13.3 The relationship between the number of plants surveyed by a study in dataset 1 and the number of identified discrete pollination systems. Each community is coded by latitudinal zone as follows: tropical (■), subtropical (*), temperate (□), and arctic (▲). Note that *x* and *y* axes are log scale.

There is a strong correlation between the number of plants surveyed by a study and the number of discrete pollination systems identified (fig. 13.3; Pearson's correlation: $r = 0.56, N = 27, P = .002$). It appears from figure 13.3 that there is some structure to the relationship between number of plant species and number of pollination systems, particularly in relation to tropical communities, which appear to have a greater number of pollination systems for a given number of plant species. We have formally tested this relationship by looking at the ratio of pollination systems to plant species for different latitudinal zones, altitudinal zones, and levels of habitat complexity, calculated as the sum (unique pollination systems)/total plant species in a community. However, there is no statistically significant difference between the mean ratio of pollination systems to plant species across latitudinal zones (general factorial general linear model [GLM]: $F_{3,27} = 1.01, P = .38$; fig. 13.4A), altitudinal zones (general factorial GLM: $F_{2,27} = 1.01, P = .38$; fig. 13.4B), or levels of habitat complexity (general factorial GLM: $F_{2,27} = 0.81, P = .46$; fig. 13.4C).

In summary, the average number of pollination systems in a community varies significantly according to latitude (tropical communities possess larger numbers of pollination systems, on average, than subtropical, temperate, or Arctic communities) and habitat complexity (complex habitats host more pollination systems than medium or simple habitats). Altitude, however, does not affect the number of pollination systems. There is some nonindependence within this dataset in that all but one of the tropical surveys were from complex habitats (tropical forest communities) and all of the complex habitats were tropical.

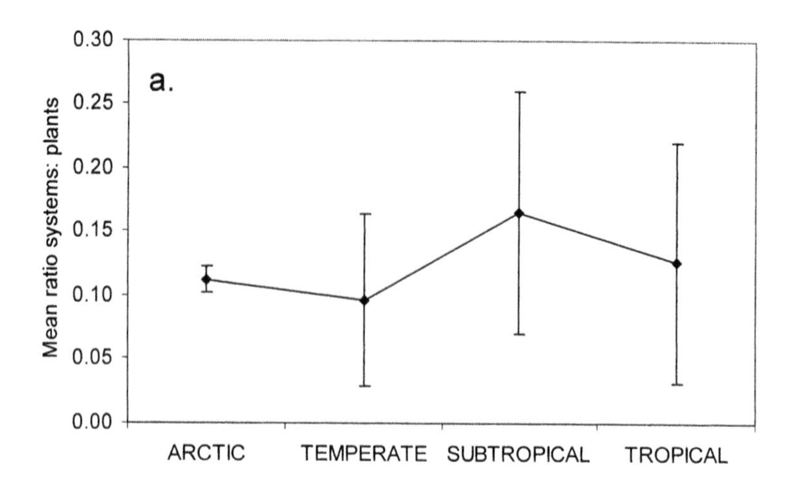

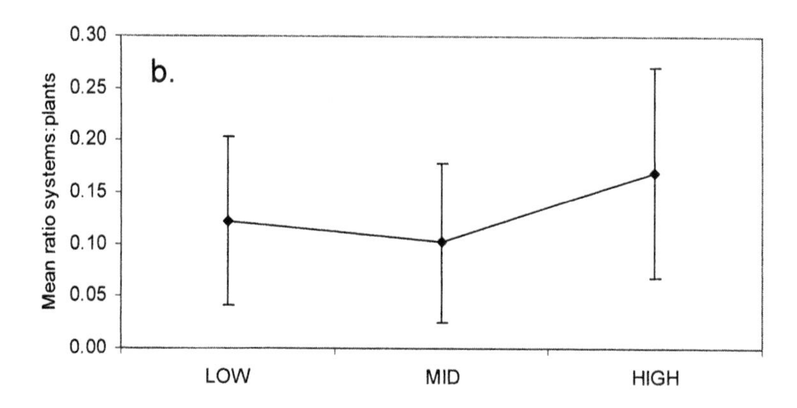

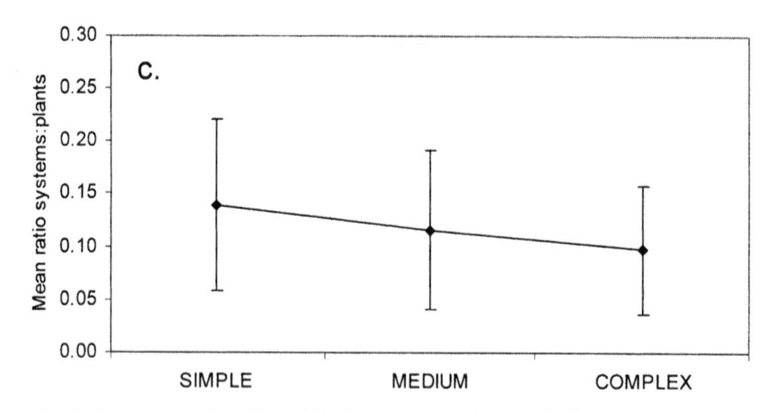

Figure 13.4 The mean ratio (± SD) of pollination systems to plant species for communities (A) in different latitudinal zones, (B) in different altitudinal zones, and (C) with different levels of habitat complexity; all results are from dataset 1.

Untangling the effect of habitat complexity from that of latitudinal zone is, therefore, impossible in this case.

Do Communities of a Given Latitudinal Zone, Altitude, or Habitat Type Show Convergence in Terms of the Types of Pollination Systems That They Support?

Ordination of the communities according to their proportional compositions of plants categorized into the various broad pollination systems also suggests that latitude and habitat complexity have greater effects on the distribution of pollination systems among communities than does altitude (fig. 13.5). Consistent with the preceding analysis of the mean number of pollination systems per community, this showed tropical communities, and those with complex habitat structure, to be distinct from other communities (figs. 13.5A, 13.5C).

Tropical communities had greater proportions of plant species specialized on bees (small, large, and euglossine), butterflies, beetles, fig wasps, wasps, and thrips than did communities from other latitudinal zones. This is indicated by the vectors associated with large proportions of plant species specialized on those insect groups (fig. 13.5D) pointing toward the cluster of tropical communities (fig. 13.5A). In contrast, tropical communities had lower proportions of plants specialized on flies and birds than did communities from higher latitudes, which is illustrated by the vectors associated with large proportions of plant species specialized on flies and birds pointing away from the cluster of tropical communities (figs. 13.5A, 13.5D). Because of the lack of independence between tropical communities considered in this review and those with complex habitat structure (fig. 13.5A, 13.5C), communities with complex habitat structure also exhibited greater proportions of plants specialized on bees (small, large, and euglossine), butterflies, beetles, fig wasps, wasps, and thrips, and lower proportions of plants specialized on flies and birds, than did communities with simpler structure (figs. 13.5C, 13.5D).

Subtropical and temperate communities were broadly similar in their proportional compositions of plants categorized into the various broad pollination systems, which is indicated by their overlap in the ordination plot (fig. 13.5A). These communities were scattered across the ordination plot in a broad band associated with the vectors describing large proportions of plants pollinated by more than one insect group, insects and birds, and birds only (figs. 13.5A, 13.5D). This suggests that plants in subtropical and temperate zones are more likely to have generalized pollination systems, or to specialize on birds rather than a particular insect group, than are those in the tropical zone.

Only two Arctic (and no Antarctic) communities were included in the analysis, making it impossible to draw firm conclusions about the relative proportions of the various pollination systems at high latitudes. However, the two Arctic communities were plotted close together on the ordination, at a position

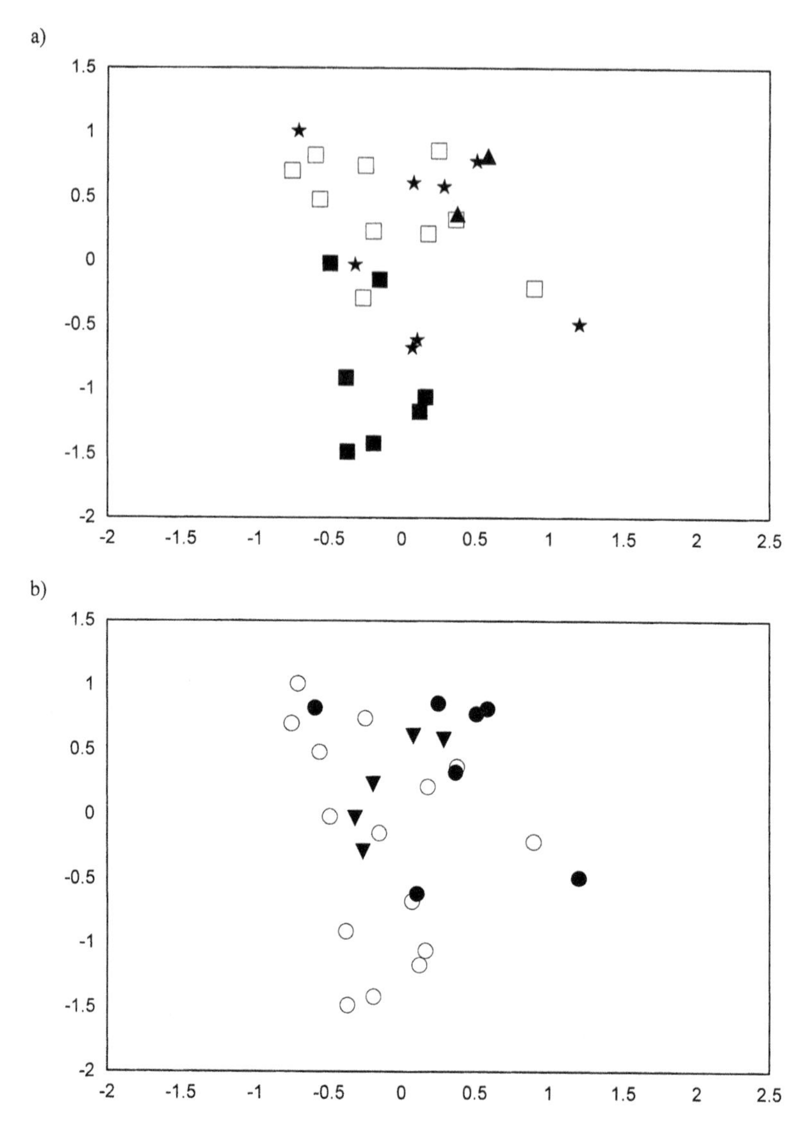

Figure 13.5 Ordination plot of the communities in dataset 1 according to the proportional compositions of the various broad pollination systems represented by their constituent plant species. The communities have been plotted in three dimensions, but only the two dimensions that encompass the greatest part of the variation among communities are shown. Communities with fewer than 10 species of categorizable plants have been excluded from the analysis. The ordination of the communities has been shown three times (A, B, and C), where communities are coded as follows: (A) latitudinal zone—tropical (■), subtropical (★), temperate (□), and arctic (▲); (B) altitudinal zone—low (○), mid (●), and high (▼); and (C) complexity of habitat structure—complex (►), medium (●), and simple (Δ). (D) The broad pollination systems have been fitted to the plot as vectors, shown in a separate plot with the same axes for clarity. Vector codes and their significance to the variation among communities are as follows: Insect, pollination by more than one insect group ($P < .01$); General, mixed bird and insect pollination ($P < .01$); Bird, bird pollination ($P < .01$); Fly, fly pollination ($P < .01$); Buttfly, butterfly pollination ($P < .01$); Beetle, beetle pollination ($P < .01$); LBEE, large-bee pollination ($P < .01$); SBEE, small-bee pollination ($P < .01$); Eugloss, euglossine-bee pollination ($P < .16$); Figwasp, fig wasp pollination ($P < .09$); Wasp, other wasp pollination ($P < .21$); and Thrips, thrips pollination ($P < .26$). Stress on three axes = 0.078.

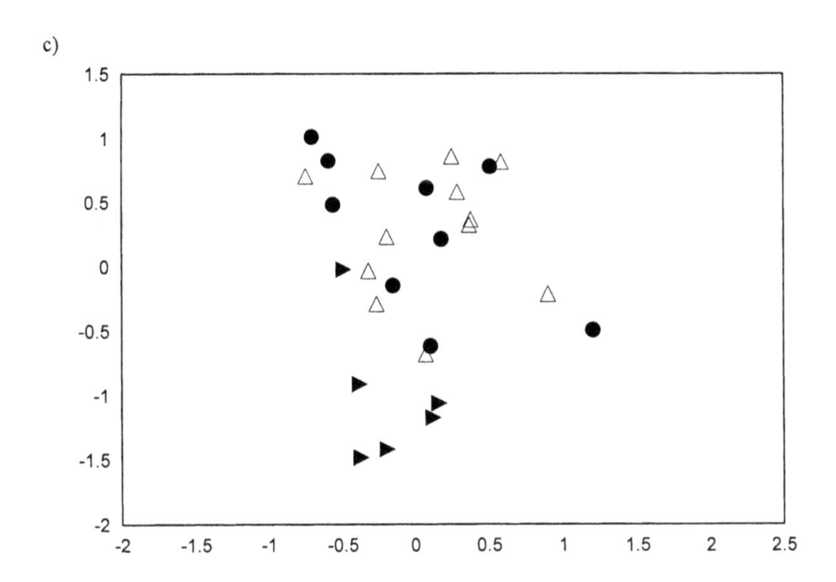

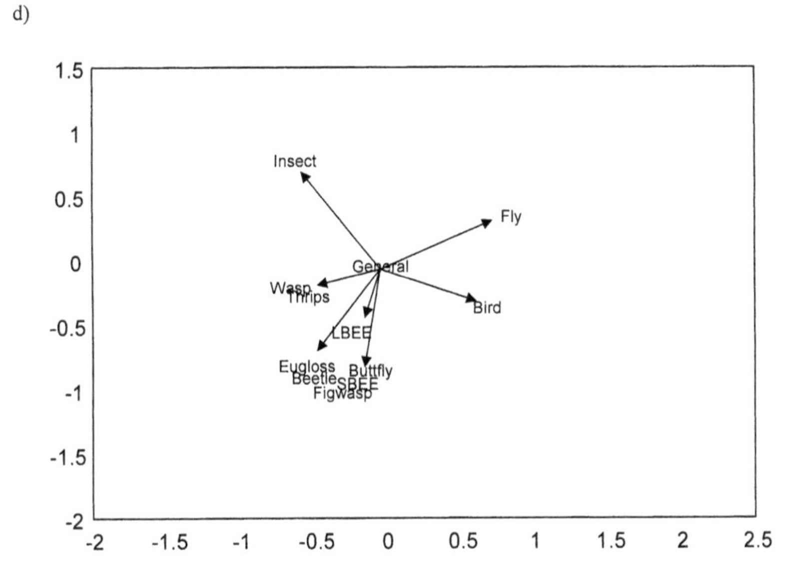

peripheral to the temperate and subtropical communities, and were associated with the vector describing large proportions of fly-pollinated plants (figs. 13.5A, 13.5D). This concurs with the findings of studies reviewed by Larson et al. (2001) that Diptera are extremely important pollinators of the Arctic flora. The vectors associated with large proportions of plant species specialized on small and large bees, butterflies, beetles, wasps, and thrips (fig. 13.5D) point away from both Arctic communities (fig. 13.5A), indicating that few plants at high latitudes are specialized on those pollinators.

Communities from the three altitudinal zones overlapped extensively in their proportional compositions of plants associated with the various pollination systems (fig. 13.5B), suggesting that any altitude effects were masked by latitude effects. There was a slight trend toward communities at low altitudes showing the weakest association with fly pollination (figs. 13.5B, 13.5D), which is consistent with the dominance of flies among insect flower visitors at higher altitudes in many studies (Müller 1880; Arroyo et al. 1982; Primack 1983; Warren et al. 1988; McCall and Primack 1992; Hingston and McQuillan 2000). However, in this review, all of the tropical communities were at low altitude, and, if these are ignored, the apparent tendency for communities at low altitude to show the least association with fly pollination does not persist (figs. 13.5A, 13.5B, 13.5D). Moreover, communities at high altitude exhibited a weaker association with fly pollination than did those at mid-altitude (figs. 13.5B, 13.5D), suggesting that altitude has little overall effect on the range and proportions of pollination systems.

In summary, existing studies suggest that communities are more predictable in the range and proportions of pollination systems that they host when compared across latitudinal zones, rather than across altitudinal zones or levels of habitat structural complexity. Although communities with complex habitat structure were distinct from those with simpler structure, this was confounded with latitude in available studies, and communities with medium and simple habitat structure overlapped extensively (fig. 13.5C). Comparisons of pollination systems at various altitudes, as well as levels of habitat structural complexity at the same latitude, are needed to shed more light on how these factors influence the range and proportions of pollination systems.

Does Ecological Specialization in Plant Pollination Systems Correlate Significantly with Latitude?

Two recent studies (Olesen and Jordano 2002; Ollerton and Cranmer 2002) have addressed this question. Using similar approaches and, to some extent, similar data, they have come to precisely opposite conclusions: Olesen and Jordano (2002) stated that tropical plants were indeed more ecologically specialized, whereas Ollerton and Cranmer (2002) concluded that any apparent trend in specialization was an artifact of systematic variation in sampling effort.

Ollerton and Cranmer (2002) used two independent datasets (described earlier as datasets 2 and 3). The first was at the scale of the plant community and comprised 27 published and unpublished surveys of plant–flower visitor interactions in 35 communities at different latitudes. The second dataset consisted of 103 published and unpublished studies of pollinators of species of asclepiads (subfamily Asclepiadoideae of the Apocynaceae sensu Endress and Bruyns 2000); this is part of the online ASCLEPOL database (http://www.uni-bayreuth.de/departments/planta2/research_wgl/pollina/as_pol_t.html). Initial analysis

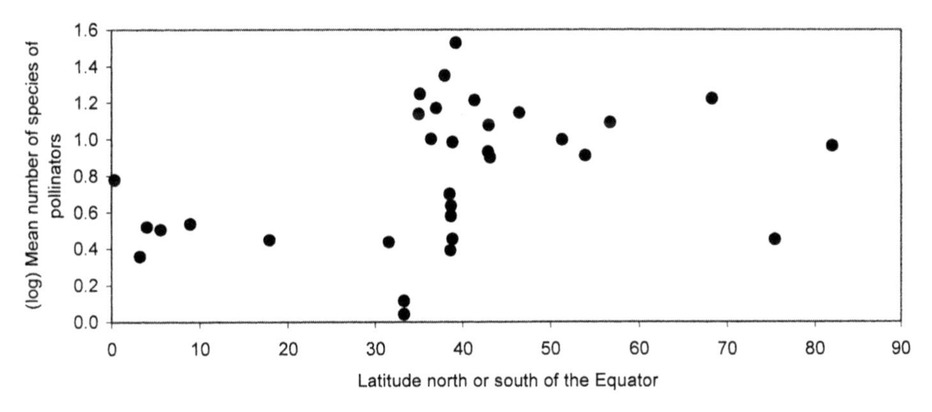

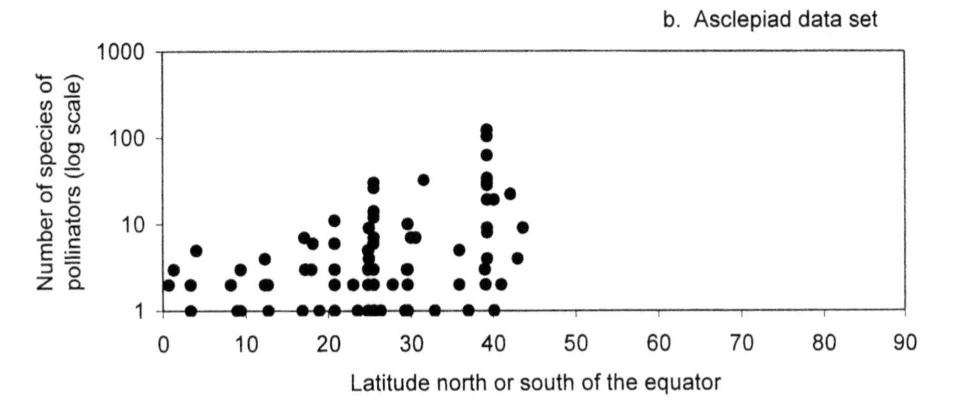

Figure 13.6 (A) Relationship between latitude and mean number of flower visitors per plant species in the community surveys (dataset 2); mean number of species of flower visitors per plant species has been log-transformed (Pearson's product moment correlation: $r = 0.33$, df = 33, $P = .05$). (B) Relationship between latitude and number of recorded asclepiad pollinators (dataset 3; Spearman rank correlation: $r = 0.33$, $N = 91$, $P = .002$). Data are from Ollerton and Cranmer (2002). Note that these correlations are artifacts of sampling effort (see text).

of both datasets revealed a strong correlation between latitude and ecological specialization: plants from lower, tropical latitudes usually had fewer recorded pollinators than plants from higher latitudes (fig. 13.6). However, more detailed analysis suggested that these patterns were an artifact of sampling effort. In both datasets, when sampling effort was taken into account, the correlation between latitude and ecological specialization of pollination systems disappeared (Ollerton and Cranmer 2002).

The study by Olesen and Jordano (2002) used 29 published and unpublished "complete" plant–pollinator networks (at a community level). A large proportion of these studies were shared in common with the community-level analysis of Ollerton and Cranmer (2002), although Olesen and Jordano (2002) addressed a broader range of questions, which related to patterns of connectance (Jordano 1987), network size, and ecological specialization from both the plant and animal perspective, in relation to species richness, altitude, and insularity, as well as latitude. In relation to geographical patterns, these authors found that network connectance was lower in highland (higher than 1500 m) than in lowland studies but only marginally different among the latitudinal zones. Levels of plant specialization were lower at higher latitudes and in lowland habitats, but higher on islands (probably because of impoverished pollinator faunas). Of particular significance to our study is their finding that plant species were more ecologically generalized at higher latitudes (although, interestingly, there was no such pattern for the animal pollinators).

Olesen and Jordano (2002) included only what they considered to be "complete" surveys of plant–pollinator interactions in a particular community. However, to be complete, a survey would have to include both diurnal and nocturnal observations, be conducted over the whole course of the flowering season, and be done in multiple years to allow for intra- and interannual variation in the identities of pollinators of particular plant species (cf. Herrera 1988; Pettersson 1991; Fishbein and Venable 1996; Lamborn and Ollerton 2000; Petanidou and Potts, chap. 10 in this volume). None of the included studies actually fulfilled these criteria, because, for the most part, such complete studies are only now being undertaken. Olesen and Jordano (2002) acknowledged some of these shortcomings but suggested that plant–pollinator networks may be more robust to assumptions of completeness than other kinds of food webs. In contrast, Ollerton and Cranmer (2002) considered that the community-level surveys represented more-or-less random samples of plants and their pollinators from those communities.

One possible explanation for the differences in the conclusions reached by Ollerton and Cranmer (2002) and Olesen and Jordano (2002) is that sampling effort was not taken into account in the latter study. In defense of their decision to omit this variable, Olesen and Jordano (2002) cite the Caribbean food web analysis of Goldwasser and Roughgarden (1997), who "simulated variation in sampling effort and found that its impact on web properties was profound" (Olesen and Jordano 2002, 2419). However, they then argue that this is unlikely to be a major problem in plant–pollinator networks because species are not recorded before they are involved in an interaction (visitor or visited). This may be true for the overall detection of interactions but is untrue for assessing levels of generalization. Work on an assemblage of grassland asclepiads in South Africa has shown that sampling effort can profoundly affect conclusions regarding the number of pollinators servicing a plant species (see also Vázquez and Aizen, chap. 9 in this volume, and Herrera 2005) and that cumulative effort curves should be utilized in this respect (Ollerton et al. 2003); few other studies have presented such curves (Petanidou and Ellis 1993; Minckley et al. 1999; Dupont and Skov 2004). There is considerable variation in the amount of sampling effort expended in community studies, and the possibility that tropical plants appear from these studies to be more ecologically specialized in their interactions with pollinators, simply because of lower sampling effort dictated by logistical problems, cannot be excluded. In this respect, there is a need for standardized sampling protocols, akin to those developed for vegetation surveys, that might include, for example, the cessation of sampling of a plant species once an asymptote in number of pollinators is reached. Standardized sampling protocols were proposed by McCall and Primack (1992) in their comparison of pollination systems in different communities, but these have not been widely adopted or agreed upon.

At a Finer Taxonomic Level, Do Plant Families Vary between Regions in Their Degree of Ecological Specialization in Pollination Systems?

One of the problems with simple comparisons of pollination systems between plant communities is that their phylogenetic composition, potentially a key determinant of specialization, is not controlled; thus, a community dominated by Asteraceae may differ from one dominated by Scrophulariaceae largely on account of the relative representation of these more generalized and more specialized families, respectively. Endress (1994) has argued that certain families have an inherent tendency to have either generalized or specialized pollination systems because of some aspect of their "Bauplan"; to the extent that this is true, one solution is to make comparisons among representatives of the same family in different regions, thus controlling, to some extent, for phylogenetic relatedness at the family level. Regional differences in the pollination systems of plants of the same family are therefore more likely to reflect either evolved or ecological differences in pollination systems. We have done this for two groups of angiosperms: the orchids (Orchidaceae) and the asclepiads (Apocynaceae subfamily Asclepiadoideae). These two groups were chosen for a number of reasons: (1) they are wholly unrelated (one belongs to the monocots [Liliopsida], the other to the eudicots [Magnoliopsida]), therefore any patterns they share in common should not be due to phylogenetic relatedness; (2) both groups in the main produce pollen masses (pollinia) rather than free pollen, which is mechanically or adhesively fixed onto the pollinator and makes pollinator identification much less ambiguous than it is for most other plant groups; (3) two of us have been studying orchids and asclepiads (SDJ and JO, respectively) for many years and have accumulated unpublished data to add to published surveys. We have used these data to compare patterns of ecological specialization in orchids and asclepiads in Southern Africa versus that in North America and Europe.

Analyses of both Orchidaceae (Johnson and Steiner 2003) and Apocynaceae

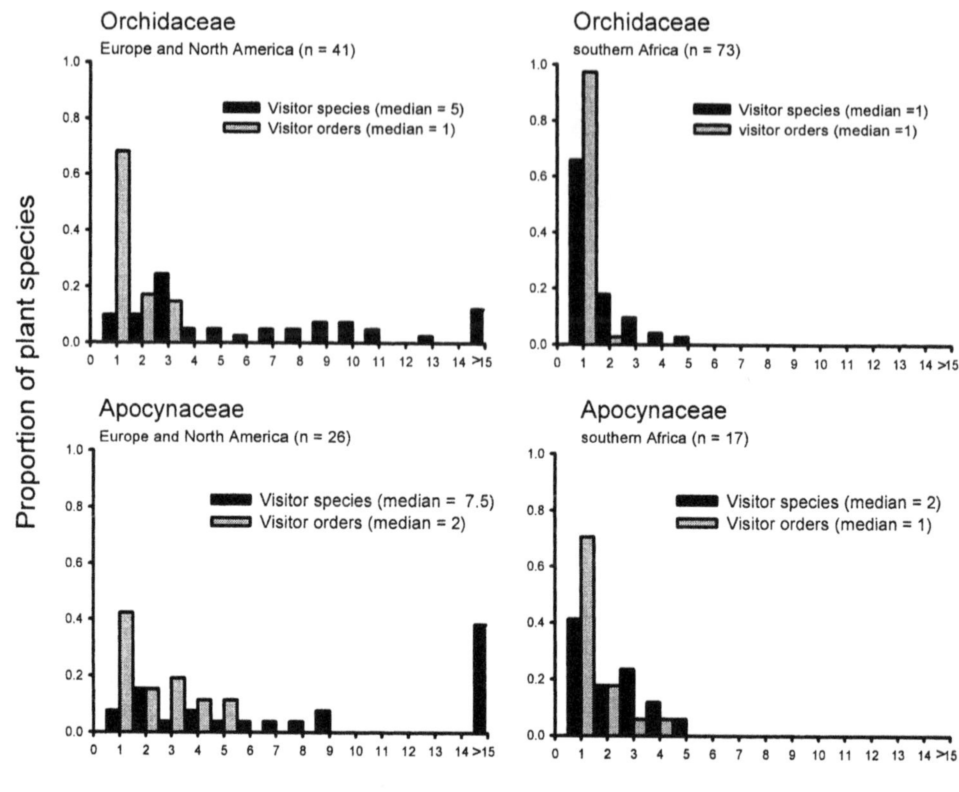

Figure 13.7 Frequency distributions of the number of pollinator species and orders recorded for plants belonging to two plant groups: the family Orchidaceae and the family Apocynaceae subfamily Asclepiadoideae (dataset 4). The median numbers of pollinator species and orders differ significantly between temperate Northern Hemisphere and southern African members of these families (Mann-Whitney U test: orchid pollinator species, $Z = 7.05$, $P < .01$; orchid pollinator orders, $Z = 2.33$, $P = .02$; asclepiad pollinator species, $Z = 3.3$, $P < .001$; asclepiad pollinator orders, $Z = 2.07$, $P = .04$). Orchid data are from Johnson and Steiner (2003); asclepiad data are from Ollerton and Cranmer (2002).

subfamily Asclepiadoideae (this study) suggest that there are significant geographical differences in the level of specificity of pollination systems in these families: plants native to southern Africa typically are significantly more ecologically specialized than their North American and European counterparts (fig. 13.7). This cannot be attributed to differences in sampling effort, because the number of hours spent recording pollinators did not differ significantly among regions for either of the two datasets (Mann Whitney U test: orchid data, $Z = 1.33$, $P = .18$; asclepiad data, $Z = 0.15$, $P = .87$).

The greater specialization evident in southern African representatives of the Orchidaceae and Apocynaceae has two possible explanations: either phenotypic specialization of floral traits or ecological specialization due to a depauperate pollinator fauna. Johnson and Steiner (2003) suggest that both explanations may apply and that some orchids show highly evolved phenotypic specialization in floral traits (e.g., oil-producing species), whereas, for others, a low diversity of pollinators within key functional groups, such as bees (fig. 13.2A), may account for some of the observed patterns at the species level in the asclepiads (Ollerton et al. 2003).

Southern African plants in general appear to have relatively specialized pollination systems. Members of the Iridaceae (a very large family in this region), for instance, are typically pollinated by a single insect taxon (Johnson and Steiner 2003). Examples of highly functionally (and sometimes ecologically) specialized pollination systems in this region include those involving long-proboscid flies, oil bees, pompilid wasps, large satyriine butterflies, cetoniid and hopliine beetles, hawkmoths and rodents (Whitehead et al. 1987; Goldblatt and Manning 1997; Johnson and Steiner 2000, 2003; Ollerton et al. 2003). Thus, the relatively generalized pollination systems that Waser et al. (1996) recorded for North American and European members of the Polemoniaceae and Orchidaceae are not necessarily representative of those found in comparable animal-pollinated plant families in other regions of the world. In seeking an explanation for the patterns in southern Africa, Johnson and Steiner (2003) speculated that a combination of high functional and low species diversity in pollinators may give rise to ecologically specialized plant pollination systems, even when flowers do not have highly modified advertising or rewards. For example, there are about 15 times more plant species than bee species in southern Africa (fig. 13.2B), whereas, in many similar-sized regions of the United States, the number of plant species exceeds the number of bee species by a factor of only 2 or 3 (fig. 13.2B). It would be interesting to test this idea with data, as they become available, from other high-plant-diversity subtropical regions, such as parts of Chile and Western Australia.

Conclusions

The analyses presented in this chapter reveal considerable geographical variation in both the diversity and the levels of specificity in plant pollination systems on a global scale. On average, tropical communities possess more types of pollination systems than do communities in other latitudinal zones; however, there does not appear to be a simple linear relationship between latitude and number of pollination systems—Arctic, temperate, and subtropical communities do not, on average, show differences in the number of pollination systems that are present. The increase in number of pollination systems from higher to lower latitudes is, therefore, a step change rather than a smooth increase. Uneven coverage of altitudes and habitats among the studies that we analyzed may have had an influence on the apparent patterns across latitudinal zones. All of the tropical communities that we considered were at low altitude and were mostly structurally complex forest habitats. Studies of high altitude and structurally simple tropical communities (e.g., grasslands) are needed to confirm that the differences between the tropics and other latitudes are indicative of tropical communities per se, not just lowland tropical forest.

Tropical communities tend to contain higher numbers of functionally specialized pollination systems. However, the long-standing assumption that plant–pollinator interactions are more ecologically specialized in the tropics is not supported by our analyses of data that are corrected for sampling effort (Ollerton and Cranmer 2002). We acknowledge that many experienced tropical biologists hold the opinion that pollination systems in the tropics are relatively specialized, but this may be due in part to different interpretations of exactly what is meant by "specialized" (see Armbruster et al. 2000, chap. 12 in this volume). Another possibility is that the range of functional specialization is greater in the tropics, extending to include more functionally specialized systems than are found in other latitudes (see Armbruster, chap. 12 in this volume), but that the effect of this is not great enough to be statistically detectable given the available data.

In relation to the questions regarding pollination niche and species diversity that were outlined at the beginning of the chapter, we interpret our datasets as follows:

1. For increased size of niche axes (i.e., utilization of pollen vectors that are not normally used) the wider range of pollinator types utilized by plants in the tropics, compared to that for other latitudinal zones, implies that this may be an important mechanism that allows species coexistence in tropical communities.

2. For decreased niche breadth, for example, by specializing on a narrower range of pollinator species, the results from datasets 2 and 3 (Ollerton and Cranmer 2002) suggest, perhaps counterintuitively, that this is not an important mechanism. However, this hypothesis is supported by the results from dataset 4 (orchid and asclepiad ecological specialization in southern Africa vs. Europe and North America). We suggest that, whereas there is no simple linear relationship between pollination niche breadth and latitude, this may be a significant factor in allowing species coexistence in some parts of the world, especially areas of extremely high plant alpha diversity such as southern Africa.

3. For increased niche overlap such that pollinators are shared between taxa, perhaps facilitated by character displacement of, for example, pollen placement (e.g., Armbruster et al. 1994, chap. 12 in this volume), our datasets are not sufficiently detailed to enable us to test this mechanism, though it should be possible in the future. It is notable, however, that Ollerton et al. (2003), working in a plant species–rich South African grassland, did find significant pollinator overlap between asclepiad species. Is the same true in other species-rich communities?

Southern Africa appears to have an excess of highly specialized and unique pollination systems, which is somewhat surprising because this is a largely temperate region. The example of southern Africa highlights the need for studies of the biogeography of species interactions to encompass not only latitudinal and altitudinal effects but also differences between both Northern and Southern and Eastern and Western Hemispheres. For example, Devy and Davidar (2003) have recently compared the proportions of pollination systems in the tree flora of tropical rainforest at Kakachi in the Western Ghats of India (Eastern Hemisphere) with that of four other sites: Lambir (Malaysia; also Eastern Hemisphere) and Venezuela, Jamaica, and La Selva, Costa Rica (all Western Hemisphere). The authors conclude that the Kakachi site is more similar to that of Lambir than to the other sites. This is confirmed by a multidimensional scaling analysis (J. Ollerton, unpublished data) of the proportional data presented by Devy and Davidar (2003). In this analysis, Lambir and Kakachi cluster together as an Eastern Hemisphere grouping, and the Venezuelan and Jamaican sites cluster as a Western Hemisphere grouping; La Selva stands isolated, albeit closer to the Eastern than Western Hemisphere sites. Thus, hemispheric differences can account for at least part, but not all, of the variation between sites. However, presently there are too few published studies from the Southern Hemisphere to attempt a truly comprehensive analysis. It may be that other biogeographic regions are similarly underrepresented in our current understanding of specialization and generalization of pollination systems.

A future agenda for the study of the biogeography of species interactions would need to include the following among its priorities: (1) standardized sampling protocols between different sites; (2) surveys of poorly explored regions, particularly in the Southern Hemisphere and including the full range of altitudinal variation and habitat complexity; and (3) quantification of the effect of phylogenetic relationships in community comparisons. The synthetic analyses published to date have been pioneering in their approaches and scope, but a dedicated research effort is now required if we are ever to truly understand pattern and process in the geographical variation in diversity and specificity of pollination systems.

Acknowledgments

Our sincere gratitude goes to the many researchers, too numerous to list individually, who collected the data that were used in our analyses. Thanks also go to Mary Price, Nick Waser, Christine Müller, Beverly Rathcke and an anonymous referee who made invaluable comments on the chapter.

References

Armbruster, W. S. 1985. Patterns of character divergence and the evolution of reproductive ecotypes of *Dalechampia scandens* (Euphorbiaceae). Evolution 39: 733–752.

Armbruster, W. S. 1986. Reproductive interactions between sympatric *Dalechampia* species: Are natural assemblages random or organized? Ecology 67: 522–533.

Armbruster, W. S. 1995. The origins and detection of plant community structure: Reproductive versus vegetative processes. Folia Geobotanica Phytotaxonomica 30: 483–497.

Armbruster, W. S., M. E. Edwards, and E. M. Debevec. 1994. Floral character displacement generates assemblage structure of Western Australian triggerplants (*Stylidium*). Ecology 75: 315–329.

Armbruster, W. S., C. B. Fenster, and M. R. Dudash. 2000. Pollination "principles" revisited: Specialization, pollination syndromes, and the evolution of flowers. Det Norske Videnskaps—Akademi. I. Matematisk Naturvidenskapelige Klasse, Skrifter, Ny Serie 39: 179–200.

Arroyo, M. T. K., R. Primack, and J. Armesto. 1982. Community studies in pollination ecology in the high temperate Andes of central Chile. I. Pollination mechanisms and altitudinal variation. American Journal of Botany 69: 82–97.

Bascompte, J., P. Jordano, C. J. Melian, and J. M. Olesen. 2003. The nested assembly of plant–animal mutualistic networks. Proceedings of the National Academy of Sciences (USA) 100: 9383–9387.

Belbin, L. 1993. PATN: Technical reference. CSIRO Division of Wildlife and Ecology, Canberra.

Crepet, W. L. 1983. The role of insect pollination in the evolution of the angiosperms. Pp. 29–50 *in* L. Real (ed.), Pollination biology. Academic Press, Orlando, FL.

Devy, M. S., and P. Davidar. 2003. Pollination systems of trees in Kakachi, a mid-elevation wet evergreen forest in Western Ghats, India. American Journal of Botany 90: 650–657.

Dupont, Y. L., and C. Skov. 2004. Influence of geographical distribution and floral traits on species richness of bees (Hymenoptera: Apoidea) visiting *Echium* species (Boraginaceae) of the Canary Islands. International Journal of Plant Sciences 165: 377–386.

Endress, M. E., and P. V. Bruyns. 2000. A revised classification of the Apocynaceae s.l. Botanical Review 66: 1–56.

Endress, P. K. 1994. Diversity and evolutionary biology of tropical flowers. Cambridge University Press, Cambridge.

Faegri, K., and L. van der Pijl. 1971. The principles of pollination ecology, 2nd ed. Pergamon Press, Oxford.

Fenster, C. B., W. S. Armbruster, P. Wilson, M. R. Dudash, and J. D. Thomson. 2004. Pollination syndromes and floral specialization. Annual Review of Ecology, Evolution, and Systematics 35: 375–403.

Fenster, C. B., and M. R. Dudash. 2001. Spatiotemporal variation in the role of hummingbirds as pollinators of *Silene virginica*. Ecology 82: 844–851.

Fishbein, M., and D. L. Venable. 1996. Diversity and temporal change in the effective pollinators of *Asclepias tuberosa*. Ecology 77: 1061–1073.

Fleming, T. H., C. T. Sahley, J. N. Holland, J. D. Nason, and J. L. Hamrick. 2001. Sonoran Desert columnar cacti and the evolution of generalized pollination systems. Ecological Monographs 71: 511–530.

Galen, C. 1999. Why do flowers vary? The functional ecology of variation in flower size and form within natural plant populations. Bioscience 49: 631–640.

Gaston, K. J., and P. H. Williams. 1996. Spatial patterns in taxonomic diversity. Pp. 202–209 *in* K. J. Gaston (ed.), Biodiversity: A biology of numbers and difference. Blackwell Science, Oxford.

Goldblatt, P., and J. C. Manning. 1997. Floristic diversity in the Cape flora of South Africa. Biodiversity and Conservation 6: 359–377.

Goldwasser, L., and J. Roughgarden. 1997. Construction and analysis of a large Caribbean food web. Ecology 74: 1216–1233.

Gómez, J. M., and R. Zamora. 1999. Generalization vs. specialization in the pollination system of *Hormathophylla spinosa* (Cruciferae). Ecology 80: 796–805.

Grimaldi, D. 1999. The co-radiations of pollinating insects and angiosperms in the Cretaceous. Annals of the Missouri Botanical Garden 86: 373–406.

Herrera, C. M. 1988. Variation in mutualisms: The spatio-temporal mosaic of a pollinator assemblage. Biological Journal of the Linnean Society 35: 95–125.

Herrera, C. M. 1996. Floral traits and adaptation to insect pollinators: A devil's advocate approach. Pp. 65–87 *in* D. G. Lloyd and S. C. H. Barrett (eds.), Floral biology. Chapman and Hall, New York.

Herrera, C. M. 2005. Plant generalization on pollinators: species property or local phenomenon? American Journal of Botany 92: 13–20.

Hillebrand, H. 2004. On the generality of the latitudinal diversity gradient. American Naturalist 163: 192–211.
Hingston, A. B., and P. B. McQuillan. 2000. Are pollination syndromes useful predictors of floral visitors in Tasmania? Austral Ecology 25: 600–609.
Howe, H. F. 1984. Constraints on the evolution of mutualisms. American Naturalist 123: 764–777.
Johnson, S. D. 1997. Pollination ecotypes of *Satyrium hallackii* (Orchidaceae) in South Africa. Botanical Journal of the Linnean Society 123: 225–235.
Johnson, S. D. 2004. An overview of plant-pollinator relationships in southern Africa. International Journal of Tropical Insect Science 24: 45–54.
Johnson, S. D., and K. E. Steiner. 1997. Long-tongued fly pollination and evolution of floral spur length in the *Disa draconis* complex (Orchidaceae). Evolution 51: 45–53.
Johnson, S. D., and K. E. Steiner. 2000. Generalization versus specialization in plant pollination systems. Trends in Ecology and Evolution 15: 190–193.
Johnson, S. D., and K. E. Steiner. 2003. Specialized pollination systems in southern Africa. South African Journal of Science 99: 345–348.
Jordano, P. 1987. Patterns of mutualistic interaction in pollination and seed dispersal: Connectance, dependence asymmetries, and coevolution. American Naturalist 129: 657–677.
Kay, K. M., and D. W. Schemske. 2003. Pollinator assemblages and visitation rates for eleven species of Neotropical *Costus* (Costaceae). Biotropica 35: 198–207.
Lamborn, E., and J. Ollerton. 2000. Experimental assessment of the functional morphology of inflorescences of *Daucus carota* (Apiaceae): Testing the "fly catcher effect." Functional Ecology 14: 445–454.
Larson, B. M. H., P. G. Kevan, and D. W. Inouye. 2001. Flies and flowers: Taxonomic diversity of anthophiles and pollinators. Canadian Entomologist 133: 439–465.
Linder, H. P. 1998 Morphology and evolution of wind pollination. Pp. 123–135 *in* S. J. Owens and P. J. Rudall (eds.), Reproductive biology in systematics, conservation, and economic botany. Royal Botanic Gardens, Kew, United Kingdom.
McCall, C., and R. B. Primack. 1992. Influence of flower characteristics, weather, time of day, and season on insect visitation rates in three plant communities. American Journal of Botany 79: 434–442.
Michener, C. D. 1979. Biogeography of the bees. Annals of the Missouri Botanical Garden 66: 277–347.
Miller, R. B. 1981. Hawkmoths and the geographic patterns of floral variation in *Aquilegia caerulea*. Evolution 35: 763–774.
Minckley, R. L., J. H. Cane, L. Kervin, and T. H. Roulston. 1999. Spatial predictability and resource specialization of bees (Hymenoptera: Apoidea) at a superabundant, widespread resource. Biological Journal of the Linnean Society 67: 119–147.
Müller, H. 1880. The fertilisers of alpine flowers. Nature 21: 275.
Olesen, J. M., and P. Jordano. 2002. Geographic patterns in plant–pollinator mutualistic networks. Ecology 83: 2416–2424.
Ollerton, J. 1996. Reconciling ecological processes with phylogenetic patterns: The apparent paradox of plant–pollinator systems. Journal of Ecology 84: 767–769.
Ollerton, J., and L. Cranmer. 2002. Latitudinal trends in plant–pollinator interactions: Are tropical plants more specialised? Oikos 98: 340–350.
Ollerton, J., S. D. Johnson, L. Cranmer, and S. Kellie. 2003. The pollination ecology of an assemblage of grassland asclepiads in KwaZulu-Natal, South Africa. Annals of Botany 92: 807–834.
Ollerton, J., and S. Liede. 1997. Pollination systems in the Asclepiadaceae: A survey and preliminary analysis. Biological Journal of the Linnean Society 62: 593–610.
Pellmyr, O. 1986. Three pollination morphs in *Cimicifuga simplex:* Inicipient speciation due to inferiority in competition. Oecologia 68: 304–307.
Petanidou, T., and W. E. Ellis. 1993. Pollinating fauna of a phryganic ecosystem: Composition and diversity. Biodiversity Letters 1: 9–22.
Pettersson, M. W. 1991. Pollination by a guild of fluctuating moth populations: Option for unspecialization in *Silene vulgaris*. Journal of Ecology 79: 591–604.

Primack, R. B. 1983. Insect pollination in the New Zealand mountain flora. New Zealand Journal of Botany 21: 317–333.

Proctor, M., P. Yeo, and A. Lack. 1996. The natural history of pollination. HarperCollins, London.

Robertson, J. L., and R. Wyatt 1990. Evidence for pollination ecotypes in the yellow-fringed orchid, *Platanthera ciliaris*. Evolution 44: 121–133.

Rohde, K. 1992. Latitudinal gradients in species diversity: The search for the primary cause. Oikos 65: 514–527.

Roubik, D. W. 1993. Direct costs of forest reproduction, bee cycling, and the efficiency of pollination modes. Journal of Biosciences 18: 537–552.

Sánchez-Lafuente, A. M. 2002. Floral variation in the generalist perennial herb *Paeonia broteroi* (Paeoniaceae): Differences between regions with different pollinators and herbivores. American Journal of Botany 89: 1260–1269.

Stone, G. N., P. Willmer, and J. A. Rowe. 1998. Partitioning of pollinators during flowering in an African *Acacia* community. Ecology 79: 2808–2827.

Thompson, J. N. 1994. The coevolutionary process. University of Chicago Press, Chicago.

Thompson, J. N., and B. M. Cunningham. 2002. Geographic structure and dynamics of coevolutionary selection. Nature 417: 735–738.

Warren, S. D., K. T. Harper, and G. M. Booth. 1988. Elevational distribution of insect pollinators. American Midland Naturalist 120: 325–330.

Waser, N. M., L. Chittka, M. V. Price, N. M. Williams, and J. Ollerton. 1996. Generalization in pollination systems, and why it matters. Ecology 77: 1043–1060.

Whitehead, V. B., J. H. Giliomee, and A. G. Rebelo. 1987. Insect pollination in the Cape flora. Pp. 52–82 *in* A. G. Rebelo (ed.), A preliminary synthesis of pollination biology in the Cape flora. National Scientific Programmes report no. 141, Pretoria.

Whitmore, T. C. 1998. An introduction to tropical rainforests, 2nd ed. Oxford University Press, Oxford.

Williams, P. H., C. J. Humphries, and K. J. Gaston. 1994. Centres of seed plant diversity: The family way. Proceedings of the Royal Society of London B 256: 67–70.

PART IV

Applications in Agriculture and Conservation

Introductory Comments by Nickolas M. Waser and Margaret M. Mayfield

Part 4 of this book is composed of four chapters, which discuss some of the ways in which knowledge of specialization and generalization informs applied pollination biology. The oldest applications of biology, which must long predate any (curiosity-driven) interest in "pure" biology, derive from the human need for food and shelter. In hunting and gathering cultures, applied biology takes the form of an understanding of the natural history of plants and animals suited for food, fiber, and other uses. With the inventions of agriculture and animal husbandry, applied biology becomes a working knowledge of how one cultivates and cares for one's living charges. This includes some understanding of pollination of crop plants by wind or animals.

An understanding of pollination can be traced to the ancients in a few cases (Baker 1983; Proctor et al. 1996); however, a clear scientific appreciation of flower visitors as pollinators dates back only to the 18th century (Waser, chap. 1 in this volume). Thus, our understanding of an essential portion of the life history of most plants has had only 20 decades in which to develop. In some regards, the development has been dramatic; in others, it can be argued that we remain at an early stage, having often taken the "easy way out" (a prime example being the application of another ancient domestication, the European honeybee *Apis mellifera,* to virtually all agricultural ends; Corbet, chap. 14 in this volume). What is worse is that even a basic appreciation of pollination has not penetrated to many farmers in developing nations, who may instead harbor erroneous folk views (including views that insects at flowers are detrimental) even as they begin to adopt other ideas from Western agriculture. We submit that an appreciation of insects at flowers has not penetrated to most citizens of the First World either! These considerations should serve as a call-to-arms to pollination biologists to redouble their efforts in applied arenas. In this context, the following chapters provide useful perspectives on human-dominated landscapes, both for those who appreciate the basic natural history and scientific context of pollination and for those who seek to educate those still lacking this appreciation.

In chapter 14, the first of part 4, Sally Corbet revises one of the early attempts at classifying pollination systems (Waser, chap. 1 in this volume) by building on recent work by Ellis and Ellis-Adam and striving for "functional" categories of plants and insects that will simplify the practical job of managers increasingly interested in developing (and conserving) insects other than honeybees to serve as pollinators. Corbet's meaning of functional is an attempt to make use of the modern understanding of insect energetics (a welcome rekindling of the "pollination energetics" of Heinrich and Raven 1972), foraging behavior (recalling the value of foraging theory, sensu Stephens and Krebs 1986), cognitive ability (which, however, she does not develop; see Chittka and Thomson 2001), and morphology, combined with knowledge of flower rewards and morphology. Thus, she narrows the great range of traits included in the "pollination syndromes" down to a few that modern knowledge suggests are key for evolution and ecology of the pollination interaction, including its degree of specialization and generalization. What remains is to test the resulting scheme, in Europe and beyond.

Conservation is a form of applied biology that is more recent than agriculture. Food, fiber, and shelter derive not only from the management of some parts of the ecosystem but, ultimately (as we increasingly recognize), from conservation of the healthy functioning of all parts. Contemplate, in the extreme, losing *all* the pollinators and, thus, potentially 90% of all flowering plants (e.g., Renner 1998), and the issue should become clear! In chapter 15, Suzanne Koptur illustrates some of these issues and uses the pine rocklands of south Florida as an example. This ancestral habitat is now reduced to traces surrounded by a matrix of human-altered habitat and a large population of humans and their camp followers, both plants and animals—this situation is the norm for much of the world and will become even more so in future. How have pollination systems responded? Koptur's case studies illustrate some anticipated changes, which include shifts in pollinator faunas and decline in pollinator diversity in smaller habitat fragments. But, the longer-term state of the system is unknown—a correct caution against making precise long-term dire predictions, which some conservation biologists do—and there are some surprising results. Although not brought to the fore, Koptur's conclusion is that specialist and generalist plants do not seem to be at different extinction risk in this setting. A possible explanation is the tendency of generalist plants to associate with specialist pollinators and vice versa, which appears to be a general feature of plant–pollinator interactions (e.g., Memmott et al. 2004; Minckley and Roulston, chap. 4 in this volume; Vázquez and Aizen, chap. 9 in this volume) and which Ashworth et al. (2004) have proposed to explain equivalent risk to generalists and specialists. This illustrates a point of contact between case-by-case studies (e.g., those of Koptur and colleagues) and recent attempts to predict "extinction dynamics" within entire pollination networks (Memmott et al. 2004; Jordano et al., chap. 8 in this volume).

Koptur ends by discussing how the surrounding habitat matrix affects pollination and presents ideas for remediation. These themes are shared in the final two chapters. First, Kwak and Bekker (chap. 16) consider remediation. They use the highly human-altered Netherlands landscape as a case study to explore the traits of plants that put them at risk of declining to extinction when they are already rare, rarity being the condition both of local populations of endangered plants and of newly established populations in restoration efforts. Can an index be calculated that predicts risk? Kwak and Bekker combine various traits into such an index, including specialization of plants in pollinator use. Rare and common plants do differ in predicted vulnerability by this index, but both groups appear similar in their degree of pollinator generalization and both share large bees as a key to seed production (see also Corbet, chap. 14 in this volume). Interestingly, Kwak and Bekker's index does poorly at predicting success in European restoration efforts, based on the scant currently available data. However, rather than react with discouragement, we suggest that the idea of simple indices be embraced and that workers be encouraged to continue their development and testing. In tropical ecosystems (and others besides!), little is known about pollination systems, and little time remains for conservation, so even crude indices may be extremely valuable for community-level assessments of risk and for directing management efforts.

Koptur's perspective of native plants and pollinators embedded in an anthropogenic habitat matrix is greatly amplified by Steffan-Dewenter et al. (chap. 17). They argue that failure to consider this context is one reason that so little is known about how pollination systems respond to change. This argument is cogently supported with their own studies of bees and bee pollination, which use an explicit landscape approach. Because so little else is available in the literature, the choice of examples reflects the pioneering nature of the work being done by the Göttingen group, not simply ego! The chapter thus serves to emphasize how early we are in these endeavors (take note, graduate students who despair of something novel to do!) and to illustrate a frame of mind that should be widely adopted in the process. Steffan-Dewenter and his colleagues cover many themes. To choose two, they provide a workable method for studies at large spatial scales, using a standard floral environment to provide some control in a nonexperimental setting (for more on this "phytometer" approach see Clements and Goldsmith 1924; Brandon and Antonovics 1996) and, thereby, obtain hard evidence for bee declines depending on landscape context; they also conclude (as did Koptur in chap. 15) that the final extinction risk to specialists versus generalists is more complex than previously imagined.

This last conclusion derives from a brief summary of recent work on coffee pollination, showing that yields increase with species diversity of native tropical bees, even though the bees are generalists and, therefore, might have been considered ecologically redundant. Thus, Steffan-Dewenter and colleagues bring us full circle back to agriculture and illustrate that two perspectives are interchangable: that of the conservation biologist viewing natural populations within a matrix of natural and anthropogenic landscape elements and that of the agricultural biologist (or farmer) viewing cultivated populations within precisely the same matrix. Hence, agriculture and conservation are alternative facets of applied biology rather than being somehow distinct, a theme echoed in each of these chapters.

Viewing agricultural pollination in the explicit context of habitat matrix has yielded other recent successes. For example, Kremen et al. (2002) found that five of six native bee species most effective at pollination of watermelon flowers were absent in fields far from natural vegetation. Similar results have been obtained for coffee pollination in Costa Rica (Ricketts 2004) and kiwifruit pollination in New Zealand (M. Mayfield, unpublished data). In all cases, isolation from natural habitat coincided with loss of native insect species with complex habitat requirements and/or short apparent foraging ranges.

Surprisingly, these studies are among the first to examine crops in a landscape setting that includes native insects and to assess the value of these native pollinators. As noted earlier, one can argue that our understanding remains superficial after 20 decades! Interestingly, even these few studies already suggest some common conclusions in relation to the diversity of landscape elements, spatial isolation of the elements containing crops, and biology of the insects. The goal of science is to discern such commonalities, and this is no less true for applied science. Time is short for conservation of native species, and so are resources—it hardly seems feasible to study each endangered species (or endangered pollination interaction) in isolation, as even current law requires in the United States (for example). Time is equally short for maintaining and improving agriculture yields in sustainable ways in complex landscapes (or, dare we dream, for convincing some of Western agriculture to move back toward such landscapes from industrialized monoculture). Therefore, we applaud the search for common features of pollination interactions, some of which emerge from the chapters in this section. And, we suggest that a focus on the degree of specialization and generalization of both pollinator and plant species will aid in recognizing commonalities and arriving at widely applicable schemes for the management of pollination systems.

References

Ashworth, L., R. Aguilar, L. Galetto, and M. A. Aizen. 2004. Why do pollination generalist and specialist plant species show similar reproductive susceptibility to habitat fragmentation? Journal of Ecology 92: 717–719.

Baker, H. G. 1983. An outline of the history of anthecology, or pollination biology. Pp. 7–28 *in* L. Real (ed.), Pollination biology, Academic Press, Orlando, FL.

Brandon, R., and J. Antonovics. 1996. The coevolution of organism and environment. Pp. 161–178 *in* R. Brandon (ed.), Concepts and methods in evolutionary biology. Cambridge University Press, Cambridge.

Chittka, L., and J. D. Thomson. 2001. Cognitive ecology of pollination. Cambridge University Press, Cambridge.

Clements, F. E., and G. W. Goldsmith. 1924. The phytometer method in ecology. Carnegie Institution of Washington, Washington, DC.

Heinrich, B., and P. H. Raven. 1972. Energetics and pollination ecology. Science 176: 597–602.

Kremen, C., N. M. Williams, and R. W. Thorp. 2002. Crop pollination from native bees at risk from agricultural intensification. Proceedings of the National Academy of Science (USA) 99: 16812–16816.

Memmott, J., N. M. Waser, and M. V. Price. 2004. Tolerance of pollination networks to species extinctions. Proceedings of the Royal Society of London B 271: 2605–2611.

Proctor, M., P. Yeo, and A. Lack. 1996. The natural history of pollination. Timber Press, Portland, OR.

Renner, S. S. 1998. Effects of habitat fragmentation on plant pollinator interactions in the tropics. Pp. 339–360 *in* D. M. Newbery, H. H. T. Prins, and N. D. Brown (eds.), Dynamics of tropical communities. Blackwell Science, Oxford.

Ricketts, T. H. 2004. Tropical forest fragments enhance pollinator activity in nearby coffee crops. Conservation Biology. 18: 1262–1271.

Stephens, D. W., and J. R. Krebs. 1986. Foraging theory. Princeton University Press, Princeton, NJ.

CHAPTER FOURTEEN

A Typology of Pollination Systems: Implications for Crop Management and the Conservation of Wild Plants

Sarah A. Corbet

Where population decline of either plant or pollinator species threatens the pollination of crops or wild flowers, or where unwelcome invasive plants spread by seed, pollination systems need to be managed. To plan effective management, it is necessary to understand which insects are effective potential pollinators of a given plant species and which plants can provide forage to support populations of a given insect species. Management would be relatively easy to plan if all pollinator–plant relationships were either completely specialized or completely generalized. If a single insect species depends on one plant species and acts as its sole pollinator, it is simply necessary to target the two species involved in the mutualism. If all pollinators can pollinate and be supported by all plant species, habitat management for any plant species would help to sustain pollinators of all others. But the real situation lies somewhere between these extremes. Except for some seed–parasite/host mutualisms, few plant species depend on a single pollinator species; it is becoming clear that few even depend on a single taxonomic group of pollinators, as reflected in the concept of pollination syndromes. Instead, most plant species have a range of potential pollinator species, just as most pollinator species have a range of potential floral hosts. Plant species differ in the nature and extent of the range of potential pollinators, and insect species vary in the nature and extent of the range of potential forage sources. Effective management requires some kind of classification of plants and pollinators that takes this diversity into account and makes it possible to assess, for a given species of crop or wildflower, what kinds of insects and how broad a spectrum of insect species are potential pollinators (or, at least, visitors). We also need a corresponding classification of insects that gives some information about the nature and diversity of plant species that can support their populations. A typology of pollination systems is needed. The most widely recognized one, based on pollination syndromes, has proved inadequate to describe real situations. A simpler, more comprehensive typology was devised long ago, and has recently been extended and tested, but has not been widely recognized in the recent literature. In this chapter, I revisit it and suggest that it deserves attention as a basis for both understanding pollination systems of intermediate levels of specialization and planning their management. I offer a tentative interpretation in terms of functional constraints on flower visiting before exploring its implications for the management of crops and the conservation of natural pollination systems.

A Typological Classification of Flowers

Attempts to classify pollinator and flower types have a long and distinguished history (Faegri and van der Pijl 1979; Waser, chap. 1 in this volume), but not all have proved robust with the shift of emphasis from teleological speculation to less subjective analysis. The concept of pollination syndromes as set out by Delpino (1868–1875) describes some (but not all) zoophilous flowers in terms of functional (rather than taxonomic) groups of flowers sharing adaptations that supposedly suit them for pollination by a particular taxonomic (not functional) group of visitors. Statistical analyses of observed visitation patterns have generally given little support to the detailed classification in terms of pollination syndromes (Ellis and Ellis-Adam 1993; Hingston and McQuillan 2000; Ollerton and Watts 2000), and this classification has several disadvantages.

A major shortcoming is the absence of a category for the many species of flowers that are visited by a wide variety of small, short-tongued insects such as flies, beetles, and small Hymenoptera (Knuth 1906, 62). The Apiaceae are the prime members of this group, but the pollination syndrome concept also fails to embrace some Asteraceae (notably the Asteroideae with yellow disc florets surrounded by a ring of ligulate florets, typically white) and other small, often white, clustered flowers with nectar accessible to short-tongued insects, such as *Sambucus nigra, Crataegus laevigata,* and *Hedera helix*. Proctor et al. (1996, 173) describe such flowers with accessible nectar and a wide range of visitor species as "catering for the mass market."

A further problem with the concept of pollination syndromes is that they are characterized and named according to the pollinator taxon to which they are supposedly adapted and, hence, according to their supposed coevolutionary history. Incorporating inferred evolutionary pathways into an operational definition is unfortunate because it is impossible to test these evolutionary pathways directly.

Recent analytical approaches (e.g., Ellis and Ellis-Adam 1993) give a more objective basis for categorization by contributing toward the development of an alternative, more comprehensive, typology-based classification that better reflects observed patterns of visitation. An ideal alternative classification should comprise functional groups of flowers, defined in terms of observable shared attributes, and corresponding functional (rather than taxonomic) groups of flower visitors, also defined in terms of observable shared attributes. Furthermore, if possible, the classification should encompass the whole range of zoophilous flowers and potential pollinators.

A candidate classification emerges from Ellis and Ellis-Adam's (1993) analysis of 29,000 published records of flower visits, which involves about 1300 species of plant and about 2600 species of insect. The study is confined to northwestern Europe and limited to insects; some pollinators important elsewhere, such as bats and birds, are not included. The records of insect visits to flowers include visits for nectar, pollen, or both and include larceny as well as legal visits. From the coevolutionary perspective, it would be desirable to segregate larceny from legal visits, and from the point of view of functional interpretation, it would be desirable to separate nectar- from pollen-collecting visits. But even in the inevitable absence of that segregation, and despite its restricted geographical and taxonomic coverage, this analysis reveals informative patterns. Ellis and Ellis-Adam (1993) arrived at a modified form of a classification proposed by Loew (1895) based on Müller's (1881) functional classification of flower types. They adopt Loew's terminology, although their categories are not identical to Loew's and are redefined in terms of functional groups (for plants) and taxonomic groups (for insects), rather than putative adaptive pathways.

Knuth (1906, 66) and Faegri and van der Pijl (1979) describe Loew's three categories of flowers as follows: (1) allotropous flowers are adapted to various short-tongued insects and include wind-pollinated flowers, pollen flowers, and those with exposed (Apiaceae) or partly concealed (Brassicaceae) nectar; (2) hemitropous flowers are "imperfectly adapted" to medium-tongued insects, having concealed nectar (*Thymus*) and/or composite inflorescences (Asteraceae); and (3) eutropous flowers are "adapted to a definite set of long-tongued insects" (table 14.1).

Loew's groups of insects that correspond to allotropous, hemitropous, and eutropous flower groups are listed by Knuth (1906, 193) as follows (table 14.2). Allotropous insects are "only slightly, if at all, adapted to flower visits and of little value for pollination," and floral rewards form a relatively small component of their diet. These are short-tongued species of Hymenoptera, Diptera, and Coleoptera and members of other orders that visit flowers only occasionally. Hemitropous insects are "partly adapted to flower visits and of moderate value for pollination." These are relatively long-tongued fossorial wasps, chrysids, solitary vespoids, and short-tongued bees among the Hymenoptera; conopids, syrphids, and bombyliids among the Diptera; all Lepidoptera, except hawkmoths (Sphingidae); and a few exotic beetles. Eutropous insects are "completely adapted flower visitors of the greatest value for pollination": insects for which floral rewards are the major component of the diet. They comprise long-tongued bees and hawkmoths. Knuth (1906, 195) found agreement between the corresponding groups of insects and flowers.

Table 14.1 Allocation of flower types to allophilous, hemiphilous, and euphilous categories

	Loew (1895) and Knuth (1906)	**Ellis and Ellis-Adam (1993)**	**Expected attributes (this chapter)**
Allophilous	Wind-pollinated flowers, pollen flowers, and flowers with exposed or partly concealed nectar	Flowers with exposed nectar, mainly or all Apiaceae	Flowers with fully exposed nectar and little or no intrafloral temperature elevation
Hemiphilous	Massed and solitary flowers with concealed nectar other than those below	Flowers with partly concealed nectar	Flowers with moderate quantities of partly concealed nectar, often with elevated intrafloral temperatures in sunshine, including cup-shaped flowers and some Asteraceae
Euphilous	Bee and bumblebee flowers; Lepidoptera flowers	Flowers with concealed nectar, massed or solitary (including Loew's bee, bumblebee, and lepidopteran flowers)	Flowers with abundant, deeply concealed nectar

Source: Data from Loew 1895; Knuth 1906–1909; Ellis and Ellis-Adam 1993.

To avoid the confusion that might result from using the same terms for both insects and plants, Faegri and van der Pijl (1979) recommend the endings -philic for categories of plants and -tropic for categories of flower-visiting animals. Preferring the -ous ending used by Davis's translation of Knuth (1906–1909) and by Ellis and Ellis-Adam (1993), I describe plants as allophilous, hemiphilous, or euphilous and insects as allotropous, hemitropous, or eutropous.

Using more sophisticated methods and a larger dataset than was available to earlier workers, Ellis and Ellis-Adam (1993) applied factor analysis to seek clusters of plant taxa based on the visiting patterns of anthophilous insects, and clusters of insect taxa based on the plant genera that they visit. For analysis they separated insects into 13 taxonomic categories: butterflies, other Lepidoptera, Apidae, other bees (Apoidea minus Apidae), Symphyta, Aculeata minus Apoidea, Parasitica, Calyptrata, Brachycera, Syrphidae, other Diptera (Nematocera plus Acalyptrata plus Phoridae), Coleoptera, and the remaining groups of visitors (mainly Heteroptera). This approach revealed three overlapping categories of insect visitors, broadly similar to those of Loew (Ellis and Ellis-Adam 1993, 207) but not identical. Their allotropous insects (966 species), listed in table 14.2, were mainly small and short-tongued; their hemitropous insects (747 species) comprised insects of medium size and tongue length; and their eutropous insects (874 species) were mainly large and long-tongued. In turn, factor analysis

Table 14.2 Allocation of insect groups to allotropous, hemitropous, and eutropous categories, with suggestions for a possible classification based on insect attributes

	Loew (1895)	Knuth (1906)	Ellis and Ellis-Adam (1993)	Expected attributes (this chapter)
Allotropous	Short-tongued Hymenoptera (except Apoidea); short-tongued Diptera (not adapted to flower feeding); Coleoptera; other orders	Unequally and only slightly adapted flower visitors of little value for pollination; broad diet to which flowers contribute only a part (p. 193)	Hymenoptera other than Apoidea; Diptera other than Brachycera and Syrphidae	Low flight threshold temperature; little or no endothermy; < 30 mg body mass if hairy, < 100 mg if not; most flies, small beetles, and short-tongued Hymenoptera
Hemitropous	Short-tongued bees; long-tongued Diptera (partly adapted to flower feeding: syrphids, bombyliids, conopids); most Lepidoptera	Partially adapted flower visitors of moderate value for pollination; hemitropous Hymenoptera also include long-tongued Fossores, chrysids, and eumenids (p. 193)	Brachycera, Syrphidae; Coleoptera; other orders	Larger insects with a moderate flight threshold temperature, including (a) robust, hairy, endothermic insects (some syrphids, short-tongued bees) and (b) slender, poorly-insulated baskers (empid flies, some Lepidoptera with low wing loading, syrphids, and some bombyliids)
Eutropous	Long-tongued bees; Sphingidae and some other Lepidoptera	Completely adapted flower visitors of the reatest value for pollination; long-tongued bees and Sphingidae (p. 194)	Apoidea including Apidae; Lepidoptera	Long-tongued, robust, facultative endotherms with good insulation and high energy requirements (long-tongued bees, sphingids and noctuids, some butterflies with high wing loading, a few syrphids and bombyliids)

Source: Data from Loew 1895; Knuth 1906–1909; Ellis and Ellis-Adam 1993.

showed that the plant genera that received large numbers of insect visits separated well. Those associated with allotropous insects were all Apiaceae, and all Apiaceae were in this group. The less distinct group associated with hemitropous insects included some Rosaceae (such as *Prunus, Crataegus, Spiraea, Rubus*) and some Asteraceae of the subfamily Asteroideae (*Leucanthemum, Achillea*) and other genera including *Salix, Calluna,* and *Ranunculus.* Many of these hemiphilous flowers, especially those that bloom in the cool weather of spring, have a form that can engender high intrafloral temperatures. They include cup-shaped flowers with pale (white or yellow), reflective petals (of the type described as solar furnaces by Kevan 1975), and *Salix* catkins, termed "hairy heat traps" by Kevan (1970, 1990; Corbet 1990). The category of flowers associated with eutropous insects comprised some Asteraceae, mainly of the subfamily Lactucoideae (nomenclature according to Stace 1997), tribes Cardueae (*Centaurea, Cirsium*), and Lactuceae (*Taraxacum, Hieracium*), together with others including *Senecio* (Asteraceae: Asteroideae) and genera in the Fabaceae (*Trifolium,*

Medicago), Dipsacaceae (*Knautia*), and Lamiaceae (*Thymus, Origanum*), as well as nectar-rich flowers such as *Silene* and *Echium* with less-dense inflorescences. Tables 14.1 and 14.2 summarize the differences between Loew's original concept and the classification resulting from Ellis and Ellis-Adam's (1993) analysis.

For analysis at the species level, Ellis and Ellis-Adam grouped plants into five functional flower types: (A) those with fully accessible nectar; (AB) those with partly accessible nectar; (B) solitary flowers with concealed nectar; (B′) flowers aggregated in capitula and with concealed nectar; and (Po) pollen flowers, including wind-pollinated flowers. Factor analysis revealed the distribution of insect visits with respect to these five flower types. As confirmed by analysis of variance, allotropous insects mainly visited flowers with exposed nectar, hemitropous insects mainly visited those with partly concealed nectar, and eutropous insects visited flowers with concealed nectar, especially those aggregated in capitula (table 14.1). Allophilous flowers are often white; hemiphilous flowers are often white or yellow; and euphilous flowers are often blue, pink, purple, or yellow.

Anemophilous and pollen flowers did not form a discrete cluster in the factor analysis plot, and Ellis and Ellis-Adam (1993) did not separate nectar visits from pollen-collecting visits in their analysis. Pollen is usually more accessible than nectar (Free 1993, 31); even in flowers with concealed nectar, pollen is often held at the mouth of the flower. Pollen collectors often represent a wider range of insect taxa than nectar collectors on a given plant species; for example, hemitropous hoverflies take pollen from flowers of bluebell, *Hyacinthoides non-scripta* (Corbet and Tiley 1999), even though they cannot reach the deep-seated nectar. In such cases, pollen-collecting insects may represent a hemitropous component in the overall visitor spectrum of euphilous flowers. In the interpretation that follows, I focus on the accessibility of nectar and the insects that collect it, paying less attention to flowers that lack nectar or to insect visitors foraging only for pollen.

The classification adopted by Ellis and Ellis-Adam (1993) does not coincide entirely with major taxonomic groups of either plants or insects. For example, both Asteraceae and Hymenoptera: Aculeata are represented in more than one category. Ellis and Ellis-Adam classified some genera of Asteraceae as euphilous and others as hemiphilous. Much earlier, Allen (1891, 51ff) drew attention to the marked differences in insect visitors between three groups of Asteraceae with different predominant flower colors. Members of the cynaroid tribe (the Cardueae of Stace 1997: thistles and knapweeds) predominantly have purple flowers and are visited by bees and other insects; Ellis and Ellis-Adam found that these flowers fell among the euphilous forms. Members of Allen's "ligulate tribe"—Stace's Lactuceae—predominantly have yellow flowers. *Taraxacum* and *Hieracium* appeared among Ellis and Ellis-Adam's euphilous flowers but were less distinctly separated from other groups than the Cardueae. Allen's "corymbiferous tribe," more or less equivalent to Stace's Asteroideae, comprises the radiate daisy-like composites, including those with a yellow disc surrounded by white ray florets such as *Leucanthemum* and *Matricaria*. These are even closer to the uncategorized basal region of Ellis and Ellis-Adam's factor analysis plot: relatively less important to eutropous insects and more important to hemitropous ones. None of the Asteraceae listed by Ellis and Ellis-Adam (1993) were predominantly visited by allotropous insects.

A plausible inference is that the nectar of Cardueae is more deeply concealed than that of Lactuceae or Asteroideae. It is not obvious how the effective corolla depth of Asteraceae should be measured: the full length of the corolla tube often exceeds the tongue length of foraging insects and the basal tube often seems too narrow to accommodate the tongue. Droplets of nectar are often seen in the corolla just above the region where the corolla flares and the stamen filaments arise. Corbet (2000) measured the effective depth to nectar in Asteraceae from the division of the perianth segments down to the region where the corolla flares. For example, *Centaurea nigra*, which includes eutropous insects in its visitor spectrum, had a mean depth to nectar of 4.8 ± 0.23 mm (sem, $N = 10$).

A Typological Classification of Insects?

A taxonomic grouping of insects, such as that arrived at by Ellis and Ellis-Adam (1993), would be an adequate surrogate for a functional grouping if the taxonomic elements fed into the analysis were homogeneous with respect to relevant functional attributes. But, just as the Asteraceae include species with both euphilous and hemiphilous flowers, some insect taxa, notably the Apoidea, include species that visit flowers of both types. Ellis and Ellis-Adam's analysis will be less clearly defined if these insect taxa are not subdivided.

In his original set of "adaptational types," Delpino distinguished between "large-bee flowers" (termed *melittophilous*) and "small-bee flowers" (*micromelittophilous*) visited by small bees "and a great variety of other small insects" (Knuth 1906, 63). Ellis and Ellis-Adam (1993) found that members of the Apidae and other members of the Apoidea predominantly visited euphilous flowers. Apoidea other than Apidae include both long-tongued and short-tongued forms and, by lumping these together in the analysis, Ellis and Ellis-Adam may have failed to recognize that different species visit flowers of different types. Members of the genus *Lasioglossum* (Halictidae) are small, slender, relatively hairless species with short tongues, are almost incapable of endothermy (Stone and Willmer 1989), and forage, at least for pollen, on allophilous (Apiaceae) as well hemiphilous (e.g., *Achillea, Rubus fruticosus,* and *Ranunculus*) and euphilous flowers (e.g., *Centaurea, Echium,* and *Taraxacum;* Westrich 1989). The nectar of euphilous flowers such as *Cirsium* and *Centaurea* must often be inaccessible to these short-tongued bees. It would be interesting to see whether an analysis with finer resolution in terms of insect taxon and visit type would separate the short-tongued bee species from the long-tongued ones.

Ellis and Ellis-Adam (1993) subdivided the order Diptera, in some cases down to the level of the family, and found that it includes forms predominantly associated with hemiphilous plant species (notably syrphids and some Brachycera) as well as taxa that focus on allophilous flowers (other Diptera). But there are insects of more than one type even within a family. Other observations show that the long-tongued, hairy-bodied bombyliids of the genus *Bombylius* (Brachycera), which are endothermic and have a high threshold temperature for flight, visit euphilous flowers such as *Hyacinthoides, Primula vulgaris,* and *Glechoma hederacea,* whereas smaller, short-tongued bombyliids of other genera visit hemiphilous and allophilous flowers (Stubbs and Drake 2001). Similarly, within the Syrphidae, the long-tongued *Rhingia campestris* takes nectar from euphilous flowers such as *Glechoma hederacea* (Gilbert 1981a), whereas most hoverflies visit hemiphilous flowers such as *Crataegus, Rubus,* and *Ranunculus* (Gilbert 1981b).

Similar considerations apply to other orders of insects. Beetles, classified as allotropous by Loew and hemitropous by Ellis and Ellis-Adam, include species that visit both allophilous and hemiphilous flowers. Flower visits by the smaller moths are probably underrecorded. Because of differences in endothermic capacity (see following discussion) and power requirements for flight among moths (Heinrich 1993), it may prove that some small-bodied, weakly-flying species predominantly visit hemiphilous flowers rather than the euphilous flowers visited by the robust, endothermic hawkmoths and noctuids. Delpino recognized a group of flowers pollinated by hawkmoths and noctuids (confusingly termed *sphingophilous*), both of which are robust and endothermic (Knuth 1906, 64). Thus, the typological classification of insects does not always correspond with the taxonomic groupings which Ellis and Ellis-Adam were obliged to use. This is particularly true when cornucopia flowers and supergeneralist insect species are involved; visitors to the cornucopian allophilous *Heracleum sphondylium* include hemitropous and eutropous insects, as well as the expected allotropous forms (Knuth 1908, 495–500; Corbet 1970), and the versatile eutropous supergeneralist honeybee *Apis mellifera* sometimes includes hemiphilous species such as *Ranunculus ficaria* in its foraging spectrum (Knuth 1908, 33).

To correspond with Ellis and Ellis-Adam's (1993) typological classification of flowers, we need a typological classification of insects based on physical or physiological attributes of each species of insect. To identify the groups of flower types, these authors analyzed plant records at the species level, rather than the generic level, but a parallel species-based analysis for insects would be much more difficult because there are more species of insects, each with fewer recorded plant hosts, and because relevant attributes such as the body mass and tongue length of each species are not readily discovered. The value of any such future typological classification of insects will depend in part on the wisdom with which insect attributes are selected. Therefore, I attempt a preliminary interpretation of the relationship between Ellis and Ellis-Adam's plant and insect groups to draw attention to some attributes of insects that may prove relevant to their preferences for allophilous, hemiphilous, and euphilous flowers.

Constraints on the Profitability of Foraging

A foraging insect can make an energetic profit only if the accessible reward per flower exceeds the energetic cost of a visit. It has been suggested that the ability of a flower-visiting insect to profitably forage for nectar depends largely on three constraints: the depth threshold (the tongue must be long enough to reach the nectar), the cost threshold (the energy content of the nectar accessible in each flower must exceed the per-flower foraging cost), and the temperature threshold (the body temperature must be high enough to permit foraging flight; Corbet et al. 1995). A further constraint which may be relevant is the cognitive ability of the insect, as recognized by Knuth (1906, 113, 191) when he described allotropous insects as "stupid" and eutropous insects as more "intelligent." Linked with this constraint is the insects' ability to learn to visit complex flowers efficiently, and the proclivity for flower constancy. Although the cognitive abilities of one group of anthophilous insects—bees—have received much attention (e.g., Chittka and Thomson 2001), they are not considered further here.

The categories of insects recognized by Loew (1895) and Ellis and Ellis-Adam (1993) differ in both body size and tongue length. I suggest that the relevance of an insect's body size to its choice of flower type may depend on the insect's energy and temperature balance, acting via the cost threshold and the temperature threshold.

Cost Threshold

When a population of insects forages on a patch of open flowers, and depletes nectar faster than it is replaced by secretion, the standing crop (the quantity of nectar in each flower) falls. When the energy content of the nectar in each flower equals or falls below the energetic cost of a foraging visit, the insects can no longer make an energetic profit and are expected to leave the patch. If several different species of insects are foraging on the patch, those with the lowest net energy gain per flower are expected to depart first; these are often those with the highest foraging costs. As the reward is further depleted, eventually the species with the highest net energy gain per flower monopolizes the patch; this is often the species with the lowest foraging costs. If foraging costs scale with body mass (Heinrich 1972), the largest insects are likely to be most vulnerable to reward depletion and the smallest may be the last to leave.

Depth Threshold

If the nectar is at the base of a deep corolla, continued depletion at a rate exceeding the secretion rate will progressively lower the meniscus. Insects with shorter tongues will stop foraging when the nectar level falls so far that they cannot reach enough nectar to meet their energy costs, until eventually the species with the longest tongue, or one with a tongue at least as long as the corolla, will monopolize the patch, maintaining the nectar at a level at which this species can just reach it and others with shorter tongues cannot. For a large insect with correspondingly high energy demands, a long tongue gives access to large nectar rewards protected from depletion by smaller insect species. The importance of tongue length is recognized by Ellis and Ellis-Adam (1993), Knuth (1906), and Loew (1895). Allophilous flowers have accessible nectar and are visited by short-tongued insects; euphilous flowers have deep-seated nectar, accessible only to long-tongued insects (except by larceny); and hemiphilous flowers have semi-concealed nectar and their visitors have tongues of moderate length.

Temperature Threshold

Many insects cannot fly unless their thoracic muscles are warm enough. Insect taxa vary in their requirement for a high thoracic muscle temperature for flight and in the relative importance of behavioral thermoregulation and endothermy (metabolic warming). In large insects, such as bumblebees and hawkmoths, the minimum threshold thoracic temperature for flight is high, often 35–40°C. Small, slender moths and flies can fly at much lower thoracic temperatures, with thresholds as low as 0–5°C. The minimum threshold temperature for flight generally rises with body mass (Heinrich 1993). Also scaling with size, at least in bees, is the capacity for endothermy (Stone and Willmer 1989). Large insects such as bumblebees and hawkmoths can warm their muscles metabolically, and this facultative endothermy (preflight warm-up) allows them to initiate flight at ambient temperatures well below the flight threshold; such insects are generally robust and well insulated. As the surface-to-volume ratio and the rate of convective cooling increase with decreasing mass, endothermy becomes less effective in small insects, and insects of body mass less than about 30 mg cannot maintain a substantial temperature difference between the thorax and the ambient temperature by endothermy (Willmer et al. 2000). The flight threshold temperature for these small insects is relatively low, but when ambient temperatures are below the flight threshold the only way they can elevate the thoracic temperature to achieve flight is by behavioral thermoregulation: adjusting their position and posture to maximize radiative gain, typically by basking in places where incident radiation is high and the rate of convective cooling is reduced by wind shelter.

Allophilous flowers—often open, flat flowers with small aliquots of unprotected nectar—are typically visited mainly by small insects: those that have a low minimum threshold temperature for flight and little or no capacity for endothermy. Such insects are often slender, with a large surface-to-volume ratio, and sparse or absent insulating hairs or scales. Euphilous flowers are visited by large, robust-bodied, well-insulated insects that have high power requirements and a correspondingly high minimum temperature threshold for flight and a good capacity for endothermic warming. Given the high energetic costs of flight and the endothermy that makes it possible, such insects require large aliquots of nectar sugar per flower for profitable foraging. Hemitropous insects are of moderate size: large enough to have a minimum threshold temperature for flight that is often above ambient temperature but too short-tongued to tap the resources of nectar-rich euphilous flowers. At least some of the larger hemitropous species are capable of effective endothermy (e.g., some syrphids; Heinrich 1993), but for many delicate, poorly insulated, nonendothermic forms, warming of the flight muscles is probably assisted by the structure of hemiphilous flowers. These are typically cup-shaped, with pale (white or yellow) reflective petals that act as parabolic reflectors and focus incident radiation at the center of the flower, possibly reducing convective cooling rates by acting as wind shields and elevating the thoracic temperature of insects basking within the flower by about 5–15°C above ambient (Kevan 1970, 1975; Heinrich 1993). Of the main hemiphilous flowers listed by Ellis and Ellis-Adam (1993), many of the spring-blooming flowers (Rosaceae, *Ranunculus*) are cup-shaped, and others bloom in full summer, when the temperature threshold is a less important constraint on foraging. Cup-shaped flowers may be particularly important to hemitropous insects at high latitudes and altitudes, where ambient temperatures are low but incident radiation may be high and where flies form a relatively high proportion of the flower visitor spectrum (Lindsey 1984; Hingston and McQuillan 2000).

Integrating the Constraints

Individual allophilous flowers have small aliquots of nectar and are visited by small insects with relatively low power requirements, and a low minimum temperature threshold, for flight. Both nectar and pollen are freely exposed, and precise positioning of the insect on the flower is probably not a prerequisite for effective pollination. The pollinator spectrum is wide (Bell 1971).

Euphilous flowers are visited by large, facultatively endothermic insects. These flowers have large aliquots of nectar, which are typically held in a deep corolla tube or spur that protects them from depletion by short-tongued insects. The protective devices that exclude small insects often dictate the position of the probing insect. Precise pollen placement on the insect's body enhances the pollination value of long-tongued visitors but may restrict the spectrum of effective pollinators.

Hemiphilous flowers are intermediate. Their visitors include insects that have flight threshold temperatures above the usual ambient temperatures and which elevate the thoracic temperature by endothermy (e.g., some syrphids) or basking, often in cup-shaped flowers (Kevan 1970). Hemiphilous flowers are larger and often contain more nectar than allophilous flowers, but, in the absence of a deep corolla, depletion by other insects may keep the standing crop of nectar too low for the more energy-demanding eutropous hawkmoths and bumblebees.

Allophilous flowers and many hemiphilous and euphilous flowers are typically massed in dense inflorescences. Although each flower may contain little nectar, an insect foraging on an inflorescence can make a substantial energy gain even if several of the flowers have been emptied, because flower-to-flower movements do not involve the energetic cost of flight. Some euphilous plant species have large, solitary flowers held so far apart that a foraging insect must fly from flower to flower. The energetic cost of visiting such a flower is high, and visits to empty flowers reduce the overall foraging profit. The cost threshold must operate as a severe constraint in such flowers, and adaptations that reduce the range of visitors capable of depleting the nectar may be more important for euphilous flowers than for flowers that are massed or that are visited by less energy-demanding insects.

Related Studies

Compartments

In a search for compartmentalization in a flower-visitor web in two English meadows, Dicks et al. (2002) found two distinct compartments. One, equivalent to the eutropous/euphilous group of Ellis and Ellis-Adam, comprised bees and butterflies visiting nectar-rich flowers including *Centaurea* and *Cirsium* (Asteraceae tribe Cardueae), *Lotus corniculatus,* and *Trifolium repens.* The other, comprising flies (Syrphidae and Brachycera) visiting *Filipendula* (Rosaceae), *Leucanthemum,* and *Achillea* (Asteraceae, subfamily Asteroideae), coincided largely with the hemitropous/hemiphilous group of Ellis and Ellis-Adam (1993). The allotropous/allophilous group was not represented; apart from four umbels of *Anthriscus sylvestris* at one site, Apiaceae were absent.

Pollination Syndromes

Because pollination syndromes supposedly represent suites of attributes evolved in relation to particular groups of pollinators, we might expect them to comprise subdivisions of the euphilous category, originally held to include flowers "more or less exclusively adapted to a definite set of insects possessing a long proboscis" which are, in turn, "completely adapted flower visitors" (Knuth 1906, 66, 194). However, the situation is more complicated. Faegri and van der Pijl (1979) point out that both cantharophilous ("beetle") and myiophilous ("fly") flowers include allophilous and hemiphilous forms. Others, such as melittophilous flowers, commonly receive visits from long-tongued lepidopterans and even hummingbirds, as well as the bees for which they are named. A classification based on animal taxa does not exactly coincide with a classification based on functional groups.

Ollerton and Watts (2000) analyzed floral features attributed to different pollination syndromes (including noninsect syndromes) by using multidimensional scaling and hierarchical cluster analysis. Among the insect groups considered, two major groups of syndromes appeared: one contained bees, moths, and butterflies; and the other, beetles, flies, and wasps. The former more or less coincides with the eutropous group of Ellis and Ellis-Adam (1993), and the latter includes those few allotropous and hemitropous insects for which syndromes have been articulated. As Ollerton and Watts (2000, 156) point out, the syndrome classification fails to include most allophilous flowers, yet the Apiaceae "have evolved and flourished despite . . . the inability of biologists to shoehorn them into syndromes."

Hingston and McQuillan (2000) recorded flower visitors in Tasmania and categorized the insects into taxonomic groups associated with the pollination syndromes described by Faegri and van der Pijl (1979). Using ordination (semi-strong hybrid multidimensional scaling) to explore relationships between plant species on the basis of their visitor profiles, they found clusters of flowers visited by butterflies, flies, beetles, wasps, and bees, but no evidence of associated syndromes of floral attributes associated with any of these groups except bees. Zygomorphic flowers with nectar guides and concealed rewards were largely visited by bees, in accord with the (long-tongued) bee pollination syndrome of Faegri and van der Pijl (1979), but at high altitudes such flowers were largely visited by flies, and many of the flowers visited by bees in Tasmania were actinomorphic with exposed rewards, which conforms with the local predominance of short-tongued bee species. Pollination syndromes were unreliable predictors of floral visitors in Tasmania.

Pollination syndromes represent finer categories than the allophilous/hemiphilous/euphilous trichotomy, and, if they are real entities, would require an even larger database than that of Ellis and Ellis-Adam (1993) to resolve them statistically, given that flowers apparently conforming to a given syndrome often receive out-of-syndrome visitors (Herrera 1996). Aigner (2001; chap. 2 in this volume) has shown how floral features that apparently adapt a flower to one of several effective pollinators may be selected for, even in a mixed pollinator climate, if they do not compromise the efficacy of other pollinators.

Which Groups of Flowers Are Morphologically and Ecologically Specialized?

The generalization/specialization spectrum might be expected to correspond to the range from small, open allophilous flowers to large, deep euphilous flowers.

There are more species of small, short-tongued insects than of large long-tongued ones in the dataset of Ellis and Ellis-Adam (1993); thus, small, open flowers have a larger pool of potentially interacting insect species, and are potentially more ecologically generalized, than large, deep flowers. Allophilous flowers are surely not ecologically specialized, but are these Apiaceae morphologically specialized in the sense of showing features that enhance the efficacy of visits by effective pollinators? For such open flowers, morphological specialization may not necessarily result in a narrowing of the pollinator spectrum. Floral adaptation to small insects with short tongues may involve further shortening of the corolla, which makes the nectar accessible to more of the many small insect species with short tongues and low energy requirements. Such open flowers do not filter their visitors by depth threshold and rarely show narrow specificity with respect to pollinators or insect visitors. Except in a few cornucopia species like *Heracleum,* the cost threshold limits the visitor spectrum by excluding large, energy-demanding eutropous insects. For small allotropous insects, net rates of energy gain can be large relative to the vector's small requirements, and nectar depletion correspondingly unimportant, in dense inflorescences such as those of the Apiaceae. Foraging success and pollinating efficacy do not depend on precise positioning behavior of the vector, and ecological generalization may be a positive advantage to the plants.

Within the family Apiaceae, Bell (1971), Ellis and Ellis-Adam (1994), and Lindsey and Bell (1985) showed specialization in the sense of some segregation of flower-visitor spectra among genera; but, with few exceptions, the ecological specialization of Apiaceae is weak in that these plants are visited, and probably also pollinated, by a wide range of insect species from several orders. This whole plant family might be regarded as morphologically specialized for ecological generalization in the sense that its members show several adaptations likely to increase the range and frequency of insect visits: exposed nectar, white petals, and flowers massed into umbellate inflorescences, often with extra-large peripheral petals.

The effective pollinators of euphilous flowers are large, hairy, endothermic insects with long tongues. For these energy-demanding insects, the cost threshold is critical, and selection may act on the plant to reduce nectar depletion by smaller insects. This could involve lengthening of the corolla, obstruction of access to the nectar by a "gate" that requires force to open it, complex handling requirements that give an advantage to insect species that can use learning to reduce their handling time, or flowering at a season (early spring) or time (night) at which temperatures are too low for efficient foraging flight by smaller species with little or no capacity for endothermy. Thus, adaptive narrowing of the pollinator spectrum can arise if some visitors contribute negatively to plant fitness (Aigner 2001). Evidently morphological specialization can result in ecological specialization in euphilous flowers.

Differences among insect species in pollinating effectiveness may be greater for euphilous flowers than for allophilous flowers. Wilson and Thomson (1991) illustrate the importance of such differences and show how depletion of pollen, as well as nectar, by poor pollinators can have negative effects on overall pollination success. Differences arise, first, because eutropous insects include bees, which forage for brood as well as for themselves; second, because the mechanics of probing for deep-seated nectar require the bee to adopt a consistent position allowing for precise pollen placement; and third because bees (Laverty 1994) and butterflies (Lewis 1993) can learn to "major" on particular species in a mixed-flower community. Possingham (1989) modeled a situation involving two species of plant and two species of foraging bees, showing how the bee species with the greater net profitability on the more profitable flower species can monopolize foraging on that species. Williams (1997) has shown that bumblebee species do assort themselves among flower species in the manner predicted by Possingham's model; such transient specialization is expected to benefit both the flower and the bee (Pellmyr 2002).

Rich nectar production, required for profitable foraging by large insects, is not necessarily obvious without careful measurement, but the deep corolla tube or spur often associated with long tongues forms a conspicuous index of ecological specialization. There is phylogenetic evidence that morphological specialization along this route has sometimes been reversed (McDade 1992; Armbruster 1993); evidently the evolutionary trajectory sometimes tends toward ecological generalization and sometimes toward ecological specialization, depending on the pollinator climate (the range of pollinators available; Grant and Grant 1965).

This natural reversal of an evolutionary trend toward ecological specialization is, perhaps, echoed in the plant breeders' response, when the nectar of red clover is not readily accessible to the relatively short-tongued honeybees. Attempts have been made to breed red clover cultivars with shorter corolla tubes (Free 1993).

Cornucopia Flowers

Ellis and Ellis-Adam's (1993) study highlighted the uneven distribution of insect visits among plant species, as also found by Dicks et al. (2002): some cornucopia plant species received visits from many insect species whereas others had a narrower visitor spectrum. Cornucopia species are plants that have a particularly wide range of visitor species because of an unusually rich floral reward (Carpenter 1983). Such species are more ecologically generalized than their relatives. This is evident in allophilous flowers such as *Heracleum sphondylium* and hemiphilous flowers such as *Rubus fruticosus,* which are partly freed from the constraint of the cost threshold that normally excludes eutropous insects from allophilous and hemiphilous flowers. In euphilous flowers, abundant nectar is a regular feature, and it is the reduced depth threshold that confers cornucopia status by giving access to shorter-tongued insects. Ellis and Ellis-Adam (1993) list *Cirsium arvense* and *Centaurea* species as euphilous cornucopia flowers. In addition to eutropous species, their wide visitor spectrum includes short-tongued hemitropous and even allotropous insects. Perhaps these visit for pollen, or perhaps the abundant nectar rises to a level within reach of short tongues, so that the depth threshold no longer limits the clientele.

Ellis and Ellis-Adam's 60 highest-ranking plant species (in terms of numbers of insect visitor species) were biennials (10), woody species (12), and perennials (38); none were annuals. A high proportion, including Apiaceae, Asteraceae, and Dipsacaceae, had flowers massed in dense inflorescences, which enhance the net rate of energy gain of foraging insects because these can walk from flower to flower instead of requiring energetically expensive flight. Many of these cornucopia flowers are common species, often occurring in disturbed habitats. The outright winner was *Heracleum* with 436 insect species recorded as visitors, followed by *Taraxacum* (375) and *Daucus* (314).

Which Groups of Insects Are Morphologically and Ecologically Specialized?

Specialization by insects is expected to improve foraging profitability. An insect might achieve this by reducing foraging costs per unit time or by increasing the reward gained per unit time.

Foraging costs can be reduced by decreasing mass, but this is also likely to reduce thermogenic capacity. This would be a disadvantage for large insects (but not for small ones, whose capacity for endothermy is negligible), especially in the cool climates of high latitudes, but perhaps less so in warm climates, where small species of solitary bees with long tongues (such as species of *Osmia, Anthophora,* and *Eucera*) are relatively numerous.

The reward gained per unit time can be increased by increasing tongue length or, more probably, adjusting tongue length to approximate corolla depth, which maximizes the rate of nectar extraction (see Harder 1983); by increasing the capacity for larceny; or by improving efficiency by learning. In social species, improvements in recruitment, or aggressive defense of the reward, may also help to ensure that a colony gains a high proportion of the available reward. Which of these changes results in ecological specialization?

With the possible exception of seed parasites, allotropous insects are unlikely to show tight evolutionary adaptations for feeding on floral rewards, because flowers contribute only a small part of the diet and open flowers are accessible to all insects, regardless of whether or not they have special tools. Such insects are commonly regarded as ecologically generalized as well as morphologically generalized, because there are more plant species with small, open flowers than with large, deep ones (Ollerton and Watts 2000), although they are ecologically specialized in the sense that the depth threshold prevents them from profitably foraging on deep flowers.

To some extent, large, eutropous insects are necessarily ecologically specialized because there are fewer large, deep flower species than small, open flower species and because the cost threshold excludes them from profitably foraging on nectar-poor flowers. Large insects with long tongues may find their resource repertoire further restricted by the requirement that they minimize handling time by foraging on flowers about the same depth as their tongue (Harder 1983), although this constraint seems to be less important for butterflies than for bees (Corbet 2000).

Nearly all social bee species are necessarily ecologically generalized in the sense that their colonies outlast the flowering season of individual flower species (Pellmyr 2002). Honeybees are extraordinarily generalized in their flower exploitation (Ellis and Ellis-Adam 1993), although they may not pollinate all the flowers on which they forage and there is little or no evidence for a coevolved "honeybee pollination syndrome" (Westerkamp 1991). Similarly, *Bombus terrestris,* which overcomes the depth threshold by robbing rather than by having a long tongue, visits a wide range of different nectar-rich flower types. It will be interesting to monitor its impact on pollination systems in Tasmania, where it has recently become established (Buttermore 1997). In contrast, *B. hortorum,* with a very long tongue, visits only very deep flowers. Among solitary bee species, some have restricted host ranges by virtue of a requirement for oil (*Macropis europaea* on *Lysimachia,* or *Rediviva* on *Diascia*) or for specific pollens, the scent of which is imprinted on the developing larvae (Dobson 1987); others are more generalized. The small, short-tongued species of *Lasioglossum* visit a wide range of forage plants, including species with small, open flowers (Westrich 1989), in the company of short-tongued flies and beetles.

Particularly interesting are the relatively small, long-tongued solitary bees such as *Anthophora* and *Osmia.* Their long tongues mean that they are seldom excluded by the depth threshold, and their small size means that they are seldom excluded by the cost threshold, although it also means that they would often be excluded by the temperature threshold in cooler climates. *Anthophora plumipes,* a solitary bee with unusually good thermoregulatory abilities for its size, overcomes this temperature problem, foraging at air temperatures down to 0°C (Stone 1994), but presumably the energetic cost of endothermic warm-up means it is more often excluded by the cost threshold than other bees of corresponding size. *Anthophora plumipes* has a limited range of flowering plant species available to it for foraging because it is active in early spring, when few nectar-rich flowers are open. This might be regarded as a type of phenological specialization.

Generalist Insects

Ellis and Ellis-Adam (1993) found that a few species of insect visited an especially wide range of plant species. The winner was the honeybee *Apis mellifera,* followed by the butterflies *Pieris napi* and *P. brassicae.* Smaller and less familiar insect species may be underrepresented in the dataset because they are less easily identified on sight, but the remarkable versatility of the honeybee in exploiting a wide range of flower types is a real phenomenon (Westerkamp 1991).

Practical Implications

Crops

Does this typology help us to generalize about the pollination of crops in ways that can inform pollinator management? De Rougemont (1989) lists the major crops grown regularly in Britain. Table 14.3 summarizes an analysis in which I took the unit as a species (thus *Brassica oleracea* appears only once, but sweet cherry, *Prunus avium,* and morello cherry, *Prunus cerasus,* are listed separately) and classified the species according to Ellis and Ellis-Adam's (1993) flower type trichotomy based on their list of genera (their fig. 4), or on whether the nectar is fully concealed or not, and on what the main visitors are (other than the ubiquitous supergeneralist honeybee; see Free 1993). There is an association between flower type, crop type, and plant family (table 14.3). All major forage crops are Fabaceae (euphilous) or Brassicaceae (hemiphilous or euphilous); nearly all culinary herbs are Lamiaceae (euphilous); and all major tree fruits and many soft fruits are Rosaceae (hemiphilous), which produce pale, cup-shaped flowers early in the year, when foraging weather is unpredictable. Herbaceous vegetables grown for the root, stem, or leaf are biennial/perennial and include an uninformative mixture of flower types from allophilous Apiaceae to euphilous Asteraceae (Lactuceae), but herbaceous crops grown for the fruit (marrow, tomato) or seed (oilseed crops, leguminous seed crops) are of particular interest. Such herbaceous crops are necessarily annual, producing a rapid yield. Annuals generally have smaller, less-nectar-rich flowers and are visited less often than perennials by large bees (Fussell and Corbet 1992; Corbet 1995; Petanidou and Smets 1995) and syrphids (Ellis and Ellis-Adam 1995), but the species selected as annual seed and fruit crops are unusual among annuals in that they have large flowers with abundant concealed nectar and produce correspondingly large seeds and fruit. Of nine such species grown regularly in Britain, all are euphilous (including four species of leguminous seed crops) or hemiphilous/euphilous (mustard, *Brassica juncea*).

As other eutropous pollinators decline (Williams 1982, 1986), many euphilous or hemiphilous crops are rescued from pollen limitation by the ubiquitous honeybees, whose value as pollinators lies in their remarkable versatility. They can exploit a wide range of flowers types that are moderately nectar rich and moderately deep, and they have excellent learning and recruitment capacities. Honeybees are moderately large (but not large enough to be very effective foragers at low ambient temperatures) and have moderately long tongues (but not long enough for rapid handling of deep flowers). They are relatively ineffective pollinators on some crops because their tongues are too short to reach the nectar except by secondary larceny (red clover, *Trifolium pratense;* field bean, *Vicia faba*), because they cannot buzz-pollinate (kiwifruit, *Actinidia deliciosa;* blueberry, *Vaccinium* spp.), because of their reaction to tripping (alfalfa, *Medicago sativa*), or be-

Table 14.3 Major British crops, categorized by flower type on the basis of lists and flower types of Ellis and Ellis-Adam (1993)

Crop type	Annual (a) or perennial (p)	Allophilous species	Hemiphilous species	Hemiphilous/ euphilous species	Euphilous species	Main family	Fraction of species in main family
Herbs	8p	1	1	0	6	Lamiaceae	6/8
Vegetables	9a, 14p	5	6	3	9	Brassicaceae	5/23
Oilseeds	4a	0	0	2	2	Brassicaceae	2/4
Forage crops	8a, 5p	0	0	3	10	Fabaceae	9/13
Soft fruits	9p	0	4	3	2	Rosaceae	4/9
Tree fruits	7p	0	7	0	0	Rosaceae	7/7

Source: Data from de Rougemont 1989.
Note: Annuals are separated from perennials, a category which here includes shrubs and trees as well as biennial and perennial forbs.

cause of their poor thermogenic ability (early-flowering orchard crops, red clover in northern Europe). Generally, where honeybees can pollinate effectively (and sometimes when they cannot), hives are introduced to the crop and the role of native pollinators remains unquantified (Kremen et al. 2002).

Where honeybees are ineffective pollinators, attempts are sometimes made to manipulate the honeybees or the crops to achieve a better match: by cultural methods (adjusting the time of flowering or the timing of honeybee introduction or, for hybrid production, adjusting the spatial arrangement of crops in the field) or by breeding to adjust the depth threshold (especially in euphilous flowers), the temperature threshold (especially for hemiphilous flowers), or the cost threshold. In red clover, attempts have been made to breed honeybees with longer tongues and clover with shorter flowers (Free 1993, 285–286). Strains of *Megachile rotundata* have been selected to fly at lower temperatures for pollination of alfalfa in northern Alberta, Manitoba, and Saskatchewan (Free 1993, 89). Where honeybees tend to desert the crop for more profitable alternative forage elsewhere, it has been suggested that the nectar secretion rate of the crop could be increased by selective breeding (e.g., Free 1993, 261).

Alternatively, poor honeybee performance is remedied by managing different species of bee. Specialists such as *Peponapis* on cucurbits or *Habropoda* on rabbiteye blueberry, *Vaccinium ashei* (Free 1993, 207, 219), may be excellent on the crop of choice but less useful on other crops, but other species of solitary bee have proved valuable on a wider range of different crops. Thus, species of *Osmia* with low temperature thresholds for foraging are managed as pollinators of various species of hemiphilous orchard fruits (e.g., Bosch and Kemp 2001).

Invasive and Declining Plant Species

Does this typology help us to draw general conclusions about invasive and declining plants? It has been suggested that alien plants invading new regions are more likely to establish and spread if the pollination system is generalized rather than specialized (Baker 1965; Woodell 1979; Pheloung et al. 1999; Preston et al. 2002a; Myers and Bazely 2003) and that ecologically specialized (euphilous)

Table 14.4 Numbers of representatives of different flower types for flowering plant species listed by Preston et al. (2002b) as those with the highest change index (increasing species) and those with the lowest change index (decreasing species) in the British flora

	Allophilous	Hemiphilous	Hemiphilous/ euphilous	Euphilous	Total
Increasing species					
Natives	0	10	3	8	21
Archaeophytes	0	4	1	2	7
Neophytes	2	12	7	26	47
Decreasing species					
Natives	2	17	5	16	40
Archaeophytes	5	12	2	12	31
Neophytes	0	3	1	6	10

Note: Nonflowering plants and anemophilous and pollen flowers are excluded (see text).

plant species are expected to be more vulnerable than ecologically generalized (allophilous) species to population decline resulting from the impact of habitat change on pollinator populations (Johnson and Steiner 2000). If open-flowered allophilous species are better colonists than deep-flowered euphilous species, and less vulnerable to decline, then plants with euphilous flowers are expected to be poorly represented among invasive species but well represented among species in decline. The publication of the *New Atlas of the British and Irish Flora* (Preston et al. 2002a) provides an opportunity to test this expectation. For each British plant species, Preston et al. (2002b) calculated an index of change proportional to the change in the numbers of 10 km squares from which the species was recorded between two recording periods (1930–1969 and 1987–1999). They list the hundred species with the highest values of the index (species showing rapid increase) and the hundred with the lowest values of the index (species showing rapid decrease). After deleting ferns, anemophilous plants, and species known to have pollen-only flowers, I classified the species on these two lists into two categories: allophilous and hemiphilous, or hemiphilous/euphilous and euphilous (table 14.4). If flower type affects invasiveness in the suggested way, we might expect euphilous species to be relatively less abundant among plants with a high change index (increasers) and relatively more abundant among those with a low change index (decreasers). Surprisingly, there is no significant association between flower type (allophilous plus hemiphilous vs. hemiphilous/euphilous plus euphilous) and increaser/decreaser category ($\chi^2 = 1.86$, df $= 1$, $P = .173$, $N = 156$).

Other attributes may outweigh the pollination system as determinants of successful invasion of new areas. Preston et al. (2002b) demonstrate a strong association between change index and plant status as native species, archaeophytes (naturalized before 1500), or neophytes (naturalized since 1500). Neophytes

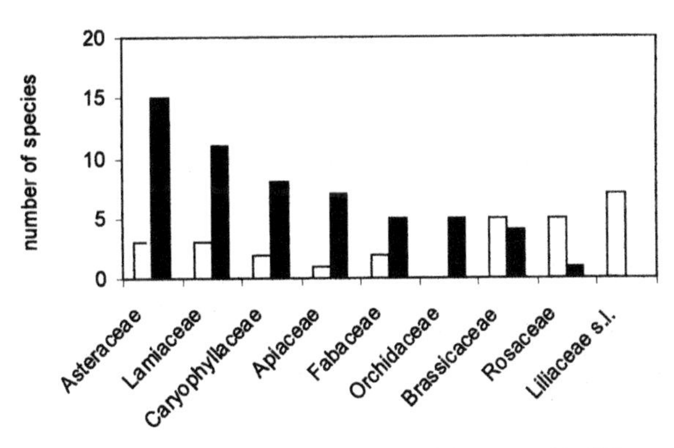

Figure 14.1 The numbers of species of increasers (white) and decreasers (black) in each of the families represented by five or more species in Preston et al.'s (2002b) lists of species with the highest and the lowest change index (increasers and decreasers, respectively) in the British flora. Nonflowering plants and anemophilous and pollen flowers are excluded (see text).

tend to have a high change index (to be "increasers"), whereas native species and archaeophytes tend to have a low index (to be "decreasers"). Analysis of the combined group of allophilous plus hemiphilous species confirmed that neophytes are disproportionately numerous among increasers, and natives and archaeophytes among decreasers ($\chi^2 = 16.60$, df = 2, $P < .001$, $N = 67$). The same was true for the combined hemiphilous/euphilous plus euphilous classes ($\chi^2 = 26.95$, df = 2, $P < .001$, $N = 89$). Among increasers, the largest category is euphilous neophytes; among decreasers, the largest category is hemiphilous natives (table 14.4).

Some plant families are more likely to decline than others. Figure 14.1 shows the numbers of species of increasers and decreasers in each of the families that have five or more species in either list. Liliaceae (with 4 of 7 species classed as euphilous/hemiphilous or euphilous) and Rosaceae (0 of 5) are well represented among increasers and Asteraceae (5 of 15), Lamiaceae (10 of 11), Caryophyllaceae (4 of 8), and Apiaceae (0 of 7) among the decreasers.

Other attributes associated with increasers include a global distribution that is Mediterranean (whereas decreasers tend to be Arctic and Boreal) and high nutrient requirements (whereas decreasers tend to be species of nutrient-poor habitats). Crawley et al. (1996) used phylogenetically independent methods to compare aliens (introduced after 500 BC) with native species in the British flora; they showed that aliens were larger plants, with larger seeds, and were more likely to be pollinated by insects than their native counterparts.

The increasers include many species that are euphilous, and the decreasers include many species that are not. Many of the neophytes with the highest change index (increasers) were originally introduced as garden plants, and many of the archaeophytes with the lowest change index (decreasers) are arable weeds, perhaps brought in with grain, or species formerly cultivated for food or medicinal use (Preston et al. 2002a). Perhaps the greater predominance of euphilous flowers among neophytes than among archaeophytes and natives depends in part on the predilection of gardeners for large flowers.

The surprising success of euphilous plants at establishing themselves, once introduced, may owe something to the ubiquity of honeybees, introduced or native, and their unusually broad floral diet. The cornucopian euphilous alien *Impatiens glandulifera* has been shown to compete for pollinators with native euphilous species (Chittka and Schürkens 2001).

Implications for Conservation

When pollination systems are threatened by environmental change, a sound categorization of types of pollinators and plants, and recognition of their relative vulnerability, will be crucial to their effective conservation. The greater focus on euphilous/eutropous species of plants and insects by conservation bodies may largely be due to their greater size, conspicuousness, and familiarity. Even if euphilous/eutropous species are not more vulnerable to decline, this attention may be justified because euphilous and hemiphilous native plant species help to support the pollinators of fruit and seed crops.

Broad-leaved annual arable field crops such as leguminous seed crops and oilseeds, as well as tree fruits and soft fruits, generally have euphilous or hemiphilous flowers and very few of the crops whose yield depends on pollination are allophilous. Honeybees are alternative pollinators for some, but not all, hemiphilous and euphilous fruit and seed crops, but the service performed by wild pollinators becomes increasingly important as honeybee colony numbers decline (Delaplane and Mayer 2000).

These considerations might reinforce the view that conservation effort should focus on eutropous/euphilous species of insects and plants if sustainable agriculture and natural biodiversity are to be maintained, but Ellis and Ellis-Adam's (1993) study draws attention to an alternative view. Their analysis highlights the importance of a few cornucopia plant species (their table 6) that support unusually large numbers of insect visitor species: *Heracleum sphondylium* (436 visitor species), *Taraxacum officinale* (375), *Daucus carota* (314), *Aegopodium podagraria* (313), *Cirsium arvense* (310), *Anthriscus sylvestris* (288), *Senecio jacobaea* (261), *Achillea millefolium* (247), *Angelica sylvestris* (245), *Knautia arvensis* (243), *Rubus fruticosus* (238), and *Jasione montana* (229). When their contribution to insect nutrition is expressed not simply as the number of insect visitor species but rather in terms of relative importance (the summation, over all visitor species, of the z-transform of an estimate of the fraction a plant species occupies of an insect species' niche width), *J. montana* and *R. fruticosus* drop out and the top 12 plant species now include *Calluna vulgaris* and *Leucanthemum vulgare.* However their importance is expressed, these top 12 cornucopia species include many allophilous (5 species of Apiaceae) and hemiphilous (*Rubus, Achillea, Leucanthemum, Jasione, Calluna*) forms; only *Knautia* and *Senecio* are euphilous.

These cornucopia plant species make a disproportionate contribution to the maintenance of insect biodiversity. Although they are generally common species and some are classed as noxious weeds, maintaining the continued abundance of cornucopia species would be a worthy management objective. None of the 60 highest-ranking plant species (in terms of numbers of visitor species) were annuals; they were all biennials, perennials, or woody species. Petanidou and Smets (1995) found nectar secretion rates to be generally higher in perennials than in annuals. Hence, Ellis and Ellis-Adam (1993) stress the importance of maintaining areas where vegetation is not subjected to repeated annual disturbance, allowing perennial herbaceous vegetation to become established.

At first sight it might appear that management should aim to enhance populations of those insect species that visit a wide range of plant species; however, not all visitors are effective pollinators, and the smaller species are less likely to have been recorded than the larger bees and butterflies. Of the highly generalist insect species in question (*Apis mellifera* visited 443 plant species, *Pieris napi* 352, and *P. brassicae* 332), honeybees are currently in decline (Delaplane and Mayer 2000), and *P. brassicae* is a pest of commercial and horticultural brassica crops. Both honeybees and pierid butterflies are regarded as poor pollinators, or even parasites, on some of the flowers they visit (Wiklund et al. 1979; Westerkamp 1991).

When loss of euphilous flowers reduces the abundance of their ecologically specialized visitors, a feedback effect is expected on their pollination and recruitment, because few alternative pollinators are available. This feedback may be less important for allophilous flowers. Loss of allophilous flower species may have a great effect on insect biodiversity, but many insect species can be lost before pollination suffers, exacerbating the decline of the plants.

Although eutropous/euphilous species of insects and flowers presently receive the most attention, allotropous/allophilous and hemitropous/hemiphilous species should not be ignored. The focus of conservation should also include cornucopia plant species, which make a disproportionate contribution to the support of pollinator communities.

The typology described here merits further exploration. It could be refined by tightening the definition of flower types, defining corresponding groups of pollinators in functional rather than taxonomic terms, and considering the extent to which it can be applied to pollen flowers, to regions outside northwestern Europe, and to pollinators other than insects.

Acknowledgments

I thank Willem Ellis, John Hobart, Manja Kwak, Jeff Ollerton, Juliet Osborne, and Martina Stang, who kindly read the manuscript and made valuable suggestions.

References

Aigner, P. 2001. Optimality modeling and fitness trade-offs: When should plants become pollinator specialists? Oikos 95: 177–184.

Allen, G. 1891. The colours of flowers as illustrated in the British flora. Macmillan, London.

Armbruster, W. S. 1993. Evolution of plant pollination systems: Hypotheses and tests with the neotropical vine *Dalechampia*. Evolution 47: 1480–1505.

Baker, H. G. 1965. Characteristics and mode of origin of weeds. Pp. 147–172 *in* H. Baker and G. Stebbins (eds.), The genetics of colonizing species. Academic Press, London.

Bell, C. R. 1971. Breeding systems and floral biology of the Umbelliferae, or evidence for specialization in unspecialized flowers. Pp. 93–106 *in* V. H. Heywood (ed.), The biology and chemistry of the Umbelliferae. Academic Press for the Linnean Society of London, London.

Bosch, J., and W. Kemp. 2001. How to manage the blue orchard bee. Sustainable Agriculture Network, National Agricultural Library, Beltsville, MD.

Buttermore, R. E. 1997. Observations of successful *Bombus terrestris* (L.) (Hymenoptera: Apidae) colonies in southern Tasmania. Australian Journal of Entomology 36: 251–254.

Carpenter, F. L. 1983. Pollination energetics in avian communities: Simple concepts and complex realities. Pp. 215–234 *in* C. E. Jones and R. J. Little (eds.), Handbook of experimental pollination biology. Van Nostrand Reinhold, New York.

Chittka, L., and S. Schürkens. 2001. Successful invasion of a floral market. Nature 411: 653.

Chittka, L., and J. D. Thomson. 2001. Cognitive ecology of pollination. Cambridge University Press, Cambridge.

Corbet, S. A. 1970. Insects on hogweed flowers: A suggestion for a student project. Journal of Biological Education 4: 133–143.

Corbet, S. A. 1990. Pollination and the weather. Israel Journal of Botany 39: 13–30.

Corbet, S. A. 1995. Insects, plants and succession: Advantages of long-term set-aside. Agriculture, Ecosystems, and Environment 53: 201–217.

Corbet, S. A. 2000. What kinds of flowers do butterflies visit? Entomologia Experimentalis et Applicata 96: 289–298.

Corbet, S. A., N. M. Saville, M. Fussell, O. E. Prŷs-Jones, and D. M. Unwin. 1995. The competition box: A graphical aid to forecasting pollinator performance. Journal of Applied Ecology 32: 707–719.

Corbet, S. A., and C. F. Tiley. 1999. Insect visitors to flowers of bluebell (*Hyacinthoides non-scripta*). Entomologist's Monthly Magazine 135: 133–141.

Crawley, M., P. Harvey, and A. Purvis. 1996. Comparative ecology of the native and alien floras of the British Isles. Philosophical Transactions of the Royal Society of London B 351: 1251–1259.

Delaplane, K. S., and D. F. Mayer. 2000. Crop pollination by bees. CABI Publishing, Wallingford, Oxon.

Delpino, F. 1868–1875. Ulteriori osservazione e considerazioni sulla dicogamia nel regno vegetale. Atti della Societa Italiana di Scienze Naturale, Milano 11: 265–332; 12: 21–141, 179–233; 16: 151–349.

de Rougemont, G. M. 1989. A field guide to the crops of Britain and Europe. Collins, London.

Dicks, L. V., S. A. Corbet, and R. F. Pywell. 2002. Compartmentalization in plant–insect flower visit webs. Journal of Animal Ecology 71: 32–43.

Dobson, H. E. M. 1987. Role of flower and pollen aromas in host-plant recognition by solitary bees. Oecologia 72: 618–623.

Ellis, W. N., and A. C. Ellis-Adam. 1993. To make a meadow it takes a clover and a bee: The entomophilous flora of NW Europe and its insects. Bijdragen tot de Dierkunde 63: 193–220.

Ellis, W. N., and A. C. Ellis-Adam. 1994. De ene schermbloem is de andere niet (Umbels are not alike). Entomologische Berichten, Amsterdam 54: 191–199.

Ellis, W. N., and A. C. Ellis-Adam. 1995. Flower visitation, plants' life forms, and ecological characteristics (Syrphidae: Parasitica). Proceedings of Experimental and Applied Entomology, Amsterdam 6: 53–58.

Faegri, K., and L. van der Pijl. 1979. The principles of pollination ecology, 3rd ed. Pergamon Press, Oxford.

Free, J. B. 1993. Insect pollination of crops. Academic Press, London.

Fussell, M., and S. A. Corbet. 1992. Forage for bumble bees and honey bees in farmland: A case study. Journal of Apicultural Research 30: 87–97.

Gilbert, F. S. 1981a. Foraging ecology of hoverflies: Morphology of the mouthparts in relation to feeding on nectar and pollen in some common urban species. Ecological Entomology 6: 245–262.

Gilbert, F. S. 1981b. Morphology and the foraging ecology of hoverflies (Diptera: Syrphidae). PhD dissertation, Cambridge University, Cambridge.

Grant, V., and K. A. Grant. 1965. Flower pollination in the phlox family. Columbia University Press, New York.

Harder, L. D. 1983. Flower handling efficiency of bumble bees: Morphological aspects of probing time. Oecologia 57: 274–280.

Heinrich, B. 1972. Energetics of temperature regulation and foraging in a bumblebee, *Bombus terricola* Kirby. Journal of Comparative Physiology 77: 49–64.

Heinrich, B. 1993. The hot-blooded insects. Springer, Berlin.

Herrera, C. M. 1996. Floral traits and plant adaptation to insect pollinators: A devil's advocate approach. Pp. 65–87 *in* S. C. H. Barrett and D. G. Lloyd (eds.), Floral biology. Chapman and Hall, New York.

Hingston, A. B., and P. B. McQuillan. 2000. Are pollination syndromes useful predictors of floral visitors in Tasmania? Austral Ecology 25: 600–609.

Johnson, S. D., and K. E. Steiner. 2000. Generalization versus specialization in plant pollination systems. Trends in Ecology and Evolution 15: 140–143.

Kevan, P. G. 1970. High Arctic insect–flower relations: The inter-relationships of arthropods and flowers at Lake Hazen, Ellesmere Island, N.W.T., Canada. PhD dissertation, University of Alberta, Canada.

Kevan, P. G. 1975. Suntracking solar furnaces in high Arctic flowers: Significance for pollination and insects. Science 189: 723–726.

Kevan, P. G. 1990. Sexual differences in temperatures of blossoms on a dioecious plant, *Salix arctica:* Significance for life in the Arctic. Arctic and Alpine Research 22: 283–289.

Knuth, P. 1906–1909. Handbook of flower pollination. Oxford University Press, Oxford.

Kremen, C., N. M. Williams, and R. W. Thorp. 2002. Crop pollination from native bees at risk from agricultural intensification. Proceedings of the National Academy of Sciences (USA) 99: 16812–16816.

Laverty, T. M. 1994. Bumble bee learning and flower morphology. Animal Behaviour 47: 531–545.

Lewis, A. C. 1993. Learning and the evolution of resources: Pollinators and flower morphology. Pp. 219–242 *in* D. R. Papaj and A. C. Lewis (eds.), Insect learning: Ecological and evolutionary perspectives. Chapman and Hall, London.

Lindsey, A. H. 1984. Reproductive biology of Apiaceae. I. Floral visitors to *Thaspium* and *Zizia* and their importance in pollination. American Journal of Botany 71: 375–387.

Lindsey, A. H., and C. R. Bell. 1985. Reproductive biology of Apiaceae. II. Cryptic specialization and floral evolution in *Thaspium* and *Zizia.* American Journal of Botany 72: 231–247.

Loew, E. 1895. Einführung in die Blütenbiologie auf historischer Grundlage. Ferdinand Dummlers Verlagsbuchhandlung, Berlin.

McDade, L. A. 1992. Pollinator relationships, biogeography, and phylogenetics. BioScience 42: 21–26.

Müller, H. 1881. Die Alpenblumen: ihre Befruchtung durch Insekten und ihre Anpassungen an dieselben. W. Engelmann, Leipzig.

Myers, J., and D. Bazely. 2003. Ecology and control of introduced plants. Cambridge University Press, Cambridge.

Ollerton, J., and S. Watts. 2000. Phenotype space and floral typology: Towards an objective assessment of pollination syndromes. Det Norske Videnskaps—Akademi. I. Matematisk Naturvidenskapelige Klasse, Skrifter, Ny Serie 39: 149–159.

Pellmyr, O. 2002. Pollination by animals. Pp. 157–184 *in* C. M. Herrera and O. Pellmyr (eds.), Plant–animal interactions. Blackwell Science, Oxford.

Petanidou, T., and E. Smets. 1995. The potential of marginal lands for bees and apiculture—nectar secretion in Mediterranean shrublands. Apidologie 26: 39–52.

Pheloung, P., P. Williams, and S. Halloy. 1999. A weed risk assessmant model for use as a biosecurity tool evaluating plant introductions. Journal of Environmental Management 57: 239–251.

Possingham, H. P. 1989. The distribution and abundance of resources encountered by a forager. American Naturalist 133: 42–60.

Preston, C., D. Pearman, and T. Dines. 2002a. New atlas of the British and Irish flora. Oxford University Press, Oxford.

Preston, C., M. Telfer, H. Arnold, and P. Rothery. 2002b. The changing flora of Britain, 1930–99. Pp. 35–45 *in* C. Preston, D. Pearman, and T. Dines (eds.), New atlas of the British and Irish flora. Oxford University Press, Oxford.

Proctor, M., P. Yeo, and D. Lack. 1996. The natural history of pollination. HarperCollins, London.

Stace, C. 1997. New flora of British Isles. Cambridge University Press, Cambridge.

Stone, G. N. 1994. Activity patterns of females of the solitary bee *Anthophora plumipes* in relation to temperature, nectar supplies and body size. Ecological Entomology 19: 177–189.

Stone, G. N., and P. G. Willmer. 1989. Warm-up rates and body temperature in bees: The importance of body size, thermal regime and phylogeny. Journal of Experimental Biology 147: 303–328.

Stubbs, A. E., and M. Drake. 2001. British soldierflies and their allies. British Entomological and Natural History Society, Reading.

Westerkamp, C. 1991. Honeybees are poor pollinators—why? Plant Systematics and Evolution 177: 71–75.

Westrich, P. 1989. Die Wildbienen Baden-Württembergs. Ulmer, Stuttgart.

Wiklund, C., T. Eriksson, and H. Lundberg. 1979. The wood white butterfly *Leptidia sinapis* and its nectar plants: A case of mutualism or parasitism? Oikos 33: 358–362.

Williams, C. S. 1997. Foraging ecology of nectar-collecting bumblebees and honeybees. PhD dissertation, University of Cambridge, Cambridge.

Williams, P. H. 1982. The distribution and decline of British bumble-bees. Journal of Apicultural Research 21: 236–245.

Williams, P. H. 1986. Environmental change and the distribution of British bumble-bees (*Bombus* Latr.). Bee World 67: 50–61.

Willmer, P. G., G. Stone, and I. Johnston. 2000. Environmental physiology of animals. Blackwell Science, Oxford.

Wilson, P., and J. D. Thomson. 1991. Heterogeneity among floral visitors leads to discordance between removal and deposition of pollen. Ecology 72: 1503–1507.

Woodell, S. R. J. 1979. The role of unspecialized pollinators in the reproductive success of Aldabran plants. Philosophical Transactions of the Royal Society of London B 286: 99–108.

CHAPTER FIFTEEN

The Conservation of Specialized and Generalized Pollination Systems in Subtropical Ecosystems: A Case Study

Suzanne Koptur

They paved paradise and put up a parking lot
—Joni Mitchell

If one is lucky enough to hike in a pristine natural area and to come upon a display of native plants in bloom, one may see native visitors pollinating flowers and enjoying the floral rewards. These visitors may even have evolved over time to best exploit the rewards and the flowers, to best export their pollen for dispersal to another individual of the same plant species. Much important research has been conducted in natural areas with minimal disturbance, and from these studies a body of ecological and evolutionary theory has grown about these striking mutualisms. This idyllic scenario is becoming the exception, however because many parts of the planet now have a disproportionately large percentage of the fauna made up of one species, *Homo sapiens*. The earth's human population has doubled in the past 40 years (surpassing six billion in 2001). Humans are prone to taking the nicest places and transforming them into places where they will live and work, often in isolation from anything natural. Even areas that superficially seem to be "pristine" often or always show the imprint of humans (McKibbin 1989); for example, nonnative plants or pollinators are likely to join the natives in the idyllic scenario just described (Brown et al. 2002; Memmott and Waser 2002).

Habitat destruction and fragmentation often shift the balance of nature in remaining habitat patches so that native organisms can no longer persist. Large predatory animals that require large areas for their home range provide the most obvious indication when they disappear, and, with the demise of predators, cascading effects of increased herbivore abundance may affect plants (Anderson 1997; Malcolm 1997; Dicke and Vet 1999; Jeffries 1999; Terborgh et al. 2001; Dyer and Letourneau 2003). Smaller animals, including insects, may hold on longer in remaining habitat patches as long as their survival requirements are met, but many groups show increased species richness with larger fragment size

(Robinson et al. 1997; Steffan-Dewenter and Tscharntke 2002; Steffan-Dewenter et al., chap. 17 in this volume). Predators and parasitoids are more strongly affected by habitat fragmentation than are lower trophic groups (Gibb and Hochuli 2002). Various phenomena accompanying fragmentation may lead to the decline or disappearance of organisms, including negative consequences of inbreeding, which results from isolation of small populations (Holsinger 1993; Hastings and Harrison 1994), and stochastic extirpation without recolonization due to greater distances from other populations (Hanski 1997). Smaller animals may have even greater effects on plants, because many of them serve as pollinators (Steffan-Dewenter et al. 2002) and seed dispersers (Bierregaard and Stouffer 1997) as well as herbivores (Rao et al. 2001) and seed predators (Donoso et al. 2003). Animals disappear more quickly than plants from landscapes affected by humans, but plants without their mutualistic animal partners may not persist long into the future.

In many situations it is not possible to preserve wild habitats, especially in the vicinity of urban areas, where human population pressures are great. Forward-thinking governments may set aside preserves, but these are often smaller and fewer than what conservation biologists might deem optimal or desirable. Plant species may be preserved in protected and/or managed habitat remnants, but, if their pollinators are lost and they cannot reproduce sexually, they may be evolutionarily dead. Habitat destruction can incur an "extinction debt" that will not be realized for decades or centuries; this is the reasoning behind using successful pollination as a measure of ecosystem health (Aizen and Feinsinger 1994), although using pollination deficits to infer pollinator declines may not be entirely straightforward (Thomson 2001).

In conquering the natural world, we humans have been largely oblivious to our dependence on pollinators for much of what we eat and use (Nabhan and Buchmann 1997) and have "forgotten pollinators" (Buchmann and Nabhan 1996). For over a decade, there have been declines in pollinators and pollination disruption has been reported worldwide (Kearns et al. 1998), though there is less direct evidence than many have presumed and such conclusions may be premature (Cane and Tepedino 2001). Long-term data are needed to track changes (Kearns 2001; Roubik 2001), and it is difficult to tell if changes are truly declines, or just supra-annual fluctations (Roubik 2001; Williams et al. 2001) or statistical artifacts (Cane 2001; Kerr 2001).

Indeed, there are some anthropogenically fragmented habitats where many of the mutualistic plant–animal relationships remain fairly intact, and not all mutualistic interactions show negative effects of habitat fragmentation or land-use intensity (Klein et al. 2001). Humans may actually enhance their own habitats in ways that can attract and sustain pollinators—to the benefit of native plant species dependent on specialized and generalized pollinators. The quality of the matrix—the space between the habitat fragments—can play a role in reducing negative effects of fragmentation (Perfecto and Vandermeer 2001). In subtropical southern Florida, extensive plantings of nonnative ornamentals provide abundant floral rewards to sustain pollinators of native plants in the urban matrix between the remaining fragments of natural habitat. Native-plant enthusiasts have promoted gardening with indigenous species, further enhancing the seemingly inhospitable between-fragment spaces for pollinator attraction and survival.

For the past decade my students and I have been studying plant–animal interactions in the South Florida pine rocklands. In this chapter I will review the effects of habitat destruction and fragmentation on native plants that remain in the natural landscape, consider the role of the matrix in ameliorating some of the negative effects of habitat fragmentation on pollinators, and discuss some measures that are being taken to conserve pollinators in the human-dominated landscape of subtropical South Florida in the United States. My hope is that this example will serve to illustrate problems and possibilities for more general maintenance of pollination systems in human-dominated landscapes.

Effects of Habitat Destruction and Fragmentation in Pine Rocklands of South Florida

The basic result of habitat destruction is that less habitat is available in which native plants can persist. I will illustrate this point by using the pine rocklands habitat from the uplands of extreme southern peninsular Florida. Pine rocklands, a fire-maintained subclimax vegetation with many endemic taxa, used to be nearly continuous albeit divided occasionally by freshwater wetlands or "transverse glades" (Snyder et al. 1990). The area covered by the rocklands ecosystems was never large (fig. 15.1A) and shrank rapidly from the mid- to late 20th century because of economic development. Rockland sites were preferred areas for clearing, building, and (after the invention of the rock plow) vegetable fields. Today, less than 2% of the original habitat outside of Everglades National Park remains, composed of a highly fragmented patchwork throughout urban and suburban Dade County (fig. 15.1B). Many of these anthropogenic fragments are protected as parks, but only some are maintained with exotic-pest-plant control and periodic fires. Other fragments are in private ownership; most of these have management problems similar to those of the parks, or precarious preservation status.

Fragments of pine rocklands also dramatically illustrate the "edge effects" resulting from increased perimeter-to-interior ratio: greater invasion by exotic species (especially weedy pest plants) that crowd out natives. The edges are greatly influenced by the surrounding inhabitants in terms of fire suppression: without periodic fires, pine rocklands undergo succession to hardwood hammock forest, losing their diverse understory of herbs and shrubs (Snyder et al. 1990; DeCoster et al. 1999). Many of these understory plants are endemic to this

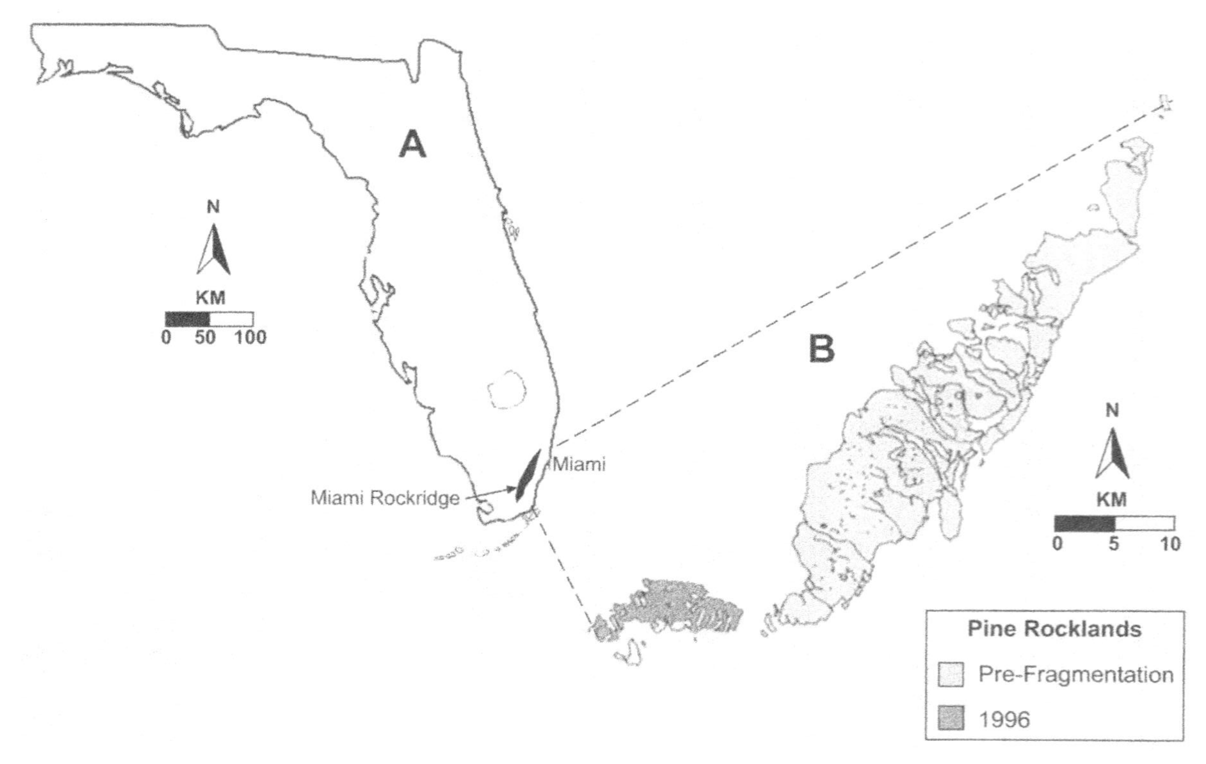

Figure 15.1 Original (*left*) and present-day (*right*) extent of pine rocklands in mainland southern Florida. There are also pine rocklands in the lower Florida Keys but they are not shown at the scale of the state map. (*Left*) Location and extent of Miami Rockridge is shown in black, and that same area is shown outlined at a much larger scale in the projection to the right. (*Right*) Prefragmentation extent (outlined) and extent in 1992 (black areas). Most remaining intact rocklands are protected within Everglades National Park; outside this park, less than 2% of the original rocklands remain.

habitat, and many are rare and becoming more so. Many former pine rocklands fragments have been degraded so completely that there is no longer a central core with native species, only a monoculture of Brazilian pepper (*Schinus terebinthifolius*) or a mixture of pest-plant species (Bradley and Gann 1999).

Exotic plants are not the only pests; exotic insects can compete with and eliminate native beneficial insect mutualists as well. Many areas in the southern United States (as well as Central America), formerly species-rich in native ants (and other insects), have become dominated by exotic ant species such as fire ants (*Solenopsis invicta*). Fire ants can limit the numbers of herbivores and pollinators with their aggressive, omnivorous foraging behavior (Fleet and Young 2000). Honeybees may be kept by beekeepers in groves adjacent to, and even in, some natural forest areas and may usurp floral resources that originally supported a diversity of native insects (Cairns 2002).

Animals kept as pets (or feral colonies maintained by kindhearted but misguided humans) can, in some cases, also have a profound impact on natural habitats. Many natural areas have networks of people who feed stray cats, capture them, neuter/spay them, and release the strays. Rather than controlling the populations, the presence of the colony serves as an "attractive nuisance," so that more cats are abandoned at the parks and populations continue to grow from the continual "immigration" of new individuals (Clarke and Pacin 2002; Castillo and Clarke 2003). The effects of domestic cats (Churcher and Lawton 1987; Schneider 2001) and other feral animals (Woodroffe et al. 1990; Schneider 2001) on wildlife are destructive and profound. Birds, lizards, and small mammals eat a variety of insects; when cats reduce their numbers, then insect populations can grow to levels that severely limit plant growth and reproduction. Some residents of Miami keep chickens that roam freely, which may travel through local parks in their search for food—eating seeds, seedlings, and small plants (and sometimes themselves providing food for resident foxes!). Goats and rabbits may similarly alter the landscape in their quest for forage and make "natural" areas less diverse and more barren, just as livestock does in midwestern U.S. forests (Dennis 1997).

Native animals may also be influenced by human interaction that in turn can affect their habitat. Sportsmen hunted the charismatic, endangered Key deer to near extinction as the Keys were exploited for tourism (Silvy 1975; Frank et al. 2003). Since their protection, Key deer have grown so numerous that populations have reached carrying capacity (Lopez 2001) and their grazing effects may have a larger impact than ever in the past (Folk et al. 1991; Koptur et al. 2002). Key deer herbivory, especially after fire, significantly reduces plant stem length and eliminates flowers on many preferred species (S. Koptur et al., unpublished data).

There is clearly need for management of pine rocklands fragments in the urbanized landscape of South Florida. County natural-areas managers prioritize activities in lands they manage and are fairly effective in controlling exotic plants via manual removal and herbicides. It is more difficult to use fire to manage fragments, especially those in close proximity to residential areas, but on occasion progress is made in this aspect of pine rocklands habitat conservation. Urban and suburban areas inhabited by humans are also managed for problem insects, especially mosquitoes, cockroaches, and ants. Pesticides that are used to control insects in adjacent areas can certainly affect survival of nontarget insects in natural area fragments in the urban landscape. Closer to humans, more pesticides are used: more than 20,000 pest control firms and 100,000 service personnel treat 12 million dwellings nationally, including most of the 280,000 retail food outlets, 480,000 commercial restaurants and kitchens, and 66,000 hotels and motels in the United States (National Pest Management Association 2001). In subtropical southern Florida, I estimate that more than 90% of homes use chemical pest control inside the dwelling, and more than 60% use some sort of chemical pest control in the yard. Termite control in dwellings is ubiquitous but usually has little broadcast effect. Certain pesticides (some used for fleas, ticks, and juvenile mosquitoes) are fairly specific; but broad-spectrum insecticides (such as those used for adult mosquito or fruit fly control) can certainly cause a decline of beneficial insects. The aerial application of pesticides to crops and forestry plantations has been shown to depress pollinator populations (Kevan 1975; Johansen 1977; Johansen et al. 1983; Kearns and Inouye 1997; Spira 2001). Coincident aerial insecticide spraying and flowering of endangered entomophilous plants puts those plants in peril (Bowlin et al. 1993; Sipes and Tepedino 1995). Even application of *Bacillus thuringensis* by organic gardeners can be detrimental to butterfly pollinators if *B. thurigensis* spores drift to weedy and native larval host plants adjacent to vegetable gardens.

Empirical Examples

Observations of pollinator–plant interactions in relatively pristine pine rocklands of the Everglades and lower Florida Keys provide a basis for comparison of the interactions of the same plants occurring in fragments of pine rocklands in suburban and urban Miami-Dade County and in the developed areas of Big Pine Key. We imagined that fragmentation of habitat would be detrimental to plant–pollinator interactions, and it seemed reasonable to hypothesize that species especially vulnerable to negative effects of fragmentation would include specialists and obligate outcrossers. Therefore, we selected to study native plant species that span a range of pollination systems, from specialized to generalized. I will choose examples from this research to illustrate that "all is not lost" for some plant species persisting in pine rocklands fragments.

Figure 15.2 Flowers of pine rocklands plants: (A) flowering shoots of the pineland clustervine, *Jacquemontia curtissii;* (B) a small individual of the pineland petunia, *Ruellia succulenta,* in full bloom; (C) the Big Pine partridge pea, *Chamaecrista keyensis,* with flowers and developing fruit; and (D) fast-moving bee (*Centris errans*) collecting oil from flowers of locustberry, *Byrsonima lucida.*

Pollinator Fauna

There are certainly differences in the pollinator fauna between pristine habitat and habitat fragments. For most of the plant species we have examined, pollinator species richness is greater in pristine habitat and in larger fragments than in medium-sized and small fragments. The composition of the fauna varies as well, as illustrated by the following examples.

The pineland clustervine, *Jacquemontia curtissii* (Convolvulaceae), has numerous white flowers with rotate, open corollas about 2–3 cm in diameter, with nectar in the center of the flower available to a wide array of visitors (fig. 15.2A);

some flower visitors also collect its white pollen. Three pristine pine rocklands sites in Everglades National Park had a total of 22 species of flower visitors, of which 19 were probable pollinators (determined by size and activities on flowers): three large (greater that 10 ha) fragments had 12 probable pollinator species, medium (3–9 ha) fragments had 11 pollinator species, and small (less than 3 ha) fragments had 6 pollinator species (Koptur and Geiger 1999). We recorded 26 species of floral visitors and observed certain visitors only in fragments, indicating that the pollinator fauna of *J. curtissii* in fragments is not simply a subset of pollinators in the intact habitat.

The pineland petunia, *Ruellia succulenta* (Acanthaceae), has large, showy flowers with a lavender funnelform corolla (fig. 15.2B), suggesting that visits are limited to insects with long proboscises. Geiger (2002) found this was not the case because numerous bees, as well as Lepidoptera visitors, crawl down the corolla tube to reach the nectar and pollinate the flowers. There were highly significant differences in the proportions of Hymenoptera (bees) and Lepidoptera (butterflies and skippers) visitors by habitat size class; bees make up an increasing proportion of the total floral visitors as one moves from small to intact/pristine size classes, and Lepidoptera are more important in the smaller fragments (Geiger 2002).

The Big Pine partridge pea (*Chamaecrista keyensis,* Fabaceae: Caesalpinioideae) has large, showy, yellow flowers (fig. 15.2C) that are buzz-pollinated by carpenter bees (*Xylocopa micans*) and two species of *Melissodes* bees (Liu and Koptur 2003); they are also visited by other, nonbuzzing bees who pick up the pollen scattered on the petals by the buzzing bees but usually do not contact the stigma in the process. *Chamaecrista keyensis* flowers received substantially more visits by *X. micans,* but fewer visits from *Melissodes* spp., in urban edge versus forest sites in the Key Deer Refuge. Unexpectedly, the buzz-pollinators made up a substantially greater proportion of the bee visits in urban edge sites than in forest sites, where nonbuzzing visitors were more common (Liu and Koptur 2003). The numbers of buzz-pollinating bees at partridge pea flowers declined after repeated aerial mosquito spraying in Big Pine Key (Liu and Koptur 2003). This aerial spraying has been observed to depress Lepidoptera populations in the Keys as well (Salvato 2001; S. Carroll and J. Loye, unpublished data).

Byrsonima lucida, the sole member of the tropical plant family Malpighiaceae native to South Florida (fig. 15.2D), has a specialized pollination system: oil is secreted as a floral reward and is collected by andrenid bees in the genus *Centris,* of which only two species occur in this area (*Centris errans* = *C. versicolor,* and *C. lanosa*). Our hand-pollination experiments show that flowers need visitation to set fruit, and plants set substantially more fruit with cross- than self-pollination (Koptur and Geiger 2000). Copious fruit production in this species is, therefore, evidence of not only visitation, but also likely deposition of pollen from other individuals. Everglades plants fruit heavily, as do plants in or near some of the larger fragments, and it is common to see *C. errans* bees at these sites. Plants in smaller fragments also set fruit, though sometimes only modestly; even plants in gardens and planted landscapes set fruit. *Centris lanosa* is the more common visitor to these plants, although both *Centris* spp. have been observed in urban areas.

Pollinator Activity in Disturbed Habitats

Native plants do exist in the urban landscape, either persisting in fragmented or semideveloped land or planted back into the landscape as garden specimens or in landscaping projects of varying size. The native plants are visited by some of the same insect species that visit them in natural environments, and by some species that are more common in disturbed situations. Plants that are both generalists and specialists in their pollinator affinities have been observed to maintain pollinator relationships in urban situations in South Florida.

The endangered crenulate leadplant (*Amorpha herbacea* var. *crenulata,* Fabaceae: Papilionoideae) needs pollinator visits for fruit set, and cross-pollinations set considerably more fruit and seed than self-pollinations (Linares 2004). *Amorpha crenulata* survives in only a few small pine rocklands fragments in Miami-Dade County, but, wherever it grows, it receives visits from a variety of native solitary bees (such as *Dianthidium curvatum floridense*) and nonnative honeybees (*Apis mellifera*). Even in sporadically mowed lots, crenulate leadplant produced abundant fruit. Planted in gardens within its native range and beyond, its striking inflorescences of tiny white flowers tipped with orange-yellow anthers are visited by native and introduced insects (figs. 15.3A, 15.3B).

The purple flowers of pine rocklands milkpeas, *Galactia* spp., are visited by nectar-collecting halictid bees which get brushed by the anthers and pollinate them (personal observation). While conducting a study of the distribution of rare milkpeas of southern Florida pine rocklands, O'Brien (1988) observed native bee pollinators (*Augochlora pura* ssp. *mosieri*) visiting remnant milkpea individuals in manicured lawns of Coral Gables.

Role of the Matrix

The characteristics of the matrix—the space between habitat fragments in a fragmented landscape—are crucial to the maintenance of plant–pollinator interactions in fragments. Those fragments that are small and/or isolated from larger areas of intact habitat may depend in particular on the matrix for support of pollinators passing through or possibly even nesting and living in the matrix. A thorough comparison of ecology of species across fragmented landscapes must also consider matrix habitat (Jules and Shahani 2003). I will consider several types of matrix habitat found between pine rocklands fragments and their potential effects on plants and insects in remnant habitat.

Figure 15.3 More pine rocklands plants and insects: (A) crenulate leadplant, *Amorpha herbacea* var. *crenulata,* plant habit; (B) inflorescence close-up of *A. herbacea;* (C) caterpillar of naturalized orange-barred sulfur (*Phoebis agarithe*) butterfly on native *Cassia bahamensis* (aka *Senna mexicana* var. *chapmanii*); and (D) flowering stem of the butterfly pea, *Centrosema virginiana.*

Concrete in the Big City

One aspect of urbanization (that is definitely not pollinator friendly) is the use of asphalt (tarmac) on roads and concrete on other horizontal surfaces to thwart the establishment and growth of any plant life. As the population of South Florida has grown, roads that were formerly unpaved became paved, then widened from two-lane, then four-lane roads, then to multilane expressways. Consequently, the area covered by asphalt has steadily increased over the past century. As areas have been developed for human habitation and other uses, more and more ground has been covered by concrete. Gardens have been eliminated from many lots for various reasons (they require care, attract unwanted animals, they look "too wild"). As in many parts of Latin America, a sign of success is a tidy, barren yard consisting of concrete (frequently painted) with a minimum of plants.

Suburban Lawns: A Golf-course Green in Every Yard?

As inhospitable as concrete is, matrix consisting of meticulously maintained lawns (turf grass) may be even more detrimental to the movements of pollinators. Turf grass science leads to the development of grass strains that are tough and easy to maintain; the goal is to make the lawn as uniform as possible. Extensive use of chemicals (fertilizers, herbicides, and pesticides/insecticides) is needed to maintain the ideal lawn. Pest control operators leave signs in lawns so that humans can avoid these areas for at least one day, but few pollinators (or pets or children) are able, or take the time, to read these signs.

Pollinator Relief in the Land of Flowers

Fortunately, a number of features of the matrix between natural habitat fragments exist that are improvements over concrete. The penchant many residents of Florida (dubbed by the Spaniards "the land of flowers") have for lush landscaping and beautiful flowers has led to an extensive array of cultivated ornamental plants that can provide pollinators with a variety of foods and shelter. Most pollinator foods are provided in flowers, usually in the forms of nectar and pollen, but certain species also provide oils (e.g., Malpighiaceae), resins (e.g., Clusiaceae), and extrafloral nectar (many families; Koptur 1992).

There are some spectacular sights involving animals and flowers to be seen on the streets of Miami. Brilliant yellow, black, and white spot-breasted orioles (*Icterus pectoralis*) visiting flowers of the sausage tree (*Kigelia pinnata,* Bignoniaceae) are the facultative pollinators of these bat-adapted flowers, the fruit of which resembles huge, pendant sausages. High up on the bare trunks of majestic Bombacaceae, squirrels drink nectar from the flowers of *Bombax malabaricus* and *Pseudobombax* sp. These visits rarely lead to fruit production because conspecific individuals of these species are few and far between. Fortunately, in big cities (especially in the subtropics), there are many opportunists who use floral rewards, sometimes, though not always, pollinating in the process.

Isolated individuals of *Byrsonima lucida* in urban garden plantings receive visits from their specialized *Centris* bee pollinators even though no other *B. lucida* are in sight. Those bees visit alternative oil sources found in frequent plantings of several ornamental species of Malpighiaceae: *Malpighia coccigera, Stigmaphyllon* spp., and *Thryallis glauca*. And when the neighbors decide to add native *B. lucida* to their gardens, fruit set is then possible for formerly isolated individuals. Perhaps this fruit set is of less fitness consequence than fruit set on individuals in native habitats, but it can serve to perpetuate this species in the matrix between natural habitat fragments.

Nonnative species cultivated for their useful fruit are readily pollinated in South Florida. Passion fruit (*Passiflora edulis*) are usually grown along fences and are most effectively pollinated by carpenter bees, but a group of honeybees working together can also effect pollination (Hardin 1987). Flower beetles visiting the purportedly wind-pollinated flowers of jackfruit (*Artocarpus heterophyllus*) growing in orchards or garden plantings may enhance fruit production and seed set in South Florida (El-Sawa 1998).

Ornamental, exotic congeners of native species can serve to attract and feed pollinators and can help pollinators make their way between fragments or between native species in the urban landscape. A very popular cultivated species, *Ruellia brittoniana*, has purple, pink, or white flowers that look very similar to the native *R. succulenta*. Ubiquitous plantings of this popular species ensure plentiful nectar for butterflies and bees, and maybe even larval food for specialist herbivore butterflies (Nymphalidae) such as the white peacock (*Anartia jatrophae*) and the malachite (*Siproeta stelenes*). Found throughout the tropics, *Lantana camara* volunteers readily as its seeds are dispersed by birds that eat the blue fruits; butterflies are nourished by its nectar and may contribute to its hybridization with native *L. depressa* in South Florida (Ramey 1999).

Even Weeds Serve a Purpose

Lantana camara is listed as a category I nonnative, invasive plant by the Florida Exotic Pest Plant Council and is one of the worst weeds in all the world (Holm 1977), forming dense thickets in 47 countries and a weed in many crops as well; yet it is frequently planted to attract butterflies in the United States and in South Florida is a well-used nectar plant for many butterfly species. Other pervasive weeds are nourishment mainstays for pollinators in the seminatural and disturbed landscape.

Devil's pitchfork (*Bidens pilosa*, Asteraceae) is a crop weed in the Old and New World tropics and a frequent resident of any disturbed ground or unmown lawn in South Florida. It is so favored by insects that one can obtain a good general collection of floral visitors for an area simply by observing its blossoms. A recent edition of a popular ecology textbook had a photo of a zebra butterfly (*Heliconius charitonius*) sipping nectar on this flower rather than any of the native plants in the area!

Brazilian pepper (*Schinus terebinthefolius,* Anacardiaceae) is a woody species with attractive red berries that facilitate its dispersal by birds into natural areas; it frequently colonizes disturbed ground to form a monospecific stand (as in the former agricultural area within Everglades National Park known as the "Hole in the Donut"). Honeybees and other insects consume the floral nectar of this pest tree, and honey production is greatly enhanced by its presence (Ewel 1982).

Another notorious pest plant, the paperbark tree (*Melaleuca quinquenervia*), has attractive white flowers full of nectar that is collected by a variety of insects, including butterflies, skippers, moths, wasps, bees, and flies. Honeybees are the most abundant pollinators (Vardaman 1994), and, although the flowers can automatically self-pollinate, seed set is enhanced by insect visitors (Vardaman 1994). The beekeeper practice of placing their bees in natural areas may therefore promote the spread of noxious pest trees (both paperbark and Brazilian pepper) that provide nectar for honeybees and in turn receive pollinator services.

Exotic Alternatives When Natives Are Absent: Benefits to Butterflies

Lepidoptera feed in different ways as either adults or larvae, and larval food plants are necessary to maintain butterflies, moths, and skippers in the landscape. Some of South Florida's rare butterflies use not only native but also exotic host plants. The Atala butterfly (*Eumaeus atala*), once thought to be extinct, lays its eggs on coontie, a native cycad (*Zamia pumila*), and the extensive coontie starch industry of the early 20th century may have led to the extirpation of this butterfly in South Florida prior to its subsequent recolonization (Smith 2000). A reintroduction program undertaken at Crandon Park utilized extensive cycad host plantings and larval relocations from colonies at Fairchild Tropical Garden (Smith 2002), where Atala larvae also feed on the cultivated cardboard palm (*Z. furfuracea*) and other cycads in the garden's extensive collection—the reason the garden is eager to farm out the larvae of this endangered butterfly species! Atala adults visit many flowers, including native palmettos, *Lantana involucrata,* and weedy *Bidens pilosa* (Smith 2000, 2002).

The Miami Blue (*Hemiargus thomasi bethunebakeri*) utilizes balloon vine (*Cardiospermum* spp.) hosts. The larvae feed on the plant and hide in the seedpods to avoid predators. Balloon vine occurs adjacent to hammocks in the lower Florida Keys (Loye and Carroll, in press), and these hammocks are often close to roads, resulting in the mowing of these areas to appease safety concerns. Consequently, the state's Department of Environmental Protection has requested that an area several feet wide be left unmown to allow the plants to fruit, to perpetuate suitable host plant for the Miami Blue.

Common butterflies also utilize both native and cultivated species for their larval hosts: the native cloudless sulfur (*Phoebis sennae*) and the naturalized orange-barred sulfur (*Phoebis agarithe*) utilize native and cultivated *Cassia* spp. as their larval host plants (Glassberg et al. 2000). These butterflies visit a variety of flowers for nectar, and their activity is greatest in areas with abundant host plants, in both natural areas and urban areas (fig. 15.3C).

General Conclusions

Urban and Anthropogenic Landscapes and Pollinator Conservation

Our results do not show a marked difference in the response of specialized versus generalized pollination systems to habitat fragmentation (table 15.1). Aizen et al. (2002) surveyed the literature and concluded that the extent of specialization does not necessarily correlate with the likelihood of a species experiencing negative effects of habitat fragmentation. Ashworth et al. (2004) noted more recently that, although pollinators are sensitive to habitat fragmentation, plants that are pollination specialists do not appear to suffer more from habitat fragmentation than do generalists, and they proposed that this is because of asymmetry in the degree of specialization of the plants and their pollinators (see also Vázquez and Aizen, chap. 9 in this volume; Petanidou and Potts, chap. 11 in this volume).

Thompson (1997) makes the case for conserving interaction biodiversity as well as species diversity. Although this may be most easily done with large preserves, there are "chronically fragmented" landscapes where this option does not exist and there is much value in small preserves (Schwartz and van Mantgem 1997). If small preserves can be managed in ways that tip the balance of nature in favor of native species (using exotic removal, fire management, and augmentation of resources in the matrix between preserves), many plant-animal interactions will also be maintained. Generalized interactions are more flexible, and it will take more care to ensure the persistence of extreme specialists; it will certainly not be possible in all cases. One way of increasing the chances of the persistence of these interactions is by "gardening for pollinators."

Gardening for Pollinators

Some naturalists have promoted butterfly gardening (e.g., Minno and Minno 1999; Glassberg et al. 2000), as have plant societies, public and private gardens, and plant-related businesses. The most important message for nonbiologists is that gardeners need to plant not only nectariferous plants but also larval food plants to encourage the butterflies to linger and proliferate. The beneficial effects on urban wildlife (specifically Lepidoptera) are noticeable. Little has yet been done, however, to promote the numbers of other pollinators. It is essential not only to include plants with floral rewards for the pollinators, but also to consider the pollinators' other needs (larval host plants and nesting sites).

The idea of gardening for pollinators was expressed in a popular article

Table 15.1 Summary of fragmentation effects (FEs) on pine rockland plants (general conclusions from work in progress)

Species (family)	Specialist/ generalist	Principle pollinators	FEs on flowering	FEs on pollen deposition	FEs on pollinators at flowers	FEs on fruit set
Amorpha crenulata (Fabaceae)	Generalist	Bees	Negative	No info	None	None
Byrsonima lucida (Malpighiaceae)	Specialist	Centris bees	Positive	None	None	Slight negative
Centrosema virginiana (Fabaceae)	Specialist	Large and medium-sized bees	Mixed	None	Medium bees more common	None
Dyschoriste angusta (Acanthaceae)	Generalist	Bees and butterflies	Mixed	None	None	No info
Evolvulus sericeus (Convolvulaceae)	Generalist	Small bees and flies	Negative	—	None	No info
Galactia spp. (Fabaceae)	Specialist	Medium and small bees	Negative	—	None	None
Jacquemontia curtissii (Convolvulaceae)	Generalist	Bees, flies, wasps, butterflies	Negative	None	Fewer species	None
Ruellia succulenta (Acanthaceae)	Generalist	Bees and butterflies	Negative	None	Butterflies more common	None

(Tasker 1996) by a newspaper columnist influenced by the "Forgotten Pollinators Campaign" (Buchman and Nabhan 1996); since that time, local interest in making pollinators welcome has been growing. The Forgotten Pollinators Campaign directed much attention to disappearing pollinators in the southwestern United States, and worldwide, and a booklet entitled *Gardening for Pollinators* was published by the Sonoran Desert Museum for guidance in the arid southwest. The humid, subtropical climate of South Florida is vastly different from the arid southwest, and some parts are considerably more urbanized; nonetheless, both areas share problems in disappearing species and declining pollinators. Although bee diversity of the desert southwestern United States dwarfs that of Florida, Florida's bee fauna is still fairly rich compared with that of the rest of the United States (Pascarella et al. 1999, 2001).

Solitary bees may find it difficult to nest in gardens that are too neatly maintained: some of these bees nest in dead twigs, which they may stuff with pieces of leaves they cut; others nest in rocky crevices, or right in the ground in sandy patches. Carpenter bees nest in wood, including wooden structures, and are often more abundant in urban edge habitats (Liu and Koptur 2003). *Centrosema virginiana,* the butterfly pea (fig. 15.3D), is pollinated primarily by these large bees, and carpenter bee activity at flowers is much greater for plants near picnic tables and park visitor facilities than those farther from wooden structures (Cardel 2004). In the Redland agricultural area of South Florida, edible passionfruit (*Passiflora edulis*) grown on fences with wooden posts, or in areas with wooden structures, receive more visits from carpenter bees; *P. edulis* on chain-link fences with only concrete structures nearby receive more honeybee visits (Hardin 1987). Therefore, it is important to have some habitat heterogeneity in a garden to promote nesting by a variety of bee pollinators. Entomologists use pollinator nest traps to study bee diversity (Pascarella et al. 1999, 2001), but nest blocks/boxes have not been yet deployed in the South Florida landscape to attract pollinators. Wasps frequently colonize nest blocks (much more than bees) in South Florida studies (J. Pascarella, personal communication).

Importance of Education in Pollinator Restoration

Insects and Gardens (Grissell 2001) gives readers an appreciation of the diversity of insects maintained by plants in a garden. This innovative work not only educates about insect biology and natural history; it also guides gardeners to a coexistence in which humans and insects can share gardens, encouraging gardeners to tolerate many types of insects (such as bees, wasps, earwigs) that may at first seem undesirable—those that benefit garden plants not only by visiting and pollinating flowers but also by eating potential pests.

The most powerful conservation education starts with children, and many activities are aimed at young people. Butterflies are lovely, and butterfly gardening is the easiest hook for most people; once hooked, they are more likely to be open to appreciating the presence and activities of other insects in the garden, the home landscape, and in natural areas. Schoolyard ecology (Berkowitz 2000) brings students (and families) in touch with the natural environment, and students who are exposed to nature activities in school are more likely to care about nonhuman life in the future. Most organizations that have conservation of flora and/or fauna as part of their mission, therefore, have a substantial educational component, for example, botanical gardens, zoos, government agencies (federal, state, and county), and nongovernment organizations. Continuing to educate people after elementary school is perhaps the most important mission of many organizations if their goals of conservation are to be realized in our complex, modern world. One example is the North American Butterfly Association, whose Miami Blue chapter conducts semiannual butterfly counts, which increase public awareness of these insects. Adult education with public programs and special events displays and activities are ways to engage members of society who might otherwise never think about the importance of pollinators.

The Florida Native Plant Society and the Tropical Audubon Society regularly have plant sales to promote creation of a habitat for wildlife. As more native plants join the home landscape, the earlier planted individuals find mates, fruit and seed are produced, and, in some cases, new populations become self-sustaining. Admittedly, the genetic structure of remnant natural populations is very likely changed with these native plantings in the matrix between natural habitat fragments as pollinators move from fragment to oases of floral rewards (some from exotic plants, some from native plants). This is a dilemma in our irreversibly altered human-dominated landscapes.

Restoration of pollinator–plant interactions by gardening for pollinators can enhance plant and pollinator diversity and help rejuvenate landscapes in which plants have lost their partners. There are many examples of pollinators that have disappeared and are presumed extinct, from localized specialists to far-ranging generalists (Buchman and Nabhan 1996). Planting projects can serve to replace floral resources lost through development and may attract and support populations of floral visitors that would otherwise decline or disappear. These may be the only means that can conserve both generalist and specialist pollinators in the face of ever-growing human populations.

Acknowledgments

I thank Joe O'Brien for help with the maps for figure 15.1, Walter M. Goldberg for photo editorial assistance, and my students and colleagues for allowing me to cite results from their yet-unpublished work. I am grateful to Nick Waser and other reviewers for their constructive comments that improved this chapter.

References

Aizen, M. A., L. Ashworth, and L. Galetto. 2002. Reproductive success in fragmented habitats: Do compatibility systems and pollination specialization matter? Journal of Vegetation Science 13: 885–892.

Aizen, M. A., and P. Feinsinger. 1994. Forest fragmentation, pollination, and plant reproduction in a Chaco dry forest, Argentina. Ecology 75: 330–351.

Anderson, R. C. 1997. Native pests: The impact of deer in highly fragmented habitats. Pp. 115–134 *in* M. W. Schwartz (ed.), Conservation in highly fragmented landscapes. Chapman and Hall, New York.

Ashworth, L., R. Aguilar, L. Galetto, and M. A. Aizen. 2004. Why do pollination generalist and specialist plant species show similar reproductive susceptibility to habitat fragmentation? Journal of Ecology 92: 717–719.

Berkowitz, A. 2000. Schoolyard ecology leader's handbook. Online URL http://www.ecostudies.org/syefest.

Bierregaard, R. O., Jr., and P. C. Stouffer. 1997. Understory birds and dynamic habitat mosaics in Amazonian rainforests. Pp. 138–155 *in* W. F. Laurance and R. O. Bierregaard Jr. (eds.), Tropical forest remnants. University of Chicago Press, Chicago.

Bowlin, W. R., V. J. Tepedino, and T. L. Griswold. 1993. The reproductive biology of *Eriogonum pelinophilum* (Polygonaceae). Pp. 296–302 *in* R. Sivinski and K. Lightfoot (eds.), Southwestern rare and endangered plants. Miscellaneous publication no. 2. New Mexico Forestry and Resources Conservation Division, Santa Fe, NM.

Bradley, K., and G. D. Gann. 1999. The status of exotic plants in the preserves of south Florida. Pp. 35–41 *in* D. T. Jones and B. W. Gamble (eds.), Florida's garden of good and evil. Proceedings of the 1998 joint symposium of the Florida Exotic Pest Plant Council and the Florida Native Plant Society. Florida Exotic Pest Plant Council and South Florida Water Management District, West Palm Beach, FL.

Brown, B. J., R. J. Mitchell, and S. A Graham. 2002. Competition for pollination between an invasive species (purple loosestrife) and a native congener. Ecology 83: 2328–2336.

Buchmann, S. L., and G. P. Nabhan. 1996. The forgotten pollinators. Island Press, Washington, DC.

Cairns, C. E. 2002. Effects of invasive africanized honey bees (*Apis mellifera scutellata*) on native stingless bee populations (Meliponinae) and traditional Mayan beekeeping in central Quintana Roo, Mexico. MS thesis, Florida International University, Miami, FL.

Cane, J. H. 2001. Habitat fragmentation and native bees: A premature verdict? Conservation Ecology 5: Online URL http://www.consecol.org/vol5/iss1/art3.

Cane, J. H., and V. J. Tepedino. 2001. Causes and extent of declines among native North American invertebrate pollinators: Detection, evidence, and consequences. Conservation Ecology 5. Online URL http://www.consecol.org/vol5/iss1/art1.

Cardel, Y. 2004. Linking herbivory and pollination: Cost and selection implications in *Centrosema virginianum* Benth. (Papilionoideae). PhD dissertation. Florida International University, Miami, Florida.

Castillo, D., and A. C. Clarke. 2003. Trap/neuter/release methods ineffective in controlling domestic cat "colonies" on public lands. Natural Areas Journal 23: 247–254.

Churcher, P. B., and J. H. Lawton. 1987. Predation by domestic cats in an English village. Journal of Zoology 212: 439–455.

Clarke, A. L., and T. Pacin. 2002. Domestic cat "colonies" in natural areas: A growing exotic species threat. Natural Areas Journal 22: 154–159.

DeCoster, J. K, W. J. Platt, and S. A. Riley. 1999. Pine savannas of Everglades National Park: An endangered ecosystem. Pp. 81–88 41 *in* D. T. Jones and B. W. Gamble (eds.), Florida's garden of good and evil. Proceedings of the 1998 joint symposium of the Florida Exotic Pest Plant Council and the Florida Native Plant Society. Florida Exotic Pest Plant Council and South Florida Water Management District, West Palm Beach, FL.

Dennis, A. 1997. Effects of livestock grazing on forest habitats. Pp. 313–341 *in* M. W. Schwartz (ed.), Conservation in highly fragmented landscapes. Chapman and Hall, New York.

Dicke, M., and L. E. M. Vet. 1999. Plant-carnivore interactions: Evolutionary and ecological consequences for plant, herbivore and carnivore. Pp. 483–20 *in* H. Olff, V. K. Brown, and R. H. Drent (eds.), Herbivores: Between plants and predators. 38th symposium of the British Ecological Society. Blackwell Science, Oxford.

Donoso, D. S., A. A. Grex, and J. A. Simonetti. 2003. Effects of forest fragmentation on the granivory of differently sized seeds. Biological Conservation 115: 3–70.

Dyer, L. A., and D. Letourneau. 2003. Top-down and bottom-up diversity cascades in detrital vs. living food webs. Ecology Letters 6: 60–68.

El-Sawa, S. 1998. Pollination and breeding of jackfruit (*Artocarpus heterophyllus* Lam.) in south Florida. MS thesis, Florida International University, Miami, FL.

Ewel, J. J. 1982. *Schinus* in successional ecosystems of Everglades National Park. South Florida Research Center Report T-676.

Fleet, R. R., and B. L. Young. 2000. Facultative mutualism between imported fire ants (*Solenopsis invicta*) and a legume (*Senna occidentalis*). Southwestern Naturalist 453: 289–298.

Folk, M. L., W. D. Klimstra, and C. R. Kruer. 1991. Habitat evaluation: National Key Deer Range. Final report to Florida Game and Fresh Water Fish Commission, Tallahassee, FL. Project no. NG88-015.

Frank, P. A., B. W. Stieglitz, J. Slack, and R. R. Lopez. 2003. The Key deer: Back from the brink. Endangered Species Bulletin 28: 20–21.

Geiger, J. H. 2002. The reproductive biology of *Ruellia succulenta* (Acanthaceae) and the effects of habitat fragmentation. MS thesis, Florida International University, Miami, FL.

Gibb, H., and D. F. Hochuli. 2002. Habitat fragmentation in an urban environment: Large and small fragments support different arthropod assemblages. Biological Conservation 106: 91–100.

Glassberg, J., M. C. Minno, and J. V. Calhoun. 2000. Butterflies through binoculars. A field, finding, and gardening guide to butterflies in Florida. Oxford University Press, Oxford.

Grissell, E. 2001. Insects and gardens. Timber Press, Portland, OR.

Hanski, I. 1997. Metapopulation dynamics: From concepts and observations to predictive models. Pp. 69–92 *in* I. A. Hanski and M. E. Gilpin (eds.), Metapopulation biology: Ecology, genetics, and evolution. Academic Press, San Diego, CA.

Hardin, L. C. 1987. The pollinators of the yellow passionfruit: Do they limit the success of *Passiflora edulis* f. *flavicarpa* as a tropical crop? MS thesis, Florida International University, Miami, FL.

Hastings, A., and S. Harrison. 1994. Metapopulation dynamics and genetics. Annual Review of Ecology and Systematics 25: 167–188.

Holm, L. G. 1977. The world's worst weeds: distribution and biology. University Press, Honolulu.
Holsinger, K. E. 1993. The evolutionary dynamics of fragmented plant populations. Pp. 198–216 *in* P. M. Kareiva, J. G. Kingsolver, and R. B. Huey (eds.), Biotic interactions and global change. Sinauer Associates, Sunderland, MA.
Jefferies, R. L. 1999. Herbivores, nutrients and trophic cascades in terrestrial environments. Pp. 301–330 *in* H. Olff, V. K. Brown, and R. H. Drent (eds.), Herbivores: Between plants and predators. 38th symposium of the British Ecological Society. Blackwell Science, Oxford.
Johansen, C. A. 1977. Pesticides and pollinators. Annual Review of Entomology 22: 177–192.
Johansen, C. A., D. F. Mayer, J. D. Eves, and C. W. Isious. 1983. Pesticides and bees. Environmental Entomology 12: 1513–1518.
Jules, E. S., and P. Shahani. 2003. A broader ecological context to habitat fragmentation: Why matrix habitat is more important than we thought. Journal of Vegetation Science 14: 459–464.
Kearns, C. A. 2001. North American dipteran pollinators: Assessing their value and conservation status. Conservation Ecology 5. Online URL http://www.consecol.org/vol5/iss1/art5.
Kearns, C. A., and D. W. Inouye. 1997. Pollinators, flowering plants, and conservation biology: Much remains to be learned about pollinators and plants. Bioscience 47: 297–307.
Kearns, C. A., D. W. Inouye, and N. M. Waser. 1998. Endangered mutualisms; the conservation of plant–pollinator interactions. Annual Review of Ecology and Systematics 29: 83–112.
Kerr, J. T. 2001. Butterfly species richness patterns in Canada: Energy, heterogeneity, and the potential consequences of climate change. Conservation Ecology 5. Online URL http://www.consecol.org/vol5/iss1/art10.
Kevan, P. G. 1975. Forest application of the insecticide Fenitrothion and its effect on wild bee pollinators (Hymenoptera: Apoidea) of lowbush blueberries (*Vaccinium* spp.) in southern New Brunswick, Canada. Biological Conservation 7: 301–309.
Klein, A. M., I. Steffan-Dewenter, D. Buchori, and T. Tscharntke. 2001. Effects of land-use intensity in tropical agroforestry systems on coffee flower-visiting and trap-nesting bees and wasps. Conservation Biology 16: 1003–1014.
Koptur, S. 1992. Extrafloral nectary-mediated interactions between insects and plants. Pp. 81–129 *in* E. Bernays (ed.), Insect–plant interactions, vol. 4. CRC Press, Boca Raton, FL.
Koptur, S., and J. H. Geiger. 1999. Pollination of an endemic morning glory in pine rockland fragments. Ecological Society of America meetings abstract. Spokane, WA: 128.
Koptur, S., and J. H. Geiger. 2000. Pollination of *Byrsonima lucida* (Malpighiaceae) in southern Florida. American Journal of Botany 87 (6 Supplement): 45.
Koptur, S., J. R. Snyder, M. S. Ross, H. Liu, and C. K. Borg. 2002. Selective grazing by Key deer after fire changes plant morphology, reproduction, and species composition in pine rockland understory. American Journal of Botany 89 (6 Supplement): 102.
Linares, L. J. 2004. Floral biology and breeding system of the Crenulate Leadplant, *Amorpha herbacea* var. *crenulata* (Fabaceae), an endangered South Florida pine rockland endemic. MS thesis, Florida International University, Miami, FL.
Liu, H., and S. Koptur. 2003. Breeding system and pollination of a narrowly endemic herb of the lower Florida Keys: Impacts of the urban wildland interface. American Journal of Botany 90: 1180–1187.
Lopez, R. R. 2001. Population ecology of Florida Key deer. PhD dissertation, Texas A&M University, College Station, TX.
Loye, J., and S. Carroll. In press. A plant piñata: Seed predation on the balloon vine by the Miami blue, silver-banded hairstreak, a eulophid wasp and the soapberry bug in the Florida Keys. Holarctic Lepidoptera.
Malcolm, J. R. 1997. Biomass and diversity of small mammals in Amazonian forest fragments. Pp. 207–221 *in* W. F. Laurance and R. O. Bierregaard Jr. (eds.), Tropical forest remnants. University of Chicago Press, Chicago.
McKibbin, W. 1989. The end of nature. Random House, New York.
Memmott, J., and N. M. Waser. 2002. Integration of alien plants into a native flower-pollinator visitation web. Proceedings of the Royal Society of London B 269: 2395–2399.

Minno, M. C., and M. Minno. 1999. Florida butterfly gardening. A complete guide to attracting, identifying, and enjoying butterflies of the Lower South. University Press of Florida, Gainesville, FL.
Nabhan, G. P., and S. L. Buchmann. 1997. Services provided by pollinators. Pp. 133–150 *in* G. C. Daily (ed.), Nature's services: Societal dependence on natural ecosystems. Island Press, Washington, DC.
National Pest Management Association 2001. Pest Control News. Online URL http://www.pestweb.com/pestcontrolnews/.
O'Brien, J. J. 1998. The distribution and habitat preferences of rare *Galactia* species (Fabaceae) and *Chamaesyce deltoidea* subspecies (Euphorbiaceae) native to southern Florida pine rockland. Natural Areas Journal 18: 208–222.
Pascarella, J. B., K. D. Waddington, and P. R. Neal. 1999. The bee fauna (Hymenoptera: Apoidea) of Everglades National Park, Florida, and adjacent areas: Distribution, phenology, and biogeography. Journal of the Kansas Entomological Society 72: 32–45.
Pascarella, J. B., K. D. Waddington, and P. R. Neal. 2001. Non-apoid flower-visiting fauna of Everglades National Park, Florida. Biodiversity Conservation 10: 551–566.
Perfecto, I., and J. Vandermeer. 2001. Quality of agroecological matrix in a tropical montane landscape: Ants in coffee plantations in southern Mexico. Conservation Biology 16: 174–192.
Ramey, V. 1999. Lantana, shrub verbena (*Lantana camara* L.). Aquaphyte Online 19 (2). Online URL http://plants.ifas.ufl.edu/aq-f99-1.html.
Rao, M., J. Terborgh, and P. Nuñez. 2001. Increased herbivory in forest isolates: Implications for plant community structure and composition. Conservation Biology 15: 624–633.
Robinson, S. K., J. D. Brawn, and J. P. Hoover. 1997. Effectiveness of small nature preserves for breeding birds. Pp. 154–179 *in* M. W. Schwartz (ed.), Conservation in highly fragmented landscapes. Chapman and Hall, New York.
Roubik, D. W. 2001. Ups and downs in pollinator populations: When is there a decline? Conservation Ecology 5. Online URL http://www.consecol.org/vol5/iss1/art2.
Salvato, M. H. 2001. Influence of mosquito control chemicals on butterflies (Nymphalidae, Lycaenidae, Hesperiidae) of the lower Florida Keys. Journal of the Lepidopterists Society 55: 8–14.
Schneider, M. F. 2001. Habitat loss, fragmentation and predator impact: Spatial implications for prey conservation. Journal of Applied Ecology 38: 720–735.
Schwartz, M. W., and P. J. van Mantgem. 1997. The value of small preserves in chronically fragmented landscapes. Pp. 379–394 *in* M. W. Schwartz (ed.), Conservation in highly fragmented landscapes. Chapman and Hall, New York.
Silvy, N. J. 1975. Population density, movements, and habitat utilization of Key deer, *Odocoileus virginianus clavium*. PhD dissertation, Southern Illinois University, Carbondale, IL.
Sipes, S. D., and V. J. Tepedino. 1995. Reproductive biology of the rare orchid, *Spiranthes diluvialis:* Breeding system, pollination, and implications for conservation. Conservation Biology 9: 929–938.
Smith, E. M. 2000. A field study and re-establishment of the butterfly *Eumaeus atala* (Lycaenidae) in Miami-Dade County, Florida. MS thesis, Florida International University, Miami, FL.
Smith, E. M. 2002. The effects of season, host plant protection, and ant predators on the survival of *Eumaeus atala* (Lycaenidae) in re-establishments. Journal of the Lepidopterists Society 56: 272–276.
Snyder, J. R., A. Herndon, and W. B. Robertson. 1990. South Florida rockland. Pp. 230–277 *in* R. L. Myers and J. J. Ewel (eds.), Ecosystems of Florida. University of Central Florida Press, Orlando, FL.
Spira, T. P. 2001. Plant–pollinator interactions: A threatened mutualism with implications for the ecology and management of rare plants. Natural Areas Journal 21: 78–88.
Steffan-Dewenter, I., U. Munzenberg, C. Burger, C. Thies, and T. Tscharntke. 2002. Scale-dependent effects of landscape context on three pollinator guilds. Ecology 83: 1421–1432.
Steffan-Dewenter, I., and T. Tscharntke. 2002. Insect communities and biotic interactions on fragmented calcareous grasslands: A mini review. Biological Conservation 104: 275–284.
Tasker, G. 1996. Birds and bees. The Miami Herald, 26 September.
Terborgh, J. L., P. Lopez, V. Nuñez, M. Ra, G. Shahabuddin, G. Orihuela, M. Riveros, R. Ascanio, G. H.

Adler, T. D. Lambert, and L. Balbas. 2001. Ecological meltdown in predator-free forest fragments. Science 294: 1923–1926.

Thompson, J. N. 1997. Conserving interaction biodiversity. Pp. 285–293 *in* S. T. A. Pickett, R. S. Ostfeld, M. Shachak, and G. E. Likens (eds.), The ecological basis of conservation. Chapman and Hall, New York.

Thomson, J. D. 2001. Using pollination deficits to infer pollinator declines: Can theory guide us? Conservation Ecology 5. Online URL http://www.consecol.org/vol5/iss1/art6.

Vardaman, S. M. 1994. The reproductive ecology of *Melaleuca quinquenervia* (Cav.) Blake. MS thesis, Florida International University, Miami, FL.

Williams, N. M., R. L. Minckley, and F. A. Silveira. 2001. Variation in native bee faunas and its implications for detecting community changes. Conservation Ecology 5. Online URL http://www.consecol.org/vol5/iss1/art7.

Woodroffe, G. L., J. H. Lawton, and W. L. Davidson. 1990. The impact of feral mink *Mustela vison* on water voles *Arvicola terrestris* in the North Yorkshire Moors National Park. Biological Conservation 51: 49–62.

CHAPTER SIXTEEN

Ecology of Plant Reproduction: Extinction Risks and Restoration Perspectives of Rare Plant Species

Manja M. Kwak and Renée M. Bekker

Introduction

In many parts of the world, vascular plant species are becoming rare and endangered as a result of habitat loss. These rare plant species have limited geographical distribution, small population sizes, or both (sensu Rabinowitz 1981). To prevent species from extinction, nature conservation aims at stabilizing and expanding the size and number of populations. Habitat restoration or the creation of new habitat often prove to be sufficient to rehabilitate plant species (Gigon 1999); however, successful restoration will only occur if there is sufficient propagule dispersal from source populations or if establishment is possible from the soil seed bank (Bakker and Berendse 1999). Additionally, newly established populations, often small in number of plants, must be pollinated to set seed in order to increase population size.

Many characteristics of plant reproductive systems may be associated with rarity; however, after a study of 84 species, Weller (1994) concluded that there is little compelling evidence that plant breeding systems or pollination biology have had a pervasive effect in determining rarity. Aizen et al. (2002) reported that no overall conclusions could be drawn with regard to susceptibility to fragmentation based on compatibility systems and pollination generalization.

Rare and common plant species form a community in which flower-visiting insects forage, may compete for visitation by the same insect species or even individuals, and, therefore, may influence one another's pollination directly (by facilitation or competition, sensu Rathcke 1983) or indirectly (by affecting fertilization and seed set after heterospecific pollen deposition). Pollination systems are under increasing threat from anthropogenic sources, including habitat fragmentation, changes in land use, modern agricultural practices, pesticides, and herbicides (Rathcke and Jules 1993; Allen-Wardell et al. 1998; Kearns et al. 1998; Kremen and Ricketts 2000).

After seed production, seeds need to be dispersed to suitable habitats for establishment. Dispersal can take place along different vectors such as wind, water, animals, and man. Nowadays, plant species adapted to long-distance dispersal are less effective due to habitat fragmentation and different cultural practices (Bonn and Poschlod 1998).

In this chapter, we relate the pollination characteristics, seed dispersal characteristics, and longevity of seeds in the soil, life form, and clonality to vulnerability for extinction and the possibility of restoration success of plants. To address the question of whether characteristics of rare plant species differ from those of common species in extinction chance and restoration potential, we ask: (1) which plant reproductive traits indicate vulnerability and (2) which plant species are actually threatened? Using many characteristics of insect-visited herbaceous and shrubby species, we analyze the reproductive mutualisms between plants and their animal pollinators and seed dispersers and calculate a vulnerability index for extinction per plant species. This index is then related to the success of plants in recently performed restoration projects.

Rarity, Pollination, and Seed Set

Human demands and activities are responsible for the deterioration of the natural habitats of our wildflower populations (Lande 1998). Numerous species are forced to live in small and isolated natural fragments either with or without formal protection. These "new" rare species are especially vulnerable to extinction, because their natural habitat and population size declined faster than they could adapt. In this respect, they differ from species that are naturally rare, such as endemics, which are often well adapted to small population size and low levels of genetic variation (Huenneke 1991).

Populations of (rare) plant species, recently established (naturally or experimentally) after a restoration project, may encounter problems similar to those of naturally occurring rare plants: small population size, low density, large distance between patches or populations, or presence of more heterospecific than conspecific flowers, all of which increase the chance of heterospecific pollinations. All these factors may lead to low interest of flower visitors, possibly resulting in a low seed set (Allee effect; Jennersten 1988; Lamont et al. 1993). Populations of plant species that are recently established after a restoration project may encounter the additional disadvantage that the "right" (efficient) pollinators are not yet available because they also need to find the new location and to establish new populations.

Rarity and Seed Dispersal

The ability to disperse in space can be described by clonality (short distance) and by the dispersal of seeds (long distance); dispersal in time is described by the longevity of seeds in the soil seed bank. A plant species' capability of vegetative expansion is often considered an escape strategy for survival if sexual reproduction fails (De Kroon and Van Groenendael 1997).

Detailed knowledge on the dispersal capacity of rare plants is scarce or absent (Tackenberg 2001). Seeds of rare plant species are often produced in low numbers; therefore, the species are not easily discovered in dispersal experiments or in seed bank samples. Even for plants with obvious morphological adaptations for potential long distance wind dispersal, such as the rare Asteraceae *Arnica montana,* the measured dispersal distance was low (about 95% of the seeds deposited at 2.5 m in a wind-tunnel experiment with a wind speed of 6.5 m/s; Strykstra et al. 1998b). A similar weak correlation between seed morphology and dispersal exists for epizoochory (dispersal of diaspores by external attachments to animals, on fur or hoofs; Bonn et al. 2000), endozoochory (dispersal by animals carrying seeds in their digestive tracts and dropping dung containing seeds), and hydrochory (dispersal by water). Bonn and Poschlod (1998) conclude that many species are animal dispersed even when they do not have obvious hooks for attachment to fur. Also, small seeds tend to float on water even if they do not have the air chambers that hydrochorous plants obviously have (Boedeltje et al. 2003).

As with dispersability, little is known about the ability of rare plant species of northwestern Europe to form persistent soil seed banks. Prevailing methods for assessing seed-bank longevity are seed burial experiments, which take a long time and a considerable number of seeds to start with (which is obviously problematic in the case of rare plant species), or soil seed-bank sampling of natural plant communities. In addition, rare plants often occur in low abundance, consequently producing few seeds in the soil, which further decreases the chances of finding seeds when random samples are taken (Thompson et al. 1997); moreover, some groups of plants, such as orchid species, will be present in soil samples from natural populations but are not detected by regular germination procedures. It might take several years for orchids to reach a detectable size (if at all) because they need mycorrhizae for their establishment (Dijk et al. 1997). Nevertheless, there is a negative correlation between seed size and seed persistence (Bekker et al. 1998; Thompson et al. 2003).

In summary, many plant species do not have any obvious adaptation to either long-distance dispersal or seed longevity, which makes them highly vulnerable for local extinction in the present dynamic and fragmented landscape (Bakker et al. 1996; Bakker and Berendse 1999; Turnbull et al. 1999).

The Dutch Situation as an Example

Most of the Netherlands consists of agricultural landscapes in which values for nature conservation are concentrated in linear elements such as field margins, forest edges, ditch banks, and road verges; for many flowering plant species, these elements are the only refugium or the most important habitat. In these linear elements, habitats for plant species are highly dynamic and fragmented. Plant communities are repeatedly destroyed and new habitats created. The

Dutch situation can have a signal function for other countries. Plant species that are rare in the Netherlands may still be common in neighboring countries but will probably become locally rare in the future. Dutch plant species are monitored intensively so that data on long-term changes in numbers and distribution are available. In addition, at various universities (Amsterdam, Utrecht, Nijmegen, and Groningen), a large knowledge base exists concerning rare plant species, including pollination and seed dispersal; insects are also intensively monitored in the Netherlands, in comparison with other countries. Finally, restoration projects are common. An evaluation of various projects is available and the data will be used in this chapter.

Materials and Methods

Vulnerability Index

Our analysis is inspired by Bond (1994), who presented a vulnerability index for plant species; here we adjust his formula to our larger set of traits. We also choose to avoid zeros in the scores: first, because of the ecological reasoning of comparing species by their optimal capacities, and second, to ensure that the formula returns rankable results. The trait rankings range from 0.1 to 0.9, with 0.1 indicating a low vulnerability and 0.9 indicating a high vulnerability for the plant's extinction. Our vulnerability index (VI) is as follows:

$$\begin{aligned} \text{VI} = {} & (\text{Life span} + \text{Seed production} + \text{Breeding system}) \\ & + (\text{Dispersal} + \text{Soil seed bank} + \text{Clonality}). \end{aligned}$$

The first part of the formula comprises traits related to the extent of deficiency of a plant's *fitness,* including life span, seed production per plant, and the breeding system. The second part of the formula expresses the extent of deficiency of *expansion,* the ability of the plant species to expand its range. It comprises seed-bank longevity, dispersability, and clonality. For each trait we will explain the classification; the assignment of two or (possibly) more levels of specialization of each trait was derived from many data sources.

Dataset and Trait Data Collection

A dataset of the Dutch flora (about 1500 species) is available, which contains information on distribution and life span (Centraal Bureau voor de Statistiek [CBS] 1987) and rarity (Van der Meijden et al. 2000). Kwak (1994) determined that only about 25–32% of the species of the Dutch flora are pollinated by wind or water; the remaining 68–75%, are pollinated by insects. The dataset used here is restricted to insect-visited herbaceous species and bushes, excluding plantains (*Plantago* species), grasses, sedges, and trees. We started with a survey of insect visitors of plants in Dutch road verges and ditches (F. Hoffmann and M. M. Kwak, unpublished data) and added data of plant species studied by Dutch colleagues, including data of plants occurring in nature reserves. Additional data have been derived from the literature (Salisbury 1942; Fenner 1992; Proctor et al. 1996; Thompson et al. 1997; Bouman et al. 2000). The dataset contained 135 plant species (30 red list and 105 unendangered, common species) from 31 families. Data on flower-visiting insects were collected during 100 m transect walks at a slow pace; only insect species longer than 2 mm were included because smaller insects are easy to miss or to misidentify. Only insects showing active feeding behavior on flowers were scored. In some cases, more detailed information will be presented because if an insect visits flowers it does not necessarily imply pollination.

Fitness Terms

Life Span

Data on *life span* of the plants were derived from the Dutch Botanical Database (CBS 1987). Species that are known to have a life span of one year were considered short lived, whereas species with a life span indication of three years or longer were considered long lived. Annual species received the high rank 0.9, biennials 0.5, and long-lived species 0.1 for vulnerability to extinction.

Seed Production

The fitness of a plant species is partly determined by its capability for sexual reproduction, that is, the number of fertile seeds produced. We expressed seed production per individual plant, based on the number of flowers and the seed number per flower produced. We made a classification with six categories: 0–250 seeds ranks 0.96 for highest vulnerability, 251–500 seeds ranks 0.80, 501–1000 seeds ranks 0.64, 1001–2500 seeds ranks 0.48, 2501–5000 seeds ranks 0.32, and more than 5001 seeds ranks 0.16.

Breeding System

Thee term *breeding system* denotes a set of flower characteristics that affect pollination. We approach this from two points of view: the specificity (or specialization) of the plant in use of pollinators and the type of pollen that plants can expect to receive.

Under pollination specificity, we distinguish four characteristics of the flowering architecture of the plant that are important features from the view of the pollinator/flower visitor. For each of the four characteristics, we scored a plant either as specialized and, therefore, more vulnerable or as having no specific adaptation and, therefore, less vulnerable.

Many plants flower in such a way that single flowers are considered the *unit visited;* however, others produce flowers in clusters like umbels (Apiaceae) or flower heads (capitula, Asteraceae). The latter was considered to reduce the failure of pollination drastically when visitors walk around foraging and simultaneously touch more flowers than they forage on.

We classified *flower morphology* as simple and accessible for many visitors (actinomorphic, i.e., radially symmetrical) versus complex and less accessible (zygomorphic, i.e., complexly bilaterally symmetrical). Many insect visitors will not be able to effectively pollinate zygomorphic species, and specialized insect species often occur in low abundance (e.g., Vázquez and Aizen, chap. 9 in this volume).

Third, we distinguished two types of *pollen presentation:* flowers that present pollen in such a way that it is distributed diffusely over a visitor's body versus those with precise presentation that deposit pollen locally on a visitor's body. The latter means that the visitor has to deliver these pollen grains in a precise way to the next stigma, and this implies a higher risk of pollination failure because insects have to behave in a precise way to pollinate.

For flowers that produce *nectar,* this was regarded as an adaptation to attract certain visitors at a cost of energy that cannot be spent on other traits. Therefore, plants with floral nectar were considered to be more specialized for pollinators and vulnerable to pollination failure.

Binary scores (symbolized here as A, B) for these four characteristics, when combined, yield five categories (i.e., AAAA, AAAB, AABB, ABBB, BBBB). We assigned these combined scores the values of 0.1 (least vulnerable), 0.3, 0.5, 0.7, and 0.9 (highly vulnerable); species that specialize in all four traits receive the highest vulnerability score on this *pollination specificity* index.

Under pollen type, we first looked at the possibility of self-pollination: the ability to deposit locally produced pollen onto the stigma of the same flower (*autodeposition*). This trait was subdivided into three classes: autodeposition, restricted autodeposition, and no autodeposition possible. Temporal or spatial differences between anthers and stigmas may prevent autodeposition. Note that self-pollination is not the same as self-fertilization. Moreover, many species are self-incompatible, and self-pollen cannot germinate or grow on self-stigmas. However, self-incompatibility was not included because of lack of data on many plant species.

Second, the *pollen purity* of outcross pollination is important for plants. Insect species that are generalists, visiting several plant species on the same foraging trip, may deposit heterospecific pollen onto stigmas. This reduces the chance for conspecific pollen to germinate successfully and, therefore, may hamper seed set. Specialist pollinators, on the other hand, deposit rather pure pollen loads. Plants might invest in the attraction of specialist pollinators, but they run the risk of low pollen deposition if their population is small or sparse or when the abundance of pollinators decreases for some demographic reason. An intermediate category was designated for plants that were visited by both specialist and generalist insect species.

We combined the scores for autodeposition and pollen purity into a matrix of final scores for pollen type, as shown in table 16.1. Plant species with generalized pollinators and the ability for autodeposition scored 0.1, and plants with spe-

Table 16.1 Matrix of values assigned for "pollen type," composed of all possible pairwise combinations of the capacity for autodeposition (which serves as a potential rescue when pollinators are rare or lacking) and pollinator type (which affects the purity of expected pollen deposition)

	Self-pollination		
Pollinator type	Autodeposition	Autodeposition but reduced seed set	No autodeposition
Generalist	0.1	0.3	0.5
Mix of generalists and specialists	0.3	0.5	0.7
Specialist	0.5	0.7	0.9

Note: A value of 0.1 connotes lowest vulnerability and 0.9 highest vulnerability.

cialized pollinators and lacking autodeposition scored 0.9 (and are considered the most vulnerable to pollination failure).

To summarize, we considered four traits under pollination specificity and two under pollen type. To generate one value for "breeding system" from these six variables we weighted their contributions equally; hence,

$$\text{Breeding system} = [(4 \times \text{Pollination specificity}) + (2 \times \text{Pollen type})]/6.$$

Expansion Terms

Dispersability

We considered four types of dispersal—anemochory (dispersal by wind), hydrochory, epizoochory, and endozoochory—which we arbitrarily considered effective over a distance of 100 m or more. This is considered the minimal distance to connect populations in a metapopulation network (Cain et al. 2000) and to have an effect in successful reestablishment of species during restoration efforts. We omitted dispersal by birds because species with such dispersal rarely occurred in our dataset.

Dispersal by wind was classified as effective (greater than 100 m) if seeds had a known low terminal velocity or were observed to fly for large distances (Ozinga et al. 2004). The possibility of floating on water for at least one week was set as the primary criterion for assigning species to be effectively dispersed by water (Praeger 1913). Dispersal by animals was considered to bridge large distances if observations such as presence in the fur or germination from dung samples could be substantiated (Tamis et al. 2004).

Those plants able to disperse over considerable distances by all four methods received the lowest ranking for vulnerability (0.1), whereas species with no adaptations for long-distance dispersal received the highest ranking (0.9).

Soil Seed-bank Longevity

We used the seed longevity index data of Bekker et al. (1998), based on data from Thompson et al. (1997), and for missing species used those of Grime et al. (1988), to apply a scale of seed longevity comprising nine classes: from 0.1 (seeds transient, do not persist longer than one year) to 0.9 (seeds persistent in the long term, for more than five years). We had to add the original seed longevity index class 0 to class 0.1 (short-living) and class 1 to class 0.9 (long-living) to adjust the ranges of this classification to the other trait ranges and avoid zeros. The reciprocal value of the index gives the ranking for vulnerability; plants with persistent seeds were considered less vulnerable to extinction than plants with transient seeds.

Clonality

We defined the criterion for ecologically effective vegetative expansion as the ability to bridge a distance of at least 10 cm from the maternal plant within a single growing season. We used mostly data from Klimes and Klimesova (1999) and we distinguished three classes for clonality: no spreading, spreading less than 10 cm, and spreading equal to or farther than 10 cm; the corresponding vulnerability ranks were 0.9, 0.5, and 0.1, respectively. Note that clonality largely coincides with a plant's life span: by our definition, annuals and biennials are not able to spread laterally.

Data Analysis

We performed a frequency analysis per trait on the total species set and observed that all classes in all traits were represented by at least 5 out of the total of 135 species. We performed a one-way analysis of variance (ANOVA) for each group of traits composing the overall VI to detect significant differences between mean values of rare versus common plant species, after checking for normality. If normality was not achieved after data transformation, we used a nonparametric Kruskal–Wallis test. Contrasts were obtained by a Tukey range test (ANOVA) or Student Newman Keuls (rank) test. Correlation coefficients were determined by factor analysis between the component terms of the VI. All tests were performed with SPSS 11.0 for Windows.

Results

Vulnerability Index

The two main components of the VI formula, deficiency in fitness and expansion, are at most only weakly correlated in either rare or common plants (figs. 16.1A, 16.1B). This result indicates that each of these terms makes an independent contribution to the vulnerability of plant species to extinction.

Values of the vulnerability indices ranged from 3.07 to 5.0 for rare plants (mean $\pm$ SE $= 3.92 \pm 0.72$) and from 1.26 to 5.0 (3.07 ± 0.30) for common species, with a significant difference in the means ($P < .0001$).

Both groups of plants contain species that have high VI values (see appendix 16.2). These species are short-lived species and have low seed production and a complex breeding system with bumblebees as the main pollinator. Moreover,

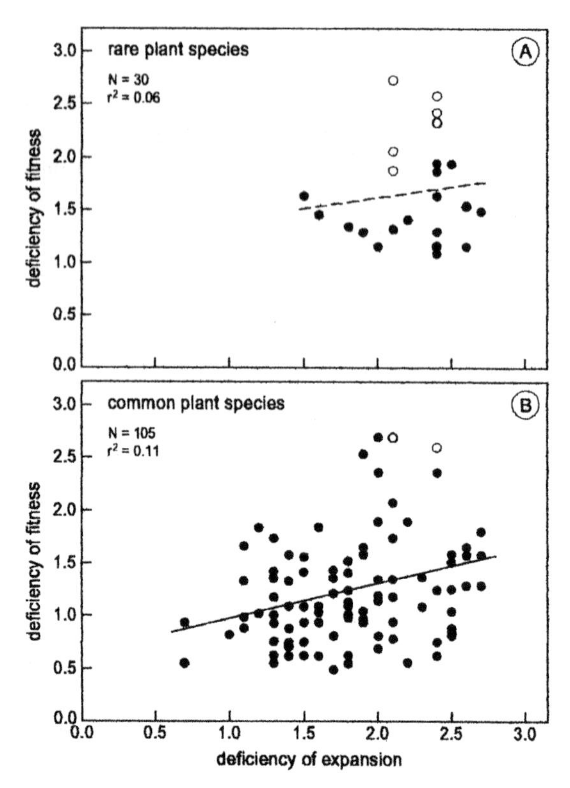

Figure 16.1 The relation between the composite terms, the deficiency in expansion, and fitness for (A) rare and (B) common plant species in the Netherlands. Significant regression only found for common plants ($P < .001$). Open circles indicate hemiparasitic plants of the Scrophulariaceae.

these species have less-persistent seeds in the soil, lower dispersability, and lower abilities for vegetative expansion. Hemiparasitic plants (family Scrophulariaceae), both rare and common species, have high vulnerability indices that vary between 3.98 and 5.0. In general, VI values correspond to expansion or contraction in species distribution observed in the Netherlands during the past century. *Impatiens glandulifera* is an interesting common plant that has a high VI value, 5.0, which has increased in the Netherlands due to weedy behavior (see Discussion).

The possible minimum and maximum VI values (0.66 and 5.46, respectively) were not reached in our dataset. The lowest value is 1.26 for the common species *Hypericum perforatum;* the highest values are 5.0 for the common species *I. glandulifera* and *Rhinanthus angustifolius* and the rare species *Rhinanthus alectorolophus.* For fitness, only the rare species *Melampyrum arvense* scored 2.76, the maximum possible value. The common *Achillea millefolium* achieved the lowest score (0.49), higher than the lowest possible value of 0.36. For expansion, one rare species (*Genista anglica*) and three common species (*Lathyrus pratensis, Lupinus polyphyllus,* and *Symphytum officinale*) scored the maximum possible value of 2.70.

Table 16.2 Values (mean ± SE) of traits of rare and common plant species and their significance

Trait	Possible range of values	Rare species (N = 30)	Common species (N = 105)	Significance
Deficiency of fitness	0.36–2.76	2.08 ± 0.06	1.84 ± 0.05	P = .030
Life span	0.1–0.9	0.27 ± 0.06	0.24 ± 0.03	P = .55
Seed production	0.16–0.96	0.69 ± 0.04	0.45 ± 0.03	P < .001
Breeding system	0.1–0.9	0.72 ± 0.04	0.56 ± 0.02	P = .001
Deficiency of expansion	0.3–2.7	1.68 ± 0.09	1.24 ± 0.05	P < .001
Seed bank longevity	0.1–0.9	0.61 ± 0.04	0.65 ± 0.02	P = .41
Dispersal	0.1–0.9	0.70 ± 0.03	0.63 ± 0.02	P = .056
Clonality	0.1–0.9	0.77 ± 0.04	0.57 ± 0.04	P = .01

Note: A value close to 0 means less vulnerable and a value close to 1 means very vulnerable. The traits in bold are composite, and their values exceed the value of 1.

Hypericum perforatum and *Ranunculus repens* scored lowest at 0.70, higher than the lowest possible value of 0.30.

Table 16.2 gives the mean values per group of rare and common plant species for the various traits composing the terms of the deficiency of the fitness and expansion of VI. Three traits—life span, seed longevity, and seed dispersal—do not differ significantly between rare and common plant species. The other components all differ in that rare plants have higher values than common plants, indicating greater vulnerability.

Insect Visitors as Fitness Components

Insect species belonging to the orders Diptera, Hymenoptera, Lepidoptera, and Coleoptera (in decreasing order of frequency) were observed as flower visitors (table 16.3). Most plant species were visited by at least two insect species. Detailed information on the frequency of insect species visiting *Succisa pratensis* is given in figure 16.2. The species is visited by 22 insect species, belonging to four orders. Five species, all common and belonging to the Eristalinae (Syrphidae), are responsible for 88% of visits, and all are pollinators. However, insect species do differ in the number of pollen grains deposited per minute: bumblebees deposited about twice as much pollen per unit time as syrphids (Kwak 1993).

Flower-visiting insects varied in body size from small (± 2 mm) to large (28 mm for cuckoo bumblebees and about 30 mm for butterflies). Tongue length, which is important to reach nectar, varied between very short (flies and small bees) to very long, having a maximum of about 16 mm for butterflies (Corbet 2000a) and 18 mm for bumblebees (Von Hagen 1994). Hairiness, which indicates pollen carrying capacity, varied between nearly naked (various fly species) to very hairy (bumblebees).

Diptera (syrphids excluded) and wasps were observed as flower visitors on 65% of the plant species investigated. The most important flower-visiting fly

Table 16.3 Percentage of insect taxa as visitors to rare and common plant species in the Netherlands

Insect taxon	All plant species (*N* = 135)	Rare plant species (*N* = 30)	Common plant species (*N* = 105)
Other Diptera + wasps	65.2	40.0	72.4
Syrphidae	62.2	40.0	68.6
All Apidae	79.3	96.7	74.3
Apis mellifera	40.7	16.7	48.6
Bombus spp.	75.6	93.3	70.5
Lepidoptera	22.2	13.3	24.8
Other[a]	10.4	3.3	12.4

Note: Almost all plant species were visited by at least two insect taxa.
[a]Includes Coleoptera and Hymenoptera.

families were Muscidae, Sarcophagidae, Calliphoridae, and Tachinidae. Syrphidae (hover flies) were the most frequent and observed on 62% of the species we investigated. For rare plants, syrphids may be critical: 40% of them had syrphid visitation. Syrphids were even more frequent on common plants, 69% of which had these hovering flies among their visitors. Syrphids, especially *Eristalis tenax* and *E. pertinax,* were the most frequent visitors of the Apiaceae and, together with bee species, were frequent on *Succisa pratensis, Valeriana officinalis,* and Asteraceae. Their behavior (active flower visiting during a long period of the day and a long active season) and morphology (hairy and large sized, able to carry a reasonable number of pollen grains, and able to reach half-hidden nectar) make them important flower visitors. They are especially frequent on a large number of wildflowers with open or half-hidden nectar.

Other Diptera, such as Muscidae and Calliphoridae (e.g., *Sarcophaga carnaria* and *Lucilla caesar*), are common on plant species with accessible nectar such as Apiaceae, *Valeriana officinalis, Achillea millefolium, Rubus fruticosus,* and *Ranunculus* species. Empidid flies were common visitors on Apiaceae such as *Anthriscus sylvestris,* but also on *Valeriana officinalis, Valeriana dioica,* yellow Asteraceae, *Rubus fruticosus,* and *Cirsium* species. Sepsidae, very small flies, were very common but, due to small body size and few hairs, were not very important pollinators on an individual basis. Small fly species, such as the tachinid *Siphona geniculata,* have a very restricted pollination value, at least for the plant species *Scabiosa columbaria.* In an experiment with flowers and flies in cages, each small fly deposited 0.7 and each *E. tenax* 145.0 pollen grains per flower head in three hours; the pollen deposition of one *E. tenax* individual therefore equaled that of nearly 200 small flies (unpublished data).

In the Netherlands, 332 bee species occur but only 67 species are considered common; this especially includes social species that are also polylectic, foraging on many plant species (Van Raaij 1998). Only 11 oligolectic species are documented for the Netherlands. About 79% of the plant species we studied received

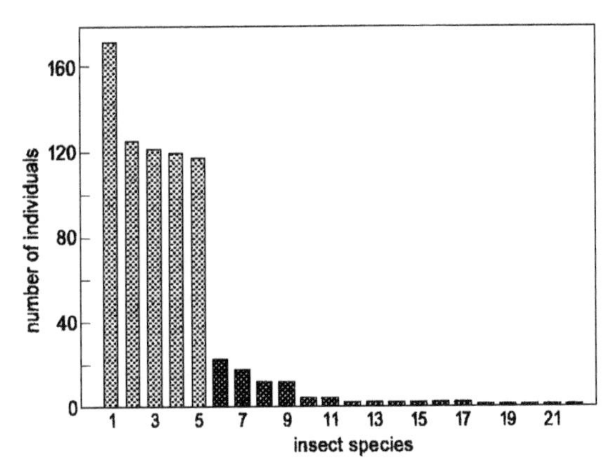

Figure 16.2 The rare plant species *Succisa pratensis* is visited by a variety of insect species numbered along the *x* axis (*N* = 740 individuals) but only five syrphid fly species (light bars) are responsible for the bulk of visitation and pollination. These syrphid species are very common in the Netherlands. Lumped data include transect observations made between September 6 and October 9, 2002, in eight *S. pratensis* populations differing in size and environment. Insect species 1, 2, 3, 4, 5, 7, 9, 11, 16, 19, 21, and 22 belong to Syrphidae; 6 belongs to Diptera; 8 and 12 belong to Hymenoptera; and 10, 13, 14, 15, 17, 18, and 20 belong to Lepidoptera.

bee visitors that included bumblebees, "wild bees," and honeybees. Bumblebees visited 93% of the rare and 71% of the common plant species, which make up the highest values for both groups of plants (table 16.3). The importance of bumblebees for rare and common plant species is similar, with 63% and 61% receiving bumblebees as important visitors, respectively, contributing substantially to the pollination of these species. Among the plants mainly pollinated by bumblebees are *Rhinanthus* spp., *Pedicularis* spp., *Melampyrum* spp., *Lamium* spp., *Gentiana pneumonanthe, Salvia pratensis, I. glandulifera,* and *Cynoglossum officinale.* Bumblebees often occurred in low numbers, but, due to their long daily period of activity, they contributed substantially to the pollination of many species. Wild bee species were observed on 20.7% of the plant species studied, but for no species were they the sole visitor. We found no plant species that had a monolectic or oligolectic bee species as its only visitor; even a species such as *Lysimachia vulgaris,* with the oligolectic bee visitor *Macropis europea* (Vogel 1986), was also visited by several syrphid fly species. However, the frequency of visits of syrphids was 10–40%, versus 60–90% for *Macropis.* An oligolectic bee species such as *Dasypoda hirtipes,* specialized on Asteraceae (Westrich 1990), was found on five ligulate Asteraceae species but never as the most frequent visitor. Other visitor species were syrphid flies, butterflies, and other bee species (Van der Muren et al. 2003). If the common species *Campanula rotundifolia* was visited by the oligolectic bee species *Melitta haemorrhoidalis,* bumblebees were absent; but if *M. haemorrhoidalis* were absent, bumblebees appeared to be the only flower visitors.

Honeybees were visitors of 55 plant species (41%) but for no plant species were they the main flower visitor. They were visitors together especially with syrphids, wild bees, and bumblebees. Honeybees were observed as regular visitors in reasonable numbers on only eight plant species, most of which are common species (e.g., *Tanacetum vulgare, Rubus fruticosus,* and *Calluna vulgaris*). Five of the 30 rare plant species (*Arnica montana, Phyteuma spicatum* subsp. *nigrum, Primula vulgaris, Knautia arvensis,* and *Scabiosa columbaria*) were visited by honeybees, but these were of minor importance considering the number of visits. In the case of *P. vulgaris,* honeybees preferred the thrum flowers (anthers just protruding past the corolla-tube opening) over the pin flowers, resulting in very low pollen deposition on the appropriate stigmas (unpublished data).

Wasps as flower visitors belong to the Symphyta (five families), Parasitica (family Ichneumonidae), Vespoidea (especially family Vespidae), and the Apoidea (family Sphecidae). Most species feed on nectar for themselves, whereas the social Vespidae (e.g., *Vespula* and *Dolichovespula*) also do so to feed larvae; pollen is not collected. Most plant species visited by wasps had accessible nectar and were also visited by flies. All the previously mentioned wasp groups visited plant species belonging to the Apiaceae, such as *Angelica sylvestris* and *Heracleum sphondylium.* Members of the Vespidae were also observed on *Eupatorium cannabium* and *Scrophularia nodosa;* the latter is often mentioned as particularly adapted to wasp pollination (Kevan and Baker 1983; Meeuse and Morris 1984; Proctor et al. 1996). Indeed several species of the Vespidae were observed as visitors, but rarely were they the only or dominant visitors. Bumblebees and, in some cases, honeybees were also frequent.

The Formicidae, ants, are flower visitors, but their role in pollination is not fully understood. Ants showed a very low mobility and remained present on the same flower for many hours. They were never the only visitors. Plant species with ant visitation were *Anthriscus sylvestris, Angelica sylvestris, Heracleum sphondylium,* and *Hieracium laevigatum.* Ants were not included as pollinators in our analysis.

Butterflies are intensively monitored in the Netherlands, and a total of 55 resident and 4 migratory species occur, of which 31 are red list species (M. Wallis de Vries, personal communication). Only 28 species are not yet threatened (Van Ommering et al. 1995), and 10 species (8 resident and 2 migratory) can be considered so common that they can be observed as flower visitors everywhere. Both groups, rare and common butterflies, have exhibited a decreasing trend in species number during the past 10 years (Plate and Van Swaay 2002). Butterflies and day-active moths (mainly *Autographa gamma*) were regular visitors but often in low numbers (e.g., figs. 16.2 and 16.3). Within our dataset, butterflies were recorded as visitors on 35 plant species (26%). No plant species were visited exclusively by butterflies; all were also visited by bees and/or flies. A few plant species attracted a large number of butterfly individuals: *Eupatorium cannabium, Cirsium arvense, Knautia arvensis, Silene latifolia, Lythrum salicaria,* and *Hieracium laevigatum.* Rare plant species occasionally visited by butterflies were *Dianthus*

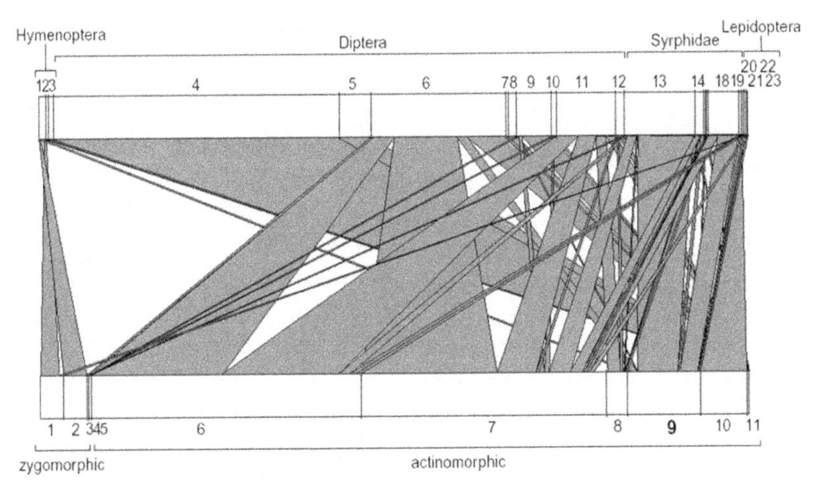

Figure 16.3 Quantitative plant–visitor web for a Dutch road-verge community containing the rare plant species *Succisa pratensis* (Dipsacaceae, plant 9), flowering during August–September (data from 2000). Each species/morphotype of plant and insect is represented by a numbered rectangle on the horizontal bar. Species identities ($N = 782$) are listed in appendix 16.1. The width of the rectangles is proportional to the species abundance at the field site. Interactions are shown as connecting trapezoids, the sizes of which reflect the proportion of all recorded visits received by that plant species. Note that, for insects, the number of individuals on a certain plant species is given, which may differ from the number of visits a plant species will have received by that species.

deltoides, Scabiosa columbaria, and *Succisa pratensis.* Rare butterfly species were not observed as flower visitors.

Many Coleoptera occur in the Netherlands but no official documentation of the number of beetle species is available. The beetle species most often observed as flower visitors were *Rhagonycha fulva* (Cantharidae) and *Alstera tabacicolor, Trichius fasciatus,* and *Phylloperta horticola* (Scarabaeidae). They were observed on 15 plant species with accessible or half-hidden nectar—members of the Apiaceae, Asteraceae, and Rosaceae. These beetles were the most important component of the "other" group in table 16.3 and were always visitors together with flies and bees. The foraging speed of beetles (number of flowers visited per time unit) is low. In some cases, they are present in more than marginal numbers, as in the case of species of the Cantharidae, which made up 14 % of visits to the plant *Anthriscus sylvestris.* They carried an average of more than 5600 *Anthriscus* pollen grains on their bodies per individual, and the average distance between two flower visits was 80 cm (unpublished data). Species of the Curculionidae visited flowers to eat and to lay eggs on the developing fruit and were not considered pollinators. Hemiptera species were not often observed as flower visitors and they are included in a group with Coleoptera ("other" in table 16.3). Plant species with Hemiptera were Apiaceae, yellow Asteraceae, and *Achillea millefolium.*

Figure 16.3 shows a "visitation web" of relations between insects and flowers observed in a road verge containing the rare plant species *Succisa pratensis.* This web depicts both the importance of a certain insect species for the plant species and the diet of the insect species. The syrphid species *Eristalis tenax/pertinax* (no. 13) is an important visitor of *S. pratensis,* but this insect species also visited *Hypericum laevigatum* and *Leontodon autumnalis.* Other insect species on *S. pratensis* are *Musca* sp 1 (no. 6) and *Rhingia campestris* (no. 18).

Discussion

Vulnerability Index

We were able to distinguish extinction risks between rare and common plants species by condensing 11 traits into 6 testable traits, forming the terms for the deficiency of fitness and expansion of an overall Vulnerability Index. Rare plant species generally had higher VI values (3.92 ± 0.72, mean ± SE) than did common species (3.07 ± 0.30); however, for some common species we also found high VI values, indicating that, among this group, vulnerable species do occur that will probably become new rare species. Our dataset could have been biased due to the fact that hemiparasitic, bumblebee-pollinated plants are included (see fig. 16.1); however, without these species the same conclusions hold, because the common species have much lower minimum VI values (1.26) than the rare species (3.07). Very low values for vulnerability (0.66–1.26) might occur in clonal, wind-pollinated species like grasses and sedges, which we deliberately excluded from our dataset.

Insect Visitors as Fitness Components

In this chapter, we stress the role that insects play in the pollination of rare and common plant species. This is the reverse of what is presented in various other papers, which focus on the role that plants play for insects (Ellis and Ellis-Adam 1993; see, for instance, the construction of flower visitation webs by Memmott 1999; Dicks et al. 2002). We mainly focus on data of herbaceous species in the remnants of "nature" in the present fragmented agricultural landscape of the Netherlands: road verges, ditches, and nature reserves. Not surprising is that the number of insect species we observed is much lower than the number of species for a certain plant species mentioned by Ellis and Ellis-Adam (1993). They summarized the anthophilous fauna of northwestern Europe, including data collected since Hermann Müller (1873–1881) and, thus, covering a long time period and a large area of Europe.

Rare plants in the Netherlands generally did not differ in insect visitors compared to common ones. In both groups of plants, species belonging to Diptera and Hymenoptera were the most important visitors, and particular plant species often were visited by insects of both orders, confirming the generalist visitation of plants (Waser et al. 1996). Even plants that are visited by specialists also had generalist visitors. For both rare and common plants, bumblebees played an important role, considering the number of visits: bumblebees often deposit more pollen grains per visit than, for instance, butterflies or syrphids (Kwak 1993; Kearns and Inouye 1994; Kwak et al. 1996; Kwak and Velterop 1997; unpublished data). Plants only or mainly visited by bumblebees can be considered specialists.

Corbet (2000b) considered the large bee/large flower compartment in pollination webs to be important but sensitive to frequent disturbance. Road verges are often disturbed, with as many as six mowings during the growing season (F. Hoffmann and M. M. Kwak, unpublished data); hence, frequent disturbance occurred—even nature reserves are often mown twice a year. Large bees have higher energy demands and must visit flowers that offer substantial energetic rewards. These plants are commonly perennials and, to a lesser extent, biennials. In addition, perennials are often self-incompatible (Bond 1994; Corbet 1995) and, thus, more sensitive to the loss of bee pollinators. An interesting common and expanding species is *Impatiens glandulifera* (Jewelweed); it has occurred in Europe since 1839, hailing from the Himalayas and India and introduced as a garden plant (Coombe 1956). It is found in the Netherlands along rivers and disturbed areas. The species has seeds with a short longevity, and, despite ballistic seed dispersal, the range of dispersal does not exceed 100 m. It has a complex flower morphology with bumblebees as the main pollinators. Chittka and Schürkens (2001) stated that this successful invader tempts bee pollinators away from native flowers, reducing the seed set of local plants in the vicinity (*Stachys palustris,* in their example). This enables *I. glandulifera* to take over, its spread facilitated by insect pollination, reducing the fitness of native flora before competition for other resources (nutrients, water, light, and space) even takes effect.

Several bumblebee species are in decline in Europe, especially long-tongued species like *Bombus hortorum* (Rasmont 1988; Williams 1989; Kwak and Tieleman 2000); however, short-tongued bumblebees can also collect pollen from certain plants, resulting in pollination, if the number of flowers is large enough to trigger this behavior (Kwak 1977, 1979, 1988).

Honeybees (*Apis mellifera*) are kept in hives by beekeepers who often transport the hives to flower-rich areas like orchards and heathland. Without the assistance of the beekeeper, colonies have problems surviving for more than one season. Honeybees were observed on 41% of the studied species, but they occurred in large numbers on only eight (6%) of the species. It is hypothesized that honeybees may compete with other flower-visiting insects and thus influence the diversity of flower-visiting insects and their abundances (Van der Goot 1981; Evertz 1995; Roubik 1996; Brugge et al. 1998), although this hypothesis is not supported by all studies (Sugden et al. 1996). The idea also exists that honeybees can replace lost pollinators (Aizen and Feinsinger 1994). Certain plant species may receive honeybees as new visitors, but, at least for some plant species, they are less effective in pollinating these flowers than the "natural" pollinator (Kwak 1980; Westerkamp 1991; Wilson and Thomson 1991). The significance of honeybees for pollination derives from the large numbers available (Corbet 1991). In general, most honeybees are found on mass-flowering species, but many wild plant species, especially rare species, do not flower in large numbers. The eight plant species that were visited regularly by honeybees indeed all occurred in large numbers.

The extinction process can take a long time, especially in perennials. Low seed set may be one of the factors, but seed set may also differ greatly between years. In general, changes in habitat or habitat loss are the most common reasons for the disappearance of plant species. We are not aware of any plant species that are decreasing from poor pollination as a consequence of the absence of the right pollinator. Because plant species receive visits from a variety of insect species, pollen will be deposited, but the quantity and quality of this deposition may decrease and influence the amount of seed set. For *Succisa pratensis,* it seems that pollination quality (conspecific pollen) is more important than pollination quantity (number of visits; F. Hoffmann and M. M. Kwak, unpublished data). In our study, we could not directly relate plant rarity with insect rarity because rare insect species were only occasionally observed: they were mostly bees and syrphids. We observed only a single case in which a rare plant species is pollinated by a rare insect species: *Phyteuma spicatum* ssp. *nigrum* (Campanulaceae) and the early bumblebee species *Bombus jonellus.* This plant species can also be pollinated by the syrphid species *Rhingia campestris,* although seed set is much lower in that case (Kwak 1993).

Besides bumblebees, syrphids appeared to be common visitors on many plant species. Few examples exist where the pollination efficiency of syrphids is compared with that of other species (Kwak 1993; Kearns and Inouye 1994; Kwak et al. 1996; Kwak and Velterop 1997, unpublished data). At least for plant species with half-hidden nectar and protruding stigmas and anthers, such as *Scabiosa columbaria* and *Succisa pratensis,* they are important as pollinators.

Restoration Perspectives

All rare plant species have been checked for their possible known prospects and measures for reestablishment in the Netherlands and Switzerland (Gigon et al. 1996; Gigon 1999; Bekker and Lammerts 2002; see appendix 16.2). One-third of the plant species could be reestablished or saved from extinction, mainly through the reintroduction of a haymaking regime and/or by sod cutting. Contradictory results for a successful rescue exist for nearly one-third of the species. The remaining set of plant species has not yet been reported as reacting successfully to restoration measures, or no information at all was available. Plant species that have been successfully restored generally seem to be the best-studied species with, in some cases, pollination studies included. Because most plant species are visited and pollinated by flies and bees, it is important to create appropriate conditions for these insects.

Concluding Thoughts

Rare and common plants differ significantly in all traits considered in our VI formula except for three: life span, seed longevity, and dispersability. Plant species of the Dutch flora, both rare and common, are primarily visited by flies and bees. Among the bees, bumblebees are very important, especially for rare species, whereas honeybees play a minor role as flower visitors of the wild flora.

For plants that we wish to reestablish in a restoration project, bumblebees may be among the first flower visitors attracted because they are able to fly relatively large distances. If we create the right conditions for bumblebees (food plants, nesting sites) in the vicinity of the area to be restored, we maximize the chance of reproductive success of these plant species.

Much effort is being made to increase connectivity in the landscape by creating corridors between remnants (Parker and Pickett 1997; Pywell et al. 2003). These corridors could be significant for the dispersal of both plants and insects; however, the importance for the dispersal of seeds is estimated to be rather low (Van Dorp 1996), especially in light of the fact that a corridor for plants can only function when it is a suitable habitat for establishment (Verhagen et al. 2001; Strykstra et al. 2002). Intrinsically, it seems impossible to connect the landscape with suitable habitats for many endangered plant species. For insects, corridors of flowering plants do increase pollen flow between patches (Kwak et al. 1998; Velterop 2000). Even non-flowering hedgerows, or an artificial structure in the landscape that does not contain flowers, increases bumblebee movement between flower patches and, thus, increases pollen flow (Cranmer 2004).

We did not consider the possible importance of dispersal by human activities such as hemerochory (dispersal through vehicles and agricultural machines, tires, trains, boots, etc.) and the decreased connectivity of habitats (through the decrease in sheep herds, hay and sod gathering, dung spreading, flooding frequency for plants with floating seeds, etc.). Still, most grassland communities in nature reserves are cut for hay by machinery nowadays instead of by scythe. This has proved to considerably increase dispersal distances for some species (Strykstra et al. 1997). Much effort is being expended to prevent rare species from going extinct in the Netherlands. Sowing seeds of desired species or distribution of seed-rich hay is also one of the current practices used to establish species after restoration practices instead of waiting until spontaneous establishment occurs. When reintroduction of plants is considered, the method of introduction needs to be carefully chosen, such as single or repeated usage of seeds, seedlings, juvenile plants, or seed-rich hay (Nunney and Campbell 1993; Oostermeijer et al. 1994; Van Groenendael et al. 1998; Strykstra et al. 1998a; Strykstra 2000). From the point of view of insects, plant species must be introduced in clusters to reach high-density patches. These high-density patches have a higher chance of being discovered by insects, thus preventing the problems met by low-density populations during flowering and pollination (Van Treuren et al. 1994).

Appendix 16.1

Table 16A.1 Legend for figure 16.3

Insect species number	Order/ family	Name	Plant species number	Name
1	Hymenoptera	*Bombus pascuorum/terrestris*	1	*Trifolium pratense*
2		Parasitic wasp	2	*Trifolium repens*
3		*Symphyta* sp.	3	*Chamerion angustifolium*
4	Diptera	*Bibio* sp.	4	*Taraxacum officinalis (agg.)*
5		*Sepsidea*	5	*Achillea millefolium*
6		*Musca* sp.1	6	*Leontodon autumnalis*
7		*Sarcophaga* sp.	7	*Hypericum laevigatum*
8		*Musca* sp.2	8	*Hypochaeris radicata*
9		*Scathophaga stercoraria*	9	*Succisa pratensis*
10		*Scathophaga* sp.	10	*Potentilla erecta*
11		Diptera, brown 3 mm	11	*Heracleum sphondylium*
12		Diptera, black 3 mm		
13	Syrphidae	*Eristalis tenax/pertinax*		
14		*Eristalis arbustorum/abussivia*		
15		*Helophilus pendulus/trivittatus*		
16		*Syrphus* sp.		
17		*Sphaerophoria* cf. *scripta*		
18		*Rhingia campestris*		
19		*Cheilosia* sp.		
20	Lepidoptera	*Pieris brassicae*		
21		*Vanessa cardui*		
22		*Thymelicus lineola*		
23		*Autographa gamma*		

Appendix 16.2

Table 16A.2 Plant species (all rare species and the top 10 and bottom 10 most common species), their families, and values for the two components of the VI, ranked in decreasing order of VI values

Rare plants	Family	Deficiency of fitness	Deficiency of expansion	VI	Restoration success and measures to gain success[a]
Rhinanthus alectorolophus	Scrophulariaceae	2.6	2.4	5.0	± Haymaking
Rhinanthus minor	Scrophulariaceae	2.5	2.4	4.9	+ Haymaking
Melampyrum arvense	Scrophulariaceae	2.8	2.1	4.9	N/A
Gentianella amarella	Gentianaceae	2.4	2.4	4.8	± Sod cut/haymaking near remnant populations
Pedicularis sylvatica	Scrophulariaceae	2.4	2.4	4.8	+ Sod cutting
Fritillaria meleagris	Liliaceae	2.0	2.5	4.5	N/A

Table 16A.2 *(continued)*

Rare plants	Family	Deficiency of fitness	Deficiency of expansion	VI	Restoration success and measures to gain success[a]
Gentianopsis ciliate	Gentianaceae	2.0	2.4	4.4	± Haymaking near remnant populations
Salvia pratensis	Labiatae	2.0	2.4	4.4	± Haymaking
Gentianella germanica	Gentianaceae	1.9	2.4	4.3	+ Haymaking
Euphrasia stricta	Scrophulariaceae	2.1	2.1	4.2	± Sod cutting
Dianthus deltoides	Caryophyllaceae	1.6	2.6	4.2	N/A
Genista anglica	Leguminoseae	1.5	2.7	4.2	N/A
Anthyllus vulneraria	Leguminoseae	1.9	2.1	4.0	± Sod cutting
Pedicularis palustris	Scrophulariaceae	1.9	2.1	4.0	± Haymaking near remnant populations
Pinguicula vulgaris	Lentibulariaceae	1.6	2.4	4.0	+ Sod cutting
Knautia arvensis	Dipsacaceae	1.2	2.6	3.8	N/A
Gentiana pneumonanthe	Gentianaceae	1.3	2.4	3.7	+ Sod cut around remnant populations
Phyteuma spicatum subspecies *nigrum*	Labiatae	1.3	2.4	3.7	N/A
Cirsium dissectum	Asteraceae	1.4	2.2	3.6	+ Sod cut/haymaking near remnant populations
Valeriana dioica	Valerianaceae	1.4	2.2	3.6	N/A
Dactylorhiza incarnata	Orchidaceae	1.2	2.4	3.6	+ Sod cut and haymaking
Phyteuma spicatum subspecies *spicatum*	Campanulaceae	1.1	2.4	3.5	N/A
Arnica montana	Asteraceae	1.1	2.4	3.5	+ Sod cut around remnant populations
Scabiosa columbaria	Dipsacaceae	1.3	2.1	3.4	+ Haymaking/grazing
Parnassia palustris	Saxifragaceae	1.3	1.9	3.2	+ Sod cut and haymaking
Succisa pratensis	Dipsacaceae	1.2	2.0	3.2	N/A
Gentiana cruciata	Gentianaceae	1.2	2.0	3.2	± Haymaking near remnant populations
Viola canina	Violaceae	1.6	1.5	3.1	± Haymaking near remnant populations
Geum rivale	Rosaceae	1.5	1.6	3.1	N/A
Primula vulgaris	Primulaceae	1.3	1.8	3.1	N/A
VI top 10 common plants					
Impatiens glandulifera	Balsaminaceae	2.6	2.4	5.0	
Rhinanthus angustifolius	Scrophulariaceae	2.6	2.4	5.0	
Melampyrum pratense	Scrophulariaceae	2.7	2.1	4.8	
Raphanus raphanistrum	Cruciferae	2.4	2.4	4.8	
Cynoglossum officinale	Boraginaceae	2.4	2.4	4.8	
Lamium purpureum	Labiatae	2.7	2.0	4.7	
Symphytum officinale	Boraginaceae	1.8	2.7	4.5	
Galeopsis tetrahit	Labiatae	2.5	1.9	4.4	

(continued)

Table 16A.2 *(continued)*

Rare plants	Family	Deficiency of fitness	Deficiency of expansion	VI	Restoration success and measures to gain success[a]
Anchusa officinalis	Boraginaceae	2.4	2.0	4.4	
Geranium robertianum	Geraniaceae	2.1	2.1	4.2	
VI bottom 10 common plants					
Achillea millefolium	Asteraceae	0.5	1.7	2.2	
Lysimachia punctata	Primulaceae	0.6	1.5	2.1	
Tanacetum vulgare	Asteraceae	0.6	1.5	2.1	
Eupatorium cannabinum	Asteraceae	0.9	1.1	2.0	
Hieracium pilosella	Asteraceae	0.6	1.4	2.0	
Ranunculus sceleratus	Ranunculaceae	0.6	1.3	1.9	
Hypericum dubium	Guttiferae	0.6	1.3	1.9	
Chamerion angustifolium	Onagraceae	0.8	1.0	1.8	
Ranunculus repens	Ranunculaceae	0.9	0.7	1.6	
Hypericum perforatum	Guttiferae	0.6	0.7	1.3	

[a]Restoration success of rare plants and the possible measures to enable successful reestablishment are shown (from Bekker and Lammerts 2002 and Gigon et al. 1996) as N/A (no information), + (good results available), and ± (contradictory results).

Acknowledgments

Many thanks go to Frank Hoffmann, Wim Ozinga, Ger Boedeltje, Jelte van Andel, Jan Bakker, Roel Strykstra, Oliver Tackenberg, Els Boerrigter, Odilia Velterop, Martina Stang, Theodora Petanidou, Gerard Oostermeijer, and Michiel Wallis de Vries for contributing data, ideas, discussions, and comments on the manuscript. The students Alie Hagedoorn, Cees van der Brandt, Alex Witte, Stefan te Velde, Alje Zandt, Hans Imberg, Theo Boudewijn, Maaike de Vlas, Maaike Smelter, Carola van der Muren, and Henk Hunneman collected parts of the data.

References

Aizen, M., L. Ashworth, and L. Galetto. 2002. Reproductive success in fragmented habitats: do compatiblity systems and pollination specialization matter? Journal of Vegetation Science 13: 885–892.

Aizen, M. A., and P. Feinsinger. 1994. Habitat fragmentation, native insect pollinators, and feral honeybees in Argentine chaco serrano. Ecological Applications 4: 378–392.

Allen-Wardell, G., P. Bernhardt, R. Bitner, A. Búrquez, S. Buchmann, J. Cane, P. Cox et al. 1998. The potential consequences of pollinator declines on the conservation of biodiversity and stability of food crop yields. Conservation Biology 12: 8–17.

Bakker, J. P., and F. Berendse. 1999. Constraints in the restoration of ecological diversity in grassland and heath land communities. Trends in Ecology and Evolution 14: 63–67.

Bakker, J. P., P. Poschlod, R. J. Strijkstra, R. M. Bekker, and K. Thompson. 1996. Seed banks and seed dispersal: Important topics in restoration ecology. Acta Botanica Neerlandica 45: 461–490.

Bekker, R. M., and E. J. Lammerts. 2002. Groene stippen voor Rode Lijstsoorten: Evaluatie van herstelmaatregelen. De Levende Natuur 103: 48–52.

Bekker, R. M., J. H. J. Schaminée, J. P. Bakker, and K. Thompson. 1998. Seed bank characteristics of Dutch plant communities. Acta Botanica Neerlandica 47: 15–26.

Boedeltje, G., J. P. Bakker, R. M. Bekker, J. M. Van Groenendael, and M. Soesbergen. 2003. Plant dispersal in a lowland stream in relation to occurrence in the species pool and three specific life-history traits. Journal of Ecology 91: 855–866.

Bond, W. 1994. Do mutualisms matter? Assessing the impact of pollinator and disperser disruption on plant extinction. Philisophical Transactions of the Royal Society of London B 344: 83–90.

Bonn, S., and P. Poschlod. 1998. Ausbreitungsbiologie der Pflanzen Mitteleuropas. Quelle and Meyer Verlag Wiesbaden.

Bonn, S., P. Poschlod, and O. Tackenberg. 2000. Diasporus: A database for diaspore dispersal. Concept and application case studies for risk assessment. Zeitschrift für Ökologie und Naturschutz 9: 85–97.

Bouman, F., D. Boesewinkel, R. Bregman, N. Devente, and G. Oostermeijer. 2000. Verspreiding van zaden. Koninklijke Nederlandse Natuurhistorische Vereniging Uitgeverij, Utrecht.

Brugge, B., E. Van der Spek, and M. M. Kwak. 1998. Honingbijen in natuurgebieden? De Levende Natuur 99: 71–76.

Cain, M. L., B. G. Milligan, and A. E. Strand. 2000. Long-distance seed dispersal in plant populations. American Journal of Botany 87: 1217–1227.

Centraal Bureau voor de Statistiek. 1987. Botanisch basisregister. Centraal Bureau voor de Statistiek, Voorburg.

Chittka, L., and S. Schürkens. 2001. Successful invasion of a floral market. Nature 411: 653.

Coombe, D. E. 1956. Biological flora of the British Isles: *Impatiens parviflora* DC. Journal of Ecology 44: 701–713.

Corbet, S. A. 1991. Applied pollination ecology. Trends in Ecology and Evolution 6: 3–4.

Corbet, S. A. 1995. Insects, plants and succession: Advantages of long-term set-aside. Agriculture, Ecosystems, and Environment 53: 201–217.

Corbet, S. A. 2000a. Butterfly nectaring flowers: Butterfly morphology and flower form. Entomologia Experimentalis et Applicata 96: 289–298.

Corbet, S. A. 2000b. Conserving compartments in pollination webs. Conservation Biology 14: 1229–1231.

Cranmer, L. 2004. The influence of linear landscape features on pollinator behaviour. PhD thesis, University College Northampton.

De Kroon, H., and J. M. Van Groenendael. 1997. The ecology and evolution of clonal plants. Backhuys Publishers, Leiden.

Dicks, L. V., S. A. Corbet, and R. F. Pywell. 2002. Compartmentalization in plant–insect flower visitor webs. Journal of Animal Ecology 71: 32–43.

Dijk, E., J. H. Willems, and J. Van Andel. 1997. Nutrient response as a key factor to the ecology of orchid species. Acta Botanica Neerlandica 46: 339–363.

Ellis, W. N., and A. C. Ellis-Adam. 1993. To make a meadow it takes a clover and a bee: The entomophilous flora of NW Europe and its insects. Bijdragen tot de Dierkunde 63: 193–220.

Evertz, S. 1995. Interspezifische Konkurrenz zwischen Honigbienen (*Apis mellifera*) und solitären Wildbienen (Hymenoptera Apoidae). Natur und Landschaft 70: 165–172.

Fenner, M. 1992. Seeds. The ecology of regeneration in plant communities. Cab International, Wallingford, UK.

Gigon, A. 1999. Case studies of success in nature conservation in Europe. Pp. 143–150 *in* A. Farina (ed.), Perspectives in ecology. Backhuys Publishers, Leiden.

Gigon, A., R. Langenauer, C. Meier, and B. Nievergelt. 1996. Blaue Listen, der erfolgreich erhaltenen oder geförderten Tier- und Pflanzenarten der Roten Listen. Technology Assessment 18/1996, Schweizerischer Wissenschaftsrat, Bern.

Grime, J. P., J. G. Hodgson, and R. Hunt. 1988. Comparative plant ecology. Unwin Hyman, London.

Huenneke, L. F. 1991. Ecological implications of genetic variation in plant populations. Pp. 31–44 *in* D. A. Falk and K. E. Holsinger (eds.), Genetics and conservation of rare plants. Oxford University Press, Oxford.

Jennersten, O. 1988. Pollination in *Dianthus deltoides* (Caryophyllaceae): Effects of habitat fragmentation on visitation and seed set. Conservation Biology 2: 359–366.

Kearns, C. A., and D. W. Inouye. 1994. Fly pollination of *Linum lewesii* (Linaceae). American Journal of Botany 81: 1091–1095.
Kearns, C. A., D. W. Inouye, and N. M. Waser. 1998. Endangered mutualisms: The conservation of plant-pollinator interactions. Annual Review of Ecology and Systematics 29: 83–112.
Kevan, P. G., and H. G. Baker. 1983. Insects as flower visitors and pollinators. Annual Review of Entomology 28: 407–453.
Klimes, L., and J. Klimesova. 1999. CLO-PLA2: A database of clonal plants in central Europe. Plant Ecology 141: 9–19.
Kremen, C., and T. Ricketts. 2000. Global perspectives on pollination disruption. Conservation Biology 14: 1226–1228.
Kwak, M. M. 1977. Pollination ecology of five hemiparasitic, large-flowered Rhinanthoideae with special reference to the pollination behaviour of nectar-thieving, short-tongued bumblebees. Acta Botanica Neerlandica 26: 97–107
Kwak, M. M. 1979. Effects of bumblebee visits on the seed set of *Pedicularis, Rhinanthus* and *Melampyrum* (Scrophulariaceae) in the Netherlands. Acta Botanica Neerlandica 28: 177–195.
Kwak, M. M. 1980. The pollination value of honeybees to the bumblebee plant *Rhinanthus*. Acta Botanica Neerlandica 29: 597–603.
Kwak, M. M. 1988. Pollination ecology and seedset in the rare annual species *Melampyrum arvense* L. (Scrophulariaceae). Acta Botanica Neerlandica 37: 153–163.
Kwak, M. M. 1993. The relative importance of syrphids and bumblebees as pollinators of three plant species. Proceedings Experimental and Applied Entomology, N.E.V. Amsterdam 4: 137–143.
Kwak, M. M. 1994. Achteruitgang leidt tot verschuivende relaties. Landschap 11: 29–39.
Kwak, M. M., and I. Tieleman. 2000. Het Hommelleven. Koninklijke Nederlandse Natuurhistorische Vereniging Uitgeverij, Utrecht.
Kwak, M. M., and O. Velterop. 1997. Flower visitation by generalists and specialists: Analysis of pollinator quality. Proceedings Experimental and Applied Entomology, N.E.V. Amsterdam 8: 85–89.
Kwak, M. M., O. Velterop, and E. J. M. Boerrigter. 1996. Insect diversity and the pollination of rare plant species. Pp. 115–124 *in* A. Matheson, S. L. Buchmann, Ch. O'Toole, P. Westrich and I. H. Williams (eds.), The conservation of bees. The Linnean Society of London and the International Bee Research Association, Academic Press, London.
Kwak, M. M., O. Velterop, and J. Van Andel. 1998. Pollen and gene flow in fragmented habitats. Applied Vegetation Science 1: 37–54.
Lamont, B. B., P. G. L. Klinkhamer, and E. T. F. Witkowski. 1993. Population fragmentation may reduce fertility to zero in *Banksia goodii:* A demonstration of the Allee effect. Oecologia 94: 446–450.
Lande, R. 1998. Anthropogenic, ecological and genetic factors in extinction and conservation. Researches on Population Ecology 40: 259–269.
Meeuse, B., and S. Morris. 1984. The sex life of flowers. Faber and Faber, London.
Memmott, J. 1999. The structure of a plant-pollinator food web. Ecology Letters 2: 276–280.
Nunney, L., and K. A. Campbell. 1993. Assessing minimum viable population size: Demography meets population genetics. Trends in Ecology and Evolution 8: 234–238.
Oostermeijer, J. B. G., M. W. Van Eck, and H. C. M. Den Nijs. 1994. Offspring fitness in relation to population size and genetic variation in the rare perennial species *Gentiana pneumonanthe* (Gentianaceae). Oecologia 97: 289–296.
Ozinga, W. A., R. M. Bekker, J. H. J. Schaminée, and J. M. van Groenendael. 2004. Dispersal potential in plant communities depends on environmental conditions. Journal of Ecology 92: 767–777.
Parker, V. T., and S. T. A. Pickett. 1997. Restoration as an ecosystem process: Implications of the modern ecological paradigm. Pp. 17–32 *in* K. M. Urbanska, N. R. Webb, and P. J. Edwards (eds.), Restoration ecology and sustainable development, Cambridge University Press, Cambridge.
Plate, C., and C. Van Swaay. 2002. Meten is weten? Vlinders 17: 21.
Praeger, R. L. 1913. Buoyancy of the seed of some Britannic plants. Scientific Proceedings of the Royal Dublin Society 14: 13–62.
Proctor M., P. Yeo, and A. Lack. 1996. The natural history of pollination. HarperCollins, London.
Pywell, R. F., J. M. Bullock, D. B. Roy, L. Warman, K. J. Walker, and P. Rothry. 2003. Plant traits as predictors of performance in ecological restoration. Journal of Applied Ecology 40: 65–77.

Rabinowitz, D. 1981. Seven forms of rarity. Pp. 205–217 *in* H. Synge (ed.), The biological aspects of rare plant conservation. John Wiley and Sons, Chichester.
Rasmont, P. 1988. Monographie ecologique et zoogeographique des bourdons de France et de Belgique (Hymenoptera, Apidae, Bombinae). PhD dissertation, University of Gembloux, Belgium.
Rathcke, B. 1983. Competition and facilitation among plants for pollination. Pp. 305–329 *in* L. Real (ed.), Pollination biology, Academic Press, Orlando, FL.
Rathcke, B. J., and E. S. Jules. 1993. Habitat fragmentation and plant–pollinator interactions. Current Science 65: 273–277.
Roubik, D. W. 1996. Measuring the meaning of honey bees. Pp. 163–172 *in* A. Matheson, S. L. Buchmann, Ch. O'Toole, P. Westrich and I. H. Williams (eds.), The conservation of bees, Linnean Society of London and International Bee Research Association, Academic Press, London.
Salisbury, E. J. 1942. The reproductive capacity of plants. Studies in quantitative biology. G. Bells and Son, London.
Strykstra, R. J. 2000. Reintroduction of plant species: S(h)ifting settings. PhD dissertation, University of Groningen, The Netherlands.
Strykstra, R. J., R. M. Bekker, and J. P. Bakker. 1998a. Assessment of dispersule availability: Its practical use in restoration management. Acta Botanica Neerlandica 47: 57–70.
Strykstra, R. J., R. M. Bekker, and J. Van Andel. 2002. Dispersal and life span spectra in plant communities: A key to safe site dynamics, species coexistence, and conservation. Ecography 25: 145–160.
Strykstra, R. J., D. M. Pegtel, and A. Bergsma. 1998b. Dispersal distance and achene quality of the rare anemochorous species *Arnica montana:* Implications for conservation. Acta Botanica Neerlandica 47: 45–56.
Strykstra, R. J., G. L. Verweij, and J. P. Bakker. 1997. Seed dispersal by mowing machinery in a Dutch brook valley system. Acta Botanica Neerlandica 46: 387–401.
Sugden E. A., R. W. Thorp, and S. L. Buchmann. 1996. Honey bee–native bee competition: Focal point for environmental change and apicultural response in Australia. Bee World 77: 26–44.
Tackenberg, O. 2001. Methoden zur Bewertung gradueller Unterschiede des Ausbreitungspotentials von Pflanzenarten. Dissertationes Botanicae, 347. Cramer, Berlin.
Tamis, W. L. M., R. Van der Meijden, J. Runhaar, R. M. Bekker, W. A. Ozinga, B. Odé, and I. Hoste. 2004. Standaardlijst van de Nederlandse flora 2003. Gorteria 30: 101–195.
Thompson, K., J. P. Bakker, and R. M. Bekker. 1997. Soil seed banks of NW Europe: Methodology, density, and longevity. Cambridge University Press, Cambridge.
Thompson, K., R. M. Ceriani, J. P. Bakker, and R. M. Bekker. 2003. Are seed dormancy and persistence in soil related? Seed Science Research 13: 97–100.
Turnbull, L. A., M. J. Crawley, and M. Rees. 1999. Are plant populations seed-limited? A review of seed sowing experiments. Oikos 88: 225–238.
Van der Goot, V. S. 1981. De zweefvliegen van Noord-west Europa en Europees Rusland, in het bijzonder van de Benelux. Koninklijke Nederlandse Natuurhistorische Vereniging, Utrecht.
Van der Meijden, R., B. Odé, C. L. G. Groen, J.-P. M. Witte, and D. Bal. 2000. Bedreigde en kwetsbare vaatplanten in Nederland. Basisrapport met voorstel voor de Rode Lijst. Gorteria 26: 85–208.
Van der Muren, C., F. Hoffmann, and M. M. Kwak. 2003. Insect diversity on yellow Asteraceae in road verges in the Netherlands. Proceedings of Experimental and Applied Entomology, N.E.V. Amsterdam 14: 115–118.
Van Dorp, D. 1996. Seed dispersal in agricultural habitats and the restoration of species-rich meadows. PhD dissertation, Agricultural University, Wageningen, The Netherlands.
Van Groenendael, J. M., N. J. Ouborg, and R. J. J. Hendriks. 1998. Criteria for the introduction of plant species. Acta Botanica Neerlandica 47: 3–13.
Van Ommering, G., I. Van Halder, C. A. M. Van Swaay, and I. Wynhoff. 1995. Bedreigde en kwetsbare dagvlinders in Nederland. Toelichting op de Rode-Lijst. IKC-N rapport no. 18, Wageningen.
Van Raaij, L. 1998. Concurrentie tussen honingbijen en andere bloembezoekende insecten. Informatie—en Kennis Centrum Landbouw, Ede.
Van Treuren, R., R. Bijlsma, N. J. Ouborg, and M. M. Kwak. 1994. Relationships between plant density, outcrossing rates, and seed set in natural and experimental populations of *Scabiosa columbaria.* Journal of Evolutionary Biology 7: 287–302.

Velterop, O. 2000. Effects of fragmentation on pollen and gene flow in insect-pollinated plant populations. PhD dissertation, University of Groningen, The Netherlands.

Verhagen, R., J. Klooker, J. P. Bakker, and R. Van Diggelen. 2001. Restoration success of low-production plant communities on former agricultural soils after top-soil removal. Applied Vegetation Science 4: 75–82.

Vogel, S. 1986. Ölblumen und ölsammelnde Bienen. Zweite Folge. Lysimachia und Macropis. Tropische und subtropische Pflanzenwelt 54: 1–312.

Von Hagen, E. 1994. Hummeln. Bestimmen, ansiedeln, vermehren, schutzen. Weltbild Verlag GmbH, Augsburg.

Waser, N. M., L. Chittka, M. V. Price, N. M. Williams, and J. Ollerton. 1996. Generalization in pollination systems, and why it matters. Ecology 77: 1043–1060.

Weller, S. G. 1994. The relationship of rarity to plant reproductive biology. Pp. 90–117 *in* M. L. Bowles and C. J. Whelan (eds.), Restoration of endangered species: Conceptual issues, planning, and implementation. Cambridge University Press, Cambridge.

Westerkamp, C. 1991. Honeybees are poor pollinators: Why? Plant Systematics and Evolution 177: 71–75.

Westrich, P. 1990. Die Wildbienen Baden-Württembergs, spezieller Teil. Die Gattungen und Arten. Ulmer, Stuttgart.

Williams, P. H. 1989. Why are there so many bumblebees at Dungeness? Botanical Journal of the Linnean Society 101: 31–44.

Wilson, P., and J. D. Thomson. 1991. Heterogeneity among floral visitors leads to discordance between removal and deposition of pollen. Ecology 72: 1503–1507.

CHAPTER SEVENTEEN

Bee Diversity and Plant–Pollinator Interactions in Fragmented Landscapes

Ingolf Steffan-Dewenter, Alexandra-Maria Klein, Volker Gaebele, Thomas Alfert, and Teja Tscharntke

The worldwide destruction and fragmentation of natural habitats and the increasing dominance of highly disturbed anthropogenic habitats are considered major threats to biodiversity and ecosystem functioning (Harrison and Bruna 1999; Naeem et al. 1999). The loss of biodiversity may threaten essential ecological interactions in ecosystems such as decomposition, parasitism, predation, or pollination (Martinez 1996); however, most research on biodiversity and ecosystem functions has focused on more general, integrating ecosystem characteristics, such as productivity, stability, or resilience of ecosystems, and has neglected biotic interactions as important drivers of ecosystem functioning (Chapin et al. 1997; Loreau et al. 2001; Lundberg and Moberg 2003). Furthermore, most research has been done with plants in small experimental patches, whereas larger spatial scales and functional animal groups have rarely been considered in the context of biodiversity and ecosystem functioning (Naeem and Wright 2003).

Pollinators do not directly affect ecosystem processes but they have the potential to change the structure and diversity of plant communities, thereby indirectly affecting ecosystem properties (Lundberg and Moberg 2003). The loss of pollinators and consequent risks for pollination as an important ecosystem service has received much attention during past years (e.g., Rathcke and Jules 1993; Kearns et al. 1998; Wilcock and Neiland 2002). The actual impact of pollinator loss on pollination services depends greatly on the degree of specialization of plant–pollinator interactions (fig. 17.1). Assuming a close one-to-one relationship between a single pollinator species and a single plant species, this should result in the extinction of the remaining mutualistic partner in case of extinction of either the pollinator or the plant species. Unbalanced relationships where several plant species rely on one pollinator species or several pollinator species on one plant species also impose a high extinction risk for either the pollinator or the plant species. At the other extreme (generalized relationships where several pollinator species visit several plant species and vice versa), plant–pollinator systems should be resilient over a broad range of pollinator loss (fig. 17.1).

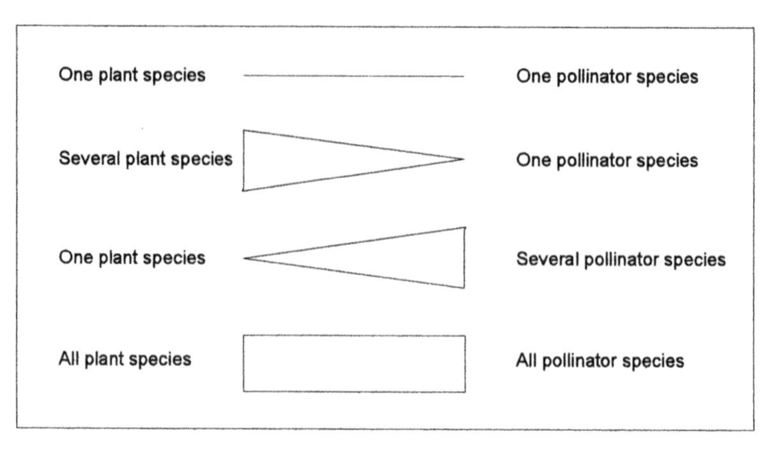

Figure 17.1 Four scenarios for the specialization of plant–pollinator interactions

The degree of specialization of plant–pollinator interactions is part of an ongoing lively discussion (Waser et al. 1996; Johnson and Steiner 2000). Whereas traditional views favor the existence of specialized pollination syndromes, more recent concepts emphasize that most plant–pollinator interactions are generalized and variable in space and time (Herrera 1988; Waser et al. 1996). One possibility is that generalization is favored in human-dominated ecosystems, whereas natural systems show a higher degree of specialization (Johnson and Steiner 2000).

Interactions between plants and pollinators play a key role in the structure of terrestrial ecosystems and have attracted much attention since Darwin's pioneering evolutionary and behavioral research (e.g., Pellmyr 2002). However, remarkably little is known about the spatial and temporal variation of plant–pollinator interactions in response to habitat fragmentation and land-use change (e.g., Rathcke and Jules 1993; Bronstein 1995; Kearns et al. 1998). This may have several explanations. First, plant–pollinator interactions have usually been studied from a narrow botanical or zoological perspective rather than a broad perspective (Bronstein 1995; Waser and Price 1998). Second, plant–pollinator interactions may have been studied at an inappropriate local spatial scale (Bronstein 1995), not taking into account the considerable foraging distances of pollinators such as bees and others, which affect genetic, ecological, and evolutionary integration on greater scales (Gathmann and Tscharntke 2002; Steffan-Dewenter and Kuhn 2003). Third, other biotic interactions that may modulate plant–pollinator interactions (e.g., between plants and herbivores or seed predators) have rarely been taken into account (but see, e.g., Jennersten and Nilsson 1993; Kéry et al. 2001; Steffan-Dewenter et al. 2001; Gómez and Zamora, chap. 7 in this volume).

In this chapter, we consider plant–pollinator interactions from a landscape perspective, and from both botanical and zoological points of view. In the first section, we attempt to summarize existing knowledge of how habitat fragmentation affects the diversity and community structure of pollinators. We present new data on species–area relationships for a diverse bee community in limestone quarries and test the hypothesis that higher trophic levels and specialized species groups respond more sensitively to habitat loss and isolation. In the second section, we ask how shifts in species richness and density of pollinators affect the seed set of rare plant populations in fragmented habitats. After reviewing existing evidence for pollination limitation, we present our own study on *Primula veris,* which is novel in that it separates subpopulation size and habitat fragment size and also considers seed predation. In the third section, we expand our perspective to a landscape scale: we argue that the analysis of multiple spatial scales is necessary for a more complete understanding of plant–pollinator interactions and we give a condensed overview of recent research on this topic from our group. In this section, we also consider biodiversity–function relationships in plant–pollinator interactions and present results from a study on coffee pollination in the tropics as rare evidence for a positive relationship. Such biodiversity effects should significantly depend on the degree of specialization or generalization of plant–pollinator interactions and may give new insights into the large-scale risks posed to these mutualistic interactions. Our main emphasis will be on the specialization and generalization of pollinators rather than of plant species. Furthermore, we focus on bees as an important pollinator group, at least in temperate ecosystems. Approximately 17,000 bee species worldwide are described and the total number has been estimated at 30,000 (Michener 2000). Bees exhibit a great diversity of habitat requirements and life histories, for example, solitary versus social organization, nesting in soil or above-ground cavities, specialized or generalized requirements for pollen (e.g., Cane and Sipes, chap. 5 in this volume), or even parasitic species that develop on the stored food of other bee species (Wcislo and Cane 1996; Michener 2000).

Bee Diversity in Fragmented Habitats

The process of habitat fragmentation mainly results in a smaller size and larger isolation of remaining habitat fragments and a reduced total area in a landscape. In most cases, species richness and population density increase with fragment or patch area: the so-called species–area and abundance–area relationships (Rosenzweig 1995; Connor et al. 2000). The relative importance of patch size and isolation is expected to differ at different levels of habitat loss. The impact of isolation is expected to increase below critical thresholds of habitat area in a landscape (Andren 1994; Bascompte and Sole 1996). An interesting but largely unknown aspect is the impact of the surrounding landscape matrix on the survival of species in habitat fragments (Ricketts 2001; Koptur, chap. 15 in this volume). Species with specialized nesting or food plant requirements, species of higher trophic levels, species with limited dispersal abilities, and species that depend on obligate mutualists are expected to suffer most from habitat fragmentation (e.g., Holt et al. 1999; Davies et al. 2000; Tscharntke et al. 2002). Studies on habitat fragmentation are biased with respect to species groups (mainly birds and small mammals) and spatial dimension (mainly small-scale experimental arrays; Debinski and Holt 2000).

For insect communities generally and bees specifically, the response to habitat fragmentation is poorly understood (Steffan-Dewenter and Tscharntke 2002). For example, Cane (2001) found only four published studies dealing with habitat fragmentation and native bees and pointed out that none of these considered complete habitats including nesting sites. Aizen and Feinsinger (1994a) found reduced taxon diversity with decreasing fragment size but no effects on total visit frequency of two spring-flowering tree species. Bolger at al. (2000) did not find significant species–area relationships for bees sampled by pitfall traps and vacuum suction in urban habitat fragments. Steffan-Dewenter (2003) analyzed species richness and abundance of trap-nesting bees on 45 orchard meadows ranging in size from 0.08 to 5.8 ha and differing in habitat connectivity and the surrounding landscape matrix. Species richness of bees showed a steep positive species–area relationship. With a z value of 0.23 (the exponent of the log-log species–area relationship) species loss in small habitat fragments was considerably greater than for most other species groups in terrestrial habitats (e.g., Rosenzweig 1995).

In another recent study in central Germany, Alfert et al. (2001) selected 24 limestone quarries ranging in size from 0.01 to 25.1 ha and in age from 1 to 120 years. Because of their high habitat heterogeneity, limestone quarries provide diverse food-plant resources and nesting sites and are expected to be valuable secondary habitats for bees in the intensively used central European landscapes (Westrich 1989). Bees were monitored by standardized transect walks. We characterized each study site via the following habitat parameters: area, age, species richness, and cover of flowering plants. Each bee species was categorized with respect to social status (solitary or social), trophic status (nest building or parasitic), nesting sites (below- or above-ground nesting sites), and food-plant requirements (oligolectic or polylectic; Cane and Sipes, chap. 5 in this volume), following Westrich (1989).

A total of 6882 individuals were identified, representing 123 wild bee species (Hymenoptera, Apidae) from 20 genera. The outstanding value of this habitat type is supported by the fact that these comprise about 41% of all bee species occurring in this part of Germany (Theunert 2003). The total number of bee species significantly increased with habitat area (fig. 17.2A) but not with habitat age ($r^2 = 0.04, P = .36$). The species–area exponent (z) was 0.33, which is unusually high (Rosenzweig 1995). The density of bees measured as number of individuals per transect area also increased with habitat area (fig. 17.2B), thus providing rare empirical evidence that insect density may also increase with habitat area (Connor

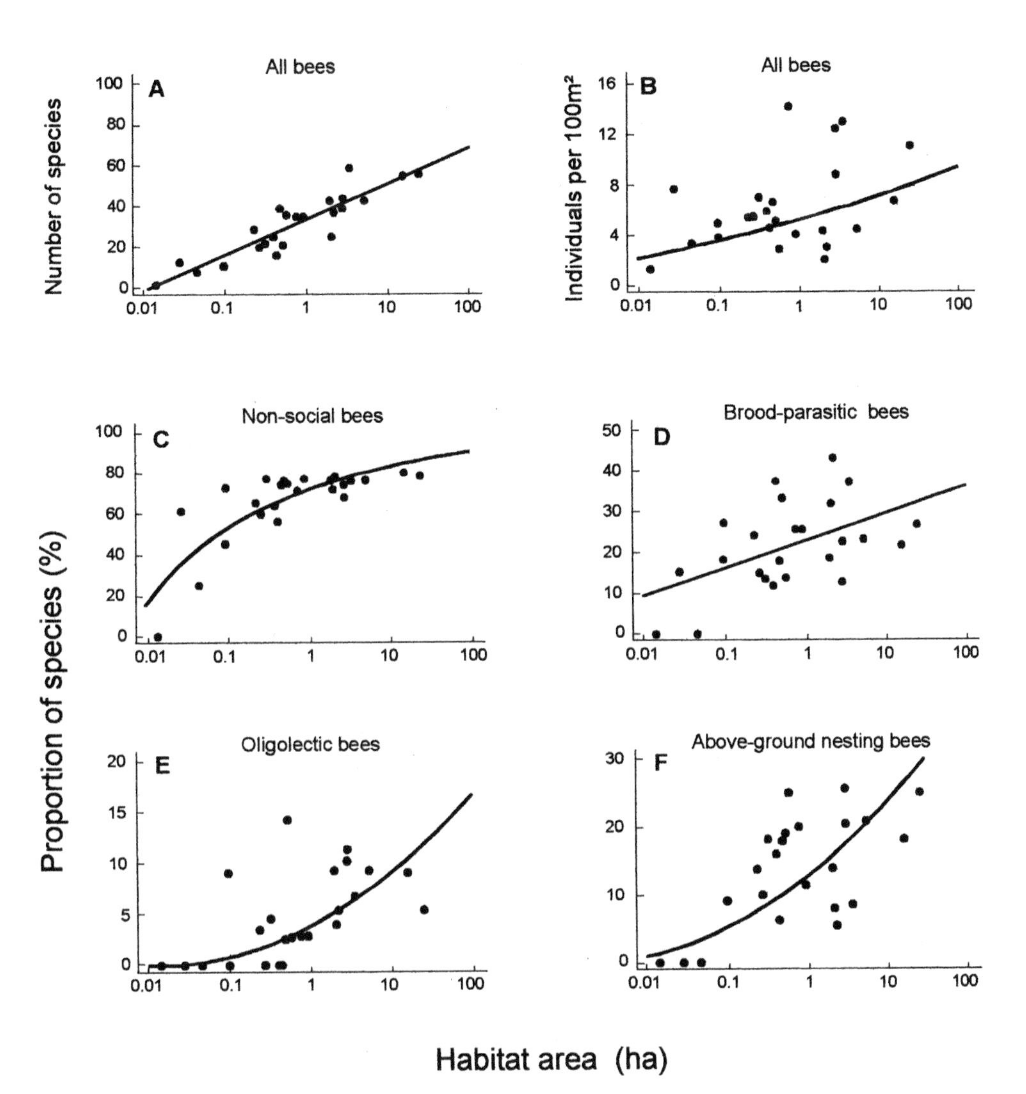

Figure 17.2 Effects of habitat area on species richness, density, and community structure of bees in limestone quarries: (A) total number of bee species in relation to habitat area ($r^2 = 0.82$, $N = 24$, $P < .001$); (B) bee density (number of individuals per 100 m²) in relation to habitat area ($r^2 = 0.25$, $N = 24$, $P = .013$); (C) proportion of nonsocial bee species in relation to habitat area ($r^2 = 0.66$, $N = 24$, $P < .001$); (D) proportion of parasitic bee species in relation to habitat area ($r^2 = 0.26$, $N = 24$, $P = .011$); (E) proportion of oligolectic bee species in relation to habitat area ($r^2 = 0.44$, $N = 24$, $P < .001$); and (F) proportion of above-ground-nesting bee species in relation to habitat area ($r^2 = 0.49$, $N = 24$, $P < .001$).

et al. 2000). Possible explanations for the bee communities' response to area come from correlations with other habitat parameters. The number of flowering plant species increased with habitat area ($r^2 = 0.61$, $P < .001$), thereby suggesting that higher resource diversity supported a greater number of different bee species. Furthermore, the percentage cover of flowering plants increased with area

($r^2 = 0.24, P = .015$), which possibly explains the higher density of bees in larger habitats.

Dividing the bee communities into species groups revealed considerable differences in responses to habitat area. Social (colony-building) bees represented about 19% of the pollen-collecting species and 73% of the individuals coming from the genera *Bombus, Halictus,* and *Lasioglossum,* whereas the remaining species were grouped as nonsocial and included solitary, colonial, and communal species. The species–area relationship was significantly steeper for nonsocial than for social bee species ($z = 0.42$ vs. 0.16) and, accordingly, the proportion of nonsocial bee species significantly increased with area (fig. 17.2C). This may be explained by the large foraging distances of at least bumblebees, the dominating group among the social species (Walter-Hellwig and Frankl 2000). An additional recent study suggests that the abundance of the three most common bumblebee species depends more on the availability of mass-flowering crops in agricultural landscapes than on the distribution of seminatural habitats (Westphal et al. 2003).

A total of 32 parasitic so-called cuckoo species from six genera were found, representing 26% of all species and 7.5% of all individuals. The slope of the species–area curve was steeper for parasitic than for pollen-collecting species ($z = 0.36$ vs. 0.29) and the proportion of brood parasitic species increased with habitat area (fig. 17.2D). These results confirm the hypothesis formulated by Holt et al. (1999) that higher trophic levels (here parasitic bees) are affected more by reduced habitat area than are lower trophic levels (here brood-provisioning bees).

Sixteen oligolectic species (13% of all species but only 3.8% of all individuals) with specialized food-plant requirements were recorded. The slope of species–area curves was similar for oligolectic and polylectic species ($z = 0.285$ vs. 0.273), but the proportion of oligoleges significantly increased with habitat area (fig. 17.2E). The results suggest that species with more specialized food-plant requirements suffer more from habitat loss than generalists.

The majority of the bee species found (67%) build their nests in the soil, 20% require above-ground nesting places (e.g., in dead wood), and 7% (mainly the bumblebee species) use both below- and above-ground nesting places. No information was available for the remaining 6%. The slope of species–area curves was not significantly different for ground-nesting species ($z = 0.317$) or above-ground cavity nesters ($z = 0.361$); however, the proportion of above-ground cavity nesters increased with area (fig. 17.2F). This may be due to the fact that above-ground nesting sites seem to be a limiting factor in most habitats; hence, these species have lower density and, thereby, should be more prone to extinction.

These results support the assumption that the destruction and fragmentation of potential bee habitats has significant effects on their species richness, population density, and community structure. The strongest impact of habitat fragmentation existed for solitary species with specialized pollen needs and for species of higher trophic levels. The reduced proportion of parasitic bees in small habitat islands suggests that not only mutualistic but also antagonistic biotic interactions might be lost. In the short term, brood-provisioning bee species could benefit from reduced parasitism, but in the long term, this may change competitive interactions and result in reduced local diversity of species and their interactions. The significant loss of oligolectic bee species in small habitats suggests that specialized plant–pollinator interactions are more threatened by habitat fragmentation than are generalized interactions.

Pollination Limitation in Fragmented Habitats

Recent reviews argue that pollination systems are under threat due to habitat fragmentation, agricultural intensification, and invasions (Rathcke and Jules 1993; Kearns et al. 1998; Richards 2001; Wilcock and Neiland 2002), but all emphasize the need for more field research. Indeed, the empirical case for pollination declines remains incomplete. The logical reasons to expect declines are that fully 91% of the estimated 240,000 plant species for which pollen vectors have been recorded are pollinated by animals (Renner 1998), and limitation of pollen receipt is a wide-spread phenomenon for animal-pollinated plant species: it occurs for 62% of 258 species surveyed in natural habitats (Burd 1994). Assuming that pollinators are often a limiting factor in intact habitats, reproductive success of plants should be even more limited by pollinator scarcity in landscapes disturbed and fragmented by human activities (Nabhan and Buchmann 1997).

To provide evidence for the existence of pollination failure due to habitat fragmentation, a study should (1) monitor the diversity and abundance of flower visitors as well as flower-visiting behavior, (2) quantify the seed set of open-pollinated flowers, and (3) perform pollination experiments to assess maximum seed set after cross-pollination and exclusion of pollinators, respectively. Evidence for pollinator limitation comes from a positive relationship between pollinator availability and fruit or seed set of open-pollinated flowers; the most direct evidence for pollinator limitation is provided by an increasing difference between seed set of open- compared to hand cross-pollinated flowers with decreasing pollinator availability or patch size. This approach also allows for the exclusion of external effects such as reduced resource availability or changed genetic structure in small habitat patches.

Several studies do show that small plant populations in fragmented habitats have reduced seed set, reduced genetic diversity and offspring fitness, and higher extinction risk (Oostermeijer et al. 1994; Fischer and Matthies 1997; Fischer and Stöcklin 1997; Morgan 1999; Hendrix and Kyhl 2000; Kéry et al. 2000; Luijten et al. 2000). However, relatively few studies include direct observations of flower visitation and experimentally test for pollinator limitation as the cause for lower seed set. Similarly, it is not well understood if such fragmentation effects are exacerbated as plant–pollinator interactions become more specialized. Reduced seed set and more severe pollinator limitation have been found in small plant populations of *Dianthus deltoides* (Jennersten 1988), *Lythrum salicaria* (Ågren 1996), *Viscaria vulgaris* (Nielsen and Ims 2000), *Gentiana campestris* (Lennartsson 2002), *Nepeta cataria* (Sih and Baltus 1987), *Calystegia collina* (Wolf and Harrison 2001), *Acacia brachybotrya* and *Eremophila glabra* (Cunningham 2000), *Primula sieboldi* (Matsumura and Washitani 2000), and for most of 16 plant species in dry subtropical forests in Argentina (Aizen and Feinsinger 1994b). However, in other studies, reproductive success did not decline in fragmented or more isolated populations (e.g., Spears 1987; Costin et al. 2001).

We recently undertook a study of how pollinator limitation and seed predation influence seed output of *Primula veris* at different spatial scales. *Primula veris* is a perennial, insect-pollinated, self-incompatible spring-flowering herb, which is endangered in central Germany where it grows on calcareous grasslands (Garve 1994). We selected 15 calcareous grassland fragments covering a gradient from 0.03 to 5.15 ha area, each of which supported several spatially separated different-sized subpopulations of the plant. We focused on subpopulations of fewer than 200 individuals, because earlier studies did not find significant variation in seed set for larger populations (Kéry et al. 2000). On each grassland, we selected a small, an intermediate, and a large subpopulation, resulting in a total of 45 subpopulations ranging in size from 6 to 450 flowering individuals. Flower visitation was observed in each subpopulation over several days. In each subpopulation, 10 inflorescences were randomly selected and permanently marked; after ripening of fruits, these inflorescences were collected and the number of seeds per fruit and percentage of fruit set per plant were measured. Additionally, we quantified the proportion of fruits destroyed by seed-feeding insects. To experimentally test for pollinator limitation, five inflorescences per subpopulation were cross-pollinated by hand. The two spatial scales under consideration (i.e., subpopulation size and habitat fragment area) were not intercorrelated ($r = -0.20, P = .20$); therefore, it was possible to independently test for effects of subpopulation size and habitat area on flower visitation, seed set, and seed predation.

Overall, flower visitation rates on *P. veris* were low: only 35 observed individuals from 8 species (mainly bumblebee queens). In contrast to expectations, flower visitation rates did not respond to habitat area ($r = -0.12, N = 45, P = .45$) but tended to increase with subpopulation size ($r = 0.28, N = 45, P = .06$). In addition, the number of observed species significantly increased only with subpopulation size ($r = 0.34, N = 45, P = .02$). We can only speculate about the reasons for low flower visitation rates: perhaps parallel flowering oil seed rape in the surrounding agricultural landscape attracted most of the social bee species, or *P. veris* is pollinated mostly by moths visiting the flowers at night (Lehtilä and Syrjänen 1995).

Our results gave no evidence for actual pollinator limitation: the proportion

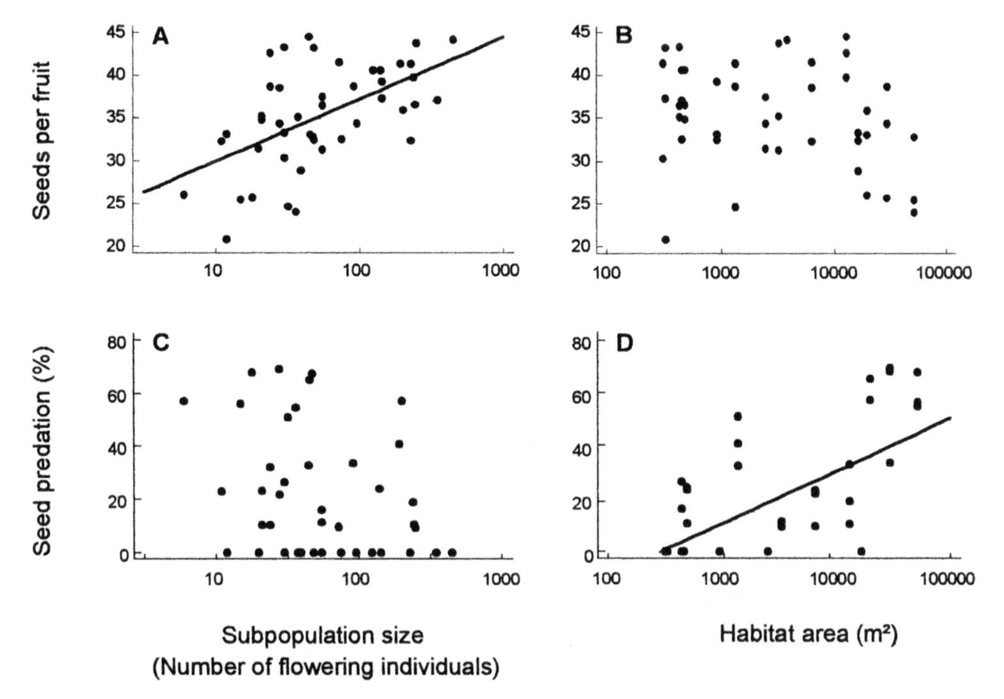

Figure 17.3 Seed set and seed predation of *Primula veris* in relation to subpopulation size and habitat area of calcareous grasslands: (A) number of seeds per fruit (only undamaged fruits) in relation to subpopulation size (number of flowering individuals; $r^2 = 0.30$, $N = 45$, $P < .0001$); (B) number of seeds per fruit (only undamaged fruits) in relation to habitat area (not significant); (C) proportion of fruits (arcsin $\sqrt{p}$ transformed) damaged by seed predators in relation to subpopulation size (number of flowering individuals; not significant); and (D) proportion of fruits (arcsin $\sqrt{p}$ transformed) damaged by seed predators in relation to habitat area ($r^2 = 0.38$, $N = 45$, $P < .0001$).

of flowers setting fruits was related neither to subpopulation size nor to habitat area. Nonetheless, the number of seeds per undamaged fruit significantly increased with subpopulation size (fig. 17.3A). This agrees with the results obtained by Kéry et al. (2000) and could be interpreted as evidence for pollinator limitation. However, for hand-pollinated inflorescences, the number of seeds per fruit also increased with subpopulation size ($r = 0.39$, $N = 45$, $P = .008$), indicating that other factors such as reduced genetic diversity, uneven distribution of flower morphs due to demographic stochasticity (Kéry et al. 2003), or less-suitable habitat conditions were responsible for reduced seed set in small subpopulations. Furthermore, habitat area had no significant effect on the number of seeds per fruit (fig. 17.3B), although this outcome was expected because this was the spatial scale at which the diversity and density of pollinators was influenced in the community-level study of bees discussed earlier in this chapter (figs. 17.2A, 17.2B).

Seed predation also played a significant role for *P. veris* and was found on 11 out of 15 study sites and in 27 out of 45 subpopulations. The main seed predator was *Phalonia ciliella* Hb. (Lepidoptera, Tortricidae), which destroyed all seeds in infested fruits. The proportion of damaged fruits did not depend on subpopulation size but significantly increased with habitat area (figs. 17.3C, 17.3D).

Our study illustrates how results may depend on the choice of spatial scale and the inclusion of both mutualistic and antagonistic interactions. If we had focused only on spatially separated subpopulations within the habitat fragments, we might have concluded that actual pollinator limitation does not seem to be important but would have had no explanation for the variation in seed predation. On the other hand, focusing only on habitat fragments and defining all subpopulations within a fragment as one population would have identified habitat area as the determinant of predispersal seed predation rate but would have obscured insights on pollinator visitation and seed set. More generally, many other factors might alter the outcome of plant–pollinator interactions such as herbivory (e.g., Strauss 1997; Herrera 2000), fungal infections (e.g., Jennersten 1985), and below-ground herbivores (Poveda et al. 2003); these have rarely been considered at multiple spatial scales in the context of habitat fragmentation (Brody 1997; Ettema and Wardle 2002; Pellmyr 2002).

It is still too early to reach conclusions with respect to general effects of habitat fragmentation on plant–pollinator interactions. Existing empirical data show that small plant populations may suffer from reduced seed set and genetic erosion, but the direct impact of pollinator limitation requires further research that should also take into account other biotic interactions, different spatial scales, and responses of plant species with different life-history traits (e.g., low or high pollinator specialization, or in different habitat types).

Effects of Landscape Context on Bees and Plant Reproduction

Important drivers such as habitat fragmentation, land-use change, climate change, and species invasions act on spatial scales much larger than single habitats; accordingly, population dynamics, species richness, and community interactions should also be affected on regional scales. However, ecologists have only recently become aware that such a landscape perspective is helpful (e.g., Kareiva and Wennergren 1995); until the past decade, they tended to consider habitat fragments in a matrix of nonhabitat and to focus on the effects of habitat area or isolation distance between neighboring habitats of the same type. But real landscapes are much more complex, and taking into account this complexity of the surrounding landscape context yields additional insights (e.g., Rickets 2001; Krauss et al. 2003, 2004; Steffan-Dewenter 2003).

The landscape approach in pollination studies takes into account that organisms such as bees depend on more than one habitat type during their life cycle (e.g., Westrich 1996) and that foraging and dispersal distances of bees are often larger than a single habitat fragment (e.g., Walter-Hellwig and Frankl 2000;

Gathmann and Tscharntke 2002; Steffan-Dewenter and Kuhn 2003). In this perspective, natural or seminatural habitats are recognized as sources for beneficial insects that spread over the surrounding landscape and supply services such as biological control of pest insects or pollination of crops (e.g., Daily 1997; Kremen et al. 2002; Thies et al. 2003). The response of different species to landscape features depends on their life-history traits, in particular foraging ranges and dispersal abilities, which emphasize the importance of considering multiple spatial scales for an understanding of ecosystems made up of interacting species (Kareiva 1990; Roland and Taylor 1997; Wiegand et al. 1999).

It is evident that ecological research on large spatial scales faces huge logistical problems. Estimating species richness in single habitat fragments with sufficient sample size and replications is already a challenge, but doing this in even larger landscapes is often impossible. Two previous solutions are the use of landscape models (Wiegand et al. 1999) or the creation of experimental fractal landscapes on a small spatial scale (With et al. 1999). Here we focus on an alternative approach: the use of large-scale experimental and observational data in a statistically meaningful number of independent real landscape units. We consider effects of landscape context on local species richness and density of pollinators, foraging distances and flower-visiting behavior of bees, and consequent effects on pollination services for wild plants and crops.

Our basic approach has been to select 15 independent circular landscape sectors in the area near Göttingen, central Germany, covering a gradient from structurally simple to complex landscapes (Steffan-Dewenter et al. 2002). In the center of each landscape unit, experimental patches were established that allowed for a standardized measurement of species richness and densities of functional groups and biotic interactions in relation to the surrounding landscape context. Landscape context was quantified within eight nested circles ranging from a 250 to a 3000 m radius. Using geographic information systems, the percentage land cover of seminatural habitats and habitat type diversity was calculated for each circular landscape sector. The parallel analysis of different spatial scales made it possible to identify "functional spatial scales," that is, the radius around experimental patches at which landscape structure most significantly influenced functional groups and their biotic interactions (Steffan-Dewenter et al. 2002; Thies et al. 2003).

To monitor species richness and density of bees independently from variation in resource quality, experimental patches with potted flowerings plants and trap nests were established in the center of each landscape unit (Steffan-Dewenter 2002; Steffan-Dewenter et al. 2002). We thereby provided both a standardized food resource for flower-visiting bees and a standardized nesting site for above-ground nesting bees (and wasps; Tscharntke et al. 1998). Consequently, variation between study sites should only be caused by differences in landscape context. Flower visitation was observed on six different plant species during summer months. The species richness and abundance of solitary wild bees increased with the proportion of seminatural habitats in the surrounding landscape. This relationship was only significant for small spatial scales but not for circular landscape sectors larger than 1000 m. In contrast, honeybees were only affected by landscape structure at a large spatial scale (landscape units with 3000 m diameter) and showed a pattern opposite that of solitary bees; thus, flower-visitor densities of honeybees were higher in structurally simple than in complex landscape units. Presumably, alternative flower resources were scarcer in simple landscapes than in complex ones, thereby enhancing the relative attractiveness of the experimental patches in simple landscapes (Steffan-Dewenter et al. 2002). This suggests that honeybees partly compensate for the loss of other more specialized pollinator species in simple landscapes but may also indicate more severe competition for the remaining food resources.

The colonization of trap nests also depended on the landscape context. The total number of trap-nesting bees and wasps significantly increased with increasing proportion of seminatural habitats in the surrounding landscape. However, this pattern was mainly determined by increasing species richness of wasps, whereas, for bees alone, only a positive trend existed ($r = 0.44$, $N = 15$, $P = .10$). In contrast with expectations, neither the abundance of bees (number of brood cells) nor the higher trophic level of natural enemies responded to landscape structure. Again landscape effects were most significant for small landscape units, indicating restricted dispersal and foraging distances of solitary bees and wasps (Steffan-Dewenter 2002).

For one focal plant, *Centaurea jacea,* we analyzed whether the reduced species richness and abundance of flower-visiting bees had negative consequences for pollination and seed set (Steffan-Dewenter et al. 2001). In contrast with expectations, the mean number of seeds per flower head did not increase with the proportion of seminatural habitats. This was presumably caused by counterbalancing effects on flower-visitor behavior, on the one hand (see following discussion), and on seed predation, on the other. Seed predation by larvae of microlepidoptera and tephritid flies played a significant role. The percentage of damaged flower heads increased from 13 to 99% along the gradient from structurally simple to complex landscapes; in addition, the proportion of damaged seeds per flower head increased with the percentage of colonized flower heads. We failed in this study to experimentally test for pollinator limitation, but indirect evidence comes from a weak positive correlation between seed set of undamaged flower heads and flower visitation rates of bees.

The foraging behavior of pollinators on single plants or patches plays an important role in pollination efficiency (e.g., Kunin 1993; Goulson 2000), but this has rarely been considered in a landscape context. In our study on *Centaurea jacea,* we found that the number of plants consecutively visited by a single bee was higher in simple landscapes with low proportions of seminatural habitats than in complex landscapes (Steffan-Dewenter et al. 2001). Simple landscapes presumably provide fewer alternative food resources, thereby changing economic constraints and increasing patch residence times (Dukas and Edelstein-Keshet 1998). As a result, a lower number of visitors provides similar flower visitation rates (see also Schulke and Waser 2001; Goverde et al. 2002). Such changes in foraging behavior possibly enhance the seed set of isolated plant populations but may incur ecological costs in terms of disrupted gene flow and increased inbreeding (Kwak et al. 1998). However, the actual gene flow via pollen dispersal in relation to changing landscape structure is difficult to assess, again because effects of foraging behavior may be counterbalanced by increased foraging distances of pollinators in structurally simple landscapes that should result in gene flow over larger distances (Schulke and Waser 2001). This may occur via shifts in pollinator community structure toward larger solitary bees, bumblebees, and honeybees, with foraging distances much larger than those of small solitary bees (Steffan-Dewenter and Tscharntke 1999; Walther-Hellwig and Frankl 2000; Gathmann and Tscharntke 2002). Foraging distances of pollen-collecting honeybees may also be significantly longer in simple landscapes than in complex ones (Steffan-Dewenter and Kuhn 2003).

We conclude that landscape context (1) affects species richness, community structure, and abundance of bees; (2) changes biotic interactions such as pollination and seed predation; and (3) modifies foraging behavior and foraging distances of pollinators. Interesting, and important, is that different species groups and interaction types are influenced at different "functional" spatial scales. Further studies are needed to assess the consequences that increasing dominance of generalist pollinators in simple landscapes have on plant–pollinator interactions.

In the studies discussed thus far, species richness and abundance of bees were closely correlated; diversity effects could not be unambiguously disentangled from pure abundance effects. Nonetheless, the results suggest that the diversity of bees may also contribute to intact pollination services in generalized, theoretically highly redundant plant–pollinator systems (Steffan-Dewenter and Tscharntke 1999; Steffan-Dewenter et al. 2001). We will expand this aspect by using a recently published case study, where we analyzed shifts of bee diversity and pollination services in tropical coffee agroforestry systems (Klein et al. 2003a, 2003b). The research was done in 2000 and 2001 in central Sulawesi (Indonesia) at the margin of the Lore-Lindu National Park. Two coffee species, lowland coffee (*Coffea canephora*) and highland coffee (*C. arabica*), which differ in pollination biology (Klein et al. 2003a), are grown in the region; here we focus on the latter species. Highland coffee has been assumed to be a self-compatible crop with no dependence on insect pollination; recently, however, Roubik (2002) published indirect evidence that honeybee pollination significantly improves coffee yields. We compared the fruit set of *C. arabica* after different pollination treatments and could show that cross-pollination by hand significantly increased fruit set compared to wind pollination or self-pollination by hand (Klein et al. 2003a).

To assess the importance of bee diversity for highland coffee yields, we selected 24 coffee agroforestry fields covering an isolation gradient from the tropical rainforest margin to a human-dominated open agricultural landscape. We assumed that undisturbed rainforests are source habitats for social bee species nesting in old trees (Liow et al. 2001). As a second possible determinant of bee diversity, we analyzed the impact of local shade management, quantified as light intensity and local plant diversity. Reduced shading is generally expected to reduce biodiversity in tropical agroforestry systems (Perfecto et al. 1996), but it may also improve local nest-site quality for soil-nesting solitary bee species and cover of flowering herbs (Klein et al. 2002). We observed flower visitation rates and fruit set of coffee plants and additionally performed experimental cross-pollination by hand to test for pollinator limitation (Klein et al. 2003b). The numbers of both all flower-visiting bee species and fruit set of coffee increased with light intensity and decreased with distance to the forest margin. Social and solitary bees showed guild-specific differences in the response to these habitat parameters: species richness of social bees only depended on the distance from the forest margin, whereas the number of solitary bee species was determined by local factors and increased with light intensity and plant diversity.

We found a significant diversity–ecological function relationship: fruit set of open-pollinated coffee flowers increased with the species richness of flower-visiting bees but, interestingly, not with the abundance of flower visitors. Furthermore, the difference between fruit set of open insect-pollinated flowers and flowers cross-pollinated by hand was negative for sites with low bee diversity and increased with bee diversity, thereby providing experimental evidence for pollinator limitation (fig. 17.4A). Thus, our study provides empirical evidence for a positive relationship between pollination as an important ecosystem service and functional group diversity of bees. Because coffee is pollinated by several unspecialized bee species, the results indicate that diversity also matters for generalized plant–pollinator relationships; complementary effects and sampling effects can generally cause such diversity effects.

Complementary effects of diverse bee assemblages include (1) reduced spatial variability and (2) reduced temporal variability of pollination services and may act on very different spatial and temporal scales. Different bee species are known to prefer low- or high-placed flowers within an individual plant (Hambäck 2001), thereby contributing to more complete pollination. In our study, the variation of fruit set (CV) between four spatially separated coffee trees per study site was reduced when bee diversity increased (fig. 17.4B).

Thus, the complementary effects of a more diverse bee community appear to have contributed to a more constant and high pollination service in coffee agro-

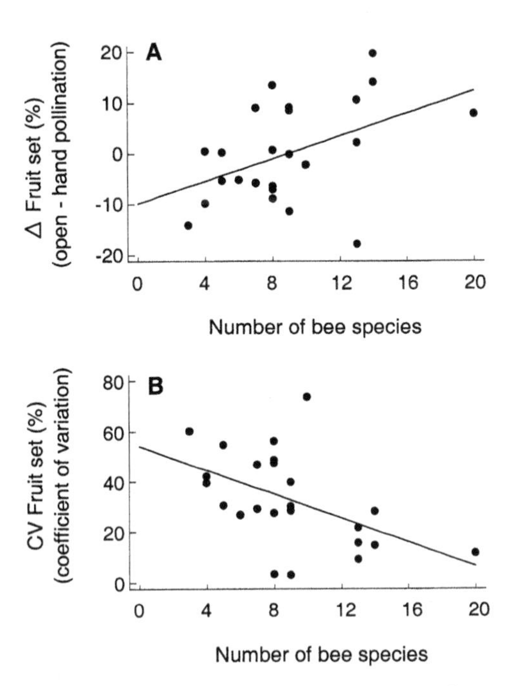

Figure 17.4 Diversity–ecological function relationship for bees and the pollination of highland coffee: (A) relationship between the difference of fruit set after open pollination minus fruit set after experimental cross-pollination, and the species richness of flower-visiting bees ($r^2 = 0.203$, $N = 24$, $P = .027$); (B) relationship between the coefficient of variation (CV) of mean fruit set after open pollination of four spatially separated coffee trees and the species richness of flower-visiting bees ($r^2 = 0.257$, $N = 24$, $P = .011$).

forestry systems. Activity patterns of bee species are also known to differ temporally: within days, between seasons, and between years (Herrera 1988; Stone et al. 1999; Kremen et al. 2002) and again diverse bee assemblages with corresponding phenological variation could act as a buffer for possible pollinator limitation.

Sampling effects suggest that within a diverse bee community there is a higher chance of a more efficient pollinator species being represented. In our coffee study, the pollination efficiency of different bee species varied significantly: single flower visits of the less abundant solitary bee community resulted in a higher probability in fruit set than for the more abundant social bee species (Klein et al. 2003b). Thus, a combination of complementary and sampling effects presumably contributed to a positive relationship between bee diversity and pollination in our generalized plant–pollinator system.

Conclusions and Future Directions

The studies we have discussed provide evidence for the decline of pollinator diversity as a result of human-induced habitat fragmentation and land-use intensification. Solitary bees, which represent the vast majority of wild bee species, are particularly threatened by reduced habitat area and increased isolation;

within this group, specialized and parasitic species are even more threatened. On a landscape scale, this translates to simplified, less-species-rich pollinator assemblages dominated by social generalist bee species. We are still far from understanding the driving forces of spatial and temporal variation in bee diversity and abundance. Bees depend on several key resources such as nesting sites, food resources, and nest-building material that are often spread over different habitat types (Westrich 1996); therefore, conservation and management activities for bees as key pollinators should take spatial scales larger than single habitat fragments into account, but further research is needed on the effects of landscape composition and the relative importance of different habitat types. Very little is known about spatial population dynamics of bees and the relative importance of bottom-up control by food or nesting resources compared with top-down control by natural enemies (Roubik 2001). The consideration of such trophic cascades is essential for understanding the mechanisms behind fragmentation- or landscape-related patterns of bee diversity and community structure.

Shifts in bee diversity and abundance associated with habitat fragmentation and land-use change are also assumed to strongly affect pollination. The limited empirical evidence suggests that plant–pollinator interactions are threatened by habitat fragmentation and land-use change, but comparative case studies for specialized and generalized plant–pollinator interactions or meta-analyses of existing data are still lacking in this context. Similarly, pollination requirements of many crop species are unknown and pollinator limitation may be much more important than at first recognized; only coarse and unsatisfying estimates of the economic value of pollination services exist (Kevan and Philips 2001). Finally, very limited insights exist for how gene flow is influenced by pollinator movement, landscape context, and the spatial arrangement of plant populations. This has profound implications for genetic erosion and inbreeding depression of plant populations in natural habitats and also for the escape of genes from genetically modified crop plants (Rieger et al. 2002).

Human-induced habitat fragmentation and shifts in landscape structure may also have evolutionary consequences for plant–pollinator interactions. We have shown that pollinator limitation presumably becomes more important in small and isolated habitat fragments and structurally simple landscapes and that species composition and foraging behavior is altered. On the other hand, herbivory and seed predation also seem to be reduced; thus, one could assume that targets and strength of natural selection are changed due to newly emerging landscape patterns. For example, this could lead to a reallocation of resources from herbivore defense to flower advertisement and rewards and could disrupt correlated evolution of mutualism- and antagonism-related traits (Herrera et al. 2002).

The threat of decreased pollination services due to loss of pollinator diversity has been considered to depend on the degree of specialization or generalization of plant–pollinator relationships (Bond 1995). Our results indicate that specialized pollinators are affected more by habitat loss and land-use change; consequently, specialized plant–pollinator relationships should be more threatened than generalized relationships. Nonetheless, our coffee study demonstrates that even generalized plant species may suffer from reduced pollinator diversity; therefore, an incomplete picture may be given when we only use the degree of specialization or generalization to judge future environmental risks for plant–pollinator interactions.

In conclusion, research on plant–pollinator interactions in changing landscapes remains a challenging topic with many possible applications. Although a beginning has been made, our understanding is still limited. Many more large-scale field experiments and multidisciplinary collaborations are called for to understand the real complexity of plant–animal interactions in their natural setting.

Acknowledgments

We are grateful to Nick Waser, Claire Kremen, and three anonymous reviewers for valuable comments on the manuscript. Financial support came from projects of the Deutsche Forschungsgemeinschaft (German Science Foundation), including the German–Indonesian Research Project STORMA ("Stability of Rainforest Margins in Indonesia"), and from the German Ministry for Research and Education.

References

Ågren, J. 1996. Population size, pollinator limitation, and seed set in the self-incompatible herb *Lythrum salicaria*. Ecology 77: 1779–1790.

Aizen, M. A., and P. Feinsinger. 1994a. Habitat fragmentation, native insect pollinators, and feral honey bees in Argentine "Chaco Serrano." Ecological Applications 4: 378–392.

Aizen, M. A., and P. Feinsinger. 1994b. Forest fragmentation, pollination, and plant reproduction in a chaco dry forest, Argentina. Ecology 75: 330–351.

Alfert T., I. Steffan-Dewenter, and T. Tscharntke. 2001. Bienen und Wespen in Kalksteinbrüchen: Flächengröße und Flächenalter. Mitteilungen der deutschen Gesellschaft für allgemeine und angewandte Entomologie 13: 579–582.

Andrén, H. 1994. Effects of habitat fragmentation on birds and mammals in landscapes with different proportions of suitable habitat. Oikos 71: 355–366.

Bascompte, J., and R. V. Sole. 1996. Habitat fragmentation and extinction thresholds in spatially explicit models. Journal of Animal Ecology 65: 465–473.

Bolger, D. T., A. V. Suarez, K. R. Crooks, S. A. Morrison, and T. J. Case. 2000. Arthropods in urban habitat fragments in southern California: Area, age, and edge effects. Ecological Applications 10: 1230–1248.

Bond, W. J. 1995. Assessing the risk of plant extinction due to pollinator and dispersal failure. Pp. 131–146 *in* J. H. Lawton and R. M. May (eds.), Extinction rates. Oxford University Press, Oxford.

Brody, A. K. 1997. Effects of pollinators, herbivores, and seed predators on flowering phenology. Ecology 78: 1624–1631.

Bronstein, J. L. 1995. The plant–pollinator landscape. Pp. 256–288 *in* L. Hannsson, L. Fahrig, and G. Merriam (eds.), Mosaic landscapes and ecological processes. Chapman and Hall, London.

Burd, M. 1994. Bateman's principle and plant reproduction: The role of pollen limitation in fruit and seed set. Botanical Review 60: 83–139.

Cane, J. H. 2001. Habitat fragmentation and native bees: A premature verdict? Conservation Ecology 5. Online URL http://www.consecol.org/vol5/iss1/art3.

Chapin, F. S., B. H. Walker, R. J. Hobbs, D. U. Hooper, J. H. Lawton, O. E. Sala, and D. Tilman. 1997. Biotic control over the functioning of ecosystems. Science 277: 500–503.

Connor, E. F., A. C. Courtney, and J. M. Yoder. 2000. Individuals–area relationships: The relationship between animal population density and area. Ecology 81: 734–748.

Costin, B. J., J. W. Morgan, and A. G. Young. 2001. Reproductive success does not decline in fragmented populations of *Leucochrysum albicans* subsp. *albicans* var. *tricolor* (Asteraceae). Biological Conservation 98: 273–284.

Cunningham, S. A. 2000. Depressed pollination in habitat fragments causes low fruit set. Proceedings of the Royal Society of London B 267: 1149–1152.

Daily, G. C. 1997. Nature's services: Societal dependence on natural ecosystems. Island Press, Washington, DC.

Davies, K. F., C. R. Margules, and J. F. Lawrence. 2000. Which traits of species predict population declines in experimental forest fragments? Ecology 81: 1450–1461.

Debinski, D. M., and R. D. Holt. 2000. A survey and overview of habitat fragmentation experiments. Conservation Biology 14: 342–355.

Dukas, R., and L. Edelstein-Keshet. 1998. The spatial distribution of colonial food provisioners. Journal of Theoretical Biology 190: 121–134.

Ettema, C. H., and D. A. Wardle. 2002. Spatial soil ecology. Trends in Ecology and Evolution 17: 177–183.

Fischer, M., and D. Matthies. 1997. Mating structure and inbreeding and outbreeding depression in the rare plant *Gentianella germanica* (Gentianaceae). American Journal of Botany 84: 1685–1692.

Fischer, M., and J. Stöcklin. 1997. Local extinctions of plants in remnants of extensively used calcareous grasslands 1950–1985. Conservation Biology 11: 727–737.

Garve, E. 1994. Atlas der gefährdeten Farn- und Blütenpflanzen in Niedersachsen und Bremen. Kartierung 1982–1992. Niedersächsisches Landesamt für Ökologie. Schriftenreihe Naturschutz und Landschaftspflege in Niedersachsen 30: 1–895.

Gathmann, A., and T. Tscharntke. 2002. Foraging ranges of solitary bees. Journal of Animal Ecology 71: 757–764.

Goulson, D. 2000. Why do pollinators visit proportionally fewer flowers in large patches? Oikos 91: 485–492.

Goverde, M., K. Schweizer, B. Baur, and A. Erhardt. 2002. Small-scale habitat fragmentation affects pollinator behaviour: Experimental evidence from calcareous grasslands. Biological Conservation 104: 293–299.

Hambäck, P. A. 2001. Direct and indirect effects of herbivory: Feeding by spittlebugs affects pollinator visitation rates and seed set of *Rudbeckia hirta.* Ecoscience 8: 45–50.

Harrison, S., and E. Bruna. 1999. Habitat fragmentation and large-scale conservation: What do we know for sure? Ecography 22: 225–232.

Hendrix, S. D., and J. F. Kyhl. 2000. Population size and reproduction in *Phlox pilosa.* Conservation Biology 14: 304–313.

Herrera, C. M. 1988. Variation in mutualisms: The spatio-temporal mosaic of a pollinator assemblage. Biological Journal of the Linnean Society 35: 95–125.

Herrera, C. M. 2000. Measuring the effects of pollinators and herbivores: Evidence for non-additivity in a perennial herb. Ecology 81: 2170–2176.

Herrera, C. M., M. Medran, P. J. Rey, A. M. Sánchez-Lafuente, M. B. Garcia, J. Guitián, and A. J. Manzaneda. 2002. Interaction of pollinators and herbivores on plant fitness suggests a pathway for correlated evolution of mutualism- and antagonism-related traits. Proceedings of the National Academy of Sciences (USA) 99: 16823–16828.

Holt, R. D., J. H. Lawton, G. A. Polis, and N. M. Martinez. 1999. Trophic rank and the species–area relationship. Ecology 80: 1495–1504.

Jennersten, O. 1985. Pollination and fungal disease transmision: Interactions between *Viscaria vulgaris, Ustilago,* and insects. Abstracts of Uppsala Dissertations from the Faculty of Science, 793, Acta Universitatis Upsaliensis, Uppsala.

Jennersten, O. 1988. Pollination in *Dianthus deltoides* (Caryophyllaceae): Effects of habitat fragmentation on visitation and seed set. Conservation Biology 2: 359–366.

Jennersten, O., and S. G. Nilsson. 1993. Insect flower visitation frequency and seed production in relation to patch size of *Viscaria vulgaris* (Caryophyllaceae). Oikos 68: 283–292.

Johnson, S. D., and K. E. Steiner. 2000. Generalization versus specialization in plant pollination systems. Trends in Ecology and Evolution 15: 140–143.

Kareiva, P. 1990. Population dynamics in spatially complex environments: Theory and data. Philosophical Transactions of the Royal Society of London B 330: 175–190.

Kareiva, P., and U. Wennergren. 1995. Connecting landscape patterns to ecosystem and population processes. Nature 373: 299–302.

Kearns, C. A., D. W. Inouye, and N. M. Waser. 1998. Endangered mutualisms: The conservation of plant-pollinator interactions. Annual Review of Ecology and Systematics 29: 83–112.

Kéry, M., D. Matthies, and M. Fischer. 2001. The effect of plant population size on the interactions between the rare plant *Gentiana cruciata* and its specialized herbivore *Maculinea rebeli*. Journal of Ecology 89: 418–427.

Kéry, M., D. Matthies, and B. Schmid. 2003. Demographic stochasticity in population fragments of the declining distylous perennial *Primula veris* (Primulaceae). Basic and Applied Ecology 4: 197–206.

Kéry, M., D. Matthies, and H.-H. Spillmann. 2000. Reduced fecundity and offspring performance in small populations of the declining grassland plants *Primula veris* and *Gentiana lutea*. Journal of Ecology 88: 17–30.

Kevan, P. G., and T. P. Phillips. 2001. The economic impacts of pollinator declines: An approach to assessing the consequences. Conservation Ecology 5. Online URL http://www.consecol.org/vol5/iss1/art8.

Klein, A.-M., I. Steffan-Dewenter, D. Buchori, and T. Tscharntke. 2002. Effects of land-use intensity in tropical agroforestry systems on coffee flower-visiting and trap-nesting bees and wasps. Conservation Biology 16: 1003–1014.

Klein, A.-M., I. Steffan-Dewenter, and T. Tscharntke. 2003a. Bee pollination and fruit set of *Coffea arabica* and *C. canephora* (Rubiaceae). American Journal of Botany 90: 153–157.

Klein, A.-M., I. Steffan-Dewenter, and T. Tscharntke. 2003b. Fruit set of highland coffee increases with the diversity of pollinating bees. Proceedings of the Royal Society of London B 270: 955–961.

Krauss, J., A.-M. Klein, I. Steffan-Dewenter, and T. Tscharntke. 2004. Effects of habitat area, isolation, and landscape diversity on plant species richness of calcareous grasslands. Biodiversity and Conservation 13: 1441–1451.

Krauss, J., I. Steffan-Dewenter, and T. Tscharntke. 2003. How does landscape context contribute to effects of habitat fragmentation on diversity and population density of butterflies? Journal of Biogeography 30: 889–900.

Kremen, C., N. M. Williams, and R. W. Thorp. 2002. Crop pollination from native bees at risk from agricultural intensification. Proceedings of the National Academy of Sciences (USA) 99: 16812–16816.

Kunin, W. E. 1993. Sex and the single mustard: Population density and pollinator behavior effects on seed-set. Ecology 74: 2145–2160.

Kwak, M. M., O. Velterop, and J. van Andel. 1998. Pollen and gene flow in fragmented habitats. Applied Vegetation Science 1: 37–54.

Lehtilä, K., and K. Syrjänen. 1995. Positive effects of pollination on subsequent size, reproduction, and survival of *Primula veris*. Ecology 76: 1084–1098.

Lennartsson, T. 2002. Extinction thresholds and disrupted plant-pollinator interactions in fragmented plant populations. Ecology 83: 3060–3072.

Liow, L. H., N. S. Sodhi, and T. Elmquist. 2001. Bee diversity along a disturbance gradient in tropical lowland forests of south-east Asia. Journal of Applied Ecology 38: 180–192.

Loreau, M., S. Naeem, P. Inchausti, J. Bengtsson, J. P. Grime, A. Hector, D. U. Hooper, M. A. Huston, D. Raffaelli, B. Schmid, D. Tilman, and D. A. Wardle. 2001. Biodiversity and ecosystem functioning: Current knowledge and future challenges. Science 294: 804–808.

Luijten, S. H., A. Dierick, J. G. B. Oostermeijer, L. E. L. Raijmann, and C. M. den Nijs. 2000. Popula-

tion size, genetic variation and reproductive success in a rapidly declining, self-incompatible perennial *Arnica montana* in the Netherlands. Conservation Biology 14: 1776–1787.
Lundberg, J., and F. Moberg. 2003. Mobile link organisms and ecosystem functioning: Implications for ecosystem resilience and management. Ecosystems 6: 87–98.
Martinez, N. D. 1996. Defining and measuring functional aspects of biodiversity. Pp. 114–148 *in* K. J. Gaston (ed.), Biodiversity: A biology of numbers and differences. Blackwell Science, Oxford.
Matsumura, C., and I. Washitani. 2000. Effects of population size and pollinator limitation on seed-set of *Primula sieboldii* populations in a fragmented landscape. Ecological Research 15: 307–322.
Michener, C. D. 2000. The bees of the world. John Hopkins University Press, Baltimore, MD.
Morgan, J. W. 1999. Effects of population size on seed production and germinability in an endangered, fragmented grassland plant. Conservation Biology 13: 266–273.
Nabhan, G. P., and S. L. Buchmann. 1997. Services provided by pollinators. Pp. 133–150 *in* G. C. Daily (ed.), Nature's services: Societal dependence on natural ecosystems. Island Press, Washington, DC.
Naeem, S., F. S. Chair, I. C. R. Chapin, P. R. Ehrlich, F. B. Golley., D. U. Hooper, J. H. Lawton, R. V. O'Neill, H. A. Mooney, O. E. Sala, A. J. Symstad, and D. Tilman. 1999. Biodiversity and ecosystem functioning: Maintaining natural life support processes. Issues in Ecology 4: 1–12.
Naeem, S., and J. P. Wright. 2003. Disentangling biodiversity effects on ecosystem functioning: Deriving solutions to a seemingly insurmountable problem. Ecology Letters 6: 567–579.
Nielsen, A., and R. A. Ims. 2000. Bumble bee pollination of the sticky catchfly in a fragmented agricultural landscape. Ecoscience 7: 157–165.
Oostermeijer, J. G. B., M. W. van Eijck, and J. C. M. den Nijs. 1994. Offspring fitness in relation to population size and genetic variation in the rare perennial plant species *Gentiana pneumonanthe* (Gentianaceae). Oecologia 97: 289–296.
Pellmyr, O. 2002. Pollination by animals. Pp. 157–184 *in* C. M. Herrera and O. Pellmyr (eds.), Plant–animal interactions: An evolutionary approach. Blackwell Science, Oxford.
Perfecto, I., R. A. Rice, R. Greenberg, and M. E. van der Voort. 1996. Shade coffee: A disappearing refuge for biodiversity. Bioscience 46: 598–608.
Poveda, K., I. Steffan-Dewenter, S. Scheu, and T. Tscharntke. 2003. Effects of below- and above-ground herbivores on plant growth, flower visitation and seed set. Oecologia 135: 601–605.
Rathcke, B. J., and E. S. Jules. 1993. Habitat fragmentation and plant–pollinator interactions. Current Science 65: 273–277.
Renner, S. S. 1998. Effects of habitat fragmentation on plant pollinator interactions in the tropics. Pp. 339–360 *in* D. M. Newbery, H. H. T. Prins, and N. Brown (eds.), Dynamics of tropical communities. Blackwell Science, Oxford.
Richards, A. J. 2001. Does low biodiversity resulting from modern agricultural practice affect crop pollination and yield? Annals of Botany 88: 165–172.
Ricketts, T. H. 2001. The matrix matters: Effective isolation in fragmented landscapes. American Naturalist 158: 87–99.
Rieger, M. A., M. Lamond, C. Preston, S. B. Powles, and R. T. Roush. 2002. Pollen-mediated movement of herbicide resistance between commercial canola fields. Science 296: 2386–2388.
Roland, J., and P. D. Taylor. 1997. Insect parasitoid species respond to forest structure at different spatial scales. Nature 386: 710–713.
Rosenzweig, M. L. 1995. Species diversity in space and time. Cambridge University Press, Cambridge.
Roubik, D. W. 2001. Ups and downs in pollinator populations: When is there a decline? Conservation Ecology 5. Online URL http://www.consecol.org/vol5/iss1/art2.
Roubik, D. W. 2002. The value of bees to the coffee harvest. Nature 417: 708.
Schulke, B., and N. M. Waser. 2001. Long-distance pollinator flights and pollen dispersal between populations of *Delphinium nuttallianum*. Oecologia 127: 239–245.
Sih, A., and M.-S. Baltus. 1987. Patch size, pollinator behavior, and pollinator limitation in catnip. Ecology 68: 1679–1690.
Spears, E. E. 1987. Island and mainland pollination ecology of *Centrosoma virginianum* and *Opuntia stricta*. Journal of Ecology 75: 351–362.

Steffan-Dewenter, I. 2002. Landscape context affects trap-nesting bees, wasps, and their natural enemies. Ecological Entomology 27: 631–637.

Steffan-Dewenter, I. 2003. Importance of habitat area and landscape context for species richness of bees and wasps in fragmented orchard meadows. Conservation Biology 17: 1036–1044.

Steffan-Dewenter, I., and A. Kuhn. 2003. Honeybee foraging in differentially structured landscapes. Proceedings of the Royal Society of London B 270: 569–575.

Steffan-Dewenter, I., U. Münzenberg, C. Bürger, C. Thies, and T. Tscharntke. 2002. Scale-dependent effects of landscape structure on three pollinator guilds. Ecology 83: 1421–1432.

Steffan-Dewenter I., U. Münzenberg, and T. Tscharntke. 2001. Pollination, seed set and seed predation on a landscape scale. Proceedings of the Royal Society London B 268: 1685–1690.

Steffan-Dewenter, I., and T. Tscharntke. 1999. Effects of habitat isolation on pollinator communities and seed set. Oecologia 121: 432–440.

Steffan-Dewenter, I., and T. Tscharntke. 2002. Insect communities and biotic interactions on fragmented calcareous grasslands—a mini review. Biological Conservation 104: 275–284.

Stone, G. N., F. Gilbert, P. Willmer, S. Potts, F. Semida, and S. Zalat. 1999. Windows of opportunity and the temporal structuring of foraging activity in a desert solitary bee. Ecological Entomology 24: 208–221.

Strauss, S. Y. 1997. Floral characters link herbivores, pollinators, and plant fitness. Ecology 78: 1640–1645.

Theunert, R. 2003. Atlas zur Verbreitung der Wildbienen (Hymenoptera: Apidae) in Niedersachsen und Bremen 1973–2002. Ökologieconsult Schriften, Hohenhameln.

Thies, C., I. Steffan-Dewenter, and T. Tscharntke. 2003. Effects of landscape context on herbivory and parasitism at different spatial scales. Oikos 101: 18–25.

Tscharntke, T., A. Gathmann, and I. Steffan-Dewenter. 1998. Bioindication using trap-nesting bees and wasps and their natural enemies: Community structure and interactions. Journal of Applied Ecology 35: 708–719.

Tscharntke, T., I. Steffan-Dewenter, A. Kruess, and C. Thies. 2002. Characteristics of insect populations on habitat fragments: A mini review. Ecological Research 17: 229–239.

Walther-Hellwig, K., and R. Frankl. 2000. Foraging distances of *Bombus muscorum, Bombus lapidarius,* and *Bombus terrestris* (Hymenoptera, Apidae). Journal of Insect Behavior 13: 239–246.

Waser, N. M., L. Chittka, M. V. Price, N. M. Williams, and J. Ollerton. 1996. Generalization in pollination systems and why it matters. Ecology 77: 1043–1060.

Waser, N. M., and M. V. Price. 1998. What plant ecologists can learn from zoology. Perspectives in Plant Ecology, Evolution, and Systematics 1: 137–150.

Wcislo, W. T., and J. H. Cane. 1996. Floral resource utilization by solitary bees (Hymenoptera: Apoidea) and exploitation of their stored foods by natural enemies. Annual Review of Entomology 41: 257–286.

Westphal, C., I. Steffan-Dewenter, and T. Tscharntke. 2003. Mass flowering crops enhance pollinator densities at a landscape scale. Ecology Letters 6: 961–965.

Westrich, P. 1989. Die Wildbienen Baden-Württembergs. Eugen Ulmer, Stuttgart.

Westrich, P. 1996. Habitat requirements of central European bees and the problems of partial habitats. Pp. 1–16 *in* A. Matheson, S. L. Buchmann, C. O'Toole, P. Westrich, and I. H. Williams (eds.), The conservation of bees. Academic Press, London.

Wiegand, T., K. A. Moloney, J. Naves, and F. Knauer. 1999. Finding the missing link between landscape structure and population dynamics: A spatially explicit perspective. American Naturalist 154: 605–627.

Wilcock, C., and R. Neiland. 2002. Pollination failure in plants: Why it happens and when it matters. Trends in Plant Science 7: 270–277.

With, K. A., S. J. Cadaret, and C. Davis. 1999. Movement responses to patch structure in experimental fractal landscapes. Ecology 80: 1340–1353.

Wolf, A. T., and S. P. Harrison. 2001. Effects of habitat size and patch isolation on reproductive success of the serpentine morning glory. Conservation Biology 15: 111–121.

PART V

Final Considerations: Pollination Compared to Other Interactions

CHAPTER EIGHTEEN

"Biological Barter": Patterns of Specialization Compared across Different Mutualisms

Jeff Ollerton

Obligate one-to-one mutualisms between species pairs are rare in practice and anomalous in theory.
—Howe (1984)

Specialization in interactions with other species is the root cause of why the world has millions of species rather than thousands.
—Thompson (1994)

It appears that we simultaneously know both a great deal and not very much at all about mutualism.
—Bronstein (1994a)

much information can be found in scattered form, and only awaits careful coordination in order to yield a rich crop of ideas. The various books and journals . . . are like a row of beehives containing an immense amount of valuable honey, which has been stored up in separate cells . . .
—Elton (1927)

Interactions between species are characterized by a continuum of evolutionary, functional, and ecological specialization (sensu Waser et al. 1996; Fenster et al. 2004; D. Vázquez et al., unpublished data; Ollerton et al., chap. 13 in this volume), ranging from obligate, two-species relationships to highly diffuse, opportunistic, multispecies associations. It should be clear from the preceding chapters that an important task of biologists is to characterize the nature of these interactions and to ask questions regarding their origin, maintenance, and stability. Although the main focus of this volume is on plant–pollinator interactions, cross-comparison with other types of interactions may yield insights that are of value to biologists interested in different types of interspecific associations, especially those searching for general laws and trends concerning mutu-

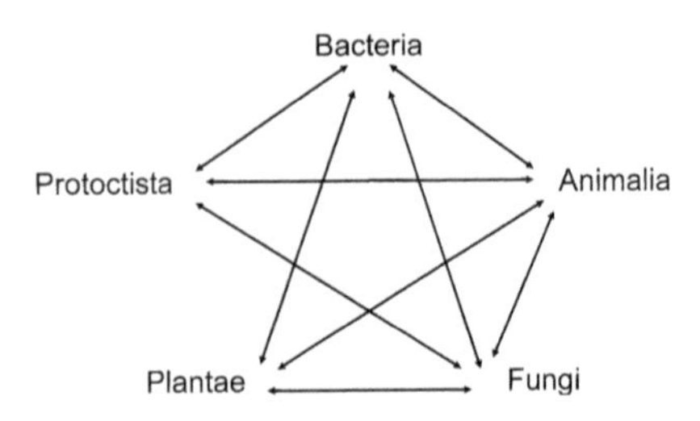

Figure 18.1 Mutualism connections between the five kingdoms of life. The connecting arrows indicate known examples of mutualistic relationships between members of each kingdom. There are in fact mutualisms between all possible pairwise combinations of kingdoms, with the exception of Plantae–Protoctista (see text). Data from Batra (1979), Boucher (1985), Smith and Douglas (1987), Reisser (1992), Douglas (1994), Jolivet (1996), Nash (1996), Paracer and Ahmadjian (2000).

alistic interactions. This chapter, therefore, addresses broader questions of patterns of specialization and generalization across mutualisms.

Plant–pollinator interactions are largely mutualistic interactions (but see Renner, chap. 6 in this volume), which are defined as relationships between species that result in reciprocal benefits. Benefits in the short term may include trophic gain, physical protection, and dispersal of gametes or propagules. One can consider the exchange of resources and services a form of "biological barter" (see following discussion); ultimately, this barter is expected to result in increased fitness for both participants (Boucher 1985).

The range of organisms involved in mutualistic interactions spans all five kingdoms (as conservatively defined) and most groups within (fig. 18.1). The absence of any known mutualisms between the kingdoms Plantae and Protoctista is an interesting anomaly and may be due to two factors relating to the autotrophic and heterotrophic members of the Protoctista. All plants and the algal protoctists are essentially autotrophic; therefore, there may be no resource basis for bartering, that is, mutual carbon exchange is unlikely to benefit either partner (see also fig. 18.2, and discussion following). The heterotrophic protoctists are not as abundant as bacteria or fungi and, therefore, are less promising partners within a mutualism; in addition, heterotrophic protoctists tend to be single celled and feed by endocytosis rather than by absorption (following production of exogenous enzymes) as in fungi and bacteria. Their solitary habits and mode of feeding may, therefore, make them less efficient than fungi and bacteria at obtaining inorganic nutrients en masse.

The partners involved in a mutualism directly exchange physical resources and services (Janzen 1985; Noë and Hammerstein 1995)—it truly is "biological barter" (fig. 18.2). Physical resources are largely concerned with nutritional gain (e.g., carbohydrates, inorganic nutrients, and water). Services range widely in their scope and include transport of propagules (e.g., seed and spore dispersal), movement of gametes (most familiarly in pollination, but also dispersal of fungal spermatia by Diptera; see Bultman and White 1988), bioluminescence, cleaning, and physical protection. Of course, as in all generalizations concerned with the natural world, the physical resources/services dichotomy is simplistic and in reality it comprises a complex, multidimensional continuum. For example, nectar contains a mix of sugars (carbohydrate resource) and amino acids (inorganic nutrient resource; see, e.g., Gardener and Gillman 2001), among other things. In addition, physical resources are occasionally nonnutritional, for example, production of resins for nest construction by pollinating bees in genera such as *Dalechampia* (Euphorbiaceae) and *Clusia* (Clusiaceae; Armbruster 1984). In such cases, the plant initially provides a physical resource that ultimately supplies a service (physical protection of the bee offspring). Occasionally, mutualistic interactions are characterized by the conferment of both services and physical resources; for example, in some ant–*Acacia* relationships, the *Acacia* provides both domatia in which the ants can build nests (a service) and food bodies on which the ants feed (a physical resource; Seigler and Ebinger 1995). A second example is plant–fungus mycorrhizal relationships in which the fungus, in addition to providing water and inorganic nutrients to the plant, also reduces the risk of it becoming infected by pathogenic fungi and may produce secondary compounds that deter herbivores (Newsham et al. 1995). Biological barter, like barter in human societies, is complex and contingent upon the abilities and requirements of both partners. Thus, the representation of mutualistic associations in the two-dimensional format of figure 18.2 loses some of the complexity of what is often a multidimensional interaction, involving the exchange of multiple resources or services. I have tried to express this by including examples where an interaction involves exchange of a minor resource or service by mentioning that a particular barter occurs "in part."

Direct mutualism (sensu Boucher et al. 1982) takes many forms, but it is clear from figure 18.2 that there is not a limitless set of permutations of resource and service barter. For instance, there are no known examples of mutualism in which partners trade the same resource or service (represented by the black cells on the diagonal in figure 18.2). This makes perfect biological sense: why should an organism barter something that it can make, obtain, or do itself? The gray cells within figure 18.2 are situations in which different resources or services could potentially be bartered but which are not documented in the literature, as far as I have been able to discover. As research progresses, such examples may come to light: suspected or confirmed mutualistic interactions new to science are discovered with almost annual predictability (Compton and Ware 1991; Grange 1991; Ellis and Midgley 1996; Ross and Newman 1996; Bultman et al. 1998; Stewart 1998; Naoki and Toapanta 2001; Brown et al. 2002; Corallini and Gaino 2003). But if they do exist, they are probably not common.

It is notable that mutualisms based on service–service bartering are rather rare and usually involve resources at some level (e.g., cleaning–physical defense associations in which the cleaner's reward includes food items as well as defense).

		Resources					
		Carbohydrate	**Nitrogen**	**Inorganic compounds**	**Proteins, amino acids, etc.**	**Complex organic compounds**	**Water**
Resources	**Carbohydrate**						
	Nitrogen	Legume-rhizobia, Actinorhizae, etc. e.g. *Lotus*-rhizobium, *Alnus*-*Frankia*. *Gunnera*-*Nostoc*					
	Inorganic compounds	Lichens, mycorrhizae, metazoan-algae, etc. *Anzia*-*Trebouxia*, *Geosiphon*-*Nostoc*, *Pinus*-fungi, *Hydra*-algae					
	Proteins, amino acids, etc.	Ant and termite fungal associations, e.g. *Cyphomyrmex*-*fungus*, *Odontotermes*-*fungus*					
	Complex organic compounds						
	Water	Mycorrhizal associations (in part)					
Services	**Gamete dispersal**	Plant-pollinator, e.g. *Asclepias*-pollinator		Plant-pollinator – (in part) calcium in nectar	Plant-pollinator – protein in pollen, amino acids in nectar	Plant-pollinator, e.g. oil, scent and resin rewards	Plant pollinator (in part) – water in nectar
	Propagule dispersal	Plant-disperser, fungus-disperser, e.g. *Viburnum*-birds, *Viola*-ants, *Phallus*-fly			Plant-disperser, fungus-disperser (in part) – proteins etc.		
	Protection	Animal gut microbes, e.g. *Lasioderma*-*Symbiotaphrina*			Insect mycetocyte (?) e.g. *Acyrthosiphon*-*Buchnera*		
	Physical defense	Plant-fungal associations, plant-ant relationships, ant-homoptera, e.g. *Lolium*-endophyte, *Passiflora*-ants, *Lasius*-homoptera					
	Bioluminescence	Fish-bioluminescence, e.g. *Leiognathus*-*Photobacterium*					
	Cleaning	Metazoan cleaning relationships (in part), e.g. *Buphagus*-mammal			Metazoan cleaning relationships (in part), e.g. *Buphagus*-mammal		

Figure 18.2 "Biological barter": examples of the range of physical resources and services offered for exchange in mutualistic relationships. The specific examples have been drawn from the studies used in the analyses in this chapter (see appendix 18.1). The black cells on the diagonal represent situations in which the same resource or service would be bartered; there are no known examples of this. The gray cells are situations in which different resources or services are bartered, but which are not documented in the literature. Data from Batra (1979), Boucher (1985), Smith and Douglas (1987), Reisser (1992), Douglas (1994), Jolivet (1996), Nash (1996), Paracer and Ahmadjian (2000).

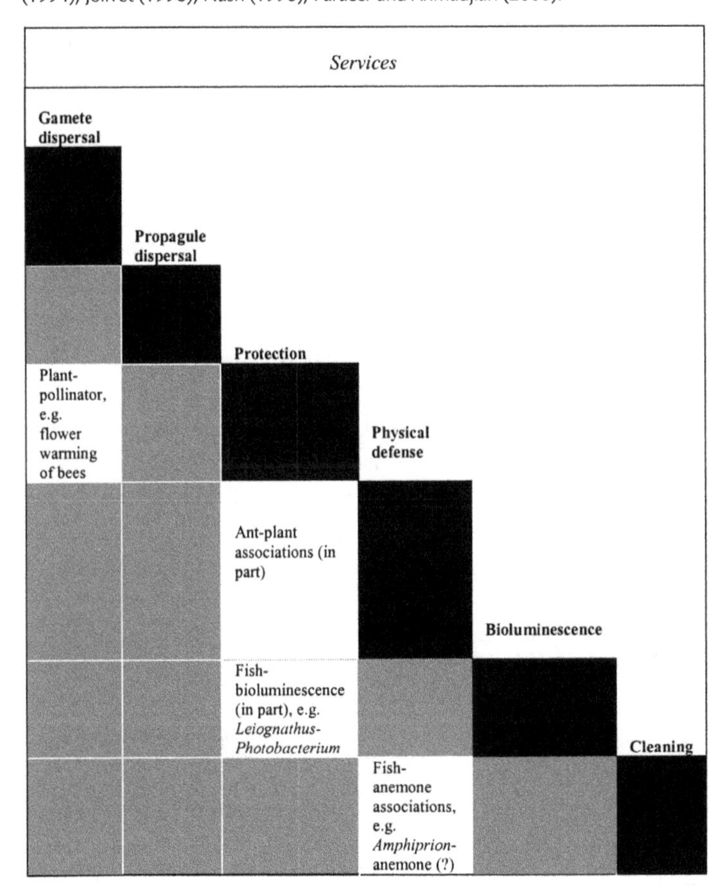

Another very obvious trend within figure 18.2 is the dominance of carbohydrate exchange in mutualistic associations, emphasizing the fundamental role of primary productivity within ecosystems.

Mutualisms are among the most ubiquitous of interactions in nature, but also the least well understood, in comparison to, say, predator–prey or parasite–host interactions. As the quote from Bronstein (1994a) suggests, whereas a great deal of effort has gone into exploring the ecology and evolution of mutualisms, their enormous variety has meant that there has only been a limited attempt to synthesize these studies to search for broad-scale generalizations about the ecology and evolution of such interactions. Indeed, it could be argued that because of their diverse nature, there are no generalizations to discover beyond the fact that these relationships benefit all participants (and even that is far from proven in many cases).

Within the published literature there exists an enormous, diffuse, scattered dataset of observations on the diversity and specificity of mutualistic relationships. The purpose of this chapter is to draw together a fraction of these data to address questions regarding the relationship between interspecific biological intimacy and ecological specialization in mutualistic interactions. In particular, I wish to address a very specific hypothesis:

As mutualistic relationships become increasingly biologically intimate through trophic, physiological, and/or physical integration, they become increasingly exclusive.

The basis of this hypothesis is that the high degree of morphological and/or biochemical specialization required to maintain very close mutualistic relationships between species precludes their usurpation by other taxa (see Borowicz and Juliano 1991 for a similar argument applied to plant–fungus interactions). Therefore, there should be a negative relationship between the number of species involved in a relationship and the level of biological intimacy displayed by that interaction. This is represented by an informal model (fig. 18.3).

It is unclear whether the predicted relationship between biological intimacy and ecological specialization should be a simple linear function (fig. 18.3, line a) or a more complex function, in which the taxonomic exclusivity of the relationship is reached after a relatively high (line b) or comparatively modest (line c) level of intimacy. The different shapes of these curves imply disparate trajectories in the evolution of ecological specialization in mutualistic interactions.

Methods

From a broad survey of the relevant literature, I have compiled a database of mutualistic interactions, which I have used to test the intimacy-specialization hypothesis. This dataset is not random (in a statistical sense) because certain classes of mutualism (and within these classes, particular systems) have been much more comprehensively studied than others; however the data are representative of the diversity of mutualisms, as presented in figure 18.2.

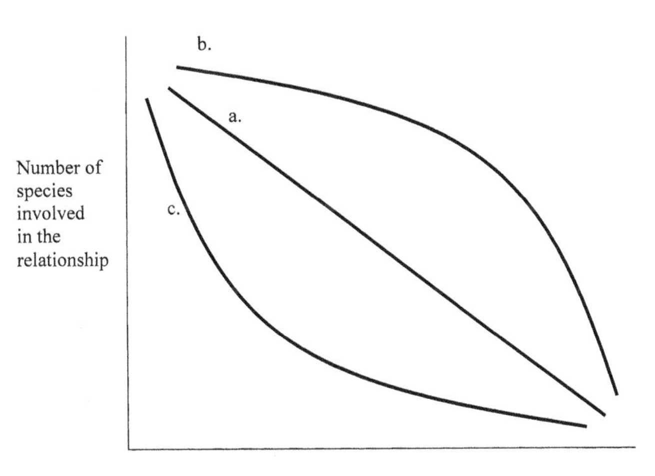

Figure 18.3 An informal graphical model of the predicted relationships between the biological intimacy of a mutualism and the number of partner species of a given mutualist (see text for a description of lines a, b, and c).

Using published sources, I compiled information regarding the ecological specificity of the relationships from studies of mutualistic associations that covered the full range of these interactions as they are currently known. The inclusion of a particular mutualistic relationship was determined by how well studied the system has been and, hence, the availability of the pertinent information (see following discussion). The dataset used for the following analyses is presented in appendix 18.1.

There is a clear problem with counting the number of species involved in an interspecific relationship: species concepts vary systematically and are particularly difficult to pin down for microorganisms. In addition, variation in observation effort between studies can be problematic (Ollerton and Cranmer 2002; Ollerton et al., chap. 13 in this volume). Therefore, to provide a degree of robustness within the data, I began by identifying "focus" genera. As the term implies, a *focus genus* is one that acts as the "focus of attention" for other species in a mutualistic interaction. For example, in a plant–mycorrhizal interaction, the focus genus is that of the plant; in a fish–fish cleaning relationship, the focus genus is that of the fish being cleaned; and so forth. I deliberately use this phraseology to avoid using the term *host,* because that word implies a large degree of physical intimacy, which in fact is one of the variables being assessed in this analysis.

For each focus genus, I recorded the number of partner genera associated with that mutualism. For example, from appendix 18.1 it can be seen that the focus genus *Asclepias* (Apocynaceae) is known to be pollinated by 150 genera of animals. The taxonomic level of genus is a more conservative one than that of species, although I accept there are still problems with comparing, for example, a "genus" of bacteria with a "genus" of flowering plants. In addition, recent molecular phylogenetic investigations have uncovered surprising levels of microbial symbiont diversity (see review by Herre et al. 1999). Nonetheless, if synthetic analyses such as this are to be conducted at all, it is necessary for us to take a robust, but honest, approach to the problems inherent in these kinds of datasets. Perhaps future analyses will incorporate less subjective measures, for example genetic diversity.

The identity of the focus genera I surveyed was determined in part by the body of work available. For example, the pollinators of the genera *Asclepias* (Apocynaceae) and *Ficus* (Moraceae) have been quite well studied compared to many other plants, and mycorrhizal associations between *Pinus* (Pinaceae) and fungi have been thoroughly documented. The criteria for including a focus genus were that (1) there was at least 10 years worth of literature available, (2) at least five species of the genus had been researched, and (3) there was some geographical diversity to these studies. To provide some balance within the dataset (i.e., to not swamp the study with plant–pollinator relationships) no more than five focus genera were included for each broad category of mutualism (e.g., mycorrhizae, endophyte, seed dispersal, pollination, etc.).

These data have been gleaned from a huge range of sources. To save space, the original references are not presented here, but a full list of sources is available upon request from the author.

In addition to recording the numbers of partner genera associated with the focus genus in that particular relationship, an "index of biological intimacy" was calculated for the relationship (independent of the number of partner genera recorded). Each mutualism was scored for the following criteria.

1. Physical dependence of the mutualism:
    - 0 = none of life cycle
    - 1 = part of life cycle
    - 2 = all of life cycle
2. Trophic dependence:
    - 0 = wholly independent
    - 1 = partially dependent
    - 2 = wholly dependent
3. Physiological integration:
    - 0 = no integration (extracorporeal)
    - 1 = integrated for part of life cycle or intercellular
    - 2 = wholly integrated/intracellular
4. Vertical transmission:
    - 0 = no vertical transmission of partner mutualist from parent to offspring
    - 1 = transmission of mutualist from parent to offspring

Of course there is no single metric of "biological intimacy" that could represent the whole diversity of mutualisms in all of their many forms; therefore, the preceding criteria were selected to represent the most biologically meaningful measures of intimacy and to provide an additive, multitrait representation. The index of biological intimacy is the sum of the scores for the individual criteria and, therefore, potentially ranges from 0 (a very low level of biological intimacy) to a maximum of 7 (a very high level of biological intimacy). As might be expected, there is considerable intercorrelation of traits (data not presented).

Comparative analyses such as this suffer from the problem that species are not statistically independent units but are linked by their phylogeny, and possible phylogenetic biases must be taken into consideration in any comparative analysis (Harvey and Pagel 1991). However, because the dataset used in these analyses spans all five kingdoms, and a wide range of genera, orders, and classes, a formal phylogenetically corrected regression is not possible—whether it is required for such a phylogenetically broad spread of taxa is arguable.

Results and Discussion

The level of biological intimacy of a mutualistic relationship has a strong effect on the degree of specialization of that relationship (fig. 18.4A). These data are strongly right skewed and nontransformable; however, there is a statistically significant, negative correlation between the index of biological intimacy and the number of partner genera associated with the focus genus (Spearman rank correlation, $r_s = -0.59$, $N = 39$, $P < .001$). Sensitivity analyses show that this is a statistically robust conclusion (see appendix 18.2).

It is clear that, as mutualistic relationships become more biologically intimate, the number of genera involved in that relationship is rapidly reduced; that is, the response function fits the line c model of figure 18.3 (concave and steeply declining).

Displaying the y axis of figure 18.4A on a log scale allows the pattern of the data to be seen more clearly (fig. 18.4B). The data points have been coded with respect to whether the mutualism would normally be considered closely symbiotic or nonsymbiotic, defined by Boucher et al. (1982, 316) as "using physiological integration as the basic criterion." In figure 18.4B, there is no clear distinction between symbiosis and other types of relationship in terms of the degree of biological intimacy of the relationship, as defined by this index: "nonsymbiotic" relationships can be as biologically intimate as symbiotic interactions. The highest scores for nonsymbiotic mutualisms were achieved by the *Cyphomyrmex*–fungus, *Odontotermes*–fungus, *Ficus*–fig wasp, *Cecropia*–ant, *Xyleborus*–fungus, *Sirex–Amylostereum,* and *Acacia*–ant mutualisms. Low scores (3 or less) for mutualisms traditionally classed as symbiotic were found for *Geosiphon–Nostoc, Sphagnum*–cyanobacteria, *Pinus*–fungus, *Lotus–Rhizobium, Alnus–Frankia,* and

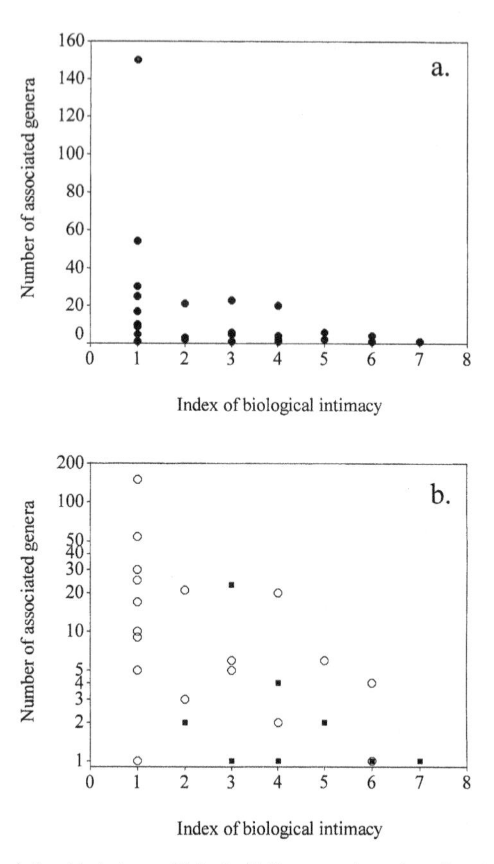

Figure 18.4 (A) The relationship between biological intimacy and number of partner genera associated with a focus genus for the mutualistic associations in appendix 18.1. Some data points represent multiple examples. (B) The relationship in (A), with the *y* axis as a log scale. The data points have been classified as representing the symbiotic (solid squares) and nonsymbiotic (open circles) interactions in appendix 18.1.

Blasia–Nostoc (see appendix 18.1). Boucher et al. (1982, 316) stated quite reasonably that the symbiotic–nonsymbiotic dichotomy was "artificial but convenient." I would go further and suggest that the symbiosis–nonsymbiosis dichotomy may be too artificial to use if we wish to study the true biological nature of mutualistic associations, especially in relation to the biological intimacy of the relationship.

The steeply declining relationship between biological intimacy and number of associated genera suggests that mutualisms can evolve to be comparatively exclusive (at a generic level) at a relatively modest level of biological intimacy (fig. 18.4A). The mutualisms at the very highest level of intimacy (index of 7) are represented by single-genus relationships (fig. 18.4B, appendix 18.1), for example, the *Azolla–Anabaena* and *Lolium–Neotyphodium* interactions. However, it is also clear that exclusivity is not a certainty at higher levels of intimacy and that, even with an index of greater than or equal to 5, relationships can be surprisingly nonspecific (e.g., the *Xyleborus*–fungus and *Cecropia*–ant mutualisms). Conversely, relationships characterized by low levels of intimacy can also be quite specialized: for example, the *Epichloe*–*Phorbia* and *Sphagnum*–cyanobacteria interactions.

The results obtained from this analysis are biological trends rather than fixed rules, but there is certainly a trend toward taxonomic exclusivity at higher levels of biological intimacy, which supports the original hypothesis set out in the beginning of this chapter.

What are the mechanisms governing the relationship between intimacy and specialization? There are broadly two factors that are important. The first is the *physical* fit between the organisms; for example, short-tongued animals cannot access nectar in long-tubed flowers, and small birds are unable to ingest large fruit (Jordano 1987). The second is the *biochemical* fit between the organisms. This is less straightforward than the former and relates to the ability of the two organisms to recognize, communicate with, and tolerate one another at a molecular level. This could be as intimate as the recognition/rejection systems in legume–rhizobia and plant–mycorrhizal relationships (Lum and Hirsch 2002); less intimate, but just as specific, olfactory signaling between a flower and its pollinators (Kite et al. 1998); or more diffuse, such as the correlation between nectar composition and pollination systems found in some plant groups (Perret et al. 2001) and identification of myrmecophilic plants by ants.

It is tempting to assume that it is the biochemical fit between organisms that governs highly specialized relationships such as microbial–plant and –animal mutualisms. However, there is clearly an important *physical* component to these interactions: bacteria and fungi are sufficiently small that they become intimately associated with the roots of the plants that they service, whereas fig wasps are tiny compared to most other insect pollinators yet need to be small to interact with *Ficus* syconia. Both the physical and biochemical "fit" of the organisms concerned are required if the relationship is to be successful.

Boucher et al. (1982) reviewed the literature on the strength of mutualistic interactions (obligate vs. facultative) and the range of partners involved in the relationship: a single partner (monophilic), a restricted range of partners (oligophilic), and a wide range of partners (polyphilic). Their findings are summarized in table 18.1, which shows that some classes of interaction within this matrix of possibilities are more probable than others. But the main finding is that, if an interaction is obligate, it usually involves a very limited set of partners, whereas facultative interactions tend to have a wider range of partners. The facultative–obligate dichotomy is inherent within the index of biological intimacy used in the analyses presented in this chapter (especially the "physical dependence" and "trophic dependence" factors; see Methods). Therefore, my find-

Table 18.1 The relationship between the strength of a mutualistic interaction and the number of partners involved, in terms of the relative rarity and commonness of those interactions in nature

	Number of partners involved		
Strength of relationship	**Monophilic**	**Oligophilic**	**Polyphilic**
Facultative	Unknown	Common	Common
Obligate	Common	Common	Rare

Source: Based on Boucher et al. 1982.

ings are in broad agreement with those of Boucher et al. (1982): at low levels of biological intimacy, generalization is the norm, and at higher levels of biological intimacy, specialization becomes more common.

Law and Koptur (1986) reviewed the earlier literature on the evolution of specificity in mutualisms. They noted that most mutualisms, of all persuasions, tend to be nonspecific, certainly in comparison to antagonistic relationships. For instance, they noted that host–parasite systems can be specific at the level of genotype (though the same is true for the legume–rhizobium symbiosis and may also apply to other, less-well-studied microbial mutualisms; see Young and Johnston 1989). Law and Koptur argued that intimacy of the relationship does not explain this disparity, because mutualisms can be, and often are, just as intimate as antagonistic relationships. To explore the circumstances under which specialized mutualisms could evolve, they developed a simple model of guild–guild coevolution. One of the surprising outcomes of this model was that it was theoretically possible for strong natural selection to favor nonspecificity at the guild level, thereby eradicating specialized phenotypes from the community. This contrasts strongly with a traditional view that evolution drives interacting species toward specialization (see also Armbruster and Baldwin 1998; Waser, chap. 1 in this volume).

Patterns of ecological specialization in pollination systems versus vertebrate seed dispersal systems were compared by Jordano (1987), who found that pollination systems of plants were, on average, a third more ecologically specialized than plant seed dispersal systems. Jordano argued that this pattern is likely to be caused by the greater requirement of a physical fit ("morphological coupling") between flowers and pollinators than between fruit and seed dispersers. This remains an intriguing possibility which has never been properly tested.

Comparisons with Nonmutualistic Interactions

Mutualism is only one category of interspecific relationship, and, in places, it blends with other types of interaction such as predation, parasitism, and commensalism. How do the patterns of specialization and biological intimacy apparent in mutualisms compare with other types of interaction? Borowicz and Juliano (1991) compared levels of host specificity between mutualistic (ectomycorrhizal), shoot parasitic, and root nectrotrophic fungi and their north temperate tree and shrub hosts. These three fungal life styles show a range of physical intimacy/integration with their hosts, with shoot parasites being more intimate than ectomycorrhizal associations, which in turn are more intimate than root necrotrophs. The authors predicted that there should be a negative correlation between the level of intimacy of the relationship and the host ranges of fungi in each group: root necrotrophs being the most broadly associating and shoot parasites the least broadly associating. Their analysis showed this not to be the case, and in fact the mutualistic ectomycorrhizal fungi were the most host specific. Therefore, the authors concluded that it was the type of species interaction (mutualistic vs. antagonistic) that determined host specificity in these fungi, rather than level of physical intimacy per se. This is in disagreement with the analysis presented in this chapter, but it may be that such patterns, if present at a smaller scale, are masked by a larger-scale analysis.

Studies which attempted to synthesize plant-microbe interactions using combinations of partners that are not normally found in nature were reviewed by Law (1985). He concluded that mutualistic relationships such as rhizobia and mycorrhizae are much less specific at a genotype level than equivalent antagonistic relationships, as expected from theory. Is this true only for biologically intimate comparisons (e.g., bacterial pathogens vs. bacterial mutualists) or is it also true for less-intimate associations (e.g., insect herbivores vs. insect pollinators)? There are certainly many examples of plants which, following their introduction to an area, can utilize the local pollinators, seed dispersers, and even mycorrhizal fungi (Richardson et al. 2000).

Thompson (1994) reviewed the literature on the relationship between physical intimacy and ecological specialization in parasitic and herbivorous interactions and concluded that a diversity of taxa, from a range of habitats and involved in different kinds of interactions, tends to be more ecologically specialized when they are more physically intimate with their hosts. This holds true for ectoparasites of birds and mammals, microlepidoptera and their larval food plants, marine limpets and the seagrasses on which they feed, and seaslugs, crabs, and amphipods in their seaweed habitats. These results further support the generality of the finding that mutualisms become much more ecologically specialized as the biological intimacy of the relationship increases and suggests that the trend may extend beyond mutualistic associations to include parasitism, herbivory, predator–prey interactions, and commensalism.

Some Evolutionary Considerations

The analyses presented in this chapter, in addition to the previous outlined studies, support the hypothesis that the taxonomic exclusivity of a mutualistic relationship is governed to some extent by the biological intimacy of that interaction. But what are the conditions under which highly exclusive interactions can evolve? For example, is it possible for an interaction to evolve from low biological intimacy to high biological intimacy (i.e., shift from, say, an index of 1 or 2 to an index of 6 or 7 in fig. 18.4)? Such extreme evolutionary shifts are known from free-living to parasitic lifestyles (e.g., *Cuscuta* compared to other Convolvulaceae; see Neyland 2001), within parasite lineages (e.g., Microsporida compared to more typical Fungi; see Keeling and Fast 2002), and even within species, in the case of gender dimorphism in deep sea anglerfish (e.g., Munk 2000). Are such extreme shifts possible within mutualistic relationships? Or is it more likely that the starting conditions of the relationship are what determines the level of intimacy and, hence, exclusivity? One way to address this question is to map organism traits, including interactions, onto independently derived (preferably molecular) phylogenies. This has been done for plant–pollinator interactions in a range of families and genera (see review by Weller and Sakai 1999) and less often for other mutualisms. The results support the notion that the functional and ecological specificity of interactions can readily evolve, perhaps even quite rapidly, and that clades do not always evolve toward specialization (Armbruster and Baldwin 1998; Machado et al. 2001; Armbruster, chap. 12 in this volume). To my certain knowledge, biological intimacy per se has never been mapped in this way and it remains a largely unexplored area.

The fitness advantage of generalizing on multiple pollinators, versus specializing on one or a few pollinators, was explored by using a simple model by Waser et al. (1996). There are clear benefits of risk-spreading when pollinators vary spatially and temporally (as they often do in nature; e.g., Herrera 1988; Lamborn and Ollerton 2000). Interactions between microbes such as fungi and algae and their animal or plant hosts are often characterized by colonization by multiple species or genotypes (Douglas 1994). Even in quite functionally specialized, biologically intimate relationships, the hosts can effectively generalize at the genotype level, and there exists evidence that this can benefit the host because different genotypes contribute different resources or services to the relationship (Newsham et al. 1995).

An extreme category of specialization involves the loss by a species of its mutualistic partner(s). In many cases, this is associated with habitat specialization by the host, for example, among those few angiosperm lineages that no longer form mycorrhizal associations (Fitter and Moyersoen 1996). In other cases, this may relate to the unpredictability of the other partner, such as obligate self-pollination in plants that grow in extreme environments (Proctor et al. 1996), or, simply, to efficiency gains in a particular environment, as in the evolution of wind pollination (Linder 1998; Wallander 2001). Loss of mutualistic partners has also been documented in an ant–plant system (Blattner et al. 2001).

Such examples may provide an insight into the relationship between specialization and the stability of mutualisms. Recent models of the evolution of mutualistic interactions have emphasized that mutualisms are inherently unstable and that they will evolve toward exploitative interactions (i.e., parasitism in its many forms) if a mutation emerges that can gain the fitness benefits of the relationship without the associated resource cost (see references in Pellmyr and Huth 1994). There certainly are a number of studies that have shown that some species can evolve as parasitic members of otherwise mutualistic interactions (see review by Yu 2001): for example, interactions between *Yucca* and yucca moths (Pellmyr 2003), *Ficus* and fig wasps (Lopez-Vaamonde et al. 2001), and some between lycaenid butterflies and their ant hosts (Fiedler 1998). Freeman and Rodriguez (1993) showed that the genetic difference between a pathogenic and a mutualistic fungus could be as small as a single gene mutation, suggesting the possibility of rapid shifts between parasitism and mutualism in these organisms. Therefore, mutualistic interactions between organisms may be very evolutionarily dynamic and fluctuate rapidly in time and space. According to Thompson's (1994) Geographic Mosaic Theory of Coevolution, this is probably true of most close biotic interactions.

Mutualistic interactions are incredibly diverse in their nature, involving organisms from all of the five kingdoms, with an almost perfectly symmetrical set of documented interactions (see beginning of chapter and fig. 18.1). Models that predict instability in mutualistic interactions, and their subsequent evolution into exploitative interactions, may hold for a subset of mutualisms, but we cannot assume that all (or even the majority?) of mutualisms can be invaded by parasitic mutations. For many mutualisms, it is extremely difficult to envision how they could be "invaded" by parasitic exploiters; in such circumstances, the mutualism itself is inherently stable. For example, most plant–pollinator interactions are ecologically generalized: flowers are visited by a number of potential pollinators, and most pollinating insects visit a wide range of flowers, which gives these relationships an element of risk-spreading stability (Waser et al. 1996 and references therein; but, for a different perspective, see Johnson and Steiner 2000). Plant–pollinator interactions have provided us with a few good examples of situations in which the interaction has been exploited by parasites (see earlier discussion) but I would argue that these are exceptional and nontypical examples.

It is well documented that not all flower visitors are pollinators. The exploitation of nectar resources without performing pollination ("nectar thieving" sensu Irwin et al. 2001; see also the review of "nectar robbery" by Maloof and Inouye 2000) appears to be mainly determined by the mechanical and behavioral fit between the flower and the visitor. Flower visitors are typically too small to contact the sexual parts of the flower, and so they take nectar without picking up or depositing pollen. But this is not an evolved interaction—the nectar robbers have probably not evolved small size to allow them to rob nectar. The situation has arisen simply as a result of the wide range of sizes of flowers and their visitors. Nectar robbery would only evolve as a specific strategy if there was a fitness cost of transferring pollen. In that case, one could imagine that a mutant which could exploit the nectar resource without contacting the anthers of the flower and picking up pollen, would be at a great advantage compared to the typical flower visitor. As far as I am aware, it has never been demonstrated that there is a fitness cost for an animal which transfers pollen, despite the often very large pollen loads that pollinators can carry. Such mutualisms are noninvasible simply because there is no opportunity for a mutant to gain a fitness advantage. It is notable that the aforementioned examples of invaded pollination mutualisms are all of relationships in which the plant offers a proportion of its ovules as the resource for the pollinator to use as larval food, and which may themselves have evolved from interactions that were originally parasitic in nature. In these examples, the fitness of the pollinator is intimately bound to the fitness of the plant.

What does this tell us about the relationship between the level of specialization of a mutualistic interaction and its evolutionary stability? It is intuitive to imagine that more evolutionarily, functionally, or ecologically generalized relationships are more stable due to risk-spreading. In the case of plant–pollinator interactions, this may be responsible for the nested pattern of interactions that appear to be commonly found in plant–pollinator and plant–seed disperser networks, in which ecologically specialized animals tend to interact with ecologically generalist plants, and specialized plants interact with generalist animals (Petanidou and Ellis 1996; Bascompte et al. 2003; Dupont et al. 2003; Jordano et al., chap. 8 in this volume; Vázquez and Aizen, chap. 9 in this volume). Ollerton et al. (2003) have suggested that long-term climatic instability has often altered the community context in which organisms find themselves, thereby filtering out specialist–specialist interactions and leaving the pattern of nestedness that we observe.

In more ecologically or functionally specialized mutualistic interactions, we may, therefore, expect to find that the relationships are less evolutionarily stable. Is there any evidence for this? Many reef animals, including corals, form mutualistic associations with photosynthetic algae (collectively termed *zooxanthellae*). These associations are the basis for some 90% of the primary productivity of reef environments (Cowen 1988). In reef-building corals, the interaction is functionally specialized and was originally also thought to be highly ecologically specialized, the photosynthetic partner being only a single species/genotype of the dinoflagellate genus *Symbiodinium* (Rowan and Powers 1991). It is now known that single species of coral, and even individual colonies, can host multiple *Symbiodinium* "species" (see review by Rowan 1998). The persistence of this relationship has been the subject of much discussion and Cowen (1988) has suggested that the slow reestablishment of reef communities following mass extinction events was constrained by the need to reestablish the coral–alga symbiosis. Rowan and Powers (1991) proposed that the relationship between the corals and algae is evolutionarily very labile, and there is little evidence for codiversification of the two groups. It is unknown if the relationship has always involved *Symbiodinium* (zooxanthellae do not fossilize), but the available evidence suggests that even an interaction as biologically intimate as this can be evolutionarily stable, and persistent to the point of reestablishing itself following mass extinction.

In another functionally specialized interaction, Piercey-Normore and DePriest (2001) documented multiple examples of switching of *Trebouxia* algal partners by lichenized fungi in the family Cladoniaceae, suggesting that although these interactions are biologically intimate (the comparable *Anzia–Trebouxia* association scores a 6; see appendix 18.1) they are nonetheless evolutionarily labile.

Other specialized examples are provided by *Ficus*–fig wasp and *Yucca*–yucca moth interactions, which are thought to have evolved 90 and 40 million years ago, respectively, again suggesting that highly specialized interactions can be remarkably persistent (Pellmyr and Leebens-Mack 1999; Machado et al. 2001). Similarly, the association between *Macaranga* (Euphorbiaceae) and *Crematogaster* ants is highly species specific and is thought to be at least seven million years old (Itino et al. 2001).

Most ectomycorrhizal fungi are generalists, capable of colonizing a range of different plant hosts. Phylogenetic analysis by Hibbett et al. (2000) has shown that ectomycorrhizal associations by basidiomycete fungi are evolutionarily unstable, with repeated evolution of these relationships and frequent reversions of fungal lineages from a mutualistic to a free-living life style. This may be related to the more generalized associations between plants and ectomycorrhizal fungi, or it may be explained by the great age of this class of mutualism (at least 200 million years; see review by Cairney 2000).

The preceding examples suggest that there is no apparent trend for either generalized or specialized mutualistic interactions to be more or less evolutionarily labile. However, as shown by the coral–zooxanthellae example (Rowan and Powers 1991; Rowan 1998), we should be cautious about assessing the degree of specialization of mutualisms, particularly those involving microbes, until molecular phylogenetic analyses of symbiont diversity have been conducted (Herre et al. 1999).

Douglas (1994, 89–90) viewed the evolution of specificity in symbiotic (in fact, microbial) interactions as "the outcome of a trade off . . . between opposing selections pressures, one to broaden the range of acceptable partners, and the other to become increasingly specialized." Douglas argued that highly intimate associations involving vertical transmission of symbionts are more likely to be specialized than less intimate associations, in which fluctuations in the availability of the partner makes generalization a less risky host strategy. However, there are clearly many other factors which may influence specificity. Indeed, interactions which are apparently specialized in the field sometimes prove to be more generalized in the laboratory, where hosts can form associations with partners with which they would not normally interact (Law 1985; Douglas 1994), suggesting a degree of lability that would only become manifest following the extinction of a host's normal symbiont or the dispersal of that host beyond its normal range (a common situation for introduced plant species; e.g., Chittka and Schürkens 2001).

Conclusions

The results presented here and the review of other relevant literature strongly support the hypothesis that biologically intimate interactions between species tend to be more ecologically specialized and taxonomically exclusive. Although the dataset used in these analyses is relatively small compared to the diversity of mutualistic interactions found in nature, it is nevertheless a representative sample. Future analyses could extend the range of mutualisms included within the dataset and more deeply explore the patterns within and between clades of taxa, and functional interactions.

Theoretical approaches to the evolution of specialization and generalization in mutualisms have begun to appear over the past 20 years or so and were reviewed by Hoeksema and Bruna (2000). Of particular interest are "virulence models," which predict that mutualistic partners should generalize because of a trade-off in ecological attributes important to the mutualism, such as the ability to colonize new hosts and vie against competitors. For example, mycorrhizal fungi, which are poor competitors, could possibly colonize the roots of young plants quite early but may be subsequently replaced by fungi that are slower colonizers but better competitors (Hoeksema and Bruna 2000; see also Hoeksema and Kummel 2003). A plant–pollinator example could involve two species of bee which show a trade-off between minimum flight temperature and aggressiveness. The less aggressive species can visit the flowers of a plant early in the morning, but it is excluded by its competitor's aggression later in the day. Although the timescales are different, in both of these examples, generalization (at some level) can be maintained by ecological trade-offs.

Howe (1984) has argued that "nonsymbiotic" mutualisms are rarely obligate, whereas "symbiotic" mutualisms (which the author argues are largely derived from host–parasite interactions) are more often highly specialized. This use of dichotomous terminology is commonplace in the mutualisms literature and is, I feel, misleading at best and divisive at worst. My analysis suggests that there is no dichotomy between "symbiotic" and "nonsymbiotic" relationships; rather there is a continuum of degrees of biological intimacy and that as the level of intimacy increases so does the exclusivity of that relationship.

The data used in these analyses were scored at the genus level for reasons of taxonomic and statistical robustness. Would the same results have been obtained had the data been analyzed at the species level? That question awaits future analyses, but my reading of the literature suggests that the pattern would be the same. Interactions involving a single focus genus and a single partner genus may be specialized at a species–species level or at a species–genus level. In the latter case, a single species of the focus genus can form an association with any of the species in the partner genus. On the other hand, interactions involving a single focus genus and multiple partner genera tend not be specialized at a species–genus level, but rather a single species can form associations with multiple genera, for example, plant–seed disperser relationships (although *Ficus*–fig wasp relationships are a notable exception to this generality). If this is indeed the case, then the greater the number of partner genera a focus genus is associated with, the greater should be the species-level generalization. Having said that, a large focus genus may contain species that are more or less specialized or generalized (for example, the genus *Xysmalobium* [Apocynaceae: Asclepiadoideae] in southern Africa; Ollerton et al. 2003). In this case, a representative sample of species from the whole range of specialized to generalized would have to be included in the analysis (perhaps another good reason for analyzing at a generic level—to give an "average" value for those taxa?). The paradox, of course, is that the level of biological intimacy is not expected to vary much within a genus, because it is controlled by biological attributes which are likely, by definition, to be similar within genera. What then controls the level of specialization exhibited by different species within a genus? Are ecological context and partner availability more important than morphological and biochemical "fit"? This promises to be a productive area for future research but requires some very detailed field and/or laboratory observations (Fox and Morrow 1981; Herrera 2005).

The benefits of mutualism to all parties involved in the interaction are often assumed rather than empirically measured (Douglas and Smith 1989; Cushman and Beattie 1991). Within the database I used for these analyses, I have no doubt that some of the interactions are not in fact mutualistic, or are at best conditionally mutualistic, varying over time and space (Bronstein 1994b). Only future analyses will reveal whether the results that I have presented are robust, though the sensitivity analyses in appendix 18.2 suggest that they are. It is worth noting that recent models of the evolution of mutualisms, using economic arguments about resource specialization, suggest that the benefits of mutualism may, in theory, be so small that they are undetectable (Schwartz and Hoeksema 1998). These emerging results, together with new findings regarding the cryptic diversity of mutualistic partners (Herre et al. 1999), are significant challenges to our current ideas about mutualisms and the ways in which we approach our research. But they also point to exciting future findings.

The quotations that open this chapter were chosen because they seem to reflect themes about the current debates regarding the commonness of highly specific plant–pollinator and other mutualistic associations (Howe 1984), the importance of specialization in generating biological diversity (Thompson 1994), how little we truly understand about the ecology and evolution of mutualisms (Bronstein 1994a), and the fact that no one researcher can hope to collect all of the data required to address these important questions, highlighting the advantages of a synthetic, comparative approach (Elton 1927). Protection and management of biodiversity is a foremost conservation priority for humanity, and synthetic analyses of available data are required to scientifically underpin judgments regarding the ability of habitats to sustain the biodiversity of species interactions.

Appendix 18.1

Table 18A.1 The dataset: A complete listing of the mutualistic associations used in the analyses in this chapter

		Components of biological intimacy[b]					
Category of association	**Genera (focus–partner[s])[a]**	**Number of genera**	**Physical dependence**	**Trophic dependence**	**Physiological integration**	**Vertical transmission**	**Total index score**
Fungus–alga	*Anzia–Trebouxia*	1	2	2	2	0	6
Fungus–cyanobacteria	*Geosiphon–Nostoc*	1	1	1	1	0	3
Metazoan–alga	*Hydra*–alga	2	1	1	2	1	5
Metazoan–alga	*Convoluta*–alga	2	1	2	2	0	5
Metazoan–alga	*Porites–Symbiodinium*	1	1	1	1	1	4
Legume–rhizobia	*Lotus–Rhizobium*	1	0	1	2	0	3
Actinorhizae	*Alnus–Frankia*	1	0	1	2	0	3
Cycad–cyanobacteria	*Cycas–Nostoc*	1	1	1	1	0	3
Angiosperm–cyanobacteria	*Gunnera–Nostoc*	1	1	1	2	0	4
Liverwort–cyanobacteria	*Blasia–Nostoc*	1	1	1	1	0	3
Moss–cyanobacteria	*Sphagnum*–cyanobacteria	2	0	1	1	0	2
Fern–cyanobacteria	*Azolla–Anabaena*	1	2	2	2	1	7
Ectomycorrhizal–conifer	*Pinus*–fungus	23	1	1	1	0	3
VA mycorrhizal–angiosperm[c]	*Festuca*–mycorrhizae	4	1	1	2	0	4
Insect–gut fungus	*Lasioderma–Symbiotaphrina*	1	2	2	2	1	7
Ant–fungus	*Cyphomyrmex*–fungus	2	2	2	0	0	4
Termite–fungus	*Odontotermes*–fungus	2	2	2	0	0	4
Ascidian–*Nephromyces*	*Molgula–Nephromyces*	1	1	2	1	0	4
Insect–mycetocyte	*Acyrthosiphon–Buchnera*	1	2	2	2	1	7
Plant–endophyte	*Lolium–Neotyphodium*	1	2	2	2	1	7
Seed dispersal	*Viburnum*–birds	25	0	1	0	0	1
Pollination	*Ficus*–fig wasp	20	1	2	1	0	4
Pollination	*Asclepias*–pollinator	150	0	1	0	0	1
Pollination	*Erythrina*–pollinator	54	0	1	0	0	1
Ant–plant seed dispersal	*Viola*–ant	9	0	1	0	0	1

Table 18A.1 *(continued)*

Category of association	Genera (focus–partner[s])[a]	Components of biological intimacy[b] Number of genera	Physical dependence	Trophic dependence	Physiological integration	Vertical transmission	Total index score
Ant–extrafloral nectary	*Passiflora*–ant	17	0	1	0	0	1
Ant–plant	*Acacia*–ant	5	0	1	0	0	1
Ant–plant	*Cecropia*–ant	6	2	2	1	0	5
Ant–food body/domatia	*Myrmecodia*–ant	3	1	1	0	0	2
Fungal spore dispersal	*Epichloe*–*Phorbia*	1	0	1	0	0	1
Fungal spore dispersal	*Phallus*–fly	30	0	1	0	0	1
Fish–biolum inescence	*Leiognathus*–*Photobacterium*	1	1	2	1	0	4
Bird–mammal cleaning	*Buphagus*–mammal	21	1	1	0	0	2
Anemone fish–anemone	*Amphiprion*–anemone	6	2	1	0	0	3
Shrimp–anemone	*Periclimenes*–anemone	5	2	1	0	0	3
Ant–hemipteran	*Lasius*–homoptera	9	0	1	0	0	1
Ant–hemipteran	*Oecophylla*–homoptera	10	0	1	0	0	1
Beetle–fungus	*Xyleborus*–fungus	4	2	2	1	1	6
Wood wasp–fungus	*Sirex*–*Amylostereum*	1	2	2	1	1	6

[a]The identity of the focus genus is given first, followed by the partner genus (if the relationship involves only a single genus of partner) or a collective term if there are two or more partner genera.

[b]Components of biological intimacy refers to the biological attributes of the partner genera in relation to the focus genus; see Methods for a description of each component of biological intimacy.

[c]VA = vesicular-arbuscular.

Appendix 18.2

Sensitivity Analyses of the Dataset

I conducted three sets of analyses designed to test the sensitivity of the dataset in appendix 18.1 to (1) sample size, (2) accuracy of recording numbers of partner genera per host genus, and (3) accuracy of scoring the index of biological intimacy.

1. I randomly selected 50% of the dataset and in each case calculated a Spearman rank correlation between the number of partner genera and the index of biological intimacy. This was done for 25 iterations; 19 of the 25 iterations (76%) returned a statistically significant ($P \leq .05$) correlation, 2 of the iterations (8%) returned a marginally statistically significant ($P \leq .10$) correlation, and 4 of the iterations (16%) returned a nonsignificant ($P > .10$) correlation.
2. To a random selection of 50% of the dataset, I added one genus to the score for number of partner genera and in each case calculated a Spearman rank

correlation between the number of partner genera and the index of biological intimacy. I ran this for 20 iterations; 20 out of 20 iterations (100%) returned a highly statistically significant ($P \leq .001$) correlation.

3. Finally, I randomly selected 50% of the dataset and added 1 to the score of biological intimacy. I then calculated a Spearman rank correlation between the number of partner genera and the index of biological intimacy for 20 iterations. Once again, 20 out of 20 iterations (100%) returned a highly statistically significant ($P \leq .001$) correlation.

The results of these sensitivity analyses suggest that the dataset I used was robust to changes in sample size (up to 50%), to minor variations in the number of recorded genera, and to inaccuracies in scoring the index of biological intimacy.

Acknowledgments

I am grateful to George Forsey (University College Northampton) for discussions during the early evolution of this work and for stimulating my interest in mutualisms beyond my personal ghetto of plant–pollinator interactions. For valuable comments on the manuscript, I thank Beverly Rathcke and an anonymous referee, plus José Maria Gómez, Jason Hoeksema, and especially Christine Müller and Dennis Hansen, who filled in a cell in figure 18.2 and introduced me to the fascinating interactions of rabbits, lizards, and cacti on the Galápagos Islands. Finally, sincere thanks go to the many, many researchers who produced the data used in these analyses.

References

Armbruster, W. S. 1984. The role of resin in angiosperm pollination: Ecological and chemical considerations. American Journal of Botany 71: 1149–1160.

Armbruster, W. S., and B. G. Baldwin. 1998. Switch from specialized to generalized pollination. Nature 394: 632–632.

Bascompte, J., P. Jordano, C. J. Melián, and J. M. Olesen. 2003. The nested assembly of plant–animal mutualistic networks. Proceedings of the National Academy of Sciences (USA) 100: 9383–9387.

Batra, L. R. 1979. Insect–fungus symbiosis: Nutrition, mutualism, and commensalism. John Wiley and Sons, New York.

Blattner, F. R., K. Weising, G. Banfer, U. Maschwitz, and B. Fiala. 2001. Molecular analysis of phylogenetic relationships among myrmecophytic *Macaranga* species (Euphorbiaceae). Molecular Phylogenetics and Evolution 19: 331–344.

Borowicz, V. A., and S. A. Juliano. 1991. Specificity in host fungus associations: Do mutualists differ from antagonists? Evolutionary Ecology 5: 385–392.

Boucher, D. H. (ed.). 1985. The biology of mutualism: Ecology and evolution. Croom Helm, London.

Boucher, D. H, S. James, and K. H. Keeler. 1982. The ecology of mutualism. Annual Review of Ecology and Systematics 13: 315–347.

Bronstein, J. L. 1994a. Our current understanding of mutualism. Quarterly Review of Biology 69: 31–51.

Bronstein, J. L. 1994b. Conditional outcomes in mutualistic interactions. Trends in Ecology and Evolution 9: 214–217.

Brown, B. L., R. P. Creed, and W. E. Dobson. 2002. Branchiobdellid annelids and their crayfish hosts: Are they engaged in a cleaning symbiosis? Oecologia 132: 250–255.

Bultman, T. L., and J. F. White. 1988. "Pollination" of a fungus by a fly. Oecologia 75: 317–319.

Bultman, T. L., J. F. White, T. I. Bowdish, and A. M. Welch. 1998. A new kind of mutualism between fungi and insects. Mycological Research 102: 235–238.

Cairney, J. W. G. 2000. Evolution of mycorrhizal systems. Naturwissenschaften 87: 467–475.

Chittka, L., and S. Schürkens. 2001. Successful invasion of a floral market. Nature 411: 653–653.

Compton, S. G., and A. B. Ware. 1991. Ants disperse the elaiosome-bearing eggs of an African stick insect. Psyche 98: 207–213.

Corallini, C., and E. Gaino. 2003. The caddisfly *Ceraclea fulva* and the freshwater sponge *Ephydatia fluviatilis:* A successful relationship. Tissue and Cell 35: 1–7.

Cowen, R. 1988. The role of algal symbiosis in reefs through time. Palaios 3: 221–227.

Cushman, J. H., and A. J. Beattie. 1991. Mutualisms: Assessing the benefits to hosts and visitors. Trends in Ecology and Evolution 6: 193–195.

Douglas, A. E. 1994. Symbiotic interactions. Oxford University Press, Oxford.

Douglas, A. E., and D. C. Smith. 1989. Are endosymbioses mutualistic? Trends in Ecology and Evolution 4: 350–352.

Dupont, Y. L., D. M. Hansen, and J. M. Olesen. 2003. Structure of a plant–pollinator network in the high altitude sub-alpine desert of Tenerife, Canary Islands. Ecography 26: 301–310.

Ellis, A. G., and J. J. Midgley. 1996. A new plant–animal mutualism involving a plant with sticky leaves and a resident hemipteran insect. Oecologia 106: 478–481.

Elton, C. 1927. Animal ecology. Methuen, London.

Fenster, C. B., W. S. Armbruster, P. Wilson, M. R. Dudash, and J. D. Thomson. 2004. Pollination syndromes and floral specialization. Annual Review of Ecology, Evolution, and Systematics 35: 375–403.

Fiedler, K. 1998. Lycaenid–ant interactions of the Maculinea type: Tracing their historical roots in a comparative framework. Journal of Insect Conservation 2: 3–14.

Fitter, A. H, and B. Moyersoen. 1996. Evolutionary trends in root-microbe symbioses. Philosophical Transactions of the Royal Society B 351: 1367–1375.

Fox, L. R., and P. A. Morrow. 1981. Specialization: species property or local phenomenon? Science 211: 887–893.

Freeman, S., and R. J. Rodriguez. 1993. Genetic conversion of a fungal plant pathogen to a nonpathogenic, endophytic mutualist. Science 260: 75–78.

Gardener, M. C., and M. R. Gillman. 2001. Analyzing variability in nectar amino acids: Composition is less variable than concentration. Journal of Chemical Ecology 27: 2545–2558.

Grange, K. R. 1991. Mutualism between the antipatharian *Antipathes fiordensis* and the ophiuroid *Astrobrachion constrictum* in New Zealand fjords. Hydrobiologia 216: 297–303.

Harvey, P. H., and M. Pagel. 1991. The comparative method in evolutionary biology. Oxford University Press, Oxford.

Herre, E. A., N. Knowlton, U. G. Mueller, and S. A. Rehner. 1999. The evolution of mutualisms: Exploring the paths between conflict and cooperation. Trends in Ecology and Evolution 14: 49–53.

Herrera, C. M. 1988. Variation in mutualisms: The spatio-temporal mosaic of a pollinator assemblage. Biological Journal of the Linnean Society 35: 95–125.

Herrera, C. M. 2005. Plant generalization on pollinators: species property or local phenomenon? American Journal of Botany 92: 13–20.

Hibbett, D. S., L. B. Gilbert, and M. J. Donoghue. 2000. Evolutionary instability of ectomycorrhizal symbioses in basidiomycetes. Nature 407: 506–508.

Hoeksema, J. D., and E. M. Bruna. 2000. Pursuing the big questions about interspecific mutualism: A review of theoretical approaches. Oecologia 125: 321–330.

Hoeksema, J. D., and M. Kummel. 2003. Ecological persistence of the plant–mycorrhizal mutualism: A hypothesis from species co-existence theory. American Naturalist 162: S40–S50.

Howe, H. F. 1984. Constraints on the evolution of mutualisms. American Naturalist 123: 764–777.

Irwin, R. E., A. K. Brody, and N. M. Waser. 2001 The impact of floral larceny on individuals, populations, and communities. Oecologia 129: 161–168.

Itino T., S. J. Davies, H. Tada, O. Hieda, M. Inoguchi, T. Itioka, S. Yamane, and T. Inoue. 2001. Cospeciation of ants and plants. Ecological Research 16: 787–793.

Janzen, D. 1985. The natural history of mutualisms. Pp. 40–99 *in* D. H. Boucher (ed.), The biology of mutualism: Ecology and evolution. Croom Helm, London.
Johnson, S. D., and K. E. Steiner. 2000. Generalization versus specialization in plant pollination systems. Trends in Ecology and Evolution 15: 140–143.
Jolivet, P. 1996. Ants and plants: An example of coevolution. Backhuys Publishers, Leiden.
Jordano, P. 1987. Patterns of mutualistic interactions in pollination and seed dispersal: Connectance, dependence asymmetries, and coevolution. American Naturalist 129: 657–677.
Keeling, P. J., and N. M. Fast. 2002. Microsporidia: Biology and evolution of highly reduced intracellular parasites. Annual Review of Microbiology 56: 93–116.
Kite, G. C., W. L. A. Hetterscheid, M. J. Lewis, P. C. Boyce, J. Ollerton, E. Cocklin, A. Diaz, and M. S. J. Simmonds. 1998. Inflorescence odours and pollinators of *Arum* and *Amorphophallus* (Araceae). Pp. 295–315 *in* S. J. Owens and P. J. Rudall (eds.), Reproductive biology in systematics, conservation and economic botany. Royal Botanic Gardens, Kew, UK.
Lamborn, E., and J. Ollerton. 2000. Experimental assessment of the functional morphology of inflorescences of *Daucus carota* (Apiaceae): Testing the "fly catcher effect." Functional Ecology 14: 445–454.
Law, R. 1985. Evolution in a mutualistic environment. Pp. 145–170 *in* D. H. Boucher (ed.), The biology of mutualism: Ecology and evolution. Croom Helm, London.
Law, R., and S. Koptur. 1986. On the evolution of non-specific mutualism. Biological Journal of the Linnean Society 27: 251–267.
Linder, H. P. 1998. Morphology and evolution of wind pollination. Pp. 123–135 *in* S. J. Owens and P. J. Rudall (eds.), Reproductive biology in systematics, conservation and economic botany. Royal Botanic Gardens, Kew, UK.
Lopez-Vaamonde, C., J. Y. Rasplus, G. D. Weiblen, and J. M. Cook. 2001. Molecular phylogenies of fig wasps: Partial co-cladogenesis of pollinators and parasites. Molecular Phylogenetics and Evolution 21: 55–71.
Lum, M. R., and A. M. Hirsch. 2002. Roots and their symbiotic microbes: Strategies to obtain nitrogen and phosphorus in a nutrient-limiting environment. Journal of Plant Growth Regulation 21: 368–382.
Machado, C. A., E. Jousselin, F. Kjellberg, S. G. Compton, and E. A. Herre. 2001. Phylogenetic relationships, historical biogeography and character evolution of fig-pollinating wasps. Proceedings of the Royal Society of London B 268: 685–694.
Maloof, J. E., and D. W. Inouye. 2000. Are nectar robbers cheaters or mutualists? Ecology 81: 2651–2661.
Munk, O. 2000. Histology of the fusion area between the parasitic male and the female in the deep-sea anglerfish *Neoceratias spinifer* Pappenheim 1914 (Teleostei, Ceratioidei). Acta Zoologica 81: 315–324.
Naoki, K., and E. Toapanta. 2001. Müllerian body feeding by Andean birds: New mutualistic relationship or evolutionary time lag? Biotropica 33: 204–207.
Nash, T. H. (ed.). 1996. Lichen biology. Cambridge University Press, Cambridge.
Newsham, K. K., A. H. Fitter, and A. R. Watkinson. 1995. Multi-functionality and biodiversity in arbuscular mycorrhizas. Trends in Ecology and Evolution 10: 407–411.
Neyland, R. 2001. A phylogeny inferred from large ribosomal subunit 26S rDNA sequences suggests that *Cuscuta* is a derived member of Convolvulaceae. Brittonia 53: 108–115.
Noë, R., and P. Hammerstein. 1995. Biological markets. Trends in Ecology and Evolution 10: 336–339.
Ollerton, J., and L. Cranmer. 2002. Latitudinal trends in plant–pollinator interactions: Are tropical plants more specialised? Oikos 98: 340–350.
Ollerton, J., S. D. Johnson, L. Cranmer, and S. Kellie. 2003 The pollination ecology of an assemblage of grassland asclepiads in KwaZulu-Natal, South Africa. Annals of Botany 92: 807–834.
Paracer, S., and V. Ahmadjian. 2000. Symbiosis: An introduction to biological associations. Oxford University Press, Oxford.
Pellmyr, O. 2003. Yuccas, yucca moths, and coevolution: A review. Annals of the Missouri Botanical Garden 90: 35–55.

Pellmyr, O., and C. J. Huth. 1994. Evolutionary stability of mutualism between yuccas and yucca moths. Nature 372: 257–260.
Pellmyr, O., and J. Leebens-Mack. 1999. Forty million years of mutualism: Evidence for Eocene origin of the yucca–yucca moth association. Proceedings of the National Academy of Sciences (USA) 96: 9178–9183.
Perret, M., A. Chautems, R. Spichiger, M. Peixoto, and V. Savolainen. 2001. Nectar sugar composition in relation to pollination syndromes in Sinningieae (Gesneriaceae). Annals of Botany 87: 267–273.
Petanidou, T., and W. N. Ellis. 1996. Interdependence of native bee faunas and floras in changing Mediterranean communities. Pp. 201–226 *in* A. Matheson, M. Buchmann, C. O'Toole, P. Westrich, and I. H. Williams (eds.), The conservation of bees. Linnean Society Symposium Series, no. 18, Academic Press, London.
Piercey-Normore, M. D., and P. T. DePriest. 2001. Algal switching among lichen symbioses. American Journal of Botany 88: 1490–1498.
Proctor, M., P. Yeo, and A. Lack. 1996. The natural history of pollination. HarperCollins, London.
Reisser, W. (ed.). 1992. Algae and symbioses: Plants, animals, fungi, viruses, interactions explored. Biopress, Bristol.
Richardson, D. M., N. Allsopp, C. M. D'Antonio, S. J. Milton, and M. Rejmanek 2000. Plant invasions: The role of mutualisms. Biological Review 75: 65–93.
Ross, A., and W. A. Newman. 1996. A new sessile barnacle symbiotic with bryozoans from Madagascar and Mauritius (Cirripedia: Balanomorpha): A unique case of co-evolution? Invertebrate Biology 115: 150–161.
Rowan, R. 1998. Diversity and ecology of zooxanthellae on coral reefs. Journal of Phycology 34: 407–417.
Rowan, R., and D. A. Powers. 1991. A molecular genetic classification of zooxanthellae and the evolution of animal–algal symbioses. Science 251: 1348–1351.
Schwartz, M. W., and J. D. Hoeksema. 1998. Specialization and resource trade: Biological markets as a model of mutualisms. Ecology 79: 1029–1038.
Seigler, D. S., and J. E. Ebinger. 1995. Taxonomic revision of the ant-acacias (Fabaceae, Mimosoideae, *Acacia,* Series Gummiferae) of the New World. Annals of the Missouri Botanical Garden 82: 117–138.
Smith, D. C., and A. E. Douglas. 1987. The biology of symbiosis. Edward Arnold, London.
Stewart, B. 1998. Can a snake star earn its keep? Feeding and cleaning behaviour in *Astrobrachion constrictum* (Farquhar) (Echinodermata: Ophiuroidea), a euryalid brittle-star living in association with the black coral, *Antipathes fiordensis* (Grange 1990). Journal of Experimental Marine Biology and Ecology 221: 173–189.
Thompson, J. N. 1994. The coevolutionary process. University of Chicago Press, Chicago.
Wallander, E. 2001. Evolution of wind-pollination in *Fraxinus* (Oleaceae): An ecophylogenetic approach. PhD dissertation, Göteborg University, Göteborg.
Waser, N. M., L. Chittka., M. V. Price, N. M. Williams, and J. Ollerton. 1996. Generalization in pollination systems, and why it matters. Ecology 77: 1043–1060.
Weller, S. G., and A. K. Sakai. 1999. Using phylogenetic approaches for the analysis of plant breeding system evolution. Annual Review of Ecology and Systematics 30: 167–199.
Young, J. P. W., and A. W. B. Johnston. 1989. The evolution of specificity in the legume–rhizobium symbiosis. Trends in Ecology and Evolution 4: 341–350.
Yu, D. W. 2001. Parasites of mutualisms. Biological Journal of the Linnean Society 72: 529–546.

Contributors

Paul A. Aigner
McLaughlin Reserve
University of California, Davis
26775 Morgan Valley Rd.
Lower Lake, CA 95457
USA

Marcelo A. Aizen
Laboratorio Ecotono
Universidad Nacional del Comahue
Bariloche
Quintral 1250, RA-8400 San Carlos De Bariloche
Rio Negro
Argentina

Thomas Alfert
Department of Agroecology
University of Göttingen
Waldweg 26
D-37073 Göttingen
Germany

W. Scott Armbruster
School of Biological Sciences
King Henry Building, King Henry I St.
University of Portsmouth
Portsmouth, PO1 2DY
UK

Norberto J. Bartoloni
Facultad de Agronomía
Universidad de Buenos Aires
Av. San Martín 4453
1417 Buenos Aires
Argentina

Jordi Bascompte
Integrative Ecology Group
Estación Biológica de Doñana
CSIC, Apdo. 1056
E-41080 Sevilla
Spain

Alicia M. Basilio
Depto Producción Animal
Facultad de Agronomía
Universidad de Buenos Aires
Av. San Martín 4453
C1417DSQ Buenos Aires
Argentina

Renée M. Bekker
Community and Conservation Ecology Group
University of Groningen
P.O. Box 14
9750 AA Haren
The Netherlands

James H. Cane
USDA Bee Biology Lab
Utah State University
Logan, UT 84322-5310
USA

Maria Clara Castellanos
Estación Biológica de Doñana
Consejo Superior de Investigaciones Científicas
Avendia de Maria Luisa s/n
E-41013 Sevilla
Spain

Sarah A. Corbet
1 St. Loy Cottages
St. Buryan
Penzance TR19 6DH
UK

Mariano Devoto
Facultad de Agronomía
Universidad de Buenos Aires
Av. San Martín 4453
1417 Buenos Aires
Argentina

Volker Gaebele
Department of Agroecology
University of Göttingen
Waldweg 26
D-37073 Göttingen
Germany

José M. Gómez
Departmento de Biología Animal y Ecología
Fac. Ciencias, Universidad de Granada
E-18071 Granada
Spain

Andrew B. Hingston
School of Geography and Environmental Studies
University of Tasmania
Private Bag 78
Hobart, Tasmania 7001
Australia

Steven D. Johnson
School of Biological and Conservation Sciences
University of KwaZulu-Natal
Private Bag X01 Scottsville
Pietermaritzburg
ZA-3209
South Africa

Pedro Jordano
Integrative Ecology Group
Estación Biológica de Doñana
CSIC, Apdo. 1056
E-41080 Sevilla
Spain

Alexandra-Maria Klein
Department of Agroecology
University of Göttingen
Waldweg 26
D-37073 Göttingen
Germany

Suzanne Koptur
Department of Biological Sciences
Florida International University
Miami, FL 33199
USA

Manja M. Kwak
Community and Conservation Ecology Group
University of Groningen
P.O. Box 14
9750 AA Haren
The Netherlands

Diego Medan
Facultad de Agronomía
Universidad de Buenos Aires
Av. San Martín 4453
1417 Buenos Aires
Argentina

Robert L. Minckley
Department of Biology
402 Hutchinson Hall
University of Rochester
Rochester, NY 14627-0211
USA

Jens M. Olesen
Department of Ecology and Genetics
University of Aarhus
Munkegade Block 540
DK-8000 Aarhus C
Denmark

Jeff Ollerton
Landscape and Biodiversity Research Group
School of Environmental Science
University of Northampton
Park Campus
Northampton NN2 7AL
UK

Theodora Petanidou
Department of Geography
University of the Aegean
Building of Geography, University Hill
GR-81100 Mytilene
Greece

Simon G. Potts
Centre for Agri-Environmental Research (CAER)
School of Agriculture
Reading University
P.O. Box 237
Reading RG6 6AR
UK

Susanne S. Renner
Systematische Botanik und Mykologie
Ludwig Maximilians University
Menzinger Str. 67
D-80638 Munich
Germany

T'ai H. Roulston
Blandy Experimental Farm
University of Virginia
400 Blandy Farm Lane
Boyce, VA 22620
USA

Sedonia Sipes
Department of Plant Biology
Southern Illinois University at Carbondale
Carbondale, IL 62901
USA

Ingolf Steffan-Dewenter
Department of Agroecology
University of Göttingen
Waldweg 26
D-37073 Göttingen
Germany

James D. Thomson
Department of Zoology
University of Toronto
25 Harbord St.
Toronto, ON M5S 3G5
Canada

Juan P. Torretta
Facultad de Agronomía
Universidad de Buenos Aires
Av. San Martín 4453
1417 Buenos Aires
Argentina

Teja Tscharntke
Department of Agroecology
University of Göttingen
Waldweg 26
D-37073 Göttingen
Germany

Diego P. Vázquez
Instituto Argentino de Investigaciones de las Zonas Áridas
Centro Regional de Investigaciones Científicas y Tecnológicas
Av. Ruiz Leal s/n
(5500) Mendoza
Argentina

Nickolas M. Waser
Department of Biology
University of California
Riverside CA 92521
USA

Paul Wilson
Department of Biology
California State University
18111 Nordhoff Street
Northridge, CA 91330-8303
USA

Andrea D. Wolfe
Department of Ecology,
Evolution, and Organismal
Biology
Ohio State University
Columbus, OH 43210
USA

Regino Zamora
Departmento de Biología
Animal y Ecología
Faculidad Ciencias,
Universidad de Granada
E-18071 Granada
Spain

Index

Note: Page numbers in italics refer to figures or tables.

CPSIA information can be obtained
at www.ICGtesting.com
Printed in the USA
LVHW03s0119150718
583502LV00004B/26/P